The DNA Damage Response: Implications on Cancer Formation and Treatment

Kum Kum Khanna · Yosef Shiloh
Editors

The DNA Damage Response: Implications on Cancer Formation and Treatment

Editors
Dr. Kum Kum Khanna
Queensland Institute of
Medical Research
Signal Transduction
Laboratory
300 Herston Road
Herston QLD 4006
Australia
kumkumK@qimr.edu.au

Dr. Yosef Shiloh
Tel Aviv University
Sackler School of Medicine
Dept. Human Molecular Genetics &
Biochemistry
69978 Tel Aviv
Ramat Aviv
Israel
yossih@post.tau.ac.il

ISBN 978-90-481-2560-9 e-ISBN 978-90-481-2561-6
DOI 10.1007/978-90-481-2561-6
Springer Dordrecht Heidelberg London New York

Library of Congress Control Number: 2009927008

Printed on acid-free paper

Springer is part of Springer Science+Business Media (www.springer.com)

Preface

The field of cellular responses to DNA damage has attained widespread recognition and interest in recent years commensurate with its fundamental role in the maintenance of genomic stability. These responses, which are essential to preventing cellular death or malignant transformation, are organized into a sophisticated system designated the "DNA damage response". This system operates in all living organisms to maintain genomic stability in the face of constant attacks on the DNA from a variety of endogenous by-products of normal metabolism, as well as exogenous agents such as radiation and toxic chemicals in the environment. The response repairs DNA damage via an intricate cellular signal transduction network that coordinates with various processes such as regulation of DNA replication, transcriptional responses, and temporary cell cycle arrest to allow the repair to take place. Defects in this system result in severe genetic disorders involving tissue degeneration, sensitivity to specific damaging agents, immunodeficiency, genomic instability, cancer predisposition and premature aging. The finding that many of the crucial players involved in DNA damage response are structurally and functionally conserved in different species spurred discoveries of new players through similar analyses in yeast and mammals. We now understand the chain of events that leads to instantaneous activation of the massive cellular responses to DNA lesions.

This book summarizes several new concepts in this rapidly evolving field, and the advances in our understanding of the complex network of processes that respond to DNA damage. The researchers who contributed chapters have profound knowledge of the recent advances in DNA damage signalling and repair and their implications for carcinogenesis, and are well positioned to anticipate upcoming developments in their respective fields. The book is divided into chapters that deal with the elaborate surveillance system and repair mechanisms used by cells to suppress mutagenic lesions to avoid cancer. It provides snapshots of what is known to date about DNA damage signalling, cell cycle checkpoints, some of the major DNA repair pathways, functional links between DNA damage, genomic instability and cancer, and how this knowledge can be mined for the new treatment modalities for cancer. Where major players in this response are potential targets for cancer therapy, the developments are discussed. We hope the book will provide the reader with a framework to understand current concepts in DNA damage signaling and repair, their critical role in the maintenance of genomic stability, how the dysfunction of these mechanisms contributes to tumorigenesis, and how those mechanisms can be exploited for therapeutic gains.

We thank all the authors who made this book possible. Their timely contributions bring us up to date on this crucial area of research on the close link between genome instability and cancer.

Australia Kum Kum Khanna
Israel Yosef Shiloh

Contents

Contributors

Byungchan Ahn Department of Life Sciences, University of Ulsan, Ulsan 680-749, Korea, bbccahn@mail.ulsan.ac.kr

Ari Barzilai Department of Neurobiology, George S. Wise Faculty of Life Sciences, Tel Aviv University, Tel Aviv 69978, Israel, barzilai@post.tau.ac.il

Vilhelm A. Bohr Laboratory of Molecular Gerontology, Gerontology Research Center, National Institute on Aging, NIH, Baltimore, MD 21224, USA, vbohr@nih.gov

Emma Bolderson Signal Transduction Laboratory, The Queensland Institute of Medical Research, Brisbane, Queensland 4006, Australia, emma.bolderson@qimr.edu.au

Keith W. Caldecott Genome Damage and Stability Centre, University of Sussex, Falmer, Brighton, BN1 9RQ, UK, k.w.caldecott@sussex.ac.uk

Wen-Hsing Cheng Department of Nutrition and Food Science, University of Maryland, College Park, MD, 20742, USA, whcheng@umd.edu

David Cortez Department of Biochemistry, Vanderbilt University School of Medicine, Nashville, TN 37232, USA, david.cortez@vanderbilt.edu

Inbal Dar Department of Neurobiology, George S. Wise Faculty of Life Sciences, Tel Aviv University, Tel Aviv 69978, Israel, darinbal@post.tau.ac.il

Daniel Durocher Samuel Lunenfeld Research Institute, Centre for Systems Biology, Mount Sinai Hospital, 600 University Avenue, Toronto, ON, Canada M5G 1X5, durocher@lunenfeld.ca

Marie Fernet INSERM U612, Bats110-112, Centre Universitaire, Orsay, 91405, France; Institut Curie-Recherche, Bats110-112, Centre Universitaire, Orsay, 91405 France, marie.fernet@curie.u-psud.fr

Thanos D. Halazonetis Department of Molecular Biology and Department of Biochemistry, University of Geneva, CH-1205, Geneva, Switzerland, Thanos.Halazonetis@unige.ch

Janet Hall INSERM U612, Bats110-112, Centre Universitaire, Orsay, 91405, France; Institut Curie-Recherche, Bats110-112, Centre Universitaire, Orsay, 91405 France, janet.hall@curie.u-psud.fr

D. Paul Harkin Centre for Cancer Research and Cell Biology, Queen's University Belfast, 97 Lisburn Road, Belfast, BT9 7BL, UK, d.harkin@qub.ac.uk

Mathew Jones Signal Transduction Laboratory, The Queensland Institute of Medical Research, Brisbane, Queensland 4006, Australia, mathew.jones@qimr.edu.au

Sachin Katyal Department of Genetics and Tumor Cell Biology, St Jude Children's Research Hospital, Memphis, TN 38105, USA, sachin.katyal@stjude.org

Kum Kum Khanna Cancer and Cell Biology Division, Signal Transduction Laboratory, The Queensland Institute of Medical Research, Brisbane, Queensland 4006, Australia, kumkum.khanna@qimr.edu.au

Anne E Kiltie Section of Experimental Oncology, Leeds Institute of Molecular Medicine, St James's University Hospital, Leeds LS9 7TF, UK, a.e.kiltie@leeds.ac.uk

Susan P. Lees-Miller Department of Biochemistry and Molecular Biology, Southern Alberta Cancer Research Institute, University of Calgary, 3330 Hospital Drive NW, Calgary, AB T2N 4N1, Canada, leesmill@ucalgary.ca

Alan R Lehmann Genome Damage and Stability Centre, University of Sussex, Falmer, Brighton BN1 9RQ, UK, a.r.lehmann@sussex.ac.uk

Peter J. McKinnon Department of Genetics and Tumor Cell Biology, St Jude Children's Research Hospital, Memphis, TN 38105, USA, peter.mckinnon@stjude.org

Sankar Mitra Department of Biochemistry and Molecular Biology, University of Texas Medical Branch, Galveston, TX 77555-1079, USA, samitra@utmb.edu

Daniel A. Mordes Department of Biochemistry, Vanderbilt University School of Medicine, Nashville, TN 37232, USA, daniel.mordes@vanderbilt.edu

Leon H.F. Mullenders Department of Toxicogenetics, Leiden University Medical Center, Leiden, Nederland, l.mullenders@lumc.nl

Julia Pagan Signal Transduction Laboratory, The Queensland Institute of Medical Research, Brisbane, Queensland 4006, Australia, julia.pagan@qimr.edu.au

Tej K. Pandita Department of Radiation Oncology, Washington University School of Medicine, 4511 Forest Park Ave., St Louis, MO 63108, USA, pandita@wustl.edu

Derek J Richard Cancer and Cell Biology Division, The Queensland Institute of Medical Research, 300 Herston Road, Herston Qld 4006, Australia, derek.richard@qimr.edu.au

Rabindra Roy Lombardi Comprehensive Cancer Center, Georgetown University, Washington, DC 20057-1468, USA, rr228@georgetown.edu

Ryo Sakasai Department of Pharmacology, University of Wisconsin School of Medicine and Public Health, Madison, WI 53706, USA, sakasai.pbc@mri.tmd.ac.jp

Kienan Savage Centre for Cancer Research and Cell Biology, Queen's University Belfast, 97 Lisburn Road, Belfast, BT9 7BL, UK, k.savage@qub.ac.uk

Toshiyasu Taniguchi Divisions of Human Biology and Public Health Sciences, Fred Hutchinson Cancer Research Center, 1100 Fairview Ave. N., C1-015, Seattle, WA 98109-1024, USA, ttaniguc@fhcrc.org

Randal S. Tibbetts Department of Pharmacology, University of Wisconsin School of Medicine and Public Health, Madison, WI 53706, USA, rstibbetts@wisc.edu

Mischa G. Vrouwe Department of Toxicogenetics, Leiden University Medical Center, Leiden, Nederland

Laura M. Williamson Department of Biochemistry and Molecular Biology, Southern Alberta Cancer Research Institute, University of Calgary, 3330 Hospital Drive, NW, Calgary, AB T2N 4N1, Canada, lbaxter@ucalgary.ca

Chris T. Williamson Department of Biochemistry and Molecular Biology, Southern Alberta Cancer Research Institute, University of Calgary, 3330 Hospital Drive, NW, Calgary, AB T2N 4N1, Canada, ctwillia@ucalgary.ca

Chapter 1
DNA Damage Sensing and Signaling

Daniel Durocher

Abstract In order to maintain genome integrity, cells have evolved a complex network of processes that detect, repair and signal DNA damage. In this chapter, I review the early steps of DNA damage signaling with a particular emphasis on how DNA lesions are sensed and how the detection of DNA damage leads to the activation of DNA damage signaling by the ATM and ATR protein kinases.

Keywords ATM · ATR · MRN · DNA damage signaling · Checkpoint

1.1 Preamble

The term "DNA damage" defines a heterogeneous group of chemical and physical alterations of the DNA double helix. DNA damage occurs as a consequence of spontaneous chemical reactions, such as base deamination; from the byproducts of normal metabolic processes, such as cellular respiration; from external mutagens and chemotherapeutics; or following aberrant or incomplete cellular processes such as aborted topoisomerase reactions and replication fork collapse [1]. Every type of DNA lesion is potentially harmful for the integrity of genetic information. DNA damage is therefore counteracted by an elaborate network of systems that detect, signal and repair DNA damage.

The first and most critical cellular response to DNA damage is the attempt to repair the lesion [1]. DNA repair likely evolved as soon as DNA became the repository of genetic information and the presence of DNA repair systems in every species, even in the smallest prokaryotic genomes, supports this idea [1–3]. DNA repair does not proceed in isolation but occurs in the context of ongoing cellular processes such as gene transcription, DNA replication, chromatin organization and cell cycle progression. As genome complexity and organization increased during evolution,

D. Durocher (✉)
Samuel Lunenfeld Research Institute, Centre for Systems Biology, Mount Sinai Hospital, 600 University Avenue, M5G 1X5, Toronto, ON, Canada
durocher@lunenfeld.ca

K.K. Khanna, Y. Shiloh (eds.), *The DNA Damage Response: Implications on Cancer Formation and Treatment*, DOI 10.1007/978-90-481-2561-6_1,

a necessity to coordinate DNA repair with these other cellular processes arose, in order to minimize the impact of DNA lesions (and that of DNA repair) on the overall integrity of the genome. The best example of this coordination is the potent slowdown of cell cycle progression in response to DNA damage. This response is termed the DNA damage checkpoint and was first hinted at by the studies of Painter and Young in 1980, which examined cell cycle progression in response to ionizing radiation (IR) in cells derived from ataxia-telangiectasia (A-T) patients [4]. Painter and Young observed, that in contrast to cells derived from "healthy" individuals, cells from A-T patients failed to arrest the cell cycle at the G2 stage in response to IR treatment [4]. They surmised that the resulting premature entry into mitosis could explain the profound radiosensitivity associated with A-T [4]. However, it was only at the end of the 1980s and the early 1990s that the concept of a DNA damage surveillance pathway emerged. Experimentally, this concept arose from the seminal discovery of the DNA damage checkpoint in yeast [5, 6]. Finally, one can consider that the field of DNA damage signaling was truly born when the Ser/Thr kinases ATM (ataxia telangiactesia-mutated), ATR (ATM and RAD3-related), DNA-PKcs (DNA-dependent protein kinase, catalytic subunit), CHK1 (checkpoint kinase 1) and CHK2 (checkpoint kinase 2) were discovered either in yeast or metazoan species, in rapid succession in the mid-nineties [7–19].

ATM, ATR and DNA-PKcs are large Ser/Thr protein kinases that are part of the PI(3) kinase-like kinase (PIKK) family [20], which also contains the TOR and SMG1 protein kinases [20–22]. The "DNA damage PIKKs" (ATM, ATR and DNA-PKcs) phosphorylate proteins on a highly similar Ser/Thr-Gln (S/T-Q) motif and are early participants in the response to DNA damage [23–25]. In fact, all three kinases are likely activated by DNA lesions [26–30] via the action of specific regulatory subunits (see below). The ATM and ATR kinases are considered to be the PIKKs that orchestrate DNA damage signaling whereas DNA-PKcs is primarily a DNA repair enzyme involved in sealing DNA double-strand breaks (DSBs) by non-homologous end-joining (NHEJ) [31].

It is important to underscore the fact that the DNA damage checkpoint is only one of many responses under the control of DNA damage signaling. Indeed, DNA repair, DNA damage-induced apoptosis, chromatin remodeling, transcriptional regulation, DNA replication progression, metabolic control and many other cellular pathways are under the influence of DNA damage signaling [32, 33]. Therefore, it is inadequate to think of DNA damage signaling as a linear pathway but rather it should be envisaged as an intricate network of responses, which we have only begun to unravel. Indeed, recent proteomics studies have identified upwards of several hundred PIKK substrates that are phosphorylated in response to DNA damage [34–36]. These substrates populate almost every significant biological process in the cell, highlighting the pleiotropic impact of DNA damage signaling on cellular physiology.

In this chapter, I will review the early steps in DNA damage signaling with a particular emphasis on the factors that might be sensing DNA lesions. In other words, I will review the current state of knowledge on how DNA lesions activate ATM and ATR. Finally, I will explore some unresolved questions in the field and will conclude this review by describing how our understanding of the early stages of DNA damage signaling might be translated in the clinical setting.

1.2 DNA Damage Sensing and the Initiation of DNA Damage Signaling

For DNA repair and DNA damage signaling to occur, a DNA lesion must first be detected or sensed. DNA damage sensing is better understood during DNA repair. Indeed, every DNA repair process is initiated by dedicated DNA damage sensors that recognize particular subsets of DNA lesions [1]. For example, the Ku heterodimer is a sensor of dsDNA ends [37, 38], while highly specialized glycosylases recognize specific base lesions during base excision repair [39]. These sensors are either DNA repair enzymes themselves or act as scaffolds for the recruitment of their associated repair systems.

Since a wide variety of DNA lesions elicit the activation of only two main signaling systems, controlled by ATM and ATR, the question of what senses DNA lesions and subsequently to initiate DNA damage signaling is considerably more challenging than it would appear at first glance. This observation raises the question of what exactly is sensed by the DNA damage signaling apparatus. Four main possibilities can be formulated: (1) the DNA damage sensors operating during DNA repair might participate in the activation of DNA damage signaling; (2) there might be "signaling-specific" sensors that turn on PIKK catalytic activity; (3) DNA damage signaling might monitor DNA repair rather than DNA lesions, and finally, (4) DNA damage signaling might monitor a common intermediate of DNA repair, rather than the primary lesions.

Apart from some specific cases, such as ATR activation in response to S(N)1 methylating agents [40, 41], DNA damage signaling is not initiated by the DNA damage sensors employed during DNA repair. Although it may at first seem odd that DNA damage signaling does not employ these highly selective and specific DNA damage sensors, coupling DNA damage signaling with the detection of every DNA lesion would have dire consequences. Indeed, the need for a selective activation of DNA damage signaling is easily understood when one considers that tens of thousands of oxidated bases are produced per cell per day. If DNA damage signaling were to be activated every time a glycosylase bound and excised a damaged base, it would present either a major impediment to cell cycle progression (due to a near permanent checkpoint) or to cell viability (via activation of the apoptotic program). In fact, DNA damage signaling must also have the built-in flexibility of being activated even when a given DNA repair process fails to initially detect a lesion. Perhaps for this reason, DNA damage signaling does not lie mechanistically downstream of DNA repair but is rather a parallel, albeit intermeshed, response.

As will be detailed in the sections below, activation of DNA damage signaling relies on specific sensors that monitor two key DNA structures: the DSB for the ATM pathway [27] and ssDNA with dsDNA/ssDNA junctions for the ATR pathway [29] (see Figs. 1.1 and 1.2 for schematic representation of their activation). These

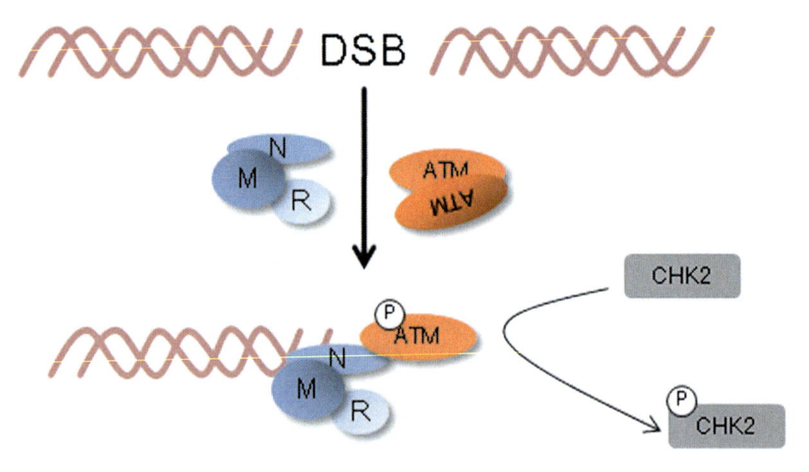

Fig. 1.1 Model of DNA damage sensing and ATM activation by MRN. In this model, the MRE11-RAD50-NBS1 (MRN) complex is first recruited to the DNA double-strand break (DSB) where it acts to convert ATM from an inactive dimer into an active monomer (which is phosphorylated) and by anchoring it to the DNA lesions. Chromatin-bound ATM is activated either by a particular DNA structure or by the C-terminus of NBS1 and then phosphorylates substrates such as components of the MRN complex (not shown) or nucleoplasmic substrates such as CHK2

two DNA structures can account for the ATM or ATR activation in response to virtually any type of DNA lesion. Indeed, ssDNA is an intermediate of many DNA repair processes such as nucleotide excision repair, BER and DSB repair. Moreover, large tracts of ssDNA are also produced by the uncoupling of the replicative helicase and the DNA polymerase at stalled replications forks [42, 43]. In parallel, DSBs not only occur as a consequence of ionizing radiation or following the action of clastogenic chemicals, but are also produced as a result of abortive DNA topoisomerase II reactions, nucleolytic processing of stalled DNA replication forks, dysfunctional telomeres or leading strand DNA replication forks encountering ssDNA gaps [44]. The recognition of ssDNA and DSBs for the initiation of DNA damage signaling therefore represents an elegant solution to the challenge of building a signaling network that monitors the myriad of types of DNA lesions that exist. Moreover, it implies that many DNA lesions must "mature" into a secondary lesion prior to PIKK activation. This latter property may provide a means to ensure that DNA damage signaling is activated solely when DNA repair fails, is slow or is overwhelmed by the sheer number of lesions.

1.3 ATM Signaling and the DNA Double-Strand Break Paradigm

The DNA double-strand break is considered the most harmful DNA lesion. A single unrepaired DSB can induce cell death [45] and DSBs are intermediates of many types of genome rearrangements [46]. Moreover, a number of DNA lesions (e.g.

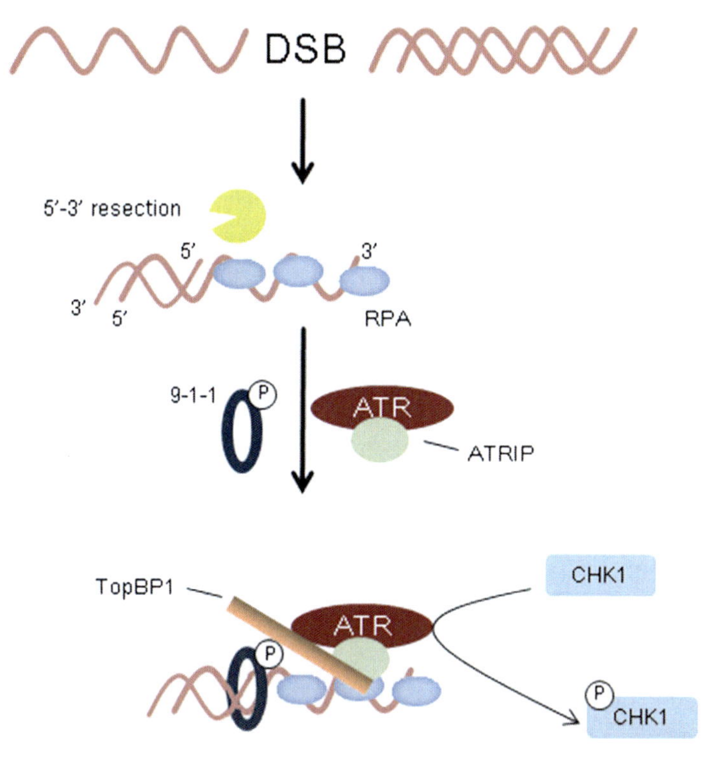

Fig. 1.2 Models of ATR activation. In this model ATR activation following a DSB depends first of the formation of ssDNA (resection) by the action of nucleases. ATR is activated following its recruitment to ssDNA coated with RPA. This recruitment occurs via a specific interaction between ATRIP and RPA. ATR recruitment is not sufficient to activate ATR but requires an interaction between TopBP1 and the kinase domain of ATR. TopBP1-dependent activation of ATR depends on the combined recruitment of ATRIP and 9-1-1 onto ssDNA. Active ATR can then phosphorylate substrates such as CHK1 (Claspin is required for this phosphorylation event but is not shown)

poisoned topoisomerase I reactions) can be converted into DSBs when replication forks encounter them, making the repair and detection of DSBs a last resort. As I elaborate below, the earliest steps of DSB signaling in eukaryotes are accomplished by the combined action of the ATM kinase and the Mre11-Rad50-NBS1 (MRN) complex.

1.3.1 ATM: The Master of DSB Signaling

ATM is the product of the gene mutated in ataxia-telangiectasia, a debilitating genetic disorder characterized by progressive neurodegeneration, immune system dysfunction, hypersensitivity to ionizing radiation and a marked predisposition to cancer [47]. The discovery of ATM was reported in 1995 by Yossi Shiloh and collaborators after a heroic positional cloning effort [11]. ATM is a large protein

(3056 amino acid residues) largely composed of HEAT repeats [48], which are predicted to form a superhelical structure with concave and convex faces [49]. The C-terminal end of ATM is composed of the PIKK-family kinase domain, which also contains, at its extreme C-terminus, a region named the "FATC" motif [50]. Another domain, termed FAT, is juxtaposed N-terminal to the kinase domain. It is not yet clear whether the FAT and FATC domains represent bona fide domains (i.e. autonomously folded protein modules) or whether they are simply regions of homology among PIKKs that reflect common ancestry. Indeed, the FAT domain is predicted to be largely composed of HEAT repeats [48]. Likewise, the FATC motif might simply be a constituent of the PIKK-family kinase domain. The current lack of structural information on ATM and the PIKK family members is perhaps one of the biggest impediments to obtaining a deep understanding of PIKK function.

In undamaged cells, ATM is in a dimeric (or multimeric) inactive form [51]. Following IR treatment, ATM activation correlates with autophosphorylation [51, 52], transition to a monomeric form [51] and localization to the chromatin that surrounds the DSB [53–56]. Active ATM is anchored on the chromatin that surrounds the DNA lesion where it phosphorylates chromatin-bound and nucleoplasmic substrates [57]. ATM phosphorylates its substrates on characteristic S/T-Q motifs [24, 25, 58, 59], but PIKKs (including ATM) can also phosphorylate some proteins on non-canonical sites [36, 60–62].

ATM is one of the earliest proteins that accumulates at DSB sites [54, 56] and is activated by DSB-inducing agents, strongly suggesting that ATM is activated by DSBs and not another type of DNA structure. In fact, studies in yeast suggest that ATM is selectively activated by minimally processed DSBs [63]. However, in a breakthrough study published in 2003, Bakkenist and Kastan observed an almost quantitative autophosphorylation of human ATM on Ser1981, even in response to a small number of DSBs [51]. This observation prompted the authors to propose that DSBs may not be the main activation signal for ATM, and suggested that ATM senses distortions in chromatin. This provocative idea that ATM is a sensor of chromatin structure rather than a DSB-activated kinase is challenged by a number of observations. Perhaps the most direct demonstration that ATM is activated by DSBs come from in vitro studies in *Xenopus* cell-free extracts [26, 64, 65] and with the reconstitution of ATM activation using recombinant mammalian proteins [27, 66]. In those studies, the addition of linear DNA (which mimics DSBs) to kinase reactions can stimulate ATM activity towards a model substrate several hundred-fold [27]. The DSB-dependent activation of ATM nevertheless requires a critical accessory factor, the MRN complex [27, 64–66].

1.3.2 ATM Activation by DSBs and MRN

The presence of mutations in the genes encoding the MRN components *MRE11* and *NBS1* results in syndromes that display a spectrum of clinical and cellular phenotypes that closely (although not entirely) match those displayed by A-T. These syndromes are the A-T-like disorder (ATLD) caused by mutations in *MRE11* [67] and the Nijmegen breakage syndrome (NBS) for mutations in *NBS1* [68–70].These observations hinted that the genes responsible for each disorder could act in the same pathway [67, 71]. Shortly after the discovery of *ATM* and *NBS1* [68–70], it was found that ATM can phosphorylate the components of the MRN complex in response to IR [72–75], which led to a model where MRN lies primarily downstream of ATM in terms of DNA damage signaling. However, it was soon recognized that in cells derived from NBS patients [76–78] or that are depleted in MRN components [79, 80], ATM is profoundly impaired in its ability to phosphorylate a number of its substrates, including SMC1 and CHK2. Collectively, these results argued that instead of acting downstream of ATM, the MRN complex rather acts upstream of ATM during DNA damage signaling.

These important findings are supported by work in budding yeast where the accumulation of the ATM ortholog (Tel1) at sites of DNA damage is entirely dependent on the MRN complex [56, 81]. In particular, Nakada et al. [81] discovered that the C-terminal region in Xrs2 (the NBS1 homolog) mediates a protein-protein interaction with Tel1 (ATM) that is necessary for the accumulation of the latter at sites of DNA damage. Subsequent work in human cells and in fission yeast corroborated this initial study and provided evidence that the NBS1-ATM interaction is conserved across eukaryotes [64, 82]. However, whereas in yeast the deletion of the critical C-terminal binding motif in Xrs2 (NBS1) largely phenocopies the deletion of *TEL1* (*ATM*), deletion of the Nbs1 C-terminus in mice only mildly affects ATM signaling [83, 84]. The reason for this discrepancy is not clear but might point to alternative pathways of ATM recruitment and/or activation. Indeed, in recent years, proteins such as ATMIN [85], p18/AIMP3 [86], Tip60 [87], hMOF [88], and hSSB1 [89] have all been involved in the regulation of ATM activity and might define NBS1-independent pathways of ATM regulation that could compensate for the loss of the NBS1-ATM interaction in the $Nbs1^{\Delta C}$ mouse.

Remarkably, the dependence on MRN for ATM activation can also be observed in cell-free systems using human- or *Xenopus*-derived proteins [26, 27, 65, 66] providing a relatively facile tool to decipher the mechanics of ATM activation. With these systems, it was found that in addition to guiding ATM to DNA ends and promoting the transition of ATM into a monomer [27, 65], the MRN complex plays at least one additional role in ATM activation [27, 65]. The nature of this other role is more difficult to address given the combined lack of structural information on ATM and the lack of true separation-of-function mutations in the MRN complex. Two broad roles have been proposed. Firstly, NBS1 could directly activate ATM, as suggested by the ability of an ATM-binding peptide derived from the NBS1 C-terminus to modestly activate ATM [65]. A second possibility is based on corroborating observations that found that mutation of the Ser1202 residue in Rad50 abolishes ATM activation by MRN [27, 65]. This mutation disrupts the adenylate kinase [90], DNA end tethering [65] and DNA unwinding [91] activities of the MRN complex. Lee and Paull [27] directly tested whether the impaired ATM activation caused by the Rad50 S1202R mutation was due to a defect in DNA end unwinding by supplementing kinase reactions containing ATM and the MR(S1202R)N complex with a dsDNA molecule containing non-complimentary DNA ends as a mimic of unwound DNA. Strikingly, whereas the MR(S1202R)N complex cannot activate ATM in the presence of dsDNA molecules with complementary ends, the partially unwound DNA molecules supported ATM activation even in the presence of MR(S1202R)N [27]. These results suggest that unwinding DNA ends might be an important function of the MRN complex during ATM activation. Lastly, it is tempting to speculate that the recently described ssDNA-binding protein hSSB1 [89] might act to promote ATM activity by stabilizing DNA structures such as unwound DNA ends.

This body of work firmly points to the MRN complex as the DSB sensor during ATM-dependent signaling. However, there is an important caveat that future studies will need to address in order to finally confirm whether MRN is indeed the DSB sensor and ATM activator. This caveat relates to another function of MRN in promoting a "feed-forward" amplification mechanism that is initiated following H2AX phosphorylation and MDC1 recruitment [92, 93]. Indeed after the initial activation of ATM, the histone variant H2AX becomes rapidly phosphorylated on its C-terminus to form so-called γ-H2AX [94, 95], an event that promotes the recruitment of MDC1 [92, 96, 97]. In turn, MDC1 can bind to ATM and MRN, providing a means to amplify the recruitment (and activation) of ATM to sites of DNA damage [92]. The potential contribution of this amplification mechanism can be best illustrated in recent experiments where the immobilization of MDC1 or MRN components on chromatin was sufficient to activate an ATM-dependent response [98]. Since most of the in vivo studies that examine ATM activation and recruitment are dependent on this feed-forward mechanism, a challenge in the future will be to design experimental systems that specifically examine the initial activation of ATM rather than relying on biological readouts that are influenced by γ-H2AX/MDC1-dependent amplification of the ATM signal.

1.3.3 Is ATM a DNA-Activated Kinase?

The experiments described above could be interpreted as indicating that ATM is either activated directly by MRN or by a particular DNA structure (e.g. unwound DNA ends). In the latter case, MRN would simply act as a "recruiter" of ATM to DSBs where it is activated by the DNA structure itself. Such a means of PIKK activation is responsible for the DNA end-dependent activation of DNA-PKcs. Indeed, DNA-PKcs can be activated directly by DSBs in vitro, under conditions of low salt [99], but its activation by DNA ends at physiological salt concentration is greatly stimulated by Ku, suggesting a model whereby the Ku70/80 heterodimer, in vivo, is instrumental for guiding and stabilizing DNA-PKcs at DNA ends, its activating structure [82, 99]. The observation that ATM can be activated by a large number of DNA ends in *Xenopus* extracts, independently of MRN [65] is certainly consistent with the possibility that ATM is directly activated by DNA.

In contrast, the observation that a peptide derived from the NBS1 C-terminus promotes ATM activation [65] tends to support an alternative model whereby MRN is a direct activator of ATM. Protein-based activation of PIKKs is in fact a common theme as well. For example, GβL binds to the kinase domain of mTOR to stabilize its interaction with its activator, Raptor [100]. In addition, ATRIP stabilizes the interaction of TopBP1 with ATR to induce its activation [101]. Further biochemical and structural characterization of ATM with DNA and/or MRN will be necessary to finally elucidate the exact mechanisms of ATM activation.

1.4 ATR Signaling: The Two-Man Rule

ATR is the second major PIKK involved in DNA damage signaling and is encoded by an essential gene in most eukaryotes [13, 102, 103]. Like ATM, ATR is a relatively large protein (2644 amino acid residues) composed chiefly of HEAT repeats as well as C-terminal PIKK-family catalytic domain [14]. ATR forms an obligate dimer with ATRIP in all species examined [104–107]. The ATR-ATRIP complex functions largely in parallel to ATM and is activated by agents that induce DSBs, such as UV lesions, alkylation damage and DNA replication stress [40, 41, 108–111]. ATR activity is particularly important to prevent the initiation of mitosis in the presence of unreplicated DNA [108, 112]. CHK1 represents the major target of ATR during the DNA damage checkpoint [111, 113] although it is clear that ATR has numerous other substrates of physiological importance [114].

1.4.1 The Role of ssDNA in ATR Activation

Pioneering studies from Garvik and Hartwell [115] in *Saccharomyces cerevisiae* led early on to the hypothesis that ssDNA was a prime candidate for a DNA damage signal. Since budding yeast DNA damage signaling relies primarily on the ATR homolog, Mec1, these studies suggested that ATR (Mec1) might be activated by ssDNA structures. This idea is supported by the fact that ATR responds to a variety of DNA lesions such as DSBs, replication stress and UV lesions, which all have, as a common intermediate, ssDNA.

More definitive evidence that ATR is activated by ssDNA came from studies in *Xenopus* cell-free extracts [116, 117], and in human and yeast cells [118]. Collectively, these studies demonstrated that ATR activation correlates with the presence of single-stranded gaps in DNA. Moreover, ATR is recruited and activated in a manner that depends on the presence of RPA, which is associated with ssDNA [117, 119] via a direct interaction between ATRIP and RPA [119].

However, the mere recruitment of ATR to ssDNA is clearly insufficient for ATR activation [29, 43, 120] indicating that additional events must occur prior to ATR being turned on. Clues as to the nature of this event came from studies in budding yeast where it was found that activation of DNA damage signaling by Mec1 (ATR) correlates with its co-localization with the 9-1-1 complex (composed of Ddc1, Rad17 and Mec3 in budding yeast; RAD9, RAD1 and HUS1 in fission yeast and vertebrates) [121, 122]. The 9-1-1 complex is analogous to the replicative sliding clamp, PCNA (proliferating cell nuclear antigen), and is loaded onto DNA by a specialized RFC (replication factor C) complex containing the human RAD17 protein (Rad24 in budding yeast). RFC^{RAD17} loads 9-1-1 at dsDNA/ssDNA junctions [123, 124] and biochemical studies indicate that in the presence of RPA, the 9-1-1 complex is loaded at the 5' primer end [123]. Together, these studies suggest that ATR is activated by primed ssDNA.

There are therefore two sensors in the ATR branch of DNA damage signaling: ATRIP-ATR, which recognizes tracts of ssDNA bound by RPA, and RFC^{RAD17} that recognizes RPA-bound primed DNA structures to load the 9-1-1 complex. A recent study by Toczyski and colleagues [125] extends this model to propose that the role of ssDNA is solely to colocalize the ATR and 9-1-1 complexes (i.e. that ATR is not activated by ssDNA per se). This conclusion was reached following the observation that the forced colocalization of Ddc2 (ATRIP) and 9-1-1 onto repetitive dsDNA arrays was sufficient to trigger a DNA damage checkpoint [125].

The necessity for the independent recruitment of two DNA damage sensors prior to ATR activation is reminiscent of the "two-man rule" [114]. The two-man rule originates from the military realm where it describes a system designed to prevent the accidental launch of nuclear weapons by a single individual. It essentially imposes a requirement that two operators complete independent sets of tasks in order to permit a weapons launch. Viewed from this angle, the combined localization of ATR and 9-1-1 might be designed to prevent the spurious activation of ATR-dependent signaling. As the ectopic activation of ATR, even in the absence of DNA breaks, leads to a permanent cell cycle arrest characteristic of cellular senescence [126], it might be critical to ensure that ATR is activated only when appropriate.

1.4.2 The Role of 9-1-1 and TopBP1 in ATR Activation

Two possible mechanisms of action could explain how the 9-1-1 complex participates in ATR activation. The first one is based on the homology to PCNA, which acts as a processivity factor during DNA replication by tethering key enzymes to the moving DNA replication fork [127]. In this model, 9-1-1 would stabilize ATR, or an activator of ATR, to the chromatin. In the second model, 9-1-1 would simply be a direct activator of ATR.

At the moment, there is support for both models. The case for 9-1-1 being a direct activator of ATR is supported by in vitro reconstitution experiments using purified yeast proteins [128]. These studies showed that the loading of 9-1-1 onto 5' ssDNA/dsDNA junctions can activate Mec1 (ATR) in complex with Ddc2 (ATRIP). Interestingly, at low-salt concentrations, the loading of 9-1-1 is no longer necessary and, under those conditions, Ddc1 (the yeast homolog of human RAD9) can alone promote Mec1 (ATR) activation [128]. However an important caveat of these studies is that all components were purified from yeast cells and thus cofactors could have been co-purified in substochiometric quantities, but at sufficient levels to impart to Ddc1 (RAD9) its Mec1 (ATR)-activating properties.

The C-terminus of RAD9, and in particular phosphorylation of the RAD9 C-terminal tail, is critical for ATR-dependent phosphorylation of CHK1 in every organism examined [129–132]. Analysis of the phosphorylated tail of RAD9 identifies a single site, Ser387 (human nomenclature) that is essential for ATR-dependent signaling [130–132]. This phosphorylated site defines an epitope that is recognized by the first two BRCT domains of TopBP1 [129–132]. This interaction is critical for ATR activation since TopBP1 is a potent activator of ATR catalytic activity in *Xenopus* cell extracts [133] and in immune kinase reactions [101]. Remarkably, a region on TopBP1 between the BRCT domains 6 and 7 is necessary and sufficient for ATR activation and has thus been dubbed the ATR-activation domain (AAD). Overexpression of the AAD in fusion to GFP [133], PCNA or histone H2B [130] is sufficient to activate ATR and bypasses the need for RAD9, indicating that the main role of 9-1-1 is to bring TopBP1 in the vicinity of the ATRIP-ATR complex. Recent studies from the Cortez laboratory further probed the mechanism of TopBP1-mediated activation of ATR and found that ATRIP collaborates with 9-1-1 to stabilize the interaction of TopBP1 with a segment of the ATR kinase domain that ultimately results in the stimulation of ATR catalytic activity [101]. This interesting study suggests that the two-man rule operates by stimulating TopBP1-dependent regulation of ATR activity. By analogy, the two men (9-1-1 and ATRIP) together turn the key (TopBP1) to initiate ATR signaling.

1.5 Unresolved Questions

After nearly two decades of research, DNA damage signaling as a field of study is reaching maturity. However, there remain a surprisingly large number of unresolved questions that are yet to be addressed satisfactorily. Below, I review a personal list of four unresolved questions that merit further investigation.

1.5.1 How is DNA Damage Sensed During DNA Damage Signaling?

Perhaps the most fundamental unanswered question in DNA damage signaling remains the elucidation of the exact mechanisms by which DSBs and ssDNA are sensed. Although we now have strong candidates for DNA damage signaling sensors (MRN for DSBs; RPA-ATRIP and RFC^{RAD17} for ssDNA) it is still unclear how they recognize lesions. For example, how does the MRN complex recognize DSBs and how does the binding of MRN to DNA lesions initiate a chain of events that ultimately leads to ATM activation? Secondly, for ssDNA and ATR, can ATRIP discriminate between RPA-bound ssDNA that is a DNA lesion (e.g. stalled replication fork) versus “normal” ssDNA, which is a rather ubiquitous structure in cells?

Studies in budding yeast suggest that the ssDNA tract must achieve a certain length prior to the accumulation of ATR at the DNA lesion [134] and work in *Xenopus* cell-free extracts also suggest that the length of ssDNA is important for ATR activation [29, 43]. How is this length-dependence achieved? In contrast, the 9-1-1 complex probably only needs a 5' ssDNA/dsDNA junction to be loaded [123, 134], also posing a question regarding the "topology" of the sensor complexes. Indeed, is ATR activation triggered by the juxtaposition of minimally overlapping populations of the ATR and 9-1-1 complexes, or as recently implied by the findings of Cortez and colleagues [101], are the ATR and 9-1-1 populations intermingled on ssDNA?

1.5.2 How are PIKK Signaling Thresholds Established?

As described above, DNA damage signaling need not to be activated every time a DNA lesion is encountered. Indeed, as most DNA lesions are rapidly repaired, the cell must "discriminate" between lesions that are on their way to being repaired from those that require DNA damage signaling. A failure to do so would result in major cell cycle delays at best and in excessive apoptosis at worst. Therefore, DNA damage signaling must have intrinsic "thresholds" that gate the extent of the response. Such an idea is self-evident when one considers the ubiquity of ssDNA during DNA replication. Moreover, recent work established that the G2/M checkpoint might require up to ten DSBs to be fully active. However, the latter study employed γ-H2AX as a marker of DSBs, which is itself a marker of activated PIKK, indicating that PIKKs can be activated locally without triggering a full DNA damage response. Therefore, DNA damage signaling is a gated response that triggers specific biological outputs when certain conditions or thresholds are met. It will be critical in the future to unravel at the molecular level how thresholds and switches are established during DNA damage signaling. Phosphorylation-dependent processes such as those that control switch-like behaviors in other signaling systems [135] are undoubtedly at play, but it seems also possible that the translocation and accumulation of DNA damage signaling proteins at sites of DNA damage into subnuclear foci might be part of this process.

1.5.3 Does DNA-PKcs Play a Signaling Role?

In mouse, the combined deletion of the *Atm* and *Prkdc* (encoding murine DNA-PKcs) genes is synthetically lethal [136]. This intriguing observation has two main interpretations: firstly, loss of DNA-PK activity may increase the number of DSBs, placing an additional burden on ATM-dependent responses. The converse (where loss of ATM would generate DNA breaks requiring DNA-PK) may also be at play. These two related scenarios suggest that ATM and DNA-PK act in parallel to enforce genome integrity. However, a second, and largely unexplored, possibility is that DNA-PK compensates for the absence of ATM by performing a signaling role. This possibility is not unprecedented as H2AX phosphorylation is both ATM- and DNA-PK- dependent [95]. Furthermore, there is anecdotal evidence that the acute inhibition of ATM with the widely employed KU-55933 inhibitor [137] often produces a more profound impact on ATM signaling (e.g. H2AX phosphorylation or 53BP1 localization to foci [138]) than what is observed in A-T cells. The existence of mouse models deficient in ATM, DNA-PKcs and other components of the non-homologous end-joining pathway should allow us to answer whether it is a repair or signaling function of DNA-PKcs that is important to maintain ATM-deficient embryos alive.

1.5.4 Disassembly of DNA Damage Sensing Complexes

Lastly, despite our increasingly advanced knowledge of the initial steps of DNA damage signaling, our understanding of how this signaling is terminated is comparatively more fragmentary. Although it is possible that DNA damage sensing complexes are passively disassembled as DNA repair occurs, there is also evidence that active processes participate in removing ATM and ATR from chromatin. A candidate for such activity exists in budding yeast where the Srs2 helicase has been proposed to strip Mec1 (ATR) from the chromatin that surrounds repaired DNA lesions [139]. Such activities could also participate in a "proofreading" process that removes accidentally localized ATM or ATR from chromatin. If such activities can be identified they will likely play important roles as modifiers of DNA damage signaling.

1.6 DNA Damage Signaling and Sensing: Clinical Perspectives

There is currently considerable interest in targeting DNA damage signaling pathways in the clinical setting. In addition to the challenge of restoring DNA damage signaling in patients with A-T and related diseases, DNA damage signaling components are being explored as drug targets for cancer therapy and as important biomarkers of DNA damage. Here I will only briefly review some of the clinical implications that pertain specifically to the early steps of DNA damage signaling.

1.6.1 Biomarkers

As radiotherapy and many cancer chemotherapeutics produce DNA damage as their mode of action, they almost invariably engage the ATM- or ATR-dependent DNA damage signaling pathways. There is therefore great interest in employing PIKK-dependent phosphorylation events as biomarkers in pre-clinical studies and in clinical trials that involve radiation therapy and agents that damage DNA. The usefulness of monitoring DNA damage resides in the ability to monitor DNA damage in situ, either at the site of the tumor or in normal tissues.

In a series of seminal papers, the activation of DNA damage signaling was found to be a hallmark of cellular senescence irrespective of whether it is triggered by telomere attrition (replicative senescence) [140] or the activation of oncogenes [141–143]. In addition to providing a new paradigm in cancer progression, these studies have identified activated DNA damage signaling as a biomarker for the identification and stratification of malignant and pre-malignant lesions [144–146] that will certainly be employed in the clinical setting in the near future.

Phosphorylation of H2AX on Ser139 (γ-H2AX) [147] is by far the most widely used biomarker of the activated DNA damage response. As mentioned above, H2AX phosphorylation in response to DSBs is primarily ATM- and DNA-PK-dependent [95] whereas it is ATR-dependent in response to replication stress [148]. H2AX phosphorylation is at least 10-fold more sensitive than techniques such as single-cell gel electophoresis (comet) assays to detect DSBs, and H2AX phosphorylation can be monitored in a variety of ways, including immunofluorescence [149], immunoblotting [149] and flow cytometry [150]. These properties make the detection of H2AX automatable and amenable to a molecular diagnostics laboratory. H2AX phosphorylation is currently being proposed as a biomarker for various clinical trials such as those examining the efficacy of next-generation topoisomerase I inhibitors [151].

However, it should be noted that H2AX phosphorylation has the disadvantage of not being entirely dependent on DNA damage. Indeed, γ-H2AX staining can be seen at low levels in a number of cell types in the absence of DNA damage [152, 153]. Thus it may be preferable to examine other robust markers of the activated DNA damage response such as phosphorylation of ATM, SMC1, KAP-1 or NBS1 [51, 62, 74, 154]. The recent identification of hundreds of possible PIKK-dependent phosphorylation events certainly provides a rich list to explore.

1.6.2 DNA Damage Signaling Inhibitors

Cancer therapy relies heavily on the use of DNA damage as a means to kill malignant cells. There is therefore great interest in developing pharmacological agents that inhibit the early steps of DNA damage signaling as a means to sensitize cancer cells to radiation or chemotherapeutics. This subject has been extensively reviewed elsewehere [155–157]. At the moment, pharmacological agents that target the ATM, DNA-PK, CHK1 and Chk2 kinases are being evaluated in pre-clinical studies or in clinical trials for tumor sensitization.

Another possible application of DNA damage signaling inhibitors is rooted in the observation of a synthetic lethality between the deletions of ATM and DNA-PK in mice. Indeed, leukemias such as mantle cell lymphoma often contain ATM-inactivating mutations [158] that might render them highly sensitive to DNA-PK inhibition. The identification of ATM mutations in a number of other types of cancers of different origins [159] raises the prospect of employing the synthetic lethality between the loss of ATM and DNA-PK activity for therapeutic purposes.

In addition to inhibition of checkpoint kinases, other proteins within the early steps of DNA damage signaling may be amenable to inhibition by small molecules. The leading candidate is the MRN complex, with its multiple enzymatic activities. In fact, a recent proof-of-principle study has identified a compound named Mirin as an inhibitor of MRN-dependent ATM activation [160], indicating that MRN inhibition is possible and could be a suitable route for the inhibition of DNA damage signaling. Another possible, yet challenging, task would be to identify strategies that rely on disrupting the key protein-protein interactions that lead to the assembly of activated PIKK complexes at sites of DNA lesions. Examples of interactions that could be targeted are the 9-1-1/TopBP1 and NBS1/ATM physical contacts.

1.6.3 Suppressors of DNA Damage Signaling Defects

The A-T, ATLD and NBS syndromes are truly debilitating diseases. One important goal of our field must be the identification of drugs or strategies that alleviate or even cure these diseases. How can our increasing knowledge of the early steps of DNA damage signaling help us with this important task?

Apart from targeting the quality control pathways that result in the degradation of missense variants of ATM (e.g. using a recently described inhibitor of nonsense-mediated mRNA decay [161]), a potentially interesting strategy to explore could be the identification of compounds that either potentiate ATM signaling or that shunt signaling towards the DNA-PK- or ATR-dependent signaling pathways. In fact, genetic studies in budding yeast [162, 163] and mice [102] instruct us that such an approach might be feasible. Such hope is primarily based on the observation that the *Rad50S* mutation in mice results in a striking suppression of ATM deficiency [164]. The mechanism of suppression by Rad50S is not well understood but since this mutation limits the ability of the MRN complex to process DSBs into ssDNA [165, 166], it is tempting to speculate that an upregulation of DNA-PK-dependent signaling, rather than ATR signaling is responsible for the suppression of ATM deficiency. No matter what the mechanism by which Rad50S leads to ATM suppression, it might be possible to screen small molecule libraries in order to identify compounds that mimic Rad50S and bypass the loss of ATM. Such compounds would obviously be of great interest for the treatment of A-T and related diseases, and screens that are designed to achieve this feat are already under way in a number of laboratories.

Acknowledgements I am indebted to Rachel Szilard who has patiently proofread this manuscript. The ideas on the bypass of the loss of ATM by activating DNA-PK were formed during discussions with Steve Jackson and Andre Nussenzweig. Finally, work on DNA damage signaling in my laboratory is supported by grants from the Canadian Institutes of Health Research (MOP89754 and MOP84297).

References

1. Friedberg EC (2003) DNA damage and repair. Nature 421: 436–440.
2. Fraser CM, Gocayne JD, White O, Adams MD, Clayton RA, Fleischmann RD, Bult CJ, Kerlavage AR, Sutton G, Kelley JM, Fritchman RD, Weidman JF, Small KV, Sandusky M, Fuhrmann J, Nguyen D, Utterback TR, Saudek DM, Phillips CA, Merrick JM, Tomb JF, Dougherty BA, Bott KF, Hu PC, Lucier TS, Peterson SN, Smith HO, Hutchison CA, 3rd, and Venter JC (1995) The minimal gene complement of Mycoplasma genitalium. Science 270: 397–403.
3. Waters E, Hohn MJ, Ahel I, Graham DE, Adams MD, Barnstead M, Beeson KY, Bibbs L, Bolanos R, Keller M, Kretz K, Lin X, Mathur E, Ni J, Podar M, Richardson T, Sutton GG, Simon M, Soll D, Stetter KO, Short JM, and Noordewier M (2003) The genome of Nanoarchaeum equitans: insights into early archaeal evolution and derived parasitism. Proc Natl Acad Sci USA 100: 12984–12988.
4. Painter RB, and Young BR (1980) Radiosensitivity in ataxia-telangiectasia: a new explanation. Proc Natl Acad Sci USA 77: 7315–7317.
5. Weinert TA, and Hartwell LH (1988) The RAD9 gene controls the cell cycle response to DNA damage in Saccharomyces cerevisiae. Science 241: 317–322.
6. Hartwell LH, and Weinert TA (1989) Checkpoints: controls that ensure the order of cell cycle events. Science 246: 629–634.
7. Hartley KO, Gell D, Smith GC, Zhang H, Divecha N, Connelly MA, Admon A, Lees-Miller SP, Anderson CW, and Jackson SP (1995) DNA-dependent protein kinase catalytic subunit: a relative of phosphatidylinositol 3-kinase and the ataxia telangiectasia gene product. Cell 82: 849–856.
8. Jackson SP, MacDonald JJ, Lees-Miller S, and Tjian R (1990) GC box binding induces phosphorylation of Sp1 by a DNA-dependent protein kinase. Cell 63: 155–165.
9. Lees-Miller SP, and Anderson CW (1989) The human double-stranded DNA-activated protein kinase phosphorylates the 90-kDa heat-shock protein, hsp90 alpha at two NH2-terminal threonine residues. J Biol Chem 264: 17275–17280.
10. Sun Z, Fay DS, Marini F, Foiani M, and Stern DF (1996) Spk1/Rad53 is regulated by Mec1-dependent protein phosphorylation in DNA replication and damage checkpoint pathways. Genes Dev 10: 395–406.
11. Savitsky K, Barshira A, Gilad S, Rotman G, Ziv Y, Vanagaite L, Tagle DA, Smith S, Uziel T, Sfez S, Ashkenazi M, Pecker I, Frydman M, Harnik R, Patanjali SR, Simmons A, Clines GA, Sartiel A, Gatti RA, Chessa L, Sanal O, Lavin MF, Jaspers NGJ, Malcolm A, Taylor R, Arlett CF, Miki T, Weissman SM, Lovett M, Collins FS, and Shiloh Y (1995) A single ataxia telangiectasia gene with a product similar to Pi 3 kinase. Science 268: 1749–1753.
12. Greenwell PW, Kronmal SL, Porter SE, Gassenhuber J, Obermaier B, and Petes TD (1995) TEL1, a gene involved in controlling telomere length in S. cerevisiae, is homologous to the human ataxia telangiectasia gene. Cell 82: 823–829.
13. Kato R, and Ogawa H (1994) An essential gene, ESR1, is required for mitotic cell growth, DNA repair and meiotic recombination in Saccharomyces cerevisiae. Nucleic Acids Res 22: 3104–3112.
14. Bentley NJ, Holtzman DA, Flaggs G, Keegan KS, DeMaggio A, Ford JC, Hoekstra M, and Carr AM (1996) The Schizosaccharomyces pombe rad3 checkpoint gene. Embo J 15: 6641–6651.
15. Matsuoka S, Huang M, and Elledge SJ (1998) Linkage of ATM to cell cycle regulation by the Chk2 protein kinase. Science 282: 1893–1897.
16. Walworth N, Davey S, and Beach D (1993) Fission yeast Chk1 protein kinase links the rad checkpoint pathway to Cdc2. Nature 363: 368–371.
17. Boddy MN, Furnari B, Mondesert O, and Russell P (1998) Replication checkpoint enforced by kinases Cds1 and Chk1. Science 280: 909–912.

18. Allen JB, Zhou Z, Siede W, Friedberg EC, and Elledge SJ (1994) The SAD1/RAD53 protein kinase controls multiple checkpoints and DNA damage-induced transcription in yeast. Genes Dev 8: 2401–2415.
19. Hari KL, Santerre A, Sekelsky JJ, McKim KS, Boyd JB, and Hawley RS (1995) The mei-41 gene of D. melanogaster is a structural and functional homolog of the human ataxia telangiectasia gene. Cell 82: 815–821.
20. Durocher D, and Jackson SP (2001) DNA-PK, ATM and ATR as sensors of DNA damage: variations on a theme? Curr Opin Cell Biol 13: 225–231.
21. Yamashita A, Ohnishi T, Kashima I, Taya Y, and Ohno S (2001) Human SMG-1, a novel phosphatidylinositol 3-kinase-related protein kinase, associates with components of the mRNA surveillance complex and is involved in the regulation of nonsense-mediated mRNA decay. Genes Dev 15: 2215–2228.
22. Brumbaugh KM, Otterness DM, Geisen C, Oliveira V, Brognard J, Li X, Lejeune F, Tibbetts RS, Maquat LE, and Abraham RT (2004) The mRNA surveillance protein hSMG-1 functions in genotoxic stress response pathways in mammalian cells. Mol Cell 14: 585–598.
23. Bannister AJ, Gottlieb TM, Kouzarides T, and Jackson SP (1993) c-Jun is phosphorylated by the DNA-dependent protein kinase in vitro; definition of the minimal kinase recognition motif. Nucleic Acids Res 21: 1289–1295.
24. O′Neill T, Dwyer AJ, Ziv Y, Chan DW, Lees-Miller SP, Abraham RH, Lai JH, Hill D, Shiloh Y, Cantley LC, and Rathbun GA (2000) Utilization of oriented peptide libraries to identify substrate motifs selected by ATM. J Biol Chem 275: 22719–22727.
25. Kim ST, Lim DS, Canman CE, and Kastan MB (1999) Substrate specificities and identification of putative substrates of ATM kinase family members. J Biol Chem 274: 37538–37543.
26. You Z, Bailis JM, Johnson SA, Dilworth SM, and Hunter T (2007) Rapid activation of ATM on DNA flanking double-strand breaks. Nat Cell Biol 9: 1311–1318.
27. Lee JH, and Paull TT (2005) ATM activation by DNA double-strand breaks through the Mre11-Rad50-Nbs1 complex. Science 308: 551–554.
28. Smith GC, Cary RB, Lakin ND, Hann BC, Teo SH, Chen DJ, and Jackson SP (1999) Purification and DNA binding properties of the ataxia-telangiectasia gene product ATM. Proc Natl Acad Sci USA 96: 11134–11139.
29. MacDougall CA, Byun TS, Van C, Yee MC, and Cimprich KA (2007) The structural determinants of checkpoint activation. Genes Dev 21: 898–903.
30. Hall-Jackson CA, Cross DA, Morrice N, and Smythe C (1999) ATR is a caffeine-sensitive, DNA-activated protein kinase with a substrate specificity distinct from DNA-PK. Oncogene 18: 6707–6713.
31. Hefferin ML, and Tomkinson AE (2005) Mechanism of DNA double-strand break repair by non-homologous end joining. DNA Repair (Amst) 4: 639–648.
32. Rouse J, and Jackson SP (2002) Interfaces between the detection, signaling, and repair of DNA damage. Science 297: 547–551.
33. Zhou BB, and Elledge SJ (2000) The DNA damage response: putting checkpoints in perspective. Nature 408: 433–439.
34. Matsuoka S, Ballif BA, Smogorzewska A, McDonald ER, 3rd, Hurov KE, Luo J, Bakalarski CE, Zhao Z, Solimini N, Lerenthal Y, Shiloh Y, Gygi SP, and Elledge SJ (2007) ATM and ATR substrate analysis reveals extensive protein networks responsive to DNA damage. Science 316: 1160–1166.
35. Stokes MP, Rush J, Macneill J, Ren JM, Sprott K, Nardone J, Yang V, Beausoleil SA, Gygi SP, Livingstone M, Zhang H, Polakiewicz RD, and Comb MJ (2007) Profiling of UV-induced ATM/ATR signaling pathways. Proc Natl Acad Sci USA 104: 19855–19860.
36. Smolka MB, Albuquerque CP, Chen SH, and Zhou H (2007) Proteome-wide identification of in vivo targets of DNA damage checkpoint kinases. Proc Natl Acad Sci USA 104: 10364–10369.

37. Mimori T, and Hardin JA (1986) Mechanism of interaction between Ku protein and DNA. J Biol Chem 261: 10375–10379.
38. Dynan WS, and Yoo S (1998) Interaction of Ku protein and DNA-dependent protein kinase catalytic subunit with nucleic acids. Nucleic Acids Res 26: 1551–1559.
39. Scharer OD, and Jiricny J (2001) Recent progress in the biology, chemistry and structural biology of DNA glycosylases. Bioessays 23: 270–281.
40. Yoshioka K, Yoshioka Y, and Hsieh P (2006) ATR kinase activation mediated by MutSalpha and MutLalpha in response to cytotoxic O6-methylguanine adducts. Mol Cell 22: 501–510.
41. Stojic L, Mojas N, Cejka P, Di Pietro M, Ferrari S, Marra G, and Jiricny J (2004) Mismatch repair-dependent G2 checkpoint induced by low doses of SN1 type methylating agents requires the ATR kinase. Genes Dev 18: 1331–1344.
42. Sogo JM, Lopes M, and Foiani M (2002) Fork reversal and ssDNA accumulation at stalled replication forks owing to checkpoint defects. Science 297: 599–602.
43. Byun TS, Pacek M, Yee MC, Walter JC, and Cimprich KA (2005) Functional uncoupling of MCM helicase and DNA polymerase activities activates the ATR-dependent checkpoint. Genes Dev 19: 1040–1052.
44. Jeggo PA, and Lobrich M (2007) DNA double-strand breaks: their cellular and clinical impact? Oncogene 26: 7717–7719.
45. Sandell LL, and Zakian VA (1993) Loss of a yeast telomere: arrest, recovery, and chromosome loss. Cell 75: 729–739.
46. Wyman C, and Kanaar R (2006) DNA double-strand break repair: all′s well that ends well. Annu Rev Genet 40: 363–383.
47. Lavin MF, and Shiloh Y (1997) The genetic defect in ataxia-telangiectasia. Annu Rev Immunol 15: 177–202.
48. Perry J, and Kleckner N (2003) The ATRs, ATMs, and TORs are giant HEAT repeat proteins. Cell 112: 151–155.
49. Kobe B, Gleichmann T, Horne J, Jennings IG, Scotney PD, and Teh T (1999) Turn up the HEAT. Structure 7: R91–97.
50. Bosotti R, Isacchi A, and Sonnhammer EL (2000) FAT: a novel domain in PIK-related kinases. Trends Biochem Sci 25: 225–227.
51. Bakkenist CJ, and Kastan MB (2003) DNA damage activates ATM through intermolecular autophosphorylation and dimer dissociation. Nature 421: 499–506.
52. Kozlov S, Gueven N, Keating K, Ramsay J, and Lavin MF (2003) ATP activates ataxia-telangiectasia mutated (ATM) in vitro. Importance of autophosphorylation. J Biol Chem 278: 9309–9317.
53. Bekker-Jensen S, Lukas C, Kitagawa R, Melander F, Kastan MB, Bartek J, and Lukas J (2006) Spatial organization of the mammalian genome surveillance machinery in response to DNA strand breaks. J Cell Biol 173: 195–206.
54. Andegeko Y, Moyal L, Mittelman L, Tsarfaty I, Shiloh Y, and Rotman G (2001) Nuclear retention of ATM at sites of DNA double strand breaks. J Biol Chem 276: 38224–38230.
55. Berkovich E, Monnat RJ, Jr., and Kastan MB (2007) Roles of ATM and NBS1 in chromatin structure modulation and DNA double-strand break repair. Nat Cell Biol 9: 683–690.
56. Lisby M, Barlow JH, Burgess RC, and Rothstein R (2004) Choreography of the DNA damage response: spatiotemporal relationships among checkpoint and repair proteins. Cell 118: 699–713.
57. Lukas C, Falck J, Bartkova J, Bartek J, and Lukas J (2003) Distinct spatiotemporal dynamics of mammalian checkpoint regulators induced by DNA damage. Nat Cell Biol 5: 255–260.
58. Banin S, Moyal L, Shieh S, Taya Y, Anderson CW, Chessa L, Smorodinsky NI, Prives C, Reiss Y, Shiloh Y, and Ziv Y (1998) Enhanced phosphorylation of p53 by ATM in response to DNA damage. Science 281: 1674–1677.
59. Canman CE, Lim DS, Cimprich KA, Taya Y, Tamai K, Sakaguchi K, Appella E, Kastan MB, and Siliciano JD (1998) Activation of the ATM kinase by ionizing radiation and phosphorylation of p53. Science 281: 1677–1679.

60. Chan DW, Ye R, Veillette CJ, and Lees-Miller SP (1999) DNA-dependent protein kinase phosphorylation sites in Ku 70/80 heterodimer. Biochemistry 38: 1819–1828.
61. Sweeney FD, Yang F, Chi A, Shabanowitz J, Hunt DF, and Durocher D (2005) Saccharomyces cerevisiae Rad9 Acts as a Mec1 Adaptor to Allow Rad53 Activation. Curr Biol 15: 1364–1375.
62. Kozlov SV, Graham ME, Peng C, Chen P, Robinson PJ, and Lavin MF (2006) Involvement of novel autophosphorylation sites in ATM activation. Embo J 25: 3504–3514.
63. Usui T, Ogawa H, and Petrini JH (2001) A DNA damage response pathway controlled by Tel1 and the Mre11 complex. Mol Cell 7: 1255–1266.
64. You Z, Chahwan C, Bailis J, Hunter T, and Russell P (2005) ATM activation and its recruitment to damaged DNA require binding to the C terminus of Nbs1. Mol Cell Biol 25: 5363–5379.
65. Dupre A, Boyer-Chatenet L, and Gautier J (2006) Two-step activation of ATM by DNA and the Mre11-Rad50-Nbs1 complex. Nat Struct Mol Biol 13: 451–457.
66. Lee JH, and Paull TT (2004) Direct activation of the ATM protein kinase by the Mre11/Rad50/Nbs1 complex. Science 304: 93–96.
67. Stewart GS, Maser RS, Stankovic T, Bressan DA, Kaplan MI, Jaspers NGJ, Raams A, Byrd PJ, Petrini JHJ, and Taylor AMR (1999) The DNA double-strand break repair gene *hMRE11* is mutated in individuals with an ataxia-telangiectasia-like disorder. Cell 99: 577–587.
68. Carney JP, Maser RS, Olivares H, Davis EM, LeBeau M, Yates JR, Hays L, Morgan WF, and Petrini JHJ (1998) The hMre11/hRad50 protein complex and Nijmegen breakage syndrome: Linkage of double strand break repair to the cellular DNA damage response. Cell 93: 477–486.
69. Matsuura S, Tauchi H, Nakamura A, Kondo N, Sakamoto S, Endo S, Smeets D, Solder B, Belohradsky BH, Kaloustian VMD, Oshimura M, Isomura M, Nakamura Y, and Komatsu K (1998) Positional cloning of the gene for Nijmegen breakage syndrome. Nature Genetics 19: 179–181.
70. Varon R, Vissinga C, Platzer M, Cerosaletti KM, Chrzanowska KH, Saar K, Beckmann G, Seemanova E, Cooper PR, Nowak NJ, Stumm M, Weemaes CMR, Gatti RA, Wilson RK, Digweed M, Rosenthal A, Sperling K, Concannon P, and Reis A (1998) Nibrin, a novel DNA double-strand break repair protein, is mutated in Nijmegen breakage syndrome. Cell 93: 467–476.
71. Shiloh Y (1997) Ataxia-telangiectasia and the Nijmegen breakage syndrome: related disorders but genes apart. Annu Rev Genet 31: 635–662.
72. Lim DS, Kim ST, Xu B, Maser RS, Lin J, Petrini JH, and Kastan MB (2000) ATM phosphorylates p95/nbs1 in an S-phase checkpoint pathway. Nature 404: 613–617.
73. Wu X, Ranganathan V, Weisman DS, Heine WF, Ciccone DN, O'Neill TB, Crick KE, Pierce KA, Lane WS, Rathbun G, Livingston DM, and Weaver DT (2000) ATM phosphorylation of Nijmegen breakage syndrome protein is required in a DNA damage response. Nature 405: 477–482.
74. Zhao S, Weng YC, Yuan SS, Lin YT, Hsu HC, Lin SC, Gerbino E, Song MH, Zdzienicka MZ, Gatti RA, Shay JW, Ziv Y, Shiloh Y, and Lee EY (2000) Functional link between ataxia-telangiectasia and Nijmegen breakage syndrome gene products. Nature 405: 473–477.
75. D'Amours D, and Jackson SP (2001) The yeast Xrs2 complex functions in S phase checkpoint regulation. Genes Dev 15: 2238–2249.
76. Kitagawa R, Bakkenist CJ, McKinnon PJ, and Kastan MB (2004) Phosphorylation of SMC1 is a critical downstream event in the ATM-NBS1-BRCA1 pathway. Genes Dev 18: 1423–1438.
77. Uziel T, Lerenthal Y, Moyal L, Andegeko Y, Mittelman L, and Shiloh Y (2003) Requirement of the MRN complex for ATM activation by DNA damage. Embo J 22: 5612–5621.
78. Cerosaletti K, and Concannon P (2004) Independent roles for nibrin and Mre11-Rad50 in the activation and function of Atm. J Biol Chem 279: 38813–38819.

79. Difilippantonio S, Celeste A, Fernandez-Capetillo O, Chen HT, Reina San Martin B, Van Laethem F, Yang YP, Petukhova GV, Eckhaus M, Feigenbaum L, Manova K, Kruhlak M, Camerini-Otero RD, Sharan S, Nussenzweig M, and Nussenzweig A (2005) Role of Nbs1 in the activation of the Atm kinase revealed in humanized mouse models. Nat Cell Biol 7: 675–685.
80. Carson CT, Schwartz RA, Stracker TH, Lilley CE, Lee DV, and Weitzman MD (2003) The Mre11 complex is required for ATM activation and the G2/M checkpoint. Embo J 22: 6610–6620.
81. Nakada D, Matsumoto K, and Sugimoto K (2003) ATM-related Tel1 associates with double-strand breaks through an Xrs2-dependent mechanism. Genes Dev 17: 1957–1962.
82. Falck J, Coates J, and Jackson SP (2005) Conserved modes of recruitment of ATM, ATR and DNA-PKcs to sites of DNA damage. Nature 434: 605–611.
83. Stracker TH, Morales M, Couto SS, Hussein H, and Petrini JH (2007) The carboxy terminus of NBS1 is required for induction of apoptosis by the MRE11 complex. Nature 447: 218–221.
84. Difilippantonio S, Celeste A, Kruhlak MJ, Lee Y, Difilippantonio MJ, Feigenbaum L, Jackson SP, McKinnon PJ, and Nussenzweig A (2007) Distinct domains in Nbs1 regulate irradiation-induced checkpoints and apoptosis. J Exp Med 204: 1003–1011.
85. Kanu N, and Behrens A (2007) ATMIN defines an NBS1-independent pathway of ATM signalling. Embo J 26: 2933–2941.
86. Park BJ, Kang JW, Lee SW, Choi SJ, Shin YK, Ahn YH, Choi YH, Choi D, Lee KS, and Kim S (2005) The haploinsufficient tumor suppressor p18 upregulates p53 via interactions with ATM/ATR. Cell 120: 209–221.
87. Sun Y, Jiang X, Chen S, Fernandes N, and Price BD (2005) A role for the Tip60 histone acetyltransferase in the acetylation and activation of ATM. Proc Natl Acad Sci USA 102: 13182–13187.
88. Gupta A, Sharma GG, Young CS, Agarwal M, Smith ER, Paull TT, Lucchesi JC, Khanna KK, Ludwig T, and Pandita TK (2005) Involvement of human MOF in ATM function. Mol Cell Biol 25: 5292–5305.
89. Richard DJ, Bolderson E, Cubeddu L, Wadsworth RI, Savage K, Sharma GG, Nicolette ML, Tsvetanov S, McIlwraith MJ, Pandita RK, Takeda S, Hay RT, Gautier J, West SC, Paull TT, Pandita TK, White MF, and Khanna KK (2008) Single-stranded DNA-binding protein hSSB1 is critical for genomic stability. Nature 453: 677–681.
90. Bhaskara V, Dupre A, Lengsfeld B, Hopkins BB, Chan A, Lee JH, Zhang X, Gautier J, Zakian V, and Paull TT (2007) Rad50 adenylate kinase activity regulates DNA tethering by Mre11/Rad50 complexes. Mol Cell 25: 647–661.
91. Moncalian G, Lengsfeld B, Bhaskara V, Hopfner KP, Karcher A, Alden E, Tainer JA, and Paull TT (2004) The rad50 signature motif: essential to ATP binding and biological function. J Mol Biol 335: 937–951.
92. Lou Z, Minter-Dykhouse K, Franco S, Gostissa M, Rivera MA, Celeste A, Manis JP, van Deursen J, Nussenzweig A, Paull TT, Alt FW, and Chen J (2006) MDC1 maintains genomic stability by participating in the amplification of ATM-dependent DNA damage signals. Mol Cell 21: 187–200.
93. Stucki M, Clapperton JA, Mohammad D, Yaffe MB, Smerdon SJ, and Jackson SP (2005) MDC1 directly binds phosphorylated histone H2AX to regulate cellular responses to DNA double-strand breaks. Cell 123: 1213–1226.
94. Burma S, Chen BP, Murphy M, Kurimasa A, and Chen DJ (2001) ATM phosphorylates histone H2AX in response to DNA double-strand breaks. J Biol Chem 276: 42462–42467.
95. Stiff T, O'Driscoll M, Rief N, Iwabuchi K, Lobrich M, and Jeggo PA (2004) ATM and DNA-PK function redundantly to phosphorylate H2AX after exposure to ionizing radiation. Cancer Res 64: 2390–2396.
96. Paull TT, Rogakou EP, Yamazaki V, Kirchgessner CU, Gellert M, and Bonner WM (2000) A critical role for histone H2AX in recruitment of repair factors to nuclear foci after DNA damage. Curr Biol 10: 886–895.

97. Stucki M, and Jackson SP (2006) gammaH2AX and MDC1: Anchoring the DNA-damage-response machinery to broken chromosomes. DNA Repair (Amst).
98. Soutoglou E, and Misteli T (2008) Activation of the cellular DNA damage response in the absence of DNA lesions. Science 320: 1507–1510.
99. Hammarsten O, and Chu G (1998) DNA-dependent protein kinase: DNA binding and activation in the absence of Ku. Proc Natl Acad Sci USA 95: 525–530.
100. Kim DH, Sarbassov DD, Ali SM, Latek RR, Guntur KV, Erdjument-Bromage H, Tempst P, and Sabatini DM (2003) GbetaL, a positive regulator of the rapamycin-sensitive pathway required for the nutrient-sensitive interaction between raptor and mTOR. Mol Cell 11: 895–904.
101. Mordes DA, Glick GG, Zhao R, and Cortez D (2008) TopBP1 activates ATR through ATRIP and a PIKK regulatory domain. Genes Dev 22: 1478–1489.
102. de Klein A, Muijtjens M, van Os R, Verhoeven Y, Smit B, Carr AM, Lehmann AR, and Hoeijmakers JH (2000) Targeted disruption of the cell-cycle checkpoint gene ATR leads to early embryonic lethality in mice. Curr Biol 10: 479–482.
103. Brown EJ, and Baltimore D (2000) ATR disruption leads to chromosomal fragmentation and early embryonic lethality. Genes Dev 14: 397–402.
104. Paciotti V, Clerici M, Lucchini G, and Longhese MP (2000) The checkpoint protein Ddc2, functionally related to S. pombe Rad26, interacts with Mec1 and is regulated by Mec1-dependent phosphorylation in budding yeast. Genes Dev 14: 2046–2059.
105. Rouse J, and Jackson SP (2000) LCD1: an essential gene involved in checkpoint control and regulation of the MEC1 signalling pathway in Saccharomyces cerevisiae. Embo J 19: 5801–5812.
106. Wakayama T, Kondo T, Ando S, Matsumoto K, and Sugimoto K (2001) Pie1, a protein interacting with Mec1, controls cell growth and checkpoint responses in Saccharomyces cerevisiae. Mol Cell Biol 21: 755–764.
107. Edwards RJ, Bentley NJ, and Carr AM (1999) A Rad3-Rad26 complex responds to DNA damage independently of other checkpoint proteins. Nat Cell Biol 1: 393–398.
108. Cliby WA, Roberts CJ, Cimprich KA, Stringer CM, Lamb JR, Schreiber SL, and Friend SH (1998) Overexpression of a kinase-inactive ATR protein causes sensitivity to DNA-damaging agents and defects in cell cycle checkpoints. Embo J 17: 159–169.
109. Wright JA, Keegan KS, Herendeen DR, Bentley NJ, Carr AM, Hoekstra MF, and Concannon P (1998) Protein kinase mutants of human ATR increase sensitivity to UV and ionizing radiation and abrogate cell cycle checkpoint control. Proc Natl Acad Sci USA 95: 7445–7450.
110. Tibbetts RS, Brumbaugh KM, Williams JM, Sarkaria JN, Cliby WA, Shieh SY, Taya Y, Prives C, and Abraham RT (1999) A role for ATR in the DNA damage-induced phosphorylation of p53. Genes Dev 13: 152–157.
111. Guo Z, Kumagai A, Wang SX, and Dunphy WG (2000) Requirement for Atr in phosphorylation of Chk1 and cell cycle regulation in response to DNA replication blocks and UV-damaged DNA in Xenopus egg extracts. Genes Dev 14: 2745–2756.
112. Nghiem P, Park PK, Kim Y, Vaziri C, and Schreiber SL (2001) ATR inhibition selectively sensitizes G1 checkpoint-deficient cells to lethal premature chromatin condensation. Proc Natl Acad Sci USA 98: 9092–9097.
113. Liu Q, Guntuku S, Cui XS, Matsuoka S, Cortez D, Tamai K, Luo G, Carattini-Rivera S, DeMayo F, Bradley A, Donehower LA, and Elledge SJ (2000) Chk1 is an essential kinase that is regulated by Atr and required for the G(2)/M DNA damage checkpoint. Genes Dev 14: 1448–1459.
114. Cimprich KA, and Cortez D (2008) ATR: an essential regulator of genome integrity. Nat Rev Mol Cell Biol 9: 616–627.
115. Garvik B, Carson M, and Hartwell L (1995) Single-stranded DNA arising at telomeres in cdc13 mutants may constitute a specific signal for the RAD9 checkpoint. Mol Cell Biol 15: 6128–6138.

116. Costanzo V, and Gautier J (2003) Single-strand DNA gaps trigger an ATR- and Cdc7-dependent checkpoint. Cell Cycle 2: 17.
117. Costanzo V, Shechter D, Lupardus PJ, Cimprich KA, Gottesman M, and Gautier J (2003) An ATR- and Cdc7-dependent DNA damage checkpoint that inhibits initiation of DNA replication. Mol Cell 11: 203–213.
118. Zou L, and Elledge SJ (2003) Sensing DNA damage through ATRIP recognition of RPA-ssDNA complexes. Science 300: 1542–1548.
119. Zou L, Liu D, and Elledge SJ (2003) Replication protein A-mediated recruitment and activation of Rad17 complexes. Proc Natl Acad Sci USA 100: 13827–13832.
120. Stokes MP, Van Hatten R, Lindsay HD, and Michael WM (2002) DNA replication is required for the checkpoint response to damaged DNA in Xenopus egg extracts. J Cell Biol 158: 863–872.
121. Melo JA, Cohen J, and Toczyski DP (2001) Two checkpoint complexes are independently recruited to sites of DNA damage in vivo. Genes Dev 15: 2809–2821.
122. Kondo T, Wakayama T, Naiki T, Matsumoto K, and Sugimoto K (2001) Recruitment of Mec1 and Ddc1 checkpoint proteins to double-strand breaks through distinct mechanisms. Science 294: 867–870.
123. Majka J, and Burgers PM (2003) Yeast Rad17/Mec3/Ddc1: A sliding clamp for the DNA damage checkpoint. Proc Natl Acad Sci USA 100: 2249–2254.
124. Ellison V, and Stillman B (2003) Biochemical Characterization of DNA Damage Checkpoint Complexes: Clamp Loader and Clamp Complexes with Specificity for 5′ Recessed DNA. PLoS Biol 1: E33.
125. Bonilla CY, Melo JA, and Toczyski DP (2008) Colocalization of sensors is sufficient to activate the DNA damage checkpoint in the absence of damage. Mol Cell 30: 267–276.
126. Toledo LI, Murga M, Gutierrez-Martinez P, Soria R, and Fernandez-Capetillo O (2008) ATR signaling can drive cells into senescence in the absence of DNA breaks. Genes Dev 22: 297–302.
127. Indiani C, and O′Donnell M (2006) The replication clamp-loading machine at work in the three domains of life. Nat Rev Mol Cell Biol 7: 751–761.
128. Majka J, Niedziela-Majka A, and Burgers PM (2006) The checkpoint clamp activates Mec1 kinase during initiation of the DNA damage checkpoint. Mol Cell 24: 891–901.
129. Furuya K, Poitelea M, Guo L, Caspari T, and Carr AM (2004) Chk1 activation requires Rad9 S/TQ-site phosphorylation to promote association with C-terminal BRCT domains of Rad4TOPBP1. Genes Dev 18: 1154–1164.
130. Delacroix S, Wagner JM, Kobayashi M, Yamamoto K, and Karnitz LM (2007) The Rad9-Hus1-Rad1 (9-1-1) clamp activates checkpoint signaling via TopBP1. Genes Dev 21: 1472–1477.
131. St Onge RP, Besley BD, Pelley JL, and Davey S (2003) A role for the phosphorylation of hRad9 in checkpoint signaling. J Biol Chem 278: 26620–26628.
132. Lee J, Kumagai A, and Dunphy WG (2007) The Rad9-Hus1-Rad1 checkpoint clamp regulates interaction of TopBP1 with ATR. J Biol Chem 282: 28036–28044.
133. Kumagai A, Lee J, Yoo HY, and Dunphy WG (2006) TopBP1 activates the ATR-ATRIP complex. Cell 124: 943–955.
134. Nakada D, Hirano Y, and Sugimoto K (2004) Requirement of the Mre11 complex and exonuclease 1 for activation of the Mec1 signaling pathway. Mol Cell Biol 24: 10016–10025.
135. Ferrell JE, Jr. (2002) Self-perpetuating states in signal transduction: positive feedback, double-negative feedback and bistability. Curr Opin Cell Biol 14: 140–148.
136. Gurley KE, and Kemp CJ (2001) Synthetic lethality between mutation in Atm and DNA-PK(cs) during murine embryogenesis. Curr Biol 11: 191–194.
137. Hickson I, Zhao Y, Richardson CJ, Green SJ, Martin NM, Orr AI, Reaper PM, Jackson SP, Curtin NJ, and Smith GC (2004) Identification and characterization of a novel and specific inhibitor of the ataxia-telangiectasia mutated kinase ATM. Cancer Res 64: 9152–9159.

138. Kolas NK, Chapman JR, Nakada S, Ylanko J, Chahwan R, Sweeney FD, Panier S, Mendez M, Wildenhain J, Thomson TM, Pelletier L, Jackson SP, and Durocher D (2007) Orchestration of the DNA-damage response by the RNF8 ubiquitin ligase. Science 318: 1637–1640.
139. Vaze MB, Pellicioli A, Lee SE, Ira G, Liberi G, Arbel-Eden A, Foiani M, and Haber JE (2002) Recovery from checkpoint-mediated arrest after repair of a double- strand break requires Srs2 helicase. Mol Cell 10: 373–385.
140. d'Adda di Fagagna F, Reaper PM, Clay-Farrace L, Fiegler H, Carr P, Von Zglinicki T, Saretzki G, Carter NP, and Jackson SP (2003) A DNA damage checkpoint response in telomere-initiated senescence. Nature 426: 194–198.
141. Bartkova J, Rezaei N, Liontos M, Karakaidos P, Kletsas D, Issaeva N, Vassiliou LV, Kolettas E, Niforou K, Zoumpourlis VC, Takaoka M, Nakagawa H, Tort F, Fugger K, Johansson F, Sehested M, Andersen CL, Dyrskjot L, Orntoft T, Lukas J, Kittas C, Helleday T, Halazonetis TD, Bartek J, and Gorgoulis VG (2006) Oncogene-induced senescence is part of the tumorigenesis barrier imposed by DNA damage checkpoints. Nature 444: 633–637.
142. Di Micco R, Fumagalli M, Cicalese A, Piccinin S, Gasparini P, Luise C, Schurra C, Garre M, Nuciforo PG, Bensimon A, Maestro R, Pelicci PG, and d'Adda di Fagagna F (2006) Oncogene-induced senescence is a DNA damage response triggered by DNA hyper-replication. Nature 444: 638–642.
143. Mallette FA, Gaumont-Leclerc MF, and Ferbeyre G (2007) The DNA damage signaling pathway is a critical mediator of oncogene-induced senescence. Genes Dev 21: 43–48.
144. Bartkova J, Horejsi Z, Koed K, Kramer A, Tort F, Zieger K, Guldberg P, Sehested M, Nesland JM, Lukas C, Orntoft T, Lukas J, and Bartek J (2005) DNA damage response as a candidate anti-cancer barrier in early human tumorigenesis. Nature 434: 864–870.
145. Gorgoulis VG, Vassiliou LV, Karakaidos P, Zacharatos P, Kotsinas A, Liloglou T, Venere M, Ditullio RA, Jr., Kastrinakis NG, Levy B, Kletsas D, Yoneta A, Herlyn M, Kittas C, and Halazonetis TD (2005) Activation of the DNA damage checkpoint and genomic instability in human precancerous lesions. Nature 434: 907–913.
146. Bartek J, Bartkova J, and Lukas J (2007) DNA damage signalling guards against activated oncogenes and tumour progression. Oncogene 26: 7773–7779.
147. Rogakou EP, Pilch DR, Orr AH, Ivanova VS, and Bonner WM (1998) DNA double-stranded breaks induce histone H2AX phosphorylation on serine 139. J Biol Chem 273: 5858–5868.
148. Ward IM, and Chen J (2001) Histone H2AX is phosphorylated in an ATR-dependent manner in response to replicational stress. J Biol Chem 276: 47759–47762.
149. Rogakou EP, Boon C, Redon C, and Bonner WM (1999) Megabase chromatin domains involved in DNA double-strand breaks in vivo. J Cell Biol 146: 905–916.
150. Olive PL (2004) Detection of DNA damage in individual cells by analysis of histone H2AX phosphorylation. Methods Cell Biol 75: 355–373.
151. Teicher BA (2008) Next generation topoisomerase I inhibitors: Rationale and biomarker strategies. Biochem Pharmacol 75: 1262–1271.
152. Ichijima Y, Sakasai R, Okita N, Asahina K, Mizutani S, and Teraoka H (2005) Phosphorylation of histone H2AX at M phase in human cells without DNA damage response. Biochem Biophys Res Commun 336: 807–812.
153. McManus KJ, and Hendzel MJ (2005) ATM-dependent DNA damage-independent mitotic phosphorylation of H2AX in normally growing mammalian cells. Mol Biol Cell 16: 5013–5025.
154. Ziv Y, Bielopolski D, Galanty Y, Lukas C, Taya Y, Schultz DC, Lukas J, Bekker-Jensen S, Bartek J, and Shiloh Y (2006) Chromatin relaxation in response to DNA double-strand breaks is modulated by a novel ATM- and KAP-1 dependent pathway. Nat Cell Biol 8: 870–876.
155. O'Connor MJ, Martin NM, and Smith GC (2007) Targeted cancer therapies based on the inhibition of DNA strand break repair. Oncogene 26: 7816–7824.

156. Ding J, Miao ZH, Meng LH, and Geng MY (2006) Emerging cancer therapeutic opportunities target DNA-repair systems. Trends Pharmacol Sci 27: 338–344.
157. Janetka JW, Ashwell S, Zabludoff S, and Lyne P (2007) Inhibitors of checkpoint kinases: from discovery to the clinic. Curr Opin Drug Discov Devel 10: 473–486.
158. Schaffner C, Idler I, Stilgenbauer S, Dohner H, and Lichter P (2000) Mantle cell lymphoma is characterized by inactivation of the ATM gene. Proc Natl Acad Sci USA 97: 2773–2778.
159. Greenman C, Stephens P, Smith R, Dalgliesh GL, Hunter C, Bignell G, Davies H, Teague J, Butler A, Stevens C, Edkins S, O'Meara S, Vastrik I, Schmidt EE, Avis T, Barthorpe S, Bhamra G, Buck G, Choudhury B, Clements J, Cole J, Dicks E, Forbes S, Gray K, Halliday K, Harrison R, Hills K, Hinton J, Jenkinson A, Jones D, Menzies A, Mironenko T, Perry J, Raine K, Richardson D, Shepherd R, Small A, Tofts C, Varian J, Webb T, West S, Widaa S, Yates A, Cahill DP, Louis DN, Goldstraw P, Nicholson AG, Brasseur F, Looijenga L, Weber BL, Chiew YE, DeFazio A, Greaves MF, Green AR, Campbell P, Birney E, Easton DF, Chenevix-Trench G, Tan MH, Khoo SK, Teh BT, Yuen ST, Leung SY, Wooster R, Futreal PA, and Stratton MR (2007) Patterns of somatic mutation in human cancer genomes. Nature 446: 153–158.
160. Dupre A, Boyer-Chatenet L, Sattler RM, Modi AP, Lee JH, Nicolette ML, Kopelovich L, Jasin M, Baer R, Paull TT, and Gautier J (2008) A forward chemical genetic screen reveals an inhibitor of the Mre11-Rad50-Nbs1 complex. Nat Chem Biol 4: 119–125.
161. Welch EM, Barton ER, Zhuo J, Tomizawa Y, Friesen WJ, Trifillis P, Paushkin S, Patel M, Trotta CR, Hwang S, Wilde RG, Karp G, Takasugi J, Chen G, Jones S, Ren H, Moon YC, Corson D, Turpoff AA, Campbell JA, Conn MM, Khan A, Almstead NG, Hedrick J, Mollin A, Risher N, Weetall M, Yeh S, Branstrom AA, Colacino JM, Babiak J, Ju WD, Hirawat S, Northcutt VJ, Miller LL, Spatrick P, He F, Kawana M, Feng H, Jacobson A, Peltz SW, and Sweeney HL (2007) PTC124 targets genetic disorders caused by nonsense mutations. Nature 447: 87–91.
162. Clerici M, Mantiero D, Lucchini G, and Longhese MP (2006) The Saccharomyces cerevisiae Sae2 protein negatively regulates DNA damage checkpoint signalling. EMBO Rep 7: 212–218.
163. Baldo V, Testoni V, Lucchini G, and Longhese MP (2008) Dominant TEL1-hy mutations compensate for Mec1 lack of functions in the DNA damage response. Mol Cell Biol 28: 358–375.
164. Morales M, Theunissen JW, Kim CF, Kitagawa R, Kastan MB, and Petrini JH (2005) The Rad50S allele promotes ATM-dependent DNA damage responses and suppresses ATM deficiency: implications for the Mre11 complex as a DNA damage sensor. Genes Dev 19: 3043–3054.
165. Alani E, Padmore R, and Kleckner N (1990) Analysis of wild-type and rad50 mutants of yeast suggests an intimate relationship between meiotic chromosome synapsis and recombination. Cell 61: 419–436.
166. Cao L, Alani E, and Kleckner N (1990) A pathway for generation and processing of double-strand breaks during meiotic recombination in S. cerevisiae. Cell 61: 1089–1101.

Chapter 2
Signaling at Stalled Replication Forks

Daniel A. Mordes and David Cortez

Abstract Completing DNA replication with a minimum of errors is critical to prevent disease and requires the coordinated action of hundreds of proteins. These include DNA repair enzymes that remove DNA lesions, thereby preventing polymerase stalling and incorporation errors. However, given the large size of the genome and the numerous chemical assaults on it from both endogenous and exogenous sources, some lesions fail to be repaired before they are encountered by elongating replication forks. This review will examine recent advances in our understanding of cellular responses to replication stress that ensure a successful resolution of this common genome integrity challenge. In particular, we will focus on the ATR signaling pathway, and describe how recent experimental approaches in multiple organisms including yeast, frogs, and humans have yielded a model for its genome maintenance activities during DNA replication.

Keywords ATR · CHK1 · DNA damage · Replication · TopBP1

2.1 Introduction

Faithfully replicating the genome is essential to prevent disease, and all organisms devote large portions of their genome and proteome to this task. Beyond the enzymes required to synthesize DNA, there are elaborate systems to detect and repair DNA lesions and to deal with the problems that arise when a replication fork encounters an unrepaired lesion. These systems include lesion bypass mechanisms, post-replicative repair, recombination activities and checkpoint signaling pathways. The goal is to ensure that replication is successfully completed with a minimum of errors. This review will focus on the cellular activities that occur when a replication fork encounters a DNA lesion or otherwise stalls with an emphasis on the detection of these problems by the ATR checkpoint-signaling pathway.

D. Cortez (✉)
Department of Biochemistry, Vanderbilt University School of Medicine, Nashville, TN 37232, USA
e-mail: david.cortez@vanderbilt.edu

K.K. Khanna, Y. Shiloh (eds.), *The DNA Damage Response: Implications on Cancer Formation and Treatment*, DOI 10.1007/978-90-481-2561-6_2,

2.2 Replication and Fork Stalling

In most eukaryotic cells the entire genome is replicated exactly once per cell cycle. Replication requires an ordered series of steps occurring in three stages: licensing of origins, initiation, and elongation. DNA replication initiates at select sites on the genome called origins. At these locations, protein complexes assemble in a highly regulated manner that leads to the formation of bi-directional DNA replication forks.

2.2.1 Initiating DNA Replication

The first step in DNA replication is the formation of a pre-replication complex (pre-RC) in the G1 phase of the cell cycle. This requires the ATP-dependent binding of a six-subunit origin recognition complex (ORC) to origins. ORC recruits the initiation factors Cdc6 and Cdt1, which together function to load the *m*ini-*c*hromosome *m*aintenance (MCM) helicase complex onto chromatin and form the pre-RC [80]. Once the MCM complex is loaded onto chromatin, the origin is 'licensed' for replication [9, 59].

Many additional factors are required for the conversion of the pre-RC to a functional replication fork. In S phase, activation of pre-RCs or 'firing of origins' requires the activity of cyclin-dependent kinase (CDK) and Cdc7/Dbf4-dependent kinase (DDK) that promote the recruitment of additional replication proteins, including CDC45, GINS, and polymerases to origins. During initiation the MCM helicase locally unwinds the DNA, followed by the loading of the single-stranded DNA binding protein *r*eplication *p*rotein *A* (RPA) [105]. The loading of RPA allows recruitment of DNA primase/polymerase α (Polα), which forms an RNA primer followed by a short DNA primer. Once the DNA primer is formed, DNA polymerase δ (Polδ) and DNA polymerase ϵ (Polϵ) begin DNA synthesis on the lagging and leading strand, respectively [73, 79]. During elongation, the activities of helicase and polymerase complexes are coordinated to complete replication of the inter-origin DNA.

2.2.2 Replication Stress

Although DNA repair activities remove the vast majority of lesions before a replication fork would encounter them, the number of chemical changes to the DNA is so vast that during every round of replication, the fork will encounter some unrepaired lesions and stall. In *E. coli*, it has been estimated that there is at least one problem per replication attempt [21]. In organisms like yeast or humans where the genome is much larger, we should expect a corresponding increase in the frequency of problems.

What happens when the replication fork encounters a DNA lesion? The answer to this question depends on the nature of the lesion. Some lesions stall the helicase and the polymerase, such as an interstrand crosslink, while others only stall the polymerase. A lesion that stalls the helicase may need to be removed before any

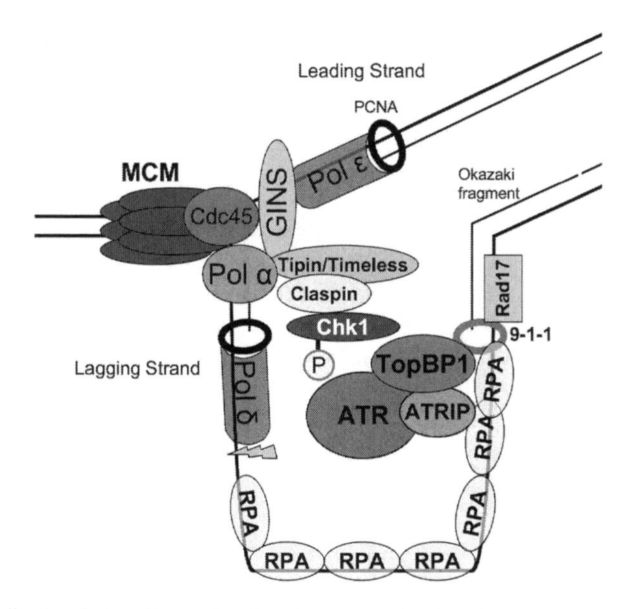

Fig. 2.1 Replication fork stalling and ATR activation. During the elongation phase of replication, the MCM helicase complex unwinds the parental DNA duplex. The replicative polymerases pol ϵ and pol δ synthesize DNA on the leading and lagging strands, respectively. Here a snapshot of the replication fork is presented as a lesion is encountered on the lagging strand. The lesion impedes the progression of the lagging strand polymerase but not the helicase or leading strand polymerase. The ssDNA gap with a 5' primer ssDNA/dsDNA junction is recognized by the RAD17-RFC complex, which loads the 9-1-1 checkpoint clamp. Independently, the ATR-ATRIP complex is recruited. Protein-protein interactions between RPA and both ATRIP and RAD9 contribute to the assembly of these complexes. 9-1-1 recruits TopBP1 and stabilizes the interaction between TopBP1 and ATR-ATRIP. TopBP1 activates the kinase activity of ATR towards its substrates. Additional proteins present at the replication fork, including the Tipin-Timeless complex and Claspin, promote CHK1 phosphorylation

further elongation of the replication fork can occur. In some cases, successfully completing replication may require rescue from a fork proceeding from a nearby origin. These lesions, however, are likely to be rare. Many types of DNA lesions, such as an intrastrand cross-links, base adducts, or abasic sites, inhibit DNA polymerases without impeding the progression of the helicase. The best data for this comes from the analysis of replication of circular plasmids in *Xenopus* egg extracts. In this system, it is clear that the helicase continues to unwind DNA while the polymerase is stalled [15, 75]. This functional uncoupling of the helicase and polymerase activities leads to the generation of abnormally long stretches of single-stranded DNA (ssDNA) (Fig. 2.1). It is worth pointing out that the replisome probably does not physically uncouple, despite how this uncoupling is usually drawn in cartoon diagrams.

In the *Xenopus* system, it is unclear whether there is a limit to how much ssDNA is generated since unwinding of an entire circular plasmid has been observed. However, in other systems it seems likely that a more limited uncoupling occurs. In *S. cerevisiae* base adducts caused by methylating agents did not cause large stretches of ssDNA that could be visualized by electron microscopy, but unrepaired ultraviolet (UV) radiation –induced lesions did generate ssDNA regions up to 3 kb [56]. In this circumstance, a ssDNA gap was generated only on one side of the fork suggesting uncoupling of leading and lagging strand synthesis. It is likely that the position of the lesion on the lagging or leading strand yields different effects. A lesion on the lagging strand might cause a small gap the size of an Okazaki fragment since re-priming naturally happens anyway. There may also be re-priming downstream of a lesion on the leading strand. In any case, the ssDNA generated when the helicase continues to unwind despite the stalling of a polymerase is a key step in the subsequent responses.

Current data indicates two major responses to the stalled polymerase. The first is a mechanism that aims to bypass the lesion through either template or polymerase switching. The mechanisms controlling lesion bypass have been reviewed elsewhere [29, 113]. The second response is a DNA damage response pathway regulated by the ATR checkpoint kinase. Both mechanisms are triggered by the ssDNA generated upon helicase and polymerase activity uncoupling [15, 16].

2.3 ATR Signaling at a Stalled Replication Fork

The DNA Damage Response (DDR) is a signaling pathway that coordinates cell-cycle transitions, DNA repair, DNA replication, and, in some cases, apoptosis and senescence. At the apex of this pathway are two closely related PIKK (Phosphoinositide 3-kinase–like kinase) protein kinases, ATM (Ataxia-Telangiectasia Mutated) and ATR (ATM and Rad3-related). ATM and ATR have similar biochemical and functional properties and target an overlapping set of substrates involved in maintaining genome stability. However, these kinases sense distinct types of DNA damage. ATM is activated in response to rare occurrences of double-stranded DNA breaks (DSBs). On the other hand, ATR responds to a broad spectrum of DNA damage, particularly those that perturb DNA replication [18]. ATR is essential for the viability of replicating cells and is activated every S-phase to regulate DNA replication [13, 14, 20].

Studies in Xenopus egg extracts using defined DNA motifs indicate that ATR senses double-stranded/single-stranded DNA junctions adjacent to stretches of ssDNA. Although both 5’ and 3’ DNA junctions can activate ATR, 5’ primer ends are more efficient in signaling checkpoint activation. The amount of single-stranded DNA correlates with the extent of ATR activation [58].

There are at least two sensors of ssDNA in the ATR signaling pathway – an ATR-interacting protein (ATRIP) and a checkpoint clamp loading complex consisting of RAD17 and four subunits of replication factor C which serves to load a checkpoint clamp consisting of RAD9-RAD1-HUS1 (9-1-1). Both of these sensors recognize the ssDNA in complex with the ssDNA binding protein RPA (replication protein A) [27,122,123]. The loaded 9-1-1 complex recruits an ATR activator, TopBP1 that directly binds and stimulates ATR-ATRIP kinase activity (Fig. 2.1). In the following sections, we will consider these important steps in ATR activation in more detail.

Table 1 Checkpoint protein orthologs

Protein function	Human	*Xenopus*	*S. cerevisiae*	*S. pombe*
PIKK kinase	ATR	ATR	Mec1	Rad3
	ATRIP	ATRIP	Ddc2	Rad26
Effector kinase	CHK1	CHK1	Chk1/ Rad53	Chk1
Clamp loader	RAD17	RAD17	Rad24	RAD17
	RFC2-5	RFC2-5	Rfc2-5	Rfc2-5
9-1-1 clamp	RAD9	RAD9	Ddc1	Rad9
	RAD1	RAD1	Rad17	Rad1
	HUS1	HUS1	Mec3	Hus1
Kinase activator	TopBP1	TopBP1	Dpb11	Cut5
CHK1 adaptor	Claspin	Claspin	Mrc1	Mrc1
Replication fork stabilizer	Timeless	Timeless	Tof1	Swi1

2.3.1 ATRIP

ATRIP was first identified as a protein that co-immunoprecipitated with ATR and co-localized with ATR to sites of DNA damage and replication stress [20]. Depletion of ATRIP causes a reduction in ATR protein levels, and vice versa, indicating the stability of ATR and ATRIP are interdependent. ATR and ATRIP form a stable complex, and depletion of ATRIP yields the same phenotype as the loss of ATR [20]. The budding and fission yeast ATR proteins (Mec1^{ATR} and Rad3^{ATR}) also bind to a yeast ATRIP (Ddc2^{ATRIP} and Rad26^{ATRIP}, respectively). (See Table 1 for a list of orthologous checkpoint proteins in yeast, frog and human systems.) Deletions of ATRIP in these organisms phenocopies deletions of ATR [26, 76, 82]. Thus, the ATR-ATRIP complex should be properly thought of as a holoenzyme complex.

ATRIP has multiple functions required for ATR signaling (Fig. 2.2A). First, it is required for the localization of ATR to sites of DNA damage. Domain mapping indicated two regions of ATRIP required for directing localization – the C-terminal ATR interaction domain and an N-terminal RPA binding domain [5]. The interaction with ATR is required not only for ATR localization, but also for ATRIP localization due to an undefined activity of ATR that promotes localization of the ATR-ATRIP complex [5, 6]. The ATRIP N-terminal RPA binding domain (hereafter referred to as the *c*heckpoint *r*ecruitment *d*omain (CRD)) binds directly to the N-terminal oligosaccharide/oligonucleotide binding (OB)-fold domain of the large RPA70 subunit (hereafter referred to as RPA70N) [4, 5]. The ATRIP CRD contains a negatively charged alpha-helix that binds in the basic cleft of the RPA70N OB fold. This interaction is conserved between Ddc2^{ATRIP} and *S. cerevisiae* RPA [4]. Disruption of this interaction greatly reduces (but does not eliminate) the ability of both the ATR-ATRIP and Mec1^{ATR}-Ddc2^{ATRIP} complexes to localize to stalled forks and other types of DNA damage. Surprisingly, disruption of this interaction only causes mild defects in ATR signaling indicating that additional means of ATR-ATRIP localization are largely sufficient to allow ATR activation [4, 5].

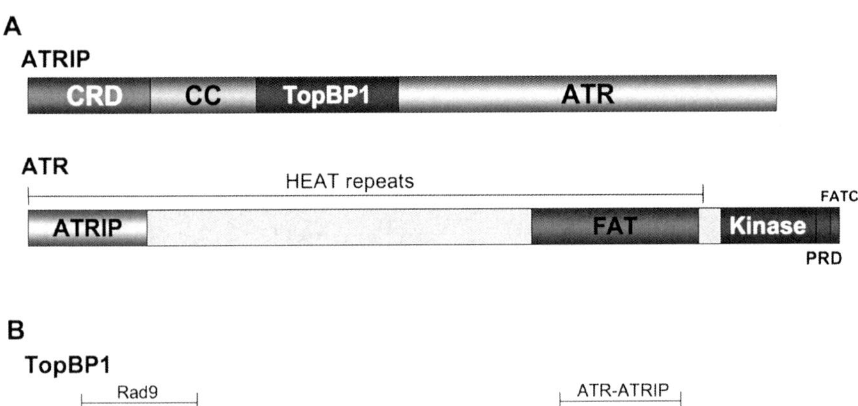

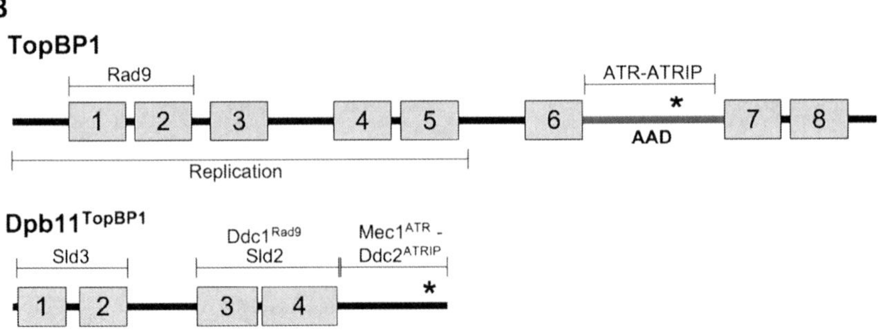

Fig. 2.2 Functional domains of ATR, ATRIP, and TopBP1. (**A**) ATRIP contains an N-terminal checkpoint recruitment domain (CRD) that promotes ATR-ATRIP localization to stalled replication forks by binding to RPA, a coiled-coil domain (CC) that mediates its oligomerization, a TopBP1-interacting region that is required for ATR activation, and a C-terminal ATR-interacting domain. ATR contains multiple α-helical HEAT repeats. The N-terminal HEAT repeats consist of an ATRIP-binding domain. The C-terminal kinase domain of ATR is flanked by FAT (FRAP, ATM, TRRAP) and FATC (FAT C-terminal) domains that share homology to related PIKKs but their function is not well defined. In between the kinase and the FATC domains is the PIKK regulatory domain (PRD) which interacts with the TopBP1 AAD and is critical for TopBP1-mediated ATR activation. (**B**) TopBP1 contains multiple BRCT (BRCA1 C-terminal) domains. The N-terminal BRCT domains interact with phosphorylated RAD9. In between the 6th and 7th BRCT domains is the ATR activation domain (AAD) that interacts with the ATR-ATRIP complex and is sufficient to stimulate the kinase activity of ATR. The asterisk indicates an ATR/ATM phosphorylation site that potentiates the ability of TopBP1 to activate ATR. The N-terminal half of TopBP1 is sufficient for its role in replication. The *S. cerevisiae* TopBP1 homolog Dpb11^{TopBP1} contains four BRCT domains that interact with the replication initiation factors Sld3 and Sld2. The third and fourth BRCT domains also interact with phosphorylated Ddc1^{RAD9}. The Dpb11^{TopBP1} C-terminal domain interacts with Mec1^{ATR}-Ddc2^{ATRIP} and is sufficient to stimulate the kinase activity of Mec1^{ATR}. The asterisk indicates a Mec1^{ATR} phosphorylation site that potentiates the ability of Dpb11^{TopBP1} to activate Mec1^{ATR}

There may be multiple mechanisms that promote ATR-ATRIP localization when the primary RPA-ATRIP interacting surfaces are mutated. One study has reported that ATRIP contains additional RPA binding domains [71]. Unfortunately, the precise locations of these RPA binding surfaces have not been mapped nor has their functional significance been tested. Many studies indicate that ATRIP localization also occurs through RPA-independent mechanisms. In yeast, Ddc2^{ATRIP} contains a DNA-binding domain that promotes the localization of Mec1^{ATR}-Ddc2^{ATRIP} to sites of DNA damage [83]. In vitro studies have shown that the ATR-ATRIP complex can associate with ssDNA in the absence of RPA [11, 101]. Additional proteins present at stalled replication forks may promote ATRIP localization. In fission yeast, the replication protein Cdc18^{Cdc6} directly interacts with Rad26^{ATRIP} and is required for the recruitment of Rad3^{ATR}-Rad26^{ATRIP} to chromatin [39]. ATR itself helps to localize ATR-ATRIP to sites of DNA damage through an unidentified mechanism [5, 6]. Finally, an alternative single-stranded DNA binding protein (hSSB1) with functions in the DNA damage response has been identified [81]. It will be important to determine whether hSSB1 can mediate the recruitment of the ATR-ATRIP complex to ssDNA.

In addition to localization, ATRIP is necessary for the DNA damage-induced increase in ATR specific activity. Neither ATRIP, RPA, or ssDNA activates ATR directly [4]. Instead ATRIP binds an ATR activating protein- TopBP1 [66]. ATRIP is necessary for the association of ATR and TopBP1. It contains a TopBP1-interacting region that is necessary for the ATR activation in vitro. The interaction between ATRIP and TopBP1 is essential for cells to survive and recover DNA synthesis following replication stress and to maintain the G2/M checkpoint after DNA damage [66]. This region of ATRIP is functionally conserved in the budding yeast ATRIP ortholog Ddc2^{ATRIP} [66]. TopBP1 will be discussed in more detail in following sections.

Finally, ATRIP post-translational modifications fine-tune the activity of the ATR-ATRIP complex. ATRIP is an ATR substrate [20, 42]. In yeast, the phosphorylation of Ddc2^{ATRIP} is an upstream indicator of Mec1^{ATR} activation [26, 76]. The functional significance of ATR-dependent ATRIP phosphorylation is unclear, but the ATRIP phosphorylation sites are within the ATRIP CRD, so it is possible that ATRIP phosphorylation regulates the association of ATR-ATRIP with RPA. ATRIP contains additional phosphorylation sites. Serine 224 matches a CDK consensus site. Its phosphorylation is both CDK2 and cell cycle-dependent and is required for ATR-ATRIP to efficiently mediate checkpoint responses to DNA damage [69]. ATRIP S239 is also phosphorylated, but the responsible kinase remains to be identified [69, 104]. Phospho-serine 239 is part of a consensus binding site for BRCT domains, which are present in many DDR-associated proteins. Serine 239 phosphorylation mediates an interaction between ATRIP and the BRCT domains of BRCA1 in vitro, and is necessary for the G2/M checkpoint after DNA damage in cells [104].

2.3.2 The 9-1-1 Complex

The 9-1-1 checkpoint clamp consists of RAD9, RAD1, and HUS1. It exhibits structural homology to the replicative DNA clamp PCNA forming a ring-like structure that circles DNA [34, 103]. DNA clamps are locked around DNA by a clamp loader in a reaction driven by ATP hydrolysis. During normal replication, the RFC (replication factor C) complex, composed of a large subunit (p140) and four small subunits (p40, p38, p37, p36), loads PCNA onto DNA. PCNA loading is directed preferentially to the 3' primer end of the double-stranded – single-stranded DNA junction where it can bind to DNA polymerases and promote elongation.

Loading of the 9-1-1 clamp requires a specialized version of the RFC complex called RAD17-RFC, where RAD17 replaces the large subunit p140 [10]. Although the 9-1-1 complex can be loaded onto both 5' and 3' DNA ends, in the presence of RPA-coated ssDNA, loading occurs preferentially at 5' primer ends [27]. During normal replication a 5'-primer end is only observed on the lagging strand at Okazaki fragments. Re-priming downstream of a lesion on the leading strand could generate the gapped DNA structure preferentially recognized by the 9-1-1 complex. Primase activity has been reported to be essential for checkpoint activation [65].

It is unclear what explains the specificity for loading the 9-1-1 complex at the 5' primer junction. Interestingly, RPA is required to generate this specificity, and RPA can bind directly to subunits of both the RFC and 9-1-1 complex [44, 123]. In fact, the same RPA70N binding surface that binds the checkpoint recruitment domain of ATRIP also binds to RAD9 [109]. Mutations that disrupt this interaction cause defects in checkpoint signaling and cellular survival to replication stress. Importantly, RPA binds to ssDNA in a specific polarity such that the RPA70N domain would be positioned next to the 5' primer junction [22, 41]. Thus, the orientation specificity of RPA on the ssDNA may serve to position or tether the 9-1-1 complex so it can present TopBP1 to ATR-ATRIP.

In S. *cerevisiae*, tethering the 9-1-1 complex and Mec1^{ATR}-Ddc2^{ATRIP} to chromatin together using fusion proteins containing sequence-specific DNA binding elements is sufficient to activate Mec1^{ATR} even in the absence of any DNA damage [12]. In this system, only a single subunit of 9-1-1 (Ddc1^{RAD9}) is required. In addition, Ddc1^{RAD9} has been demonstrated to directly activate Mec1^{ATR}-Ddc2^{ATRIP} [60]. However, there is also evidence that Ddc1^{RAD9} may recruit Dpb11^{TopBP1} to promote checkpoint signaling [78]. In vertebrate systems, 9-1-1 is not sufficient to activate ATR. Instead, it must recruit the TopBP1 protein, which is a direct ATR activator.

2.3.3 TopBP1

TopBP1 has dual functions in the initiation of DNA replication and checkpoint signaling. TopBP1 (Topoisomerase-II-binding protein 1) was initially identified in a yeast two-hybrid screen for interacting factors of a C-terminal region of DNA topoisomerase-II beta, yet the functional significance of this interaction remains unknown [110]. TopBP1 contains eight BRCT (BRCA1 C-terminal repeat) domains that often function in tandem as phospho-protein binding domains (Fig. 2.2B). For example, the BRCT domains of BRCA1 bind directly to a phospho-serine motif in BACH1/BRIP1 [117]. The TopBP1 N-terminal pair of BRCT domains (1&2) interacts with a phospho-serine site on RAD9 [23, 50] and this interaction concentrates TopBP1 at the stalled replication fork.

The role of TopBP1 in replication has been best characterized for its budding yeast ortholog Dpb11^{TopBP1} (Fig. 2.2B). Dpb11^{TopBP1} was initially identified as a high-copy suppressor of mutations in polymerase ϵ [3]. Dpb11^{TopBP1} is essential for cell proliferation and progression through S-phase. Dpb11^{TopBP1} acts as a molecular bridge between components of the polymerase and helicase complexes. It recruits DNA polymerase ϵ and polymerase α/primase to origins of replication [62]. It genetically and physically interacts with the GINS tetramer, which forms part of the CMG (Cdc45/MCM2-7/GINS) helicase complex [68]. The Dpb11^{TopBP1} C-terminal pair of BRCT domains binds to a CDK site on Sld2, which associates with the polymerase ϵ complex. The Dpb11^{TopBP1} N-terminal pair of BRCT domains interacts with a CDK phosphorylation site on Sld3, a protein necessary for the interaction of the MCM helicase complex and its co-factor Cdc45 [68, 95]. Dpb11^{TopBP1} is also required for the association of GINS with replication origins to initiate replication [93]. A fusion protein between the C-terminus of Dpb11^{TopBP1} and Sld3 is sufficient to bypass the need for the Dpb11^{TopBP1} N-terminal pair of BRCT domains in replication [118]. Thus, the main role of Dpb11^{TopBP1} in replication involves recognizing the helicase complex at origins and facilitating the recruitment of the replicative polymerase.

TopBP1 is required for replication in vertebrate systems including human cells. In *Xenopus* egg extracts TopBP1 is essential for the initiation of DNA replication rather than the elongation phase of replication. Specifically, TopBP1 is required for the chromatin loading of DNA polymerases and Cdc45 [36]. The replication and checkpoint functions of TopBP1 are physically separable, where the N-terminal half of TopBP1 is sufficient for its role in DNA replication [37, 112]. The human orthologs of the Sld2 and Sld3 proteins have not yet been identified but may include RecQL4. RecQL4 binds TopBP1 and is important for the initiation of replication [85].

TopBP1 has an essential function in the DNA damage response. In response to replication stress, TopBP1 co-localizes with PCNA at stalled DNA replication forks [61, 111]. TopBP1 also forms DNA damage-induced foci that co-localize with many DDR proteins, including BRCA1, 53BP1, NBS1, γH2AX, RPA, and ATR [54, 111]. TopBP1 is required for ATR-dependent checkpoint signaling in response to DNA damage and replication stress [37, 48, 54, 112]. Like the loss of ATR or CHK1, depletion of TopBP1 also leads to increased fragile site expression and chromosomal breaks under conditions of replication stress [45].

TopBP1 has a direct role in the activation of ATR. TopBP1 robustly stimulates the kinase activity of the ATR-ATRIP complex towards multiple substrates [48]. A region of TopBP1 in between the 6th and 7th BRCT domains called the ATR Activation Domain (AAD) is sufficient for this effect. Addition of recombinant TopBP1 AAD to *Xenopus* egg extracts induces CHK1 phosphorylation in the absence of DNA damage [48]. Additionally, overexpression of the TopBP1 AAD in human cells results in pan-nuclear phosphorylation of the ATR substrates MCM2 and γH2AX without causing DNA damage [4, 48, 98]. Thus, the AAD is able to bypass the requirements for other checkpoint protein and regulatory steps in ATR activation if expressed at high enough levels. Disruption of the TopBP1 AAD abolishes CHK1 phosphorylation after replication stress but does not effect DNA replication [48, 112]. Therefore, TopBP1, in addition to being a DNA replication protein, is a general activator of ATR.

TopBP1 is recruited to sites of DNA damage and replication stress through an interaction between its N-terminal pair of BRCT domains and a constitutive phosphorylation site on the C-terminal tail of RAD9 [23, 50]. This interaction stabilizes the binding of TopBP1 to the ATR-ATRIP complex [50]. Localization of the TopBP1 AAD to chromatin is the essential function of 9-1-1 complex in checkpoint signaling. When the TopBP1 AAD is fused to PCNA or histone H2B, both RAD9 and RAD17 are dispensable for checkpoint signaling [23].

A study performed in the *Xenopus* egg extract system has indicated an exception to this model. ATR phosphorylation of the RAD1 subunit of the 9-1-1 complex is TopBP1-dependent but independent of the RAD9 C-terminal tail, which interacts with TopBP1 [57]. The 5th BRCT repeat of TopBP1 is important for its ability to form DNA damage-induced foci [111]. Thus, there may be an alternative mechanism for localization of TopBP1 to sites of replication stress, perhaps through an interaction with ATR-ATRIP or replication proteins, such as RecQL4. RAD9-dependent concentration of TopBP1 may be dispensable for the phosphorylation of certain chromatin-bound ATR substrates, but necessary for the sustaining ATR in its active form and the phosphorylation of other substrates, especially CHK1. In support of this idea, another study in *Xenopus* egg extracts showed that an antibody directed against the C-terminus of TopBP1 was sufficient to inhibit its checkpoint function and blocked the phosphorylation of CHK1 but did not affect phosphorylation of Claspin [112]. Furthermore, at least in *S. pombe*, phosphorylation of the most proximal components of the ATR signaling pathway may not require the TopBP1-dependent ATR activation step. Although Cut5^{TopBP1} is required for the phosphorylation of CHK1 and Cds1^{Chk2}, it is not necessary for the phosphorylation of the 9-1-1 complex and Rad26^{ATRIP} [35].

Post-translation modification of TopBP1 also regulates its ability to activate ATR. TopBP1 is phosphorylated at multiple sites in response to DNA damage [111]. Phosphorylation of an ATR/ATM consensus site in the TopBP1 AAD potentiates TopBP1-mediated ATR activation [37, 115]. There are several additional TopBP1 post-translational modifications including sumoylation [40] and AKT-dependent phosphorylation [52], but it is not clear if these regulate the DNA damage response functions of TopBP1.

The conservation of the TopBP1-dependent ATR-activation mechanism in *S. cerevisiae* has been questioned since there is no sequence homology between the TopBP1 AAD and any region of Dpb11^{TopBP1}. Furthermore, the Ddc1^{RAD9} protein was shown to directly activate Mec1^{ATR}-Ddc2^{ATRIP}[60]. Thus, the need for a TopBP1 activation step is unclear. However, recent data suggest that Dpb11^{TopBP1} does function as a direct Mec1^{ATR}-Ddc2^{ATRIP} activator. First, cells harboring the *dbp11-1* allele are checkpoint compromised [3, 106]. Second, the TopBP1 binding site in ATRIP is conserved in Ddc2^{ATRIP} and mutations in this region of Ddc2^{ATRIP} compromise checkpoint signaling [66]. Third, Dpb11^{TopBP1} binds the phosphorylated Ddc1^{RAD9} C-terminal tail [78]. Finally, we have shown that the C-terminal tail of Dpb11^{TopBP1} binds directly to Mec1^{ATR}-Ddc2^{ATRIP}

complexes and activates $Mec1^{ATR}$ activity in vitro [67, 72]. Interestingly, like TopBP1, $Dpb11^{TopBP1}$ is phosphorylated and this phosphorylation appears to amplify the ability of $Dpb11^{TopBP1}$ to activate $Mec1^{ATR}$-$Ddc2^{ATRIP}$ (D. Mordes and D. Cortez, unpublished data). It is still unclear why a $Dpb11^{TopBP1}$ activation mechanism for $Mec1^{ATR}$ is needed since $Ddc1^{RAD9}$ by itself can activate. One possibility is that in *S. cerevisiae* $Dpb11^{TopBP1}$ serves primarily to amplify the activation signal. In other organisms, the direct RAD9 ATR activation pathway may have been lost, thereby increasing the importance of the Dpb11/TopBP1 proteins.

Besides its role as an ATR activator, TopBP1 may have additional upstream and downstream functions in the ATR-checkpoint signaling pathway. The primase activity Polα is required for CHK1 phosphorylation presumably because it creates the 5' primer junction important for 9-1-1 recruitment [38, 65]. TopBP1 is required for the chromatin association of Polα and the 9-1-1 complex [77]. Hence, TopBP1 may be needed for the formation of the primed-ssDNA structure that signals the recruitment and loading of the 9-1-1 complex onto chromatin. This would place TopBP1 both upstream and downstream of the 9-1-1 complex. It is also possible that TopBP1 can promote ATR signaling by promoting the recruitment of specific substrates to ATR. In response to DNA damage, TopBP1 associates with a BRCA1-BACH1 complex [33], which may facilitate the recruitment of BRCA1 to sites of replication stress and its phosphorylation by activated ATR [97].

2.3.4 CHK1

The CHK1 kinase is a major target of ATR. Its activity is responsible for many of the cellular effects of ATR activation (Fig. 2.3). Similar to the loss ATR, CHK1 deficiency in mice causes embryonic lethality and apoptosis in embryonic stem cells. CHK1(-/-) cells fail to delay the cell cycle before entry into mitosis in response to DNA damage and replication stress [53, 92]. ATR phosphorylates CHK1 at two

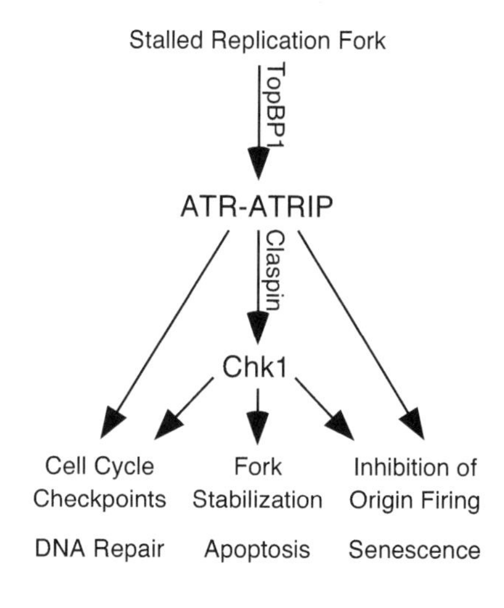

Fig. 2.3 The ATR signaling pathway. Stalled replication forks can result in TopBP1-dependent activation of the ATR-ATRIP complex. The key ATR substrate is the CHK1 effector kinase. CHK1 phosphorylation requires the adapator protein Claspin. Activated ATR and activated CHK1 phosphorylate numerous downstream substrates involved in multiple cellular responses that preserve genomic integrity and prevent the progression of genetically unstable cells

ATR consensus sites on serine 317 and serine 345 [53, 119]. Phosphorylation of CHK1 increases its kinase activity and is critical for its checkpoint function.

The activation of CHK1 requires an adaptor protein called Claspin [46]. Unlike TopBP1, which is required for the phosphorylation of all ATR substrates, Claspin is specifically required for CHK1 phosphorylation after replication stress [54]. In response to double-strand breaks, both Claspin and BRCA1 are necessary for optimal CHK1 phosphorylation [51, 114]. Claspin has been proposed to monitor DNA replication and recruit CHK1 to stalled replication forks. The chromatin association of Claspin depends upon the replication protein, Cdc45, but does not require RPA, ATR or RAD17 [49]. This suggests that Claspin may recognize components of stalled replication forks distinct from the signal sensed by ATR and RAD17. Claspin is phosphorylated in response to replication stress and this phosphorylation is required for the binding of Claspin to CHK1 [47]. Sustained CHK1 phosphorylation requires ATR phosphorylation of RAD17. Phosphorylated RAD17 interacts with Claspin, which may be a mechanism to maintain Claspin at stalled replication forks [107].

The requirement for an adaptor protein that regulates the phosphorylation of the effector kinase specifically in response to stalled replication forks is conserved in yeast. In *S. cerevisiae*, Mec1^{ATR} regulates two effector kinases, Rad53 and Chk1, that have some distinct and some overlapping functions. Although the Rad53 checkpoint kinase has greater sequence homolog to human CHK2 than CHK1, it is more functionally equivalent to human CHK1. In response to DNA damage and replication stress, Mec1^{ATR} phosphorylates Rad53 leading to increased Rad53 kinase activity [84, 91]. Mrc1^{Claspin} is necessary for cell-cycle arrest and efficient Rad53 activation after replication stress but not DNA damage [1].

Replication-stress induced Mrc1^{Claspin} phosphorylation occurs in a Mec1^{ATR}-dependent and Rad53-dependent manner and is required for cell-cycle checkpoints [1, 74]. However, Mrc1^{Claspin} also has a function in DNA replication that is independent of its checkpoint function. The absence of Mrc1^{Claspin} results in slow S-phase progression and activation of the DNA damage response. Mrc1^{Claspin} associates with origins during the initiation of replication, moves with replication forks, and associates with stalled replication forks [74]. Although Mrc1^{Claspin} is not required to prevent the collapse of stalled replication forks, it is necessary to restart replication from stalled forks [99]. Taken together, this suggests that Mrc1^{Claspin} is well-positioned at replication forks to serve as a signal for the checkpoint response when helicase-polymerase uncoupling occurs, and to maintain the integrity of replication forks.

2.3.5 Regulation of DNA Replication by ATR

The role of the ATR-CHK1 pathway is regulating replication has been most clearly defined in yeast systems. Activated Rad53 inhibits the firing of late origins of replication [86, 89]. This active inhibition of the initiation of replication at origins is primarily responsible for the slowing of DNA synthesis by the intra-S-phase checkpoint at least in *S. cerevisiae*. The loss of Rad53 causes late origins to fire earlier in S-phase, even in the absence of DNA damaging agents [89]. This suggests that Rad53 also regulates the normal temporal pattern of DNA replication. Although origins of replications are not as well defined in higher eukaryotes, the ATR-CHK1 pathway also regulates the timing of origin firing in vertebrates [28, 63, 64, 88].

Besides its function in cell cycle checkpoints, the Mec1^{ATR} pathway is also important to maintain the stability of stalled replication forks. Mec1^{ATR} and Rad53 are required for the progression of replication forks under stressful conditions [96]. In the absence of Rad53, replication stress results in the collapse of stalled replication forks and the generation of abnormal replication structures [55]. Maintenance of some replisome components, including polymerase α and ϵ, at stalled replication forks requires Mec1^{ATR} but not Rad53 [19]. Recently, the other effector kinase Chk1 has been implicated in the stability of stalled replication forks in the absence of Rad53 [87]. It remains to be determined how Mec1^{ATR}, Rad53, and Chk1 preserve replication fork integrity and allow for the resumption of DNA replication. A recent proteomic screen for Mec1^{ATR}- and Rad53-dependent phosphorylation sites identified multiple replication proteins including Ctf4 and Tof1 as potential substrates of the Mec1^{ATR} and/or Rad53 [90]. Future research will be focused on defining the functional consequences of these post-translational modifications.

Rad53 also promotes viability after DNA damage by increasing nucleotide synthesis. Rad53 activates the kinase Dun1, which causes an increased transcription of ribonucleotide reductase (RNR) genes [2]. Additionally, Dun1 targets the RNR inhibitor Sml1 for degradation [121]. In the absence of DNA damage, the regulation of nucleotide levels is the essential function of the Mec1^{ATR}-Rad53 pathway. Overexpression of ribonucleotide reductase or deletion of sml1 suppresses the lethality but not the DNA damage sensitivity of $\Delta rad53$ or $\Delta mec1^{ATR}$ yeast [24, 120]. A similar pathway operates in fission yeast [108], and vertebrate cells contain a p53-inducible RNR subunit [70, 94]. However, it is unclear whether human RNR is regulated by a protein inhibitor or whether the essential function of ATR in human cells can be rescued by increasing nucleotide levels.

The Tipin-Timeless complex also mediates CHK1 activation and the intra-S-phase checkpoint [17, 100, 102, 113]. Deletion of Timeless causes early embryonic lethality [31]. Timeless forms a complex with ATR-ATRIP and CHK1 and its depletion inhibits CHK1 phosphorylation [102]. The Tipin-Timeless complex interacts with replisome components including RPA [32, 100] and the MCM complex [17, 32]. The Tipin-Timeless complex in *S. cerevisiae* (Tof1-Swi1) is also present in replisomes and regulates fork stability in response to replication stress [43]. Thus, Tipin-Timeless is well positioned to respond to replication fork challenges.

2.4 ATR Signaling and Cancer

The DNA damage response acts as an anti-cancer barrier. Markers of DDR activation are present in premalignant lesions in a variety of tissues types [7, 30]. DDR activation in these cells is thought to arise from aberrant replication. Overexpression or mutations of oncogenes or loss of some tumor suppressors can result in replication stress, characterized by inappropriate origin firing, abnormal fork progression and double-stranded DNA breaks that leads to activation of the DDR [8, 25]. In these circumstances, the DDR induces cell cycle checkpoints, apoptosis or cellular senescence (Fig. 2.3). Consistent with this analysis, ectopic activation of ATR is sufficient to drive cells into senescence in multiple cell lines [98]. Overcoming the DDR activation allows the continued proliferation of genetically unstable cells, which eventually progress to malignant lesions.

Although the genome instability resulting from loss of some DDR activities promotes tumorigenesis, it may also provide a therapeutic opportunity. Most mutations that allow DDR bypass occur downstream in the pathway at the level of p53. The upstream components of the pathway are rarely mutated except in inherited caner predisposition syndromes. This may be because the genetically unstable cells in the tumor experience high levels of replication stress and require some activities of the DDR including ATR for successful replication and cell division. Decreasing the upstream activities of the ATR pathway in these cells would be expected to have pleiotropic effects to such an extent that the cells may be unable to successfully complete replication. In other words, the ATR pathway may be a target for novel cancer therapeutics. Eliminating all ATR function would be cytotoxic even in normal cells, but drugs usually yield partial loss of function phenotypes. The elevated levels of replication stress in cancer cells may present an opportunity for a synthetic lethality in these cells. An ATR or CHK1 inhibitor would be expected to have some activity as a mono-therapy. However, its most successful application may be in combination with other agents that increase the DNA damage burden in cells.

2.5 Conclusions and Future Directions

The concept of exploiting the replication stress found in tumors for treatment is attractive; however, the utility of targeting the ATR pathway for cancer therapy remains unproven. Further advances in defining the normal cellular responses to replication stress and the defects that exist in cancer cells will be important to both identify useful drug targets and define the patients most likely to benefit from these types of treatments.

Critical questions about the replication stress responses remain. A structural description of ATR and its binding partners would help in the design of small molecules to disrupt the pathway. Markers of ATR signaling that can be assayed in patient samples will be critical to evaluate any potential ATR-directed therapies. Advances in finding ATR substrates are encouraging, but the functional consequences of most ATR-dependent phosphorylation events remain unknown. Our understanding of how ATR promotes the maintenance and recovery of a stalled replisome is still poorly defined. We should also expect advances in understanding how lesion bypass, template switching, and recombination act cooperatively with the DNA damage response to ensure DNA replication is completed as faithfully as possible. The replication stress response is a universal requirement in all cellular organisms to maintain viability and prevent disease. Therefore, a continued exploitation of all model systems to define the mechanisms underlying genome maintenance during replication will be essential to support translational research activities.

Acknowledgements Research on ATR, the DNA damage response, and cancer in the Cortez laboratory is sponsored by the National Cancer Institute, the Susan G. Komen Foundation and the Ingram Charitable Fund. The authors would like to acknowledge their many colleagues for helpful discussions, and apologize to those authors whose important results were omitted due to space constraints.

References

1. Alcasabas, A.A., A.J. Osborn, J. Bachant, F. Hu, P.J. Werler, K. Bousset, K. Furuya, J.F. Diffley, A.M. Carr, and S.J. Elledge (2001) Mrc1 transduces signals of DNA replication stress to activate Rad53. Nat Cell Biol. 3:958–965.
2. Allen, J.B., Z. Zhou, W. Siede, E.C. Friedberg, and S.J. Elledge (1994) The SAD1/RAD53 protein kinase controls multiple checkpoints and DNA damage-induced transcription in yeast. Genes Dev. 8:2401–2415.
3. Araki, H., S.H. Leem, A. Phongdara, and A. Sugino (1995) Dpb11, which interacts with DNA polymerase II(epsilon) in Saccharomyces cerevisiae, has a dual role in S-phase progression and at a cell cycle checkpoint. Proc Natl Acad Sci USA. 92:11791–11795.
4. Ball, H.L., M.R. Ehrhardt, D.A. Mordes, G.G. Glick, W.J. Chazin, and D. Cortez (2007) Function of a conserved checkpoint recruitment domain in ATRIP proteins. Mol Cell Biol. 27:3367–3377.
5. Ball, H.L., J.S. Myers, and D. Cortez (2005) ATRIP Binding to RPA-ssDNA Promotes ATR-ATRIP Localization but Is Dispensable for Chk1 Phosphorylation. Mol Biol Cell. 16: 2372–2381.
6. Barr, S.M., C.G. Leung, E.E. Chang, and K.A. Cimprich (2003) ATR kinase activity regulates the intranuclear translocation of ATR and RPA following ionizing radiation. Curr Biol. 13:1047–1051.
7. Bartkova, J., Z. Horejsi, K. Koed, A. Kramer, F. Tort, K. Zieger, P. Guldberg, M. Sehested, J.M. Nesland, C. Lukas, T. Orntoft, J. Lukas, and J. Bartek (2005) DNA damage response as a candidate anti-cancer barrier in early human tumorigenesis. Nature. 434:864–870.
8. Bartkova, J., et al. (2006) Oncogene-induced senescence is part of the tumorigenesis barrier imposed by DNA damage checkpoints. Nature. 444:633–637.
9. Bell, S.P. and A. Dutta (2002) DNA replication in eukaryotic cells. Annu Rev Biochem. 71:333–374.
10. Bermudez, V.P., L.A. Lindsey-Boltz, A.J. Cesare, Y. Maniwa, J.D. Griffith, J. Hurwitz, and A. Sancar (2003) Loading of the human 9-1-1 checkpoint complex onto DNA by the checkpoint clamp loader hRad17-replication factor C complex in vitro. Proc Natl Acad Sci USA. 100:1633–1638.
11. Bomgarden, R.D., D. Yean, M.C. Yee, and K.A. Cimprich (2004) A novel protein activity mediates DNA binding of an ATR-ATRIP complex. J Biol Chem. 279:13346–13353.
12. Bonilla, C.Y., J.A. Melo, and D.P. Toczyski (2008) Colocalization of sensors is sufficient to activate the DNA damage checkpoint in the absence of damage. Mol Cell. 30:267–276.
13. Brown, E.J. and D. Baltimore (2000) ATR disruption leads to chromosomal fragmentation and early embryonic lethality. Genes Dev. 14:397–402.
14. Brown, E.J. and D. Baltimore (2003) Essential and dispensable roles of ATR in cell cycle arrest and genome maintenance. Genes Dev. 17:615–628.
15. Byun, T.S., M. Pacek, M.C. Yee, J.C. Walter, and K.A. Cimprich (2005) Functional uncoupling of MCM helicase and DNA polymerase activities activates the ATR-dependent checkpoint. Genes Dev. 19:1040–1052.

16. Chang, D.J., P.J. Lupardus, and K.A. Cimprich (2006) Monoubiquitination of proliferating cell nuclear antigen induced by stalled replication requires uncoupling of DNA polymerase and mini-chromosome maintenance helicase activities. J Biol Chem. 281:32081–32088.
17. Chou, D.M. and S.J. Elledge (2006) Tipin and Timeless form a mutually protective complex required for genotoxic stress resistance and checkpoint function. Proc Natl Acad Sci USA. 103:18143–18147.
18. Cimprich, K.A. and D. Cortez (2008) ATR: an essential regulator of genome integrity. Nat Rev Mol Cell Biol. 9:616–627.
19. Cobb, J.A., L. Bjergbaek, K. Shimada, C. Frei, and S.M. Gasser (2003) DNA polymerase stabilization at stalled replication forks requires Mec1 and the RecQ helicase Sgs1. Embo J. 22:4325–4336.
20. Cortez, D., S. Guntuku, J. Qin, and S.J. Elledge (2001) ATR and ATRIP: partners in checkpoint signaling. Science. 294:1713–1716.
21. Cox, M.M., M.F. Goodman, K.N. Kreuzer, D.J. Sherratt, S.J. Sandler, and K.J. Marians (2000) The importance of repairing stalled replication forks. Nature. 404:37–41.
22. de Laat, W.L., E. Appeldoorn, K. Sugasawa, E. Weterings, N.G. Jaspers, and J.H. Hoeijmakers (1998) DNA-binding polarity of human replication protein A positions nucleases in nucleotide excision repair. Genes Dev. 12:2598–2609.
23. Delacroix, S., J.M. Wagner, M. Kobayashi, K. Yamamoto, and L.M. Karnitz (2007) The Rad9-Hus1-Rad1 (9-1-1) clamp activates checkpoint signaling via TopBP1. Genes Dev. 21:1472–1477.
24. Desany, B.A., A.A. Alcasabas, J.B. Bachant, and S.J. Elledge (1998) Recovery from DNA replicational stress is the essential function of the S-phase checkpoint pathway. Genes Dev. 12:2956–2970.
25. Di Micco, R., M. Fumagalli, A. Cicalese, S. Piccinin, P. Gasparini, C. Luise, C. Schurra, M. Garre, P.G. Nuciforo, A. Bensimon, R. Maestro, P.G. Pelicci, and F. d′Adda di Fagagna (2006) Oncogene-induced senescence is a DNA damage response triggered by DNA hyper-replication. Nature. 444:638–642.
26. Edwards, R.J., N.J. Bentley, and A.M. Carr (1999) A Rad3-Rad26 complex responds to DNA damage independently of other checkpoint proteins. Nat Cell Biol. 1:393–398.
27. Ellison, V. and B. Stillman (2003) Biochemical characterization of DNA damage checkpoint complexes: clamp loader and clamp complexes with specificity for 5′ recessed DNA. PLoS Biol. 1:E33.
28. Feijoo, C., C. Hall-Jackson, R. Wu, D. Jenkins, J. Leitch, D.M. Gilbert, and C. Smythe (2001) Activation of mammalian Chk1 during DNA replication arrest: a role for Chk1 in the intra-S phase checkpoint monitoring replication origin firing. J Cell Biol. 154:913–923.
29. Friedberg, E.C. (2005) Suffering in silence: the tolerance of DNA damage. Nat Rev Mol Cell Biol. 6:943–953.
30. Gorgoulis, V.G., L.V. Vassiliou, P. Karakaidos, P. Zacharatos, A. Kotsinas, T. Liloglou, M. Venere, R.A. Ditullio, Jr., N.G. Kastrinakis, B. Levy, D. Kletsas, A. Yoneta, M. Herlyn, C. Kittas, and T.D. Halazonetis (2005) Activation of the DNA damage checkpoint and genomic instability in human precancerous lesions. Nature. 434:907–913.
31. Gotter, A.L., T. Manganaro, D.R. Weaver, L.F. Kolakowski, Jr., B. Possidente, S. Sriram, D.T. MacLaughlin, and S.M. Reppert (2000) A time-less function for mouse timeless. Nat Neurosci. 3:755–756.
32. Gotter, A.L., C. Suppa, and B.S. Emanuel (2007) Mammalian TIMELESS and Tipin are evolutionarily conserved replication fork-associated factors. J Mol Biol. 366:36–52.
33. Greenberg, R.A., B. Sobhian, S. Pathania, S.B. Cantor, Y. Nakatani, and D.M. Livingston (2006) Multifactorial contributions to an acute DNA damage response by BRCA1/BARD1-containing complexes. Genes Dev. 20:34–46.
34. Griffith, J.D., L.A. Lindsey-Boltz, and A. Sancar (2002) Structures of the human Rad17-replication factor C and checkpoint Rad 9-1-1 complexes visualized by glycerol spray/low voltage microscopy. J Biol Chem. 277:15233–15236.

35. Harris, S., C. Kemplen, T. Caspari, C. Chan, H.D. Lindsay, M. Poitelea, A.M. Carr, and C. Price (2003) Delineating the position of rad4+/cut5+ within the DNA-structure checkpoint pathways in Schizosaccharomyces pombe. J Cell Sci. 116:3519–3529.
36. Hashimoto, Y. and H. Takisawa (2003) Xenopus Cut5 is essential for a CDK-dependent process in the initiation of DNA replication. Embo J. 22:2526–2535.
37. Hashimoto, Y., T. Tsujimura, A. Sugino, and H. Takisawa (2006) The phosphorylated C-terminal domain of Xenopus Cut5 directly mediates ATR-dependent activation of Chk1. Genes Cells. 11:993–1007.
38. Hekmat-Nejad, M., Z. You, M.C. Yee, J.W. Newport, and K.A. Cimprich (2000) Xenopus ATR is a replication-dependent chromatin-binding protein required for the DNA replication checkpoint. Curr Biol. 10:1565–1573.
39. Hermand, D. and P. Nurse (2007) Cdc18 enforces long-term maintenance of the S phase checkpoint by anchoring the Rad3-Rad26 complex to chromatin. Mol Cell. 26:553–563.
40. Holway, A.H., C. Hung, and W.M. Michael (2005) Systematic, RNA-interference-mediated identification of mus-101 modifier genes in Caenorhabditis elegans. Genetics. 169: 1451–1460.
41. Iftode, C. and J.A. Borowiec (2000) 5′ –> 3′ molecular polarity of human replication protein A (hRPA) binding to pseudo-origin DNA substrates. Biochemistry. 9:11970–11981.
42. Itakura, E., K. Umeda, E. Sekoguchi, H. Takata, M. Ohsumi, and A. Matsuura (2004) ATR-dependent phosphorylation of ATRIP in response to genotoxic stress. Biochem Biophys Res Commun. 323:1197–1202.
43. Katou, Y., Y. Kanoh, M. Bando, H. Noguchi, H. Tanaka, T. Ashikari, K. Sugimoto, and K. Shirahige (2003) S-phase checkpoint proteins Tof1 and Mrc1 form a stable replication-pausing complex. Nature. 424:1078–1083.
44. Kim, H.S. and S.J. Brill (2001) Rfc4 interacts with Rpa1 and is required for both DNA replication and DNA damage checkpoints in Saccharomyces cerevisiae. Mol Cell Biol. 21:3725–3737.
45. Kim, J.E., S.A. McAvoy, D.I. Smith, and J. Chen (2005) Human TopBP1 ensures genome integrity during normal S phase. Mol Cell Biol. 25:10907–10915.
46. Kumagai, A. and W.G. Dunphy (2000) Claspin, a novel protein required for the activation of Chk1 during a DNA replication checkpoint response in Xenopus egg extracts. Mol Cell. 6:839–849.
47. Kumagai, A. and W.G. Dunphy (2003) Repeated phosphopeptide motifs in Claspin mediate the regulated binding of Chk1. Nat Cell Biol. 5:161–165.
48. Kumagai, A., J. Lee, H.Y. Yoo, and W.G. Dunphy (2006) TopBP1 activates the ATR-ATRIP complex. Cell. 124:943–955.
49. Lee, J., A. Kumagai, and W.G. Dunphy (2003) Claspin, a Chk1-regulatory protein, monitors DNA replication on chromatin independently of RPA, ATR, and Rad17. Mol Cell. 11: 329–340.
50. Lee, J., A. Kumagai, and W.G. Dunphy (2007) The Rad9-Hus1-Rad1 checkpoint clamp regulates interaction of TopBP1 with ATR. J Biol Chem. 282:28036–28044.
51. Lin, S.Y., K. Li, G.S. Stewart, and S.J. Elledge (2004) Human Claspin works with BRCA1 to both positively and negatively regulate cell proliferation. Proc Natl Acad Sci USA. 101:6484–6489.
52. Liu, K., J.C. Paik, B. Wang, F.T. Lin, and W.C. Lin (2006) Regulation of TopBP1 oligomerization by Akt/PKB for cell survival. Embo J. 25:4795–4807.
53. Liu, Q., S. Guntuku, X.S. Cui, S. Matsuoka, D. Cortez, K. Tamai, G. Luo, S. Carattini-Rivera, F. DeMayo, A. Bradley, L.A. Donehower, and S.J. Elledge (2000) Chk1 is an essential kinase that is regulated by Atr and required for the G(2)/M DNA damage checkpoint. Genes Dev. 14:1448–1459.
54. Liu, S., S. Bekker-Jensen, N. Mailand, C. Lukas, J. Bartek, and J. Lukas (2006) Claspin operates downstream of TopBP1 to direct ATR signaling towards Chk1 activation. Mol Cell Biol. 26:6056–6064.

55. Lopes, M., C. Cotta-Ramusino, A. Pellicioli, G. Liberi, P. Plevani, M. Muzi-Falconi, C.S. Newlon, and M. Foiani (2001) The DNA replication checkpoint response stabilizes stalled replication forks. Nature. 412:557–561.
56. Lopes, M., M. Foiani, and J.M. Sogo (2006) Multiple mechanisms control chromosome integrity after replication fork uncoupling and restart at irreparable UV lesions. Mol Cell. 21:15–27.
57. Lupardus, P.J. and K.A. Cimprich (2006) Phosphorylation of Xenopus Rad1 and Hus1 defines a readout for ATR activation that is independent of Claspin and the Rad9 carboxy terminus. Mol Biol Cell. 17:1559–1569.
58. MacDougall, C.A., T.S. Byun, C. Van, M.C. Yee, and K.A. Cimprich (2007) The structural determinants of checkpoint activation. Genes Dev. 21:898–903.
59. Machida, Y.J., J.L. Hamlin, and A. Dutta (2005) Right place, right time, and only once: replication initiation in metazoans. Cell. 123:13–24.
60. Majka, J., A. Niedziela-Majka, and P.M. Burgers (2006) The checkpoint clamp activates Mec1 kinase during Initiation of the DNA damage checkpoint. Mol Cell. 24:891–901.
61. Makiniemi, M., T. Hillukkala, J. Tuusa, K. Reini, M. Vaara, D. Huang, H. Pospiech, I. Majuri, T. Westerling, T.P. Makela, and J.E. Syvaoja (2001) BRCT domain-containing protein TopBP1 functions in DNA replication and damage response. J Biol Chem. 276:30399–30406.
62. Masumoto, H., A. Sugino, and H. Araki (2000) Dpb11 controls the association between DNA polymerases alpha and epsilon and the autonomously replicating sequence region of budding yeast. Mol Cell Biol. 20:2809–2817.
63. Maya-Mendoza, A., E. Petermann, D.A. Gillespie, K.W. Caldecott, and D.A. Jackson (2007) Chk1 regulates the density of active replication origins during the vertebrate S phase. Embo J. 26:2719–2731.
64. Merrick, C.J., D. Jackson, and J.F. Diffley (2004) Visualization of altered replication dynamics after DNA damage in human cells. J Biol Chem. 279:20067–20075.
65. Michael, W.M., R. Ott, E. Fanning, and J. Newport (2000) Activation of the DNA replication checkpoint through RNA synthesis by primase. Science. 289:2133–2137.
66. Mordes, D.A., G.G. Glick, R. Zhao, and D. Cortez (2008) TopBP1 activates ATR through ATRIP and a PIKK regulatory domain. Genes Dev. 22:1478–1489.
67. Mordes, D.A., E.A. Nam, and D. Cortez (2008) Dpb11 activates the Mec1-Ddc2 complex. Proc Natl Acad Sci USA.105:18730–18734.
68. Moyer, S.E., P.W. Lewis, and M.R. Botchan (2006) Isolation of the Cdc45/Mcm2-7/GINS (CMG) complex, a candidate for the eukaryotic DNA replication fork helicase. Proc Natl Acad Sci USA. 103:10236–10241.
69. Myers, J.S., R. Zhao, X. Xu, A.J. Ham, and D. Cortez (2007) Cyclin-dependent kinase 2 dependent phosphorylation of ATRIP regulates the G2-M checkpoint response to DNA damage. Cancer Res. 67:6685–6690.
70. Nakano, K., E. Balint, M. Ashcroft, and K.H. Vousden (2000) A ribonucleotide reductase gene is a transcriptional target of p53 and p73. Oncogene. 19:4283–4289.
71. Namiki, Y. and L. Zou (2006) ATRIP associates with replication protein A-coated ssDNA through multiple interactions. Proc Natl Acad Sci USA. 103:580–585.
72. Navadgi-Patil, V.M. and P.M. Burgers (2008) Yeast DNA replication protein Dpb11 activates the Mec1/ATR checkpoint Kinase. J Biol Chem. 283:35853–35859.
73. Nick McElhinny, S.A., D.A. Gordenin, C.M. Stith, P.M. Burgers, and T.A. Kunkel (2008) Division of labor at the eukaryotic replication fork. Mol Cell. 30:137–144.
74. Osborn, A.J. and S.J. Elledge (2003) Mrc1 is a replication fork component whose phosphorylation in response to DNA replication stress activates Rad53. Genes Dev. 17:1755–1767.
75. Pacek, M. and J.C. Walter (2004) A requirement for MCM7 and Cdc45 in chromosome unwinding during eukaryotic DNA replication. Embo J. 23:3667–3676.
76. Paciotti, V., M. Clerici, G. Lucchini, and M.P. Longhese (2000) The checkpoint protein Ddc2, functionally related to S. pombe Rad26, interacts with Mec1 and is regulated by Mec1-dependent phosphorylation in budding yeast. Genes Dev. 14:2046–2059.

77. Parrilla-Castellar, E.R. and L.M. Karnitz (2003) Cut5 is required for the binding of Atr and DNA polymerase alpha to genotoxin-damaged chromatin. J Biol Chem. 278: 45507–45511.
78. Puddu, F., M. Granata, L. Di Nola, A. Balestrini, G. Piergiovanni, F. Lazzaro, M. Giannattasio, P. Plevani, and M. Muzi-Falconi (2008) Phosphorylation of the budding yeast 9-1-1 complex is required for Dpb11 function in the full activation of the UV-induced DNA damage checkpoint. Mol Cell Biol. 28:4782–4793.
79. Pursell, Z.F., I. Isoz, E.B. Lundstrom, E. Johansson, and T.A. Kunkel (2007) Yeast DNA polymerase epsilon participates in leading-strand DNA replication. Science. 317:127–130.
80. Randell, J.C., J.L. Bowers, H.K. Rodriguez, and S.P. Bell (2006) Sequential ATP hydrolysis by Cdc6 and ORC directs loading of the Mcm2-7 helicase. Mol Cell. 21:29–39.
81. Richard, D.J., et al. (2008) Single-stranded DNA-binding protein hSSB1 is critical for genomic stability. Nature. 453:677–681.
82. Rouse, J. and S.P. Jackson (2000) LCD1: an essential gene involved in checkpoint control and regulation of the MEC1 signalling pathway in Saccharomyces cerevisiae. Embo J. 19:5801–5812.
83. Rouse, J. and S.P. Jackson (2002) Lcd1p recruits Mec1p to DNA lesions in vitro and in vivo. Mol Cell. 9:857–869.
84. Sanchez, Y., B.A. Desany, W.J. Jones, Q. Liu, B. Wang, and S.J. Elledge (1996) Regulation of RAD53 by the ATM-like kinases MEC1 and TEL1 in yeast cell cycle checkpoint pathways [see comments]. Science. 271:357–360.
85. Sangrithi, M.N., J.A. Bernal, M. Madine, A. Philpott, J. Lee, W.G. Dunphy, and A.R. Venkitaraman (2005) Initiation of DNA replication requires the RECQL4 protein mutated in Rothmund-Thomson syndrome. Cell. 121:887–898.
86. Santocanale, C. and J.F. Diffley (1998) A Mec1- and Rad53-dependent checkpoint controls late-firing origins of DNA replication. Nature. 395:615–618.
87. Segurado, M. and J.F.X. Diffley (2008) Separate roles for the DNA damage checkpoint protein kinases in stabilizing DNA replication forks. Genes Dev. 22:1816–1827.
88. Shechter, D., V. Costanzo, and J. Gautier (2004) ATR and ATM regulate the timing of DNA replication origin firing. Nat Cell Biol. 6:648–655.
89. Shirahige, K., Y. Hori, K. Shiraishi, M. Yamashita, K. Takahashi, C. Obuse, T. Tsurimoto, and H. Yoshikawa (1998) Regulation of DNA-replication origins during cell-cycle progression. Nature. 395:618–621.
90. Smolka, M.B., C.P. Albuquerque, S.H. Chen, and H. Zhou (2007) Proteome-wide identification of in vivo targets of DNA damage checkpoint kinases. Proc Natl Acad Sci USA. 104:10364–10369.
91. Sun, Z., D.S. Fay, F. Marini, M. Foiani, and D.F. Stern (1996) Spk1/Rad53 is regulated by Mec1-dependent protein phosphorylation in DNA replication and damage checkpoint pathways. Genes Dev. 10:395–406.
92. Takai, H., K. Tominaga, N. Motoyama, Y.A. Minamishima, H. Nagahama, T. Tsukiyama, K. Ikeda, K. Nakayama, M. Nakanishi, and K. Nakayama (2000) Aberrant cell cycle checkpoint function and early embryonic death in Chk1(-/-) mice. Genes Dev. 14:1439–1447.
93. Takayama, Y., Y. Kamimura, M. Okawa, S. Muramatsu, A. Sugino, and H. Araki (2003) GINS, a novel multiprotein complex required for chromosomal DNA replication in budding yeast. Genes Dev. 17:1153–1165.
94. Tanaka, H., H. Arakawa, T. Yamaguchi, K. Shiraishi, S. Fukuda, K. Matsui, Y. Takei, and Y. Nakamura (2000) A ribonucleotide reductase gene involved in a p53-dependent cell-cycle checkpoint for DNA damage. Nature. 404:42–49.
95. Tanaka, S., T. Umemori, K. Hirai, S. Muramatsu, Y. Kamimura, and H. Araki (2007) CDK-dependent phosphorylation of Sld2 and Sld3 initiates DNA replication in budding yeast. Nature. 445:328–332.
96. Tercero, J.A. and J.F. Diffley (2001) Regulation of DNA replication fork progression through damaged DNA by the Mec1/Rad53 checkpoint. Nature. 412:553–557.

97. Tibbetts, R.S., D. Cortez, K.M. Brumbaugh, R. Scully, D. Livingston, S.J. Elledge, and R.T. Abraham (2000) Functional interactions between BRCA1 and the checkpoint kinase ATR during genotoxic stress. Genes Dev. 14:2989–3002.
98. Toledo, L.I., M. Murga, P. Gutierrez-Martinez, R. Soria, and O. Fernandez-Capetillo (2008) ATR signaling can drive cells into senescence in the absence of DNA breaks. Genes Dev. 22:297–302.
99. Tourriere, H., G. Versini, V. Cordon-Preciado, C. Alabert, and P. Pasero (2005) Mrc1 and Tof1 promote replication fork progression and recovery independently of Rad53. Mol Cell. 19:699–706.
100. Unsal-Kacmaz, K., P.D. Chastain, P.P. Qu, P. Minoo, M. Cordeiro-Stone, A. Sancar, and W.K. Kaufmann (2007) The human Tim/Tipin complex coordinates an Intra-S checkpoint response to UV that slows replication fork displacement. Mol Cell Biol. 27:3131–3142.
101. Unsal-Kacmaz, K., A.M. Makhov, J.D. Griffith, and A. Sancar (2002) Preferential binding of ATR protein to UV-damaged DNA. Proc Natl Acad Sci U S A. 99:6673–6678.
102. Unsal-Kacmaz, K., T.E. Mullen, W.K. Kaufmann, and A. Sancar (2005) Coupling of human circadian and cell cycles by the timeless protein. Mol Cell Biol. 25:3109–3116.
103. Venclovas, C. and M.P. Thelen (2000) Structure-based predictions of Rad1, Rad9, Hus1 and Rad17 participation in sliding clamp and clamp-loading complexes. Nucleic Acids Res. 28:2481–2493.
104. Venere, M., A. Snyder, O. Zgheib, and T.D. Halazonetis (2007) Phosphorylation of ATR-interacting protein on Ser239 mediates an interaction with breast-ovarian cancer susceptibility 1 and checkpoint function. Cancer Res. 67:6100–6105.
105. Walter, J. and J. Newport (2000) Initiation of eukaryotic DNA replication: origin unwinding and sequential chromatin association of Cdc45, RPA, and DNA polymerase alpha. Mol Cell. 5:617–627.
106. Wang, H. and S.J. Elledge (1999) DRC1, DNA replication and checkpoint protein 1, functions with DPB11 to control DNA replication and the S-phase checkpoint in Saccharomyces cerevisiae. Proc Natl Acad Sci USA. 96:3824–3829.
107. Wang, X., L. Zou, T. Lu, S. Bao, K.E. Hurov, W.N. Hittelman, S.J. Elledge, and L. Li (2006) Rad17 phosphorylation is required for claspin recruitment and Chk1 activation in response to replication stress. Mol Cell. 23:331–341.
108. Woollard, A., G. Basi, and P. Nurse (1996) A novel S phase inhibitor in fission yeast. Embo J. 15:4603–4612.
109. Xu, X., S. Vaithiyalingam, G.G. Glick, D.A. Mordes, W.J. Chazin, and D. Cortez (2008) The basic cleft of RPA70N binds multiple checkpoint proteins, including RAD9, to regulate ATR signaling. Mol Cell Biol. 28:7345–7353.
110. Yamane, K., M. Kawabata, and T. Tsuruo (1997) A DNA-topoisomerase-II-binding protein with eight repeating regions similar to DNA-repair enzymes and to a cell-cycle regulator. Eur J Biochem. 250:794–799.
111. Yamane, K., X. Wu, and J. Chen (2002) A DNA damage-regulated BRCT-containing protein, TopBP1, is required for cell survival. Mol Cell Biol. 22:555–566.
112. Yan, S., H.D. Lindsay, and W.M. Michael (2006) Direct requirement for Xmus101 in ATR-mediated phosphorylation of Claspin bound Chk1 during checkpoint signaling. J Cell Biol. 173:181–186.
113. Yang, W. and R. Woodgate (2007) What a difference a decade makes: Insights into translesion DNA synthesis. Proc Natl Acad Sci. 104:15591–15598.
114. Yoo, H.Y., S.Y. Jeong, and W.G. Dunphy (2006) Site-specific phosphorylation of a checkpoint mediator protein controls its responses to different DNA structures. Genes Dev. 20:772–783.
115. Yoo, H.Y., A. Kumagai, A. Shevchenko, A. Shevchenko, and W.G. Dunphy (2007) Ataxia-telangiectasia mutated (ATM)-dependent activation of ATR occurs through phosphorylation of TopBP1 by ATM. J Biol Chem. 282:17501–17506.
116. Yoshizawa-Sugata, N. and H. Masai (2007) Human Tim/Timeless-interacting protein, Tipin, is required for efficient progression of S phase and DNA replication checkpoint. J Biol Chem. 282:2729–2740.

117. Yu, X., C.C. Chini, M. He, G. Mer, and J. Chen (2003) The BRCT domain is a phosphoprotein binding domain. Science. 302:639–642.
118. Zegerman, P. and J.F. Diffley (2007) Phosphorylation of Sld2 and Sld3 by cyclin-dependent kinases promotes DNA replication in budding yeast. Nature. 445:281–285.
119. Zhao, H. and H. Piwnica-Worms (2001) ATR-mediated checkpoint pathways regulate phosphorylation and activation of human Chk1. Mol Cell Biol. 21:4129–4139.
120. Zhao, X., E.G. Muller, and R. Rothstein (1998) A suppressor of two essential checkpoint genes identifies a novel protein that negatively affects dNTP pools. Mol Cell. 2:329–340.
121. Zhao, X. and R. Rothstein (2002) The Dun1 checkpoint kinase phosphorylates and regulates the ribonucleotide reductase inhibitor Sml1. Proc Natl Acad Sci USA. 99:3746–3751.
122. Zou, L. and S.J. Elledge (2003) Sensing DNA damage through ATRIP recognition of RPA-ssDNA complexes. Science. 300:1542–1548.
123. Zou, L., D. Liu, and S.J. Elledge (2003) Replication protein A-mediated recruitment and activation of Rad17 complexes. Proc Natl Acad Sci USA. 100:13827–13832.

Chapter 3
An Oncogene-Induced DNA Replication Stress Model for Cancer Development

Thanos D. Halazonetis

Abstract Given that cells progressing through mitosis with unrepaired DNA double-strand breaks (DSBs) are very likely to die, the notion that in cancer cells DNA DBSs would be continuously generated appeared unlikely. Yet, recent experimental findings suggest that in precancerous lesions and cancers, the activation of oncogenes leads to DNA replication stress and, subsequently to the formation of DNA DSBs. In precancerous lesions, the continuous generation of DNA DSBs activates the DNA damage checkpoint, raising a p53-dependent barrier to tumorigenesis. However, the DNA DSBs also lead to genomic instability. Thus, cancer development lies in the balance between the tumor-suppressing DNA damage checkpoint and the tumor-promoting oncogene-induced genomic instability.

Keywords Genomic instability · DNA DSBs · Oncogenes · Replication stress · p53 · ARF · Senescence

3.1 Introduction

Cancer is a progressive disease, often starting as a benign "precancerous" lesion that over time increases in mass and acquires the potential to invade the host tissues and spread throughout the body. In the past 25 years, there has been considerable progress in understanding cancer development and progression at the molecular level. Here, I will review key concepts of the field and then discuss a model for cancer development that tries to explain the key findings.

T.D. Halazonetis (✉)
Department of Molecular Biology and Department of Biochemistry, University of Geneva, CH-1205, Geneva, Switzerland
e-mail: Thanos.Halazonetis@unige.ch

K.K. Khanna, Y. Shiloh (eds.), *The DNA Damage Response: Implications on Cancer Formation and Treatment*, DOI 10.1007/978-90-481-2561-6_3,

3.2 Key Developments and Concepts

3.2.1 Identification of Oncogenes and Tumor Suppressors

The heritable capacity of cancer cells to grow beyond their normal confines implies the presence of genetic changes that can be transmitted from parent to daughter cells. The discovery of oncogenes and tumor suppressor genes has validated this prediction [1–15]. Typically, more than one oncogene are required to transform primary cells, suggesting that oncogenes may have distinct functions. For example, the oncogenes myc and ras together, but not on their own, transform primary cells; myc immortalizes the cells, whereas ras contributes features of the transformed phenotype, such as increased proliferation and growth in soft agar [2].

Transformation of primary cells may also be achieved by combining activation of an oncogene, such as *ras*, with inactivation of a tumor suppressor, such as *p53* or *p16INK4a* [9–11,13,14]. Indeed, in many cases, loss of a tumor suppressor gene is equivalent to activation of an oncogene, since the two genes may function within the same signaling pathway, but have opposite activities. For example, *p16INK4a* inhibits the cyclin-dependent kinases Cdk4 and Cdk6, whereas the proto-oncogene *cyclin D1* activates them [12].

Based on the findings described above, it has been proposed that cancer development requires the acquisition of properties that can be conferred by deregulation of oncogenes and tumor suppressors. Key properties that need to be acquired for cancer development have been listed as: self-sufficiency in growth signals, insensitivity to anti-growth signals, limitless replicative potential, evasion of apoptosis, tissue invasion and metastasis, and sustained angiogenesis [16–18]. In its most simple form, this model postulates that each of these properties is acquired by activation of a single oncogene or loss of a single tumor suppressor and that the sequence by which these properties are acquired is not critical.

3.2.2 Genomic Instability as a Characteristic of Most Human Cancers and Its Underlying Genetic Basis

Theoretical analysis of the frequencies by which mutations arise in normal human cells, suggests that humans should rarely develop cancer, because it would be difficult for a single cell to accumulate all the mutations necessary to become a cancer cell [19–24]. Accordingly, it has been suggested that precancerous cells have a higher mutation rate (genomic instability) than normal cells. Two mutually nonexclusive possibilities have been considered: the genomic instability is transient, occuring at a specific stage of cancer development; or the genomic instability has a genetic basis and is, therefore, permanent. Either model is consistent with the multitude of genomic aberrations present in virtually all human cancers.

A transient surge in genomic instability during cancer evolution could arise in response to telomere erosion [25–28]. Both human and mouse models suggest that telomere attrition, which inevitably occurs during cancer development, leads to chromosomal aberrations, especially, in the presence of DNA damage checkpoint defects. Then, as cancer cells re-express telomerase or employ other mechanisms to maintain telomeric length, the genome of these cells is stabilized. In human precancerous lesions, telomere attrition is common and, therefore, likely to contribute to genomic instability [29–40]. Nevertheless, genomic instability can also be observed in the absence of telomere attrition and there is also strong evidence for a heritable and permanent increase in genomic instability in most human cancers.

The heritable component of genomic instability has been attributed to mutations in genes controlling genomic integrity, such as DNA repair and cell cycle checkpoint genes. Inactivation of such caretaker genes may facilitate the acquisition of mutations in oncogenes and tumor suppressors that are needed for tumorigenesis [41–45]. At least two types of heritable genomic instability are recognized: microsatellite instability (MIN) and chromosomal instability (CIN) [46, 47]. MIN, a prototypical example of genomic instability arising from mutations targeting caretaker genes, is characterized by mutations in simple repeated sequences, microsatellites, and is due to mutations in DNA mismatch repair genes [48, 49]. CIN is characterized by chromosomal aberrations, including deletions, duplications and translocations. A significant effort is being undertaken to identify the caretaker genes, whose deregulation is responsible for CIN. In experimental models, several such genes were identified, but none of them are frequently mutated in sporadic human cancers [50–54]. Further, large scale sequencing analysis of all well-annotated coding regions in breast and colorectal cancers has failed to identify other genes that could account for CIN in a large fraction of human cancers [55, 56].

Perhaps one notable exception in the search for genes that could explain the CIN in human cancers is the tumor suppressor *p53*. *p53* is the most frequently inactivated gene in human cancer [6–8]. Its protein product is activated by a variety of cellular stresses, but most notably, by DNA damage through a checkpoint pathway that involves the kinases ataxia-telangiectasia mutated (ATM) and Chk2 [57–73]. Loss of *p53* function in cells exposed to DNA damaging agents leads to aneuploidy, presumably mainly because in the absence of *p53* function, cells with DNA damage are not eliminated by *p53*-dependent apoptosis. Based on these properties *p53* was considered a caretaker gene and named "guardian of the genome" [74]. However, if *p53* were a caretaker gene, then restoring *p53* function in advanced cancers would be tolerated, since these cancers have already accumulated all the mutations necessary for acquisition of the transformed phenotype. But, in both tissue culture and animal mouse cancer models, restoring *p53* function, leads to cell death and tumor regression, respectively [75–81]. Further, as discussed below, the onset of genomic instability in precancerous lesions precedes the acquisition of *p53* mutations.

3.2.3 Identification of a Pathway, Involving ARF and p53, by Which Oncogenes Induce Apoptosis or Senescence

The early studies on oncogenes identified activities that favor tumor development, such as induction of cell proliferation and growth in soft agar. Then in a landmark study, the oncogene *myc* was shown to induce apoptosis. Remarkably, the oncogenic and apoptotic functions of *myc* could not be separated; *myc* mutants either retained or lost both functions [82–84]. Similar observations were then made with other oncogenes. In a variation of the concept that oncogenes have both tumor-promoting and suppressing functions, the oncogene *ras* was shown to induce senescence, a cellular state characterized by irreversible cell cycle arrest [85–87].

The pro-apoptotic and pro-senescent functions of oncogenes were shown to be mediated by a signaling pathway that involves *p53* and the *INK4a/ARF* locus [85–89]. The latter locus encodes two protein products, p16INK4a and ARF, that share no common amino acid sequence [90–92]. Before the discovery of ARF, *ras*-induced senescence was shown to be suppressed by inactivating either *p53* or *p16INK4a*; this was achieved by mutations or deletions targeting the *p53* gene, by deletion of the *INK4a/ARF* locus or by viral oncogenes, such as adenoviral *E1A* and papillomavirus *E7*, that inhibit the p16INK4a-Rb pathway [85–87]. Thus, these studies suggest that two independent pathways, involving p53 and p16INK4a-Rb, respectively, are important for the induction of senescence.

With the discovery of ARF, as the second protein product of the *INK4a/ARF* locus, a pathway by which oncogenes can activate p53 was identified. This pathway involves induction of ARF expression by oncogenes, inhibition of the Mdm2 ubiquitin ligase by ARF and increased p53 protein levels (since Mdm2 targets p53 for degradation in the proteasome) [88,89]. Interestingly, the mechanism by which ARF activates p53 is distinct from the mechanism by which DNA damage activates p53, since p53 activated by ARF was not phosphorylated at the N-terminus (at least when adenoviral E1A was used to induce ARF expression), unlike p53 activated by DNA damage [88,89].

These findings have significantly influenced our understanding of cancer development. They have suggested the existence of a specific pathway by which oncogenes activate p53 through ARF. However, how oncogenes induce ARF expression is less clear. It appears that distinct oncogenes utilize distinct mechanisms. For example, activated *ras* enhances Jun activity, which in turn upregulates expression of the DMP1 transcription factor, which in turn upregulates ARF. However, *myc* and *E2F1* upregulate ARF without upregulating DMP1 [93].

The key concepts presented above suggest a model for cancer development, whose key features are an oncogene-induced ARF-p53 pathway for tumor suppression and, separately, inactivation of caretaker genes for induction of genomic instability. While this model can explain the high frequency of *p53* mutations in human cancer, two key aspects of the model remain unresolved. What are the mechanisms by which oncogenes activate ARF? With the exception of a small subset of cancers, why haven't the caretaker genes, whose inactivation is responsible for CIN, been identified?

3.3 A New Model to Explain Genomic Instability and Tumor Suppression in Human Cancers

3.3.1 Identification of DNA DSBs in Human Cancers and in Cells Expressing Activated Oncogenes

With the identification and characterization of proteins involved in the DNA damage response, it has recently become possible to visualize DNA DSBs, indirectly by immunofluorescence. Phosphorylation of histone H2AX and recruitment of 53BP1 and the Mre11-Nbs1-Rad50 complex to large chromatin domains encompassing about 1 Mb of genomic DNA flanking the DNA DSB provide ample signal to monitor even a single DNA DSB [94–97]. Immunofluorescence analysis monitoring 53BP1 recruitment to such chromatin domains (53BP1 irradiation-induced foci), revealed the presence of DNA DSBs in non-irradiated cancer cell lines (about 10–20 53BP1 foci per cell). Since cancer cells are typically not defective in DNA DSB repair, these results suggested a continuous cycle of DNA DSB formation and repair. Interestingly, in the small panel of cancer cell lines examined, there was a perfect correlation between the presence of large numbers of 53BP1 foci and *p53* mutations, suggesting that the presence of DNA DSBs in these cell lines could be selecting for loss of p53 function [98, 99].

Growth of human cells in tissue culture can be a stressful event for the cells, since it is hard to reproduce in tissue culture the conditions present in the human body. Thus, identifying DNA DSBs in human cancer cell lines does not necessarily mean that DNA DSBs are present in human cancers. To determine whether and at what stage of cancer development DNA DSBs are present in human cancers, tissue sections from human patients representing various stages of cancer development were analysed. In the earliest lesions (precancerous lesions), before *p53* mutations were acquired, 53BP1 foci and a cellular response to the DNA DSBs, in the form of phosphorylated histone H2AX, ATM, Chk2 and p53, were evident [100–102]. In more advanced lesions (cancers), DNA DSBs were again present, but the cells also had *p53* mutations and/or loss of expression of various DNA damage response proteins, such as 53BP1, MDC1 or Chk2, leading to inactivation of the DNA DSB checkpoint. Interestingly, the presence of DNA damage was not observed in normal tissues, even tissues with high proliferation rates [100–102].

What could be causing the induction of DNA DSBs in the precancerous lesions and cancers? Deregulation of a caretaker gene or eroded telomeres could be responsible, but neither of these are known to affect all precancerous lesions [29–56]. Activated oncogenes could be the driving force, since this is a feature that distinguishes all precancerous lesions from normal tissues. Analysis of various model systems suggests that activation of oncogenes can lead to formation of DNA DSBs and to a DNA damage checkpoint response. First, overexpression of mutant *ras* was shown to induce chromosomal breaks in NIH3T3 fibroblasts [103]. Then, overexpression of other oncogenes, such as *myc*, *cyclin E*, *mos* and *E2F1*, was shown to induce histone H2AX, ATM, Chk2 and p53 phosphorylation in various cell types and animal models, indicating the presence of a DNA damage response [104–112].

In these models, the majority of cells overexpressing the oncogene showed signs of a DNA damage response, suggesting that oncogene activation was sufficient for DNA damage induction.

3.3.2 The DNA Damage Checkpoint as an Important Mediator of Oncogene-Induced Senescence and/or Apoptosis and a Barrier to Tumor Development

As mentioned above, human precancerous lesions have DNA DSBs and activation of the ATM, Chk2 and p53 checkpoint pathway. In various cell and animal models, DNA DSBs induce cell cycle arrest, apoptosis and/or senescence in an ATM, Chk2 and p53-dependent manner [58–73]. The question then arises whether apoptosis and senescence are present in human precancerous lesions and, if so, whether they subside during cancer progression, when *p53* mutations are acquired. It turns out that human precancerous lesions do exhibit p53-dependent senescence and/or apoptosis [101, 113–125], albeit with interesting tissue-specific variations [101,119]. Precancerous lesions in the lung are characterized by increased apoptosis, which subsides dramatically at later stages of cancer, when *p53* mutations are acquired. In precancerous melanocytic lesions (dysplastic nevi), there is a high level of senescence, which again subsides as the lesions progress to melanoma. Precancerous lesions of the colon (adenomas) also exhibit senescence, which again subsides in carcinomas, when *p53* mutations are acquired. The colon is a particularly interesting tissue, because the proliferation index in the adenomas is lower than that in the normal tissue, even though adenomas have taken a step towards cancer progression. This is because the normal colon has a very high proliferation index and the induction of senescence in adenomas leads to inhibition of cell proliferation. Thus, the colon example illustrates that cancer development is not a process characterized by step-wise acquisition of tumorigenic features, but a "two step forward, one step back" process with a tumorigenesis barrier in the precancerous stage limiting, via senescence or apoptosis, the emergence of tumorigenic properties.

All evidence points to p53 as a key protein imposing the tumorigenesis barrier in precancerous lesions [100, 101, 113–127]. The question that remains to be answered is whether, in precancerous lesions, p53 is activated by the DNA damage checkpoint pathway, by the DNA damage-independent ARF pathway or by both pathways. Initially, oncogene-induced apoptosis and senescence were proposed to be mediated exclusively by the ARF-p53 pathway, because in cells undergoing adenoviral E1A-induced apoptosis, p53 lacked the ATM-dependent phosphorylation that is characteristic of cells with DNA damage [88, 89]. However, unlike E1A, numerous cellular oncogenes induce DNA DSBs and activate the ATM-Chk2-p53 DNA damage checkpoint pathway leading to p53 phosphorylation [103–112]. Importantly, in human precancerous lesions, DNA DSBs are present, the ATM-Chk2-p53 pathway is activated and its activation coincides with induction of senescence and apoptosis [100, 101, 113–123]. In the face of these results, it is hard to argue that ARF is the only pathway leading to p53-dependent senescence and apoptosis in human precancerous lesions.

A key experiment to demonstrate that the DNA damage checkpoint pathway is important for oncogene-induced senescence and apoptosis is to suppress ATM activity and show that senescence and apoptosis are abrogated. This experiment has been performed in both cell-based and animal model systems by several laboratories and, in every case, ATM function was shown to be critical [103–112, 119–125]. In some, but not all, settings, inhibition of ATM was not sufficient for complete escape from oncogene-induced senescence and this required, in addition, inhibition of the p16INK4a-Rb pathway [124, 125].

Similar experiments to test the role of ARF in oncogene-induced senescence and apoptosis have not always given consistent results. In some models, ARF is required for oncogene-induced senescence and apoptosis, while in others it is dispensable [88, 89, 128, 129]. In some models, escape from oncogene-induced senescence has been attributed to inactivation of ARF, but the entire INKA4a/ARF locus was deleted, so p16INK4a and, possibly p15INK4b, were also affected.

The spectrum of mutations targeting DNA damage checkpoint genes in human cancers also provides a measure of the importance of this pathway in human cancer development. The *atm* gene is among the most frequently mutated kinase-encoding genes in human cancer [130]; mutations in the *chk2* gene predispose to breast and other human cancers [131]; and 53BP1 expression is suppressed in about 3–5% of melanomas and lung carcinomas [101]. Of course, the most common mechanism leading to inactivation of the DNA DSB checkpoint pathway in human cancer is *p53* mutations [6–8]. The high frequency of *p53* mutations can be explained because amino acid substitutions in the p53 DNA binding domain generate dominant negative mutants and because inactivation of p53 preserves the G2 DNA damage checkpoint, upon which cancer cells must rely to ensure that they enter mitosis after the DNA DSBs have been repaired.

Mutations targeting *ARF* are also frequent in human cancer, but they are typically deletions targeting not only *ARF*, but also *p16INK4a*, and often *p15INK4b*, which is physically only a few Kb away from the *INK4a/ARF* locus [12–15, 90–92]. Further, point mutations that target exclusively *p16INK4a* outnumber those targeting exclusively *ARF* by a factor of about 20:1 [90]. Finally, transforming viruses (adenoviruses, papillomaviruses, and the SV40 polyomavirus) all target p53 and Rb, but none target ARF. Thus, the preponderance of the evidence supports the concept that in humans, and possibly other organisms, the DNA damage checkpoint pathway presents the major barrier to tumorigenesis, without excluding a role for the ARF-p53 pathway.

3.3.3 DNA Replication Stress Induces DNA DSBs and Genomic Instability in Cancer

The results reviewed so far raise the question what is the mechanism by which activated oncogenes induce DNA DSBs in human precancerous lesions and cancers. One possibility is that oncogenes compromise DNA replication leading to DNA replication stress, a loosely-defined term that refers to stalling or collapse of replication forks, but used here to refer only to collapsed replication forks (because only when the replication fork dissociates from the replicating DNA, there is the potential for DNA DSB formation and genomic instability). Indeed, in human cells, overexpression of *cyclin E* or other oncogenes leads to collapse of DNA replication forks and DNA DSB formation, as demonstrated by the presence of prematurely terminated DNA replication forks, foci of single-stranded DNA and activation of the ATR-Chk1 DNA replication stress checkpoint pathway [119–125,132–134] (Fig. 3.1). In yeast, overexpression of G1 cyclins also leads to the formation of DNA DSBs during DNA replication, the mechanism probably involving compromised licensing of DNA replication origins and subsequent collapse of DNA replication forks [135,136].

Monitoring DNA replication stress by direct methods, such as DNA combing analysis, is possible for cells grown in tissue culture, but not for human tissues. However, the presence of DNA replication stress can be monitored indirectly. Collapse of DNA replication forks leads to the formation of DNA DSBs that need to be repaired. Such repair could lead to gene conversion, if the repair is mediated by homologous recombination that utilizes the allele from the other parent. Alternatively, the repair could be associated with small deletions. Both gene conversion and deletions lead to changes in allele ratios (loss of heterozygosity, LOH), which leave a permanent signature on the genomic DNA of the cell and which can be easily monitored in human clinical specimens. Interestingly, under conditions of DNA replication stress, DNA DSBs form at specific genomic regions, called common fragile sites [137–140]. Thus, the presence of LOH specifically at common fragile sites is a molecular signature for DNA replication stress. In human precancerous lesions and in human cells overexpressing oncogenes, there is LOH that involves predominantly the common fragile sites [100, 101, 124, 141–146]. Thus, the induction of DNA DSBs after oncogene activation is due, at least in part, to DNA replication stress.

The induction of genomic instability at common fragile sites in precancerous models and in human cells overexpressing oncogenes occurs rapidly, within a few weeks [101, 124]. This suggests that genomic instability is a feature of human cancer from its very beginning and is driven, at least in part, by activated oncogenes. Further, while changes in allele ratios can occur in the absence of chromosomal

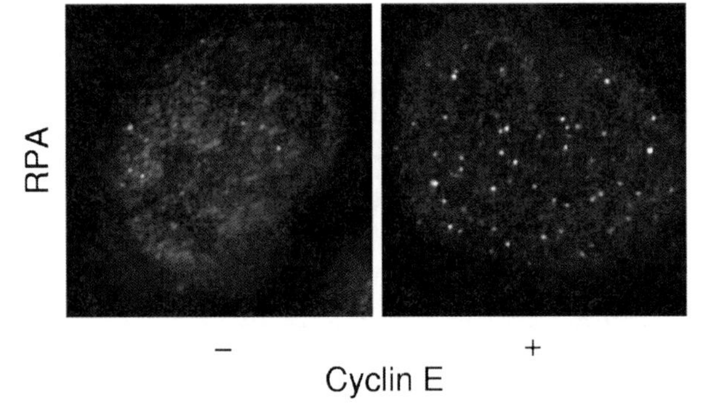

Fig. 3.1 DNA replication stress in response to cyclin E overexpression (100), as monitored by formation of replication protein A (RPA) foci. This specific experiment was performed by Lorenzo Costantino at the University of Geneva

instability, it is likely that the DNA DSBs that give rise to LOH in human precancerous lesions have the potential to induce CIN, when there is an error in template selection during homologous-mediated repair. Thus, activated oncogenes, rather than inactivation of caretaker genes, may be responsible for CIN in precancerous lesions and cancer. This could also explain why, in cell fusion experiments, CIN is a dominant trait [46, 47].

DNA replication stress targets preferentially the common fragile sites, but other sites in the genome are also targeted. Thus, even though DNA replication stress establishes a barrier to tumorigenesis by activating p53, it may also drive cancer progression by inducing genomic instability. Interestingly, in oral precancerous lesions, LOH at the common fragile site FRA3B predicted much better progression to cancer than any other molecular and histopathological marker, including *p53* mutations and LOH at the *INK4a/ARF* locus [141, 145, 146].

3.4 A Model for Cancer Development

The observations reviewed here allow us to formulate a model for cancer development (Fig. 3.2). In its most simple form, this model states that precancerous lesions arise when genetic mutations lead to oncogene activation. At this stage of cancer development, the activated oncogenes promote cell proliferation, but in the process also induce DNA replication stress and, subsequently, DNA DSBs. As a result the DNA damage checkpoint pathway is activated and through p53-induced apoptosis and senescence curtails tumor development. However, the balance between the

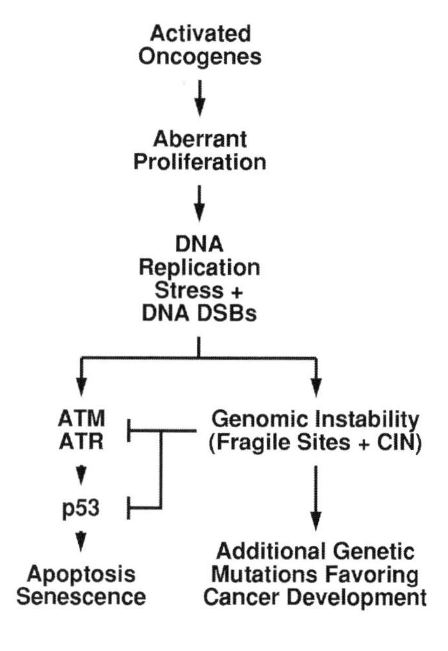

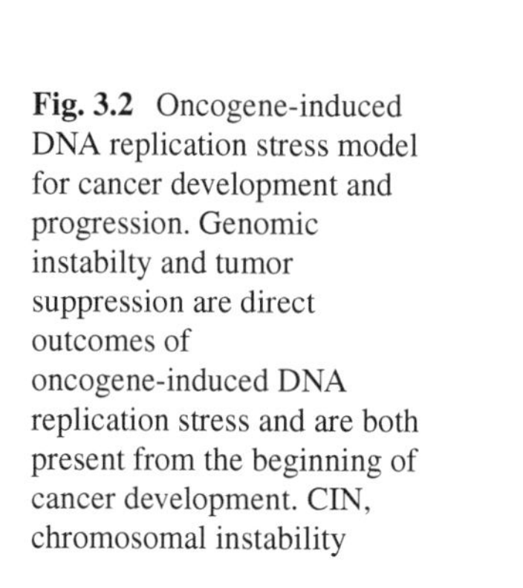
Fig. 3.2 Oncogene-induced DNA replication stress model for cancer development and progression. Genomic instabilty and tumor suppression are direct outcomes of oncogene-induced DNA replication stress and are both present from the beginning of cancer development. CIN, chromosomal instability

oncogene-stimulated cell proliferation and the DNA damage checkpoint-induced apoptosis and senescence is precarious. The continuous induction of DNA DSBs promotes genomic instability and the acquisition of mutations that drive progression of the precancerous lesions to cancer. Once these mutations target p53, then the tumor suppressor effect of the DNA damage checkpoint is relieved and progression to cancer can ensue.

The oncogene-induced DNA replication stress model explains why DNA DSB response genes, such as *p53*, are tumor suppressor genes. Further, the model explains the widespread presence of genomic instability in cancer, even though in most cancers high frequency mutations targeting genes that play a direct role in genomic stability, such as, for example, DNA repair genes, are not found [55, 56].

References

1. Bishop JM (1987) The molecular genetics of cancer. Science 235: 305–311
2. Land H, Parada LF, Weinberg RA (1983) Tumorigenic conversion of primary embryo fibroblasts requires at least two cooperating oncogenes. Nature 304: 596–602
3. Ruley HE (1983) Adenovirus early region 1A enables viral and cellular transforming genes to transform primary cells in culture. Nature 304: 602–606
4. Rassoulzadegan M, Cowie A, Carr A, Glaichenhaus N, Kamen R, Cuzin F (1982) The roles of individual polyoma virus early proteins in oncogenic transformation. Nature 300: 713–718
5. Rassoulzadegan M, Gaudray P, Canning M, Trejo-Avila L, Cuzin F (1981) Two polyoma virus gene functions involved in the expression of the transformed phenotype in FR 3T3 rat cells. I. Localization of a transformation maintenance function in the proximal half of the large T coding region. Virology 114: 489–500
6. Vogelstein B, Lane D, Levine AJ (2000) Surfing the p53 network. Nature 408: 307–310
7. Levine AJ, Perry ME, Chang A, Silver A, Dittmer D, Wu M, Welsh D (1994) The 1993 Walter Hubert Lecture: the role of the p53 tumour-suppressor gene in tumorigenesis. Br J Cancer 69: 409–416
8. Soussi T, Lozano G (2005) p53 mutation heterogeneity in cancer. Biochem Biophys Res Commun 331: 834–842
9. Eliyahu D, Raz A, Gruss P, Givol D, Oren M (1984) Participation of p53 cellular tumour antigen in transformation of normal embryonic cells. Nature 312: 646–649
10. Parada LF, Land H, Weinberg RA, Wolf D, Rotter V (1984) Cooperation between gene encoding p53 tumour antigen and ras in cellular transformation. Nature 312: 649–651
11. Jenkins JR, Rudge K, Currie GA (1984) Cellular immortalization by a cDNA clone encoding the transformation-associated phosphoprotein p53. Nature 312: 651–654
12. Ortega S, Malumbres M, Barbacid M (2002) Cyclin D-dependent kinases, INK4 inhibitors and cancer. Biochim Biophys Acta 1602: 73–87
13. Serrano M, Gomez-Lahoz E, DePinho RA, Beach D, Bar-Sagi D (1195) Inhibition of ras-induced proliferation and cellular transformation by p16INK4. Science 267: 249–252
14. Serrano M, Lee H, Chin L, Cordon-Cardo C, Beach D, DePinho RA (1996) Role of the INK4a locus in tumor suppression and cell mortality. Cell 85: 27–37
15. Lukas J, Parry D, Aagaard L, Mann DJ, Bartkova J, Strauss M, Peters G, Bartek J (1995) Retinoblastoma-protein-dependent cell-cycle inhibition by the tumour suppressor p16. Nature 375: 503–506
16. Hanahan D, Weinberg RA (2000) The hallmarks of cancer. Cell 100: 57–70
17. Evan GI, Vousden KH (2001) Proliferation, cell cycle and apoptosis in cancer. Nature 411: 342–348

18. Hahn WC, Counter CM, Lundberg AS, Beijersbergen RL, Brooks MW, Weinberg RA (1999) Creation of human tumour cells with defined genetic elements. Nature 400: 464–468
19. Beckman RA, Loeb LA (2006) Efficiency of carcinogenesis with and without a mutator mutation. Proc Natl Acad Sci USA 103: 14140–14145
20. Sieber O, Heinimann K, Tomlinson I (2005) Genomic stability and tumorigenesis. Semin Cancer Biol 15: 61–66
21. Nowak MA, Komarova NL, Sengupta A, Jallepalli PV, Shih IeM, Vogelstein B, Lengauer C (2002) The role of chromosomal instability in tumor initiation. Proc Natl Acad Sci USA 99: 16226–16231.
22. Loeb LA (2001) A mutator phenotype in cancer. Cancer Res 61: 3230–3239
23. Sweezy MA, Fishel R (1994) Multiple pathways leading to genomic instability and tumorigenesis. Ann NY Acad Sci 726: 165–177
24. Loeb LA (1991) Mutator phenotype may be required for multistage carcinogenesis. Cancer Res 51: 3075–3079
25. Maser RS, DePinho RA (2002) Connecting chromosomes, crisis, and cancer. Science 297: 565–569
26. De Lange T (2005) Telomere-related genome instability in cancer. Cold Spring Harb Symp Quant Biol 70: 197–204
27. Romanov SR, Kozakiewicz BK, Holst CR, Stampfer MR, Haupt LM, Tlsty TD (2001) Normal human mammary epithelial cells spontaneously escape senescence and acquire genomic changes. Nature 409: 633–637
28. O'Hagan RC, Chang S, Maser RS, Mohan R, Artandi SE, Chin L, DePinho RA (2002) Telomere dysfunction provokes regional amplification and deletion in cancer genomes. Cancer Cell 2: 149–155
29. Meeker AK, Hicks JL, Iacobuzio-Donahue CA, Montgomery EA, Westra WH, Chan TY, Ronnett BM, De Marzo AM (2004) Telomere length abnormalities occur early in the initiation of epithelial carcinogenesis. Clin Cancer Res 10: 3317–3326
30. Miracco C, Margherita De Santi M, Schurfeld K, Santopietro R, Lalinga AV, Fimiani M, Biagioli M, Brogi M, De Felice C, Luzi P, Andreassi L (2002) Quantitative in situ evaluation of telomeres in fluorescence in situ hybridization-processed sections of cutaneous melanocytic lesions and correlation with telomerase activity. Br J Dermatol 146: 399–408
31. Michaloglou C, Vredeveld LC, Soengas MS, Denoyelle C, Kuilman T, van der Horst CM, Majoor DM, Shay JW, Mooi WJ, Peeper DS (2005) BRAFE600-associated senescence-like cell cycle arrest of human naevi. Nature 436: 720–724
32. Kammori M, Nakamura K, Kanauchi H, Obara T, Kawahara M, Mimura Y, Kaminishi M, Takubo K (2002) Consistent decrease in telomere length in parathyroid tumors but alteration in telomerase activity limited to malignancies: preliminary report. World J Surg 26: 1083–1087
33. van Heek NT, Meeker AK, Kern SE, Yeo CJ, Lillemoe KD, Cameron JL, Offerhaus GJ, Hicks JL, Wilentz RE, Goggins MG, De Marzo AM, Hruban RH, Maitra A (2002) Telomere shortening is nearly universal in pancreatic intraepithelial neoplasia. Am J Pathol 161: 1541–1547
34. Vukovic B, Park PC, Al-Maghrabi J, Beheshti B, Sweet J, Evans A, Trachtenberg J, Squire J (2003) Evidence of multifocality of telomere erosion in high-grade prostatic intraepithelial neoplasia (HPIN) and concurrent carcinoma. Oncogene 22: 1978–1987
35. Meeker AK, Hicks JL, Platz EA, March GE, Bennett CJ, Delannoy MJ, De Marzo AM (2002) Telomere shortening is an early somatic DNA alteration in human prostate tumorigenesis. Cancer Res 62: 6405–6409
36. Meeker AK, Hicks JL, Gabrielson E, Strauss WM, De Marzo AM, Argani P (2004) Telomere shortening occurs in subsets of normal breast epithelium as well as in situ and invasive carcinoma. Am J Pathol 164: 925–935
37. Chin K, de Solorzano CO, Knowles D, Jones A, Chou W, Rodriguez EG, Kuo WL, Ljung BM, Chew K, Myambo K, Miranda M, Krig S, Garbe J, Stampfer M, Yaswen P, Gray JW,

Lockett SJ (2004) In situ analyses of genome instability in breast cancer. Nat Genet 36: 984–988
38. Hansel DE, Meeker AK, Hicks J, De Marzo AM, Lillemoe KD, Schulick R, Hruban RH, Maitra A, Argani P (2006) Telomere length variation in biliary tract metaplasia, dysplasia, and carcinoma. Mod Pathol 19: 772–779
39. Maida Y, Kyo S, Forsyth NR, Takakura M, Sakaguchi J, Mizumoto Y, Hashimoto M, Nakamura M, Nakao S, Inoue M (2006) Distinct telomere length regulation in premalignant cervical and endometrial lesions: implications for the roles of telomeres in uterine carcinogenesis. J Pathol 210: 214–223
40. Kawai T, Hiroi S, Nakanishi K, Meeker AK (2007) Telomere length and telomerase expression in atypical adenomatous hyperplasia and small bronchioloalveolar carcinoma of the lung. Am J Clin Pathol 127: 254–262
41. Hartwell LH, Kastan MB (1994) Cell cycle control and cancer. Science 266: 1821–1828
42. Setlow RB (1978) Repair deficient human disorders and cancer. Nature 271: 713–717
43. Hartwell L (1992) Defects in a cell cycle checkpoint may be responsible for the genomic instability of cancer cells. Cell 71: 543–546
44. Kinzler KW, Vogelstein B (1997) Cancer-susceptibility genes. Gatekeepers and caretakers. Nature 386: 761–763
45. Cahill DP, Kinzler KW, Vogelstein B, Lengauer C (1999) Genetic instability and darwinian selection in tumours. Trends Cell Biol 9: M57-60
46. Lengauer C, Kinzler KW, Vogelstein B (1998) Genetic instabilities in human cancers. Nature 396: 643–649
47. Lengauer C, Kinzler KW, Vogelstein B (1997) Genetic instability in colorectal cancers. Nature 386: 623–627
48. Fishel R, Lescoe MK, Rao MR, Copeland NG, Jenkins NA, Garber J, Kane M, Kolodner R (1993) The human mutator gene homolog MSH2 and its association with hereditary nonpolyposis colon cancer. Cell 75: 1027–1038
49. Leach FS, Nicolaides NC, Papadopoulos N, Liu B, Jen J, Parsons R, Peltomaki P, Sistonen P, Aaltonen LA, Nystrom-Lahti M, Guan XY, Zhang J, Meltzer PS, Yu JW, Kao FT, Chen DJ, Cerosaletti KM, Fournier RE, Todd S, Lewis T, Leach RJ, Naylor SL, Weissenbach J, Mecklin JP, Jarvinen H, Petersen GM, Hamilton SR, Green J, Jass J, Watson P, Lynch HT, Trent JM, de la Chapelle A, Kinzler KW, Vogelstein B (1993) Mutations of a mutS homolog in hereditary nonpolyposis colorectal cancer. Cell 75: 1215–1225
50. Rajagopalan H, Lengauer C (2004) Aneuploidy and cancer. Nature 432: 338–341
51. Cahill DP, Lengauer C, Yu J, Riggins GJ, Willson JK, Markowitz SD, Kinzler KW, Vogelstein B (1998) Mutations of mitotic checkpoint genes in human cancers. Nature 392: 300–303
52. Jallepalli PV, Waizenegger IC, Bunz F, Langer S, Speicher MR, Peters JM, Kinzler KW, Vogelstein B, Lengauer C (2001) Securin is required for chromosomal stability in human cells. Cell 105: 445–457
53. Rajagopalan H, Jallepalli PV, Rago C, Velculescu VE, Kinzler KW, Vogelstein B, Lengauer C (2004) Inactivation of hCDC4 can cause chromosomal instability. Nature 428: 77–81
54. Wang Z, Cummins JM, Shen D, Cahill DP, Jallepalli PV, Wang TL, Parsons DW, Traverso G, Awad M, Silliman N, Ptak J, Szabo S, Willson JK, Markowitz SD, Goldberg ML, Karess R, Kinzler KW, Vogelstein B, Velculescu VE, Lengauer C (2004) Three classes of genes mutated in colorectal cancers with chromosomal instability. Cancer Res 64: 2998–3001
55. Sjoblom T, Jones S, Wood LD, Parsons DW, Lin J, Barber TD, Mandelker D, Leary RJ, Ptak J, Silliman N, Szabo S, Buckhaults P, Farrell C, Meeh P, Markowitz SD, Willis J, Dawson D, Willson JK, Gazdar AF, Hartigan J, Wu L, Liu C, Parmigiani G, Park BH, Bachman KE, Papadopoulos N, Vogelstein B, Kinzler KW, Velculescu VE (2006) The consensus coding sequences of human breast and colorectal cancers. Science 314: 268–274
56. Wood LD, Parsons DW, Jones S, Lin J, Sj öblom T, Leary RJ, Shen D, Boca SM, Barber T, Ptak J, Silliman N, Szabo S, Dezso Z, Ustyanksky V, Nikolskaya T, Nikolsky Y, Karchin

R, Wilson PA, Kaminker JS, Zhang Z, Croshaw R, Willis J, Dawson D, Shipitsin M, Willson JK, Sukumar S, Polyak K, Park BH, Pethiyagoda CL, Pant PV, Ballinger DG, Sparks AB, Hartigan J, Smith DR, Suh E, Papadopoulos N, Buckhaults P, Markowitz SD, Parmigiani G, Kinzler KW, Velculescu VE, Vogelstein B (2007) The genomic landscapes of human breast and colorectal cancers. Science 318: 1108–1113

57. Kastan MB, Bartek J (2004) Cell-cycle checkpoints and cancer. Nature 432: 316–323.
58. Kastan MB, Zhan Q, el-Deiry WS, Carrier F, Jacks T, Walsh WV, Plunkett BS, Vogelstein B, Fornace AJ Jr (1992) A mammalian cell cycle checkpoint pathway utilizing p53 and GADD45 is defective in ataxia-telangiectasia. Cell 71: 587–597
59. Canman CE, Lim DS, Cimprich KA, Taya Y, Tamai K, Sakaguchi K, Appella E, Kastan MB, Siliciano JD (1998) Activation of the ATM kinase by ionizing radiation and phosphorylation of p53. Science 281: 1677–1679
60. Banin S, Moyal L, Shieh S, Taya Y, Anderson CW, Chessa L, Smorodinsky NI, Prives C, Reiss Y, Shiloh Y, Ziv Y (1998) Enhanced phosphorylation of p53 by ATM in response to DNA damage. Science 281: 1674–1677
61. Khanna KK, Keating KE, Kozlov S, Scott S, Gatei M, Hobson K, Taya Y, Gabrielli B, Chan D, Lees-Miller SP, Lavin MF (1998) ATM associates with and phosphorylates p53: mapping the region of interaction. Nat Genet 20: 398–400
62. Tibbetts RS, Brumbaugh KM, Williams JM, Sarkaria JN, Cliby WA, Shieh SY, Taya Y, Prives C, Abraham RT (1999) A role for ATR in the DNA damage-induced phosphorylation of p53. Genes Dev 13: 152–157
63. Matsuoka S, Huang M, Elledge SJ (1998) Linkage of ATM to cell cycle regulation by the Chk2 protein kinase. Science 282: 1893–1897
64. Hirao A, Kong YY, Matsuoka S, Wakeham A, Ruland J, Yoshida H, Liu D, Elledge SJ, Mak TW (2000) DNA damage-induced activation of p53 by the checkpoint kinase Chk2. Science 287: 1824–1827
65. Chehab NH, Malikzay A, Appel M, Halazonetis TD (2000) Chk2 /hCds1 functions as a DNA damage checkpoint in G (1) by stabilizing p53. Genes Dev 14: 278–288
66. Chao C, Herr D, Chun J, Xu Y (2006) Ser18 and 23 phosphorylation is required for p53-dependent apoptosis and tumor suppression. EMBO J 25: 2615–2622
67. Lotem J, Sachs L (1993) Regulation by bcl-2, c-myc, and p53 of susceptibility to induction of apoptosis by heat shock and cancer chemotherapy compounds in differentiation-competent and-defective myeloid leukemic cells. Cell Growth Differ 4: 41–47
68. Clarke AR, Purdie CA, Harrison DJ, Morris RG, Bird CC, Hooper ML, Wyllie AH (1993) Thymocyte apoptosis induced by p53-dependent and independent pathways. Nature 362: 849–852
69. Lowe SW, Schmitt EM, Smith SW, Osborne BA, Jacks T (1993) p53 is required for radiation-induced apoptosis in mouse thymocytes. Nature 362: 847–849
70. Di Leonardo A, Linke SP, Clarkin K, Wahl GM (1994) DNA damage triggers a prolonged p53-dependent G1 arrest and long-term induction of Cip1 in normal human fibroblasts. Genes Dev 8: 2540–2551
71. te Poele RH, Okorokov AL, Jardine L, Cummings J, Joel SP (2002) DNA damage is able to induce senescence in tumor cells in vitro and in vivo. Cancer Res 62: 1876–1883
72. Han Z, Wei W, Dunaway S, Darnowski JW, Calabresi P, Sedivy J, Hendrickson EA, Balan KV, Pantazis P, Wyche JH (2002) Role of p21 in apoptosis and senescence of human colon cancer cells treated with camptothecin. J Biol Chem 277: 17154–17160
73. von Zglinicki T, Saretzki G, Ladhoff J, d'Adda di Fagagna F, Jackson SP (2005) Human cell senescence as a DNA damage response. Mech Ageing Dev 126: 111–117
74. Lane DP (1992) Cancer. p53, guardian of the genome. Nature 358: 15–16
75. Kastan MB (2007) Wild-type p53: tumors can't stand it. Cell 128: 837–840
76. Sharpless NE, DePinho RA (2007) Cancer biology: gone but not forgotten. Nature 445: 606–607
77. Baker SJ, Markowitz S, Fearon ER, Willson JK, Vogelstein B (1990) Suppression of human colorectal carcinoma cell growth by wild-type p53. Science 249: 912–915

78. Zambetti GP, Olson D, Labow M, Levine AJ (1992) A mutant p53 protein is required for maintenance of the transformed phenotype in cells transformed with p53 plus ras cDNAs. Proc Natl Acad Sci USA 89: 3952–3956
79. Martins CP, Brown-Swigart L, Evan GI (2006) Modeling the therapeutic efficacy of p53 restoration in tumors. Cell 127: 1323–1334
80. Xue W, Zender L, Miething C, Dickins RA, Hernando E, Krizhanovsky V, Cordon-Cardo C, Lowe SW (2007) Senescence and tumour clearance is triggered by p53 restoration in murine liver carcinomas. Nature 445: 656–660
81. Ventura A, Kirsch DG, McLaughlin ME, Tuveson DA, Grimm J, Lintault L, Newman J, Reczek EE, Weissleder R, Jacks T (2007) Restoration of p53 function leads to tumour regression in vivo. Nature 445: 661–665
82. Evan GI, Wyllie AH, Gilbert CS, Littlewood TD, Land H, Brooks M, Waters CM, Penn LZ, Hancock DC (1992) Induction of apoptosis in fibroblasts by c-myc protein. Cell 69: 119–128
83. Rao L, Debbas M, Sabbatini P, Hockenbery D, Korsmeyer S, White E (1992) The adenovirus E1A proteins induce apoptosis, which is inhibited by the E1B 19–kDa and Bcl-2 proteins. Proc Natl Acad Sci USA 89: 7742–7746
84. Fukasawa K, Rulong S, Resau J, Pinto da Silva P, Woude GF (1995) Overexpression of mos oncogene product in Swiss 3T3 cells induces apoptosis preferentially during S-phase. Oncogene 10: 1–8
85. Serrano M, Lin AW, McCurrach ME, Beach D, Lowe SW (1997) Oncogenic ras provokes premature cell senescence associated with accumulation of p53 and p16INK4a. Cell 88: 593–602
86. Fukasawa K, Vande Woude GF (1997) Synergy between the Mos /mitogen-activated protein kinase pathway and loss of p53 function in transformation and chromosome instability. Mol Cell Biol 17: 506–518
87. Mallette FA, Gaumont-Leclerc MF, Ferbeyre G (2007) The DNA damage signaling pathway is a critical mediator of oncogene-induced senescence. Genes Dev 21: 43–48
88. de Stanchina E, McCurrach ME, Zindy F, Shieh SY, Ferbeyre G, Samuelson AV, Prives C, Roussel MF, Sherr CJ, Lowe SW (1998) E1A signaling to p53 involves the p19 (ARF) tumor suppressor. Genes Dev 12: 2434–2442.
89. Zindy F, Eischen CM, Randle DH, Kamijo T, Cleveland JL, Sherr CJ, Roussel MF (1998) Myc signaling via the ARF tumor suppressor regulates p53-dependent apoptosis and immortalization. Genes Dev 12: 2424–2433
90. Kim WY, Sharpless NE (2006) The regulation of INK4 /ARF in cancer and aging. Cell 127: 265–275
91. Quelle DE, Zindy F, Ashmun RA, Sherr CJ (1995) Alternative reading frames of the INK4a tumor suppressor gene encode two unrelated proteins capable of inducing cell cycle arrest. Cell 83: 993–1000
92. Kamijo T, Zindy F, Roussel MF, Quelle DE, Downing JR, Ashmun RA, Grosveld G, Sherr CJ (1997) Tumor suppression at the mouse INK4a locus mediated by the alternative reading frame product p19ARF. Cell 91: 649–659
93. Sreeramaneni R, Chaudhry A, McMahon M, Sherr CJ, Inoue K (2005) Ras-Raf-Arf signaling critically depends on the Dmp1 transcription factor. Mol Cell Biol 25: 220–232
94. Schultz LB, Chehab NH, Malikzay A, Halazonetis TD (2000) p53 binding protein 1 (53BP1) is an early participant in the cellular response to DNA double-strand breaks. J Cell Biol 151: 1381–1390
95. Rogakou EP, Boon C, Redon C, Bonner WM (1999) Megabase chromatin domains involved in DNA double-strand breaks in vivo. J Cell Biol 146: 905–916
96. Rogakou EP, Pilch DR, Orr AH, Ivanova VS, Bonner WM (1998) DNA double-stranded breaks induce histone H2AX phosphorylation on serine 139. J Biol Chem 273: 5858–5868
97. Nelms BE, Maser RS, MacKay JF, Lagally MG, Petrini JH (1998) In situ visualization of DNA double-strand break repair in human fibroblasts. Science 280: 590–592

98. DiTullio RA Jr, Mochan TA, Venere M, Bartkova J, Sehested M, Bartek J, Halazonetis TD (2002) 53BP1 functions in an ATM-dependent checkpoint pathway that is constitutively activated in human cancer. Nat Cell Biol 4: 998–1002
99. Schultz LB, Chehab NH, Malikzay A, DiTullio RA Jr, Stavridi ES, Halazonetis TD (2000) The DNA damage checkpoint and human cancer. Cold Spring Harb Symp Quant Biol 65: 489–498
100. Bartkova J, Horejsi Z, Koed K, Kramer A, Tort F, Zieger K, Guldberg P, Sehested M, Nesland JM, Lukas C, Orntoft T, Lukas J, Bartek J (2005) DNA damage response as a candidate anti-cancer barrier in early human tumorigenesis. Nature 434: 864–870
101. Gorgoulis VG, Vassiliou LV, Karakaidos P, Zacharatos P, Kotsinas A, Liloglou T, Venere M, Ditullio RA Jr, Kastrinakis NG, Levy B, Kletsas D, Yoneta A, Herlyn M, Kittas C, Halazonetis TD (2005) Activation of the DNA damage checkpoint and genomic instability in human precancerous lesions. Nature 434: 907–913.
102. Halazonetis TD (2004) Constitutively active DNA damage checkpoint pathways as the driving force for the high frequency of p53 mutations in human cancer. DNA Repair (Amst) 3: 1057–1062
103. Denko NC, Giaccia AJ, Stringer JR, Stambrook PJ (1994) The human Ha-ras oncogene induces genomic instability in murine fibroblasts within one cell cycle. Proc Natl Acad Sci USA 91: 5124–5128
104. Felsher DW, Bishop JM (1999) Transient excess of MYC activity can elicit genomic instability and tumorigenesis. Proc Natl Acad Sci USA 96: 3940–3944
105. Mai S, Fluri M, Siwarski D, Huppi K (1996) Genomic instability in MycER-activated Rat1A-MycER cells. Chromosome Res 4: 365–371
106. Berkovich E, Ginsberg D (2003) ATM is a target for positive regulation by E2F-1. Oncogene 22: 161–167
107. Karlsson A, Deb-Basu D, Cherry A, Turner S, Ford J, Felsher DW (2003) Defective double-strand DNA break repair and chromosomal translocations by MYC overexpression. Proc Natl Acad Sci USA 100: 9974–9979
108. Frame FM, Rogoff HA, Pickering MT, Cress WD, Kowalik TF (2006) E2F1 induces MRN foci formation and a cell cycle checkpoint response in human fibroblasts. Oncogene 25: 3258–3266
109. Ray S, Atkuri KR, Deb-Basu D, Adler AS, Chang HY, Herzenberg LA, Felsher DW (2006) MYC can induce DNA breaks in vivo and in vitro independent of reactive oxygen species. Cancer Res 66: 6598–6605
110. Abulaiti A, Fikaris AJ, Tsygankova OM, Meinkoth JL (2006) Ras induces chromosome instability and abrogation of the DNA damage response. Cancer Res 66: 10505–10512
111. Fikaris AJ, Lewis AE, Abulaiti A, Tsygankova OM, Meinkoth JL (2006) Ras triggers ataxia-telangiectasia-mutated and Rad-3-related activation and apoptosis through sustained mitogenic signaling. J Biol Chem 281: 34759–34767
112. Abulaiti A, Fikaris AJ, Tsygankova OM, Meinkoth JL (2006) Ras induces chromosome instability and abrogation of the DNA damage response. Cancer Res 66: 10505–10512
113. Campisi J (2005) Suppressing cancer: the importance of being senescent. Science 309: 886–887
114. Sharpless NE, DePinho RA (2005) Cancer: crime and punishment. Nature 436: 636–637
115. Michaloglou C, Vredeveld LC, Soengas MS, Denoyelle C, Kuilman T, van der Horst CM, Majoor DM, Shay JW, Mooi WJ, Peeper DS (2005) BRAFE600-associated senescence-like cell cycle arrest of human naevi. Nature 436: 720–724
116. Chen Z, Trotman LC, Shaffer D, Lin HK, Dotan ZA, Niki M, Koutcher JA, Scher HI, Ludwig T, Gerald W, Cordon-Cardo C, Pandolfi PP (2005) Crucial role of p53-dependent cellular senescence in suppression of Pten-deficient tumorigenesis. Nature 436: 725–730
117. Braig M, Lee S, Loddenkemper C, Rudolph C, Peters AH, Schlegelberger B, Stein H, Dorken B, Jenuwein T, Schmitt CA (2005) Oncogene-induced senescence as an initial barrier in lymphoma development. Nature 436: 660–665

118. Collado M, Gil J, Efeyan A, Guerra C, Schuhmacher AJ, Barradas M, Benguria A, Zaballos A, Flores JM, Barbacid M, Beach D, Serrano M (2005) Tumour biology: senescence in premalignant tumours. Nature 436: 642
119. Bartkova J, Rezaei N, Liontos M, Karakaidos P, Kletsas D, Issaeva N, Vassiliou LV, Kolettas E, Niforou K, Zoumpourlis VC, Takaoka M, Nakagawa H, Tort F, Fugger K, Johansson F, Sehested M, Andersen CL, Dyrskjot L, Orntoft T, Lukas J, Kittas C, Helleday T, Halazonetis TD, Bartek J, Gorgoulis VG (2006) Oncogene-induced senescence is part of the tumorigenesis barrier imposed by DNA damage checkpoints. Nature 444: 633–637
120. Rogoff HA, Pickering MT, Frame FM, Debatis ME, Sanchez Y, Jones S, Kowalik TF (2004) Apoptosis associated with deregulated E2F activity is dependent on E2F1 and Atm/ Nbs1/ Chk2. Mol Cell Biol 24: 2968–2977
121. Powers JT, Hong S, Mayhew CN, Rogers PM, Knudsen ES, Johnson DG (2004) E2F1 uses the ATM signaling pathway to induce p53 and Chk2 phosphorylation and apoptosis. Mol Cancer Res 2: 203–214
122. Pusapati RV, Rounbehler RJ, Hong S, Powers JT, Yan M, Kiguchi K, McArthur MJ, Wong PK, Johnson DG (2006) ATM promotes apoptosis and suppresses tumorigenesis in response to Myc. Proc Natl Acad Sci USA 103: 1446–1451
123. Hong S, Pusapati RV, Powers JT, Johnson DG (2006) Oncogenes and the DNA damage response: Myc and E2F1 engage the ATM signaling pathway to activate p53 and induce apoptosis. Cell Cycle 5: 801–803
124. Di Micco R, Fumagalli M, Cicalese A, Piccinin S, Gasparini P, Luise C, Schurra C, Garre' M, Nuciforo PG, Bensimon A, Maestro R, Pelicci PG, d'Adda di Fagagna F (2006) Oncogene-induced senescence is a DNA damage response triggered by DNA hyper-replication. Nature 444: 638–642
125. Mallette FA, Gaumont-Leclerc MF, Ferbeyre G (2007) The DNA damage signaling pathway is a critical mediator of oncogene-induced senescence. Genes Dev 21: 43–48
126. Christophorou MA, Ringshausen I, Finch AJ, Swigart LB, Evan GI (2006) The pathological response to DNA damage does not contribute to p53-mediated tumour suppression. Nature 443: 214–217
127. Efeyan A, Garcia-Cao I, Herranz D, Velasco-Miguel S, Serrano M (2006) Tumour biology: Policing of oncogene activity by p53. Nature 443: 159
128. Tolbert D, Lu X, Yin C, Tantama M, Van Dyke T (2002) p19 (ARF) is dispensable for oncogenic stress-induced p53-mediated apoptosis and tumor suppression in vivo. Mol Cell Biol 22: 370–377
129. Rogoff HA, Pickering MT, Debatis ME, Jones S, Kowalik TF (2002) E2F1 induces phosphorylation of p53 that is coincident with p53 accumulation and apoptosis. Mol Cell Biol 22: 5308–5318
130. Greenman C, Stephens P, Smith R, Dalgliesh GL, Hunter C, Bignell G, Davies H, Teague J, Butler A, Stevens C, Edkins S, O'Meara S, Vastrik I, Schmidt EE, Avis T, Barthorpe S, Bhamra G, Buck G, Choudhury B, Clements J, Cole J, Dicks E, Forbes S, Gray K, Halliday K, Harrison R, Hills K, Hinton J, Jenkinson A, Jones D, Menzies A, Mironenko T, Perry J, Raine K, Richardson D, Shepherd R, Small A, Tofts C, Varian J, Webb T, West S, Widaa S, Yates A, Cahill DP, Louis DN, Goldstraw P, Nicholson AG, Brasseur F, Looijenga L, Weber BL, Chiew YE, DeFazio A, Greaves MF, Green AR, Campbell P, Birney E, Easton DF, Chenevix-Trench G, Tan MH, Khoo SK, Teh BT, Yuen ST, Leung SY, Wooster R, Futreal PA, Stratton MR (2007) Patterns of somatic mutation in human cancer genomes. Nature 446: 153–158
131. Nevanlinna H, Bartek J (2006) The CHEK2 gene and inherited breast cancer susceptibility. Oncogene 25: 5912–5919
132. Mailand N, Diffley JF (2005) CDKs promote DNA replication origin licensing in human cells by protecting Cdc6 from APC /C-dependent proteolysis. Cell 122: 915–926
133. Vaziri C, Saxena S, Jeon Y, Lee C, Murata K, Machida Y, Wagle N, Hwang DS, Dutta A (2003) A p53-dependent checkpoint pathway prevents rereplication. Mol Cell 11: 997–1008

134. Deb-Basu D, Aleem E, Kaldis P, Felsher DW (2006) CDK2 is required by MYC to induce apoptosis. Cell Cycle 5: 1342–1347
135. Lengronne A, Schwob E (2002) The yeast CDK inhibitor Sic1 prevents genomic instability by promoting replication origin licensing in late G (1). Mol Cell 9: 1067–1078
136. Tanaka S, Diffley JF (2002) Deregulated G1-cyclin expression induces genomic instability by preventing efficient pre-RC formation. Genes Dev 16: 2639–2649
137. Arlt MF, Durkin SG, Ragland RL, Glover TW (2006) Common fragile sites as targets for chromosome rearrangements. DNA Repair 5: 1126–1135
138. Casper AM, Durkin SG, Arlt MF, Glover TW (2004) Chromosomal instability at common fragile sites in Seckel syndrome. Am J Hum Genet 75: 654–660
139. Casper AM, Nghiem P, Arlt MF, Glover TW (2002) ATR regulates fragile site stability. Cell 111: 779–789
140. Osborn AJ, Elledge SJ, Zou L (2002) Checking on the fork: the DNA-replication stress-response pathway. Trends Cell Biol 12: 509–516
141. Mao L, Lee JS, Fan YH, Ro JY, Batsakis JG, Lippman S, Hittelman W, Hong WK (1996) Frequent microsatellite alterations at chromosomes 9p21 and 3p14 in oral premalignant lesions and their value in cancer risk assessment. Nat Med 2: 682–685
142. Emilion G, Langdon JD, Speight P, Partridge M (1996) Frequent gene deletions in potentially malignant oral lesions. Br J Cancer Mar; 73 (6): 809–813
143. Wistuba II, Behrens C, Virmani AK, Mele G, Milchgrub S, Girard L, Fondon JW 3rd, Garner HR, McKay B, Latif F, Lerman MI, Lam S, Gazdar AF, Minna JD (2000) High resolution chromosome 3p allelotyping of human lung cancer and preneoplastic /preinvasive bronchial epithelium reveals multiple, discontinuous sites of 3p allele loss and three regions of frequent breakpoints. Cancer Res 60: 1949–1960
144. Maitra A, Wistuba II, Washington C, Virmani AK, Ashfaq R, Milchgrub S, Gazdar AF, Minna JD (2001) High-resolution chromosome 3p allelotyping of breast carcinomas and precursor lesions demonstrates frequent loss of heterozygosity and a discontinuous pattern of allele loss. Am J Pathol 159: 119–130
145. Lee JJ, Hong WK, Hittelman WN, Mao L, Lotan R, Shin DM, Benner SE, Xu XC, Lee JS, Papadimitrakopoulou VM, Geyer C, Perez C, Martin JW, El-Naggar AK, Lippman SM (2000) Predicting cancer development in oral leukoplakia: ten years of translational research. Clin Cancer Res 6: 1702–1710
146. Rosin MP, Lam WL, Poh C, Le ND, Li RJ, Zeng T, Priddy R, Zhang L (2002) 3p14 and 9p21 loss is a simple tool for predicting second oral malignancy at previously treated oral cancer sites. Cancer Res 62: 6447–6450

Chapter 4
Cellular Responses to Oxidative Stress

Inbal Dar and Ari Barzilai

Abstract Oxygen is essential for life yet at the same time has been shown to be a harmful cause of cellular deterioration. To avoid its deleterious effects many intricate cellular mechanisms have evolved to neutralize oxygen reactive metabolites. Every mulitcellular organism needs different types of cells with different energy needs. The level of oxygen consumption and oxidative phosphorylation of each cell type and tissue correspond to its energy needs. Oxidative stress occurs when the production of reactive oxygen species (ROS), a normal product of cellular metabolism, exceeds the ability of the cell to repair the damage caused by ROS. For that purpose, cellular oxygen concentrations are maintained with a narrow "nomoxic" range to circumvent the risk of oxidative damage from excess O_2. High levels of ROS can lead to the accumulation of damage to various cellular macromolecules including DNA, protein and lipids. Here we provide a definition for the redox state and oxidative stress. We also describe, in brief, the most prevalent oxygen radicals and non-radicals ROS. Additionally, we depict the antioxidant defense mechanisms. Since oxidative stress is a common denominator of many neurodegenerative diseases we provide information about the vulnerability of the brain to oxidative stress and the effect of ROS on brain aging.

Keywords Oxidative stress · Redox state · Reactive oxygen species (ROS) · DNA damage · Neurodegenerative disease

4.1 Introduction

4.1.1 Cellular Redox State

In general, a chemical reaction in which a substance gains electrons is defined as a reduction [59]. Oxidation is a process in which a loss of electrons occurs. When

---

A. Barzilai (✉)
Department of Neurobiology, George S. Wise Faculty of Life Sciences, Tel Aviv University, Tel Aviv 69978, Israel
e-mail: barzilai@post.tau.ac.il

K.K. Khanna, Y. Shiloh (eds.), *The DNA Damage Response: Implications on Cancer Formation and Treatment*, DOI 10.1007/978-90-481-2561-6_4,

a reductant donates its electrons, it causes another substance to be reduced, and, when an oxidant accepts electrons, it causes another substance to be oxidized [32]. In biology, a reducing agent acts via donation of electrons, usually by donation of hydrogen or removal of oxygen. An oxidation process is always accompanied by a reduction process in which there is usually a loss of oxygen, while in an oxidation process there is a net gain in oxygen [11, 50, 32, 59]. Such reactions, called redox reactions, are the basis for numerous biochemical pathways and cellular chemistry, biosynthesis, and regulation [62]. They are also important for understanding biological oxidation and radical/antioxidant effects. While reductant and oxidant are chemical terms, in biological environments they should be termed antioxidant and pro-oxidant, respectively [11].

The redox potential is defined as the ratio between oxidant and reductant. Since it is a thermodynamic parameter it can be determined under appropriate thermodynamic conditions. In biological systems it cannot be simply calculated according to the Nernst equation:

$$E = E^{o}\frac{RT}{nF}\ln\frac{\text{oxidant}}{\text{reductant}}$$

E is the potential at any given condition, E^{o} is the potential at equilibrium, R is the universal gas constant, T is the absolute temperature in Kelvins, n is the number of electrons transferred and F is the Faraday constant.

The redox potential can be determined by the Nernst equation only in reversible systems, where all factors affecting the system are known and can be controlled. In biological systems equilibrium does not exist and the systems are not fully reversible. Instead, a relative definition that describes the steady state approximation of redox was suggested and termed redox state. It was suggested that this parameter is the total reduction potential and reducing capacity of all redox couples found in biological fluids, organelles, cells and tissues. Under normal conditions the redox state of biological systems is kept within a narrow range. The redox state represents the overall capability of the biological system to donate electrons (oxidation potential) and the overall concentration of the reducing equivalents responsible for this ability.

4.1.2 Oxidative Stress

Oxidative stress occurs when the production of reactive oxygen species (ROS), a normal product of cellular metabolism, is greater than the ability of the cell to repair the damage resulting from ROS production. For that purpose, cellular oxygen concentrations are maintained within a narrow "nomoxic" range to circumvent the risk of oxidative damage from excess O_2 (hyperoxia) and of metabolic demise from insufficient O_2 (hypoxia) [60]. pO_2 ranges from 90 to below 3 Torr in mammalian organs under normoxic conditions with arterial p_2 of 100 Torr or ~14% [49]. Thus, "normoxia" for cells is an adjustable variable that is dependent on the specific localization of the cell in organs and the functional status of the specific tissue. The fact that oxygen is dangerous to the very life forms for which it has become an essential component of energy production is referred to as the "Oxygen Paradox." The first defense against oxygen toxicity is the sharp gradient of oxygen tension, seen in all mammals, from environmental level of 20% to tissue concentration of only 0.5–5% oxygen. These relatively low tissue levels of oxygen prevent most oxidative damage from ever occurring. Removal of this line of defense from cultured cells, that under normal culturing procedures are exposed to conditions of O_2-rich room air (20% O_2), can result in the activation of specific O_2-sensitive signal transduction pathways that alter cellular phenotype and function. It has been reported that cardiac fibroblasts isolated from adult murine ventricle, cultured in 10% or 21% O_2 (high O_2, relative to the pO_2 to which they are adjusted in vivo), compared with 3% O_2, exhibit reversible growth inhibition and a phenotype indicative of differentiation [56]. The role of the abortive redox state in genome instability disorders and its effects on DNA damage response were described in detail in previous reviews [5, 6]. It has been estimated that around 2×10^4 DNA damaging events occur in every cell of the human body every day [2]. A significant portion of the damage is caused by ROS. As early as 1952, Conger and Fairchild [13] demonstrated that increased oxygen pressure could lead to the accumulation of chromosomal aberrations. The effect of excessive production of ROS and/or the inadequacy of the antioxidant cellular defense systems to neutralize them, is commonly referred to as oxidative stress. Oxidation-reduction (redox) based regulation of gene expression appears to be a fundamental regulatory mechanism in cell biology. In contrast to the conventional idea that reactive species mostly serve as a trigger for oxidative damage of biological structures, we now know that low, physiologically relevant concentrations of ROS, can regulate a variety of key molecular mechanisms [61, 1, 67]. Thus, maintaining homeostatic levels of oxygen is critically important for proper cellular functioning. One common product of nucleic acid damage by oxidation is 8-hydroxyguanosine (8OHG). Indeed, 8OHG immunoreactivity is widely used as a marker for evaluating the effect of oxidative stress on nucleic acids. The production of this molecule can be induced by various environmental factors. Furthermore, 80HG is known to permanently damage cytoplasmic RNA and mitochondrial DNA, thereby contributing to neurodegeneration.

4.1.3 The Oxygen Molecule (O_2)

Approximately 90% of the oxygen that enters the cell is used by the mitochondrial respiratory chain for the production of energy. During this process, four electrons are added to each O_2 molecule, resulting in the formation of two molecules of water. It is estimated that 1–4% of the O_2 taken into cells form partially reduced O_2 species, ROS (Fig. 4.1). Some ROS contain unpaired electrons and are therefore referred to as free radicals. The diatomic oxygen molecule itself qualifies as a radical, because it possesses two unpaired electrons, each located at a different orbital. Since both electrons have the same quantum spin number, O_2 itself has relatively low reactivity, in contrast to other radicals which can be highly reactive. Another radical derived

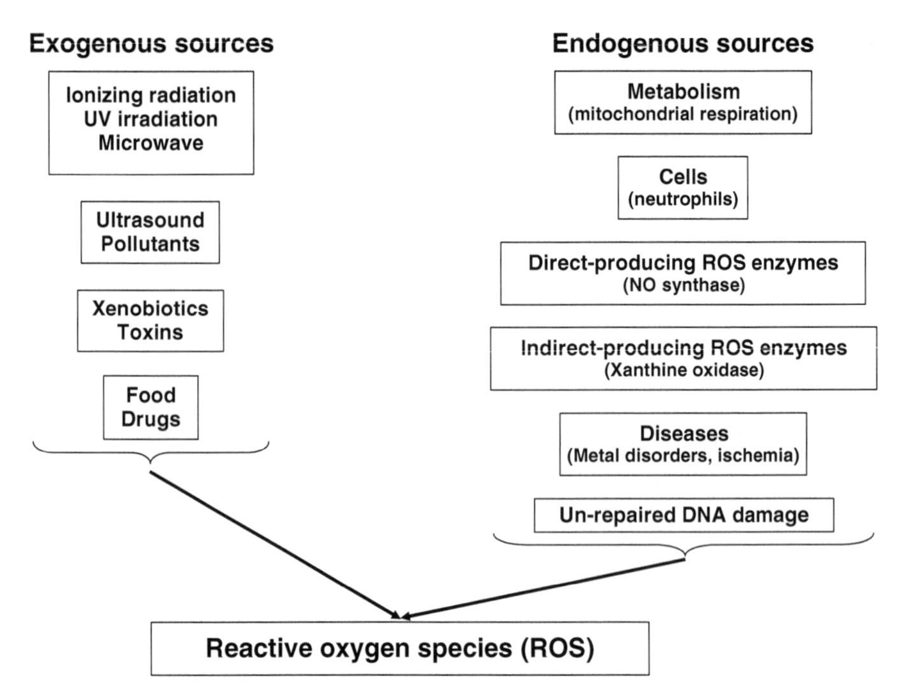

Fig. 4.1 Exogenous and endogenous sources of ROS. Each cell in our body is exposed to a variety of ROS that are generated from exogenous as well as endogenous sources. The relative contribution of each source to oxidative stress is highly dependent on cell type, its metabolic rate and the environment that the organism functions

from oxygen is singlet oxygen, designated as $^1O_2$. This is an excited form of oxygen in which one of the electrons jumps to a superior orbital following absorption of energy. For O_2 to oxidize a molecule directly, it would have to accept a pair of electrons with a spin opposite to that of the oxygen. This situation is quite rare. In biological systems, O_2 can accept an electron and form one of the following species: superoxide anion ($\cdot O_2-$), hydroxyl radical ($\cdot OH$), or hydrogen peroxide (H_2O_2). These molecules possess various degrees of reactivity with nonradical compounds. Table 4.1 summarises various types of ROS which participates in the generation of oxidative stress.

4.1.4 Oxygen Radicals

A radical (often, but unnecessarily called a free radical) is an atom or group of atoms that have one or more unpaired electrons. Radicals can have positive, negative or neutral charge. They are formed as necessary intermediates in a variety of normal biochemical reactions, but when generated in excess or not appropriately controlled, radicals can wreak havoc on a broad range of macromolecules. A prominent feature of radicals is that they have extremely high chemical reactivity, which explains not only their normal biological activities, but how they inflict damage on cells. There are many types of radicals, but in biological systems most significant are those derived from oxygen. Oxygen has two unpaired electrons in separate orbitals in its outer shell. This electronic structure makes oxygen especially susceptible to radical formation.

4.1.4.1 The Superoxide Anion Radical ($\cdot O_2-$)

The acceptance of a single electron by O_2 generates cellular $\cdot O_2-$. The mitochondrial respiratory chain is the major source of $\cdot O_2-$ [45, 65].The superoxide anion radical is abundant and can reach an intracellular concentration of about 10−11 M [23]. $\cdot O_2-$ is not highly reactive with biological molecules however once formed it quickly undergoes dismutation to generate hydrogen peroxide, which is highly reactive. This reaction is markedly accelerated by a family of enzymes, the superoxide dismutases (SODs). $\cdot O_2-$ can react with H+ to form $HO_2\cdot$ (hydroperoxy radical) which is much more reactive than $\cdot O_2-$. Other enzymes can generate superoxide anions. A notable example is NADPH oxidase, primarily located in phagocytes, neutrophils and monocytes, this enzyme produces large amounts of $\cdot O_2-$ and other reactive oxidants that are used for fighting invading microorganisms (reviewed in [4]).

4.1.5 The Hydroxyl Radical (·OH)

The hydroxyl radical is an extremely reactive oxidant [26]. It is also a short-lived molecule with an estimated half-life of nanoseconds at 37°C, during which it can travel only a few Ångstroms. Despite its short life span, ·OH is capable of inducing considerable damage to nuclear and mitochondrial DNA. This radical alone can cause over an 100 types DNA modifications [42]. In addition, ·OH can lead to lipid peroxidation and oxidation of carbohydrates and proteins. The brain is particularly susceptible to lipid peroxidation since it is rich in lipids and contains high levels of iron. The hydroxyl radical is a major product of IR due to radiation-induced dissociation of water molecules.

4.1.6 The Peroxyl Radical (L·)

Lipid peroxidation is a process in which lipids undergo oxidation and peroxyl radicals are formed. The hydroxyl radical is not the only radical that can initiate lipid peroxidation: $^1O_2$ and the peroxynitrite anion can also do so [21, 30]. Lipid peroxidation, and decomposition of fatty acids can lead to the formation of toxic products, such as lipid hydroperoxides (LOO·) and peroxyl radicals. LOO· in turn can attack adjacent polyunsaturated fatty acids (PUFA) and reinitiate the process. Thus, this complex, self-propagating process, can lead to oxidation of most or all of the cellular lipids, making it highly destructive.

4.2 Non-Radical ROS

4.2.1 Hydrogen Peroxide (H_2O_2)

This is one of the most stable ROS and acts as a messenger in cellular signaling pathways (reviewed in [35]). In addition to SOD, there are several other cellular systems that generate H_2O_2, including monoamine oxidase (MAO), diamine and polyamine oxidase, and glycolate oxidase. H_2O_2 is quite stable and under normal conditions is not toxic up to a cellular concentration of about 10^{-8} M [33] However, H_2O_2 is highly diffusible through cell membranes and organelles and it is this capacity to travel long distances from its site of generation that makes it hazardous. In the presence of transition metals such as Fe^{2+} or Cu^+, H_2O_2 it can be converted to the highly reactive hydroxyl radical, either by Fenton or Harber–Weiss reactions [33, 71, 26]. H_2O_2 is detoxified by a set of enzymes that includes the selenium-dependent glutathione peroxidase (GPx) and catalase.

4.2.2 Nitric Oxide (NO) and Generation of the Peroxinitrite Anion (ONOO−)

NO is synthesized from l-arginine [48] by any of the three NO synthase isoforms. NO is quite stable and benign for a free radical, with a lifetime of several seconds. Under normal conditions NO has many physiological functions as a neuronal messenger and modulator of smooth muscle contraction. However, when its intracellular level is increased it can induce a cascade of events that can eventually lead to cell death. NO can interact with $\cdot O_2-$ to generate the peroxynitrite anion (ONOO−) [52]. This molecule accounts for much of the NO toxicity. The reactivity of ONOO− is roughly the same as that of ·OH and $NO_2\cdot$. Its toxicity is derived from its ability to directly nitrate and hydroxylate the aromatic rings of amino acid residues [7] and to react with sulfahydryls [52], lipids [51], proteins [43] and DNA [36]. Peroxynitrite anion can also effect cellular energy status by inactivating key mitochondrial enzymes [53], and it may trigger calcium release from the mitochondria [47]. The peroxynitrite anion and perhaps a few other reactive nitrogen species (RNS) with the exception of NO, can nitrate tyrosine residues [25], potentially leading to protein dysfunction [34]. This strong ubiquitous activity can have devastating effects on cellular physiology and viability.

4.2.3 Cellular Defense Mechanisms Against Oxidative Stress

In the aerobic world our cells are constantly exposed to ROS and under certain conditions the levels of these reactive molecules are dangerously elevated. To neutralize their toxic effects, mammalian cells had to developed several defense mechanisms (reviewed in [40, 10, 14, 18, 22, 39, 46, 9, 15]). Among the various defense

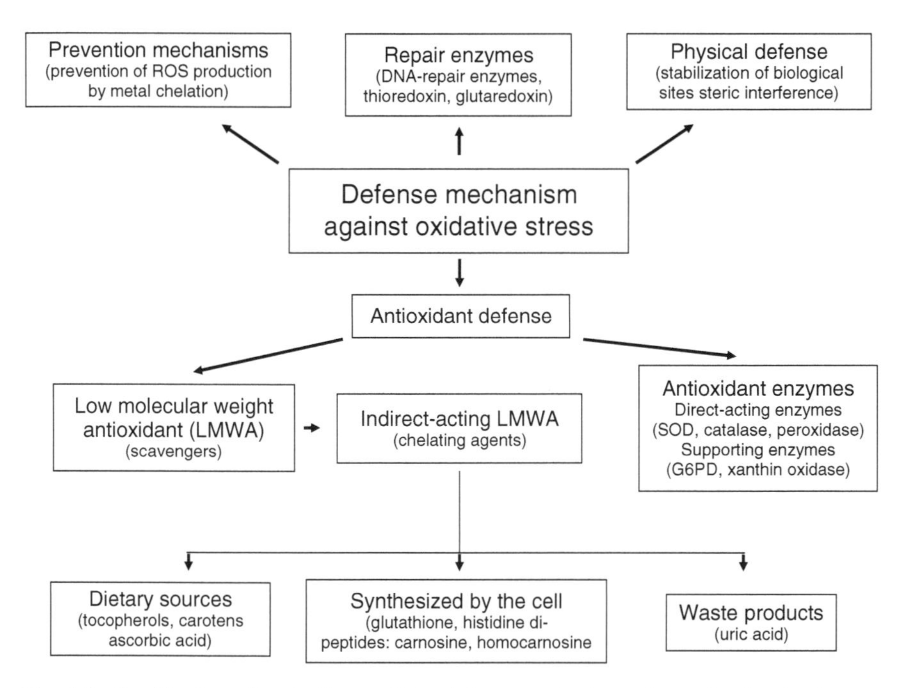

Fig. 4.2 Classification of antioxidant cellular defense mechanisms. Increasing levels of atmospheric oxygen has prompted cells to evolve several mechanisms for antioxidant defense. These mechanisms include: reduced oxygen tension in various tissue, prevention of ROS production, repair enzymes, low molecular weight antioxidants, antioxidant enzymes, dietary sources and waste products. Altogether, these mechanisms endow the cell with sufficient protection against oxidative stress-induced damage

mechanisms, antioxidant-based mechanisms are extremely important because of the variety of antioxidant compounds and the ability to directly remove pro-oxidants thus ensuring maximum protection for specific biological sites. This system apparently developed throughout the evolutionary process in response to the changing concentration of oxygen in the atmosphere. The first line of defense is based on direct interaction with the ROS and their detoxification by low molecular weight antioxidants or protein scavengers (Fig. 4.2). Typical low molecular weight antioxidants are vitamin E (α-tocopherol), vitamin C (ascorbic acid), uric acid, glutathione, β-carotene and ubiquinone. In addition, molecules such as Histidine-related compounds, and Melatonin. Lipoic acid can also serve as antioxidants (reviewed in [63]). The DNA repair system, for example, can identify a DNA-oxidized adduct [eg, 8-hydroxy-2-deoxyguanosine, thiamine glycol, and apurinic and apyramidenic (AP) sites] [3], remove it, and incorporate an undamaged base [3, 16]. Molecules that can donate hydrogen atoms to damaged molecules are also considered repair compounds; one such example is the donation of a hydrogen atom by ascorbic or tocopherol to a fatty acid radical – a product of radical-attack,. Physical defense of biological sites such as membranes is also an important mechanism that allows the cell to cope with oxidative stress. Compounds such as tocopherols can provide enhanced stability to cellular membranes, and steric interference can prevent ROS from approaching a target. The prominent reducing molecule, glutathione which can function alone or in association with enzymatic activities, appears in cells in both oxidized (GSSG) and reduced (GSH) form. Antioxidant proteins such as ferritin and ceruloplasmin, remove two transition metals, iron and copper, which enhance oxidative processes.

The second line of defense includes antioxidant enzymes. A prominent example are the SODs, which dismutate the dangerous superoxide anion via the reaction $\cdot O_2 - + \cdot O_2 \rightarrow H_2O_2 + O_2$. Proteins in this family differ in their structure and cofactors. Cu-Zn SOD contains 2 subunits, each of which possesses an active site and is widely distributed in eukaryotic cells localized in the cytoplasm, whereas Mn- SOD can be found in prokaryotic cells and eukaryotic mitochondria [26]. Other types of SOD exist, such as extracellular SOD (EC-SOD) and Fe-SOD in plants. These enzymes possess different structures, molecular masses, and reaction rate constants. The end product of the dismutation reaction – H_2O_2 – can be removed by the activity of the enzyme catalase and members of the peroxidase family including glutathione peroxidase [12]. Catalase is a unique enzyme with a very high Km for its substrate and can remove H_2O_2 present in high concentrations. The enzyme consists of 4 protein subunits, each of which contains ferric ions of the heme group that undergo oxidation following interaction with the first molecule of H_2O_2 to produce Fe^{4+} in a structure called compound 1. In contrast to catalase, peroxidases possesse high affinity for and can remove H_2O_2 even when it is present in low concentration [12, 26]; however, the electron donors in these reactions are small molecules, such as glutathione or ascorbic acid (in plants). Thus, the removal of H_2O_2 is an "expensive" reaction from the cell's point of view, as it consumes valuable molecules in the cellular environment; 2 molecules of glutathione are consumed for the removal of 1 molecule of H_2O_2. Another important class of antioxidant enzymes includes the glutathione-*S*-transferases, a large family of multifunctional dimeric enzymes that conjugate reduced GSH to electrophilic centers in hydrophobic organic compounds, including lipid peroxides (reviewed in [51]). Auxiliary enzymes in this process are glutathione reductase, which regenerates GSH from GSSG using NADPH, and glucose-6-phosphate dehydrogenase (G6PD), which recycles NADPH using energy derived from cellular metabolism. The next layer of defense involves direct repair of damaged macromolecules. Several enzymes are capable of reducing oxidized sulfhydryl groups of proteins in situ, e.g. bidisulfide reductase and methionine sulfoxide reductase [39]. Thioredoxin and glutaredoxin, are ubiquitous proteins, have a variety of activities as hydrogen donors and act as major protein oxidoreductases [31, 70]. Oxidized proteins that cannot be repaired are usually degraded via the proteasome pathway.

The low-molecular-weight antioxidant (LMWA) group contains numerous compounds capable of preventing oxidative damage by direct and indirect interaction with ROS [37, 38]. The indirect mechanism involves the chelation of transition metals that prevents them from participating in the metal-mediated Haber-Weiss reaction [72, 58]. The directacting molecules share a similar chemical trait that allows them to donate electrons to the oxygen radical so that they can scavenge the radical and prevent it from attacking the biological target. Scavengers possess many advantages over the group of enzymatic antioxidants. Since scavengers are small molecules, they can penetrate cellular membranes and can be localized in close proximity to the biological target. Furthermore, the cell can regulate their concentrations and they can be regenerated intracellularly.

4.2.4 Brain Vulnerability to Oxidative Stress

Among our organs, the brain is the most vulnerable to oxidative stress. The amount of oxygen consumed by the brain far exceeds its size relative to the rest of the organs. Several reasons account for this vulnerability:

Metabolism: the brain is a highly active organ. The neurons, as excitable cells, require high energy levels to maintain their excitability. The neurons, as the building blocks of brain functionality, receive and process enormous amount of information. To enable normal brain functionality, the brain consumes high levels of oxygen and as a consequence intensively produces high levels of reactive oxygen metabolites. Paradoxically, the brain has relatively low antioxidant capacity and low repair mechanism activity.

Structure: Neuronal cells are mostly non-replicating cells with a unique morphology. To enable them to transmit information over long distances their elongated axons have to be insulated by layers of fat known as myelin. As a consequence they have a high membrane surface to cytoplasm ratio [19, 54]. The high concentrations of polyunsaturated fatty acids in the membrane lipids of the brain are the source for the decomposition reactions termed "lipid peroxidation," in which a single initiating free radical can precipitate the destruction of adjacent molecules [69].

Transition metals: transition metals such as iron and copper are capable of catalyzing redox reactions that lead to the formation of highly reactive oxygen metabolites such as hydroxyl radicals via the Haber-Weiss reaction. The high levels of transition metals, such as iron, which are located at specific sites, such as the substantia nigra in the midbrain, expose the brain to highly toxic redox metabolites [27]. The presence of ascorbic acid in close proximity to the binding sites of the ferric ions may result in the pro-oxidant rather than antioxidant activity of ascorbic acid [24].

4.2.5 Oxidative Stress in Neurodegenerative Diseases

In 1954, a pioneering work was published by Daniel Gilbert and Rebecca Gersham, [20] suggesting that free radicals are important players in biological environments and responsible for deleterious processes in the cell. Soon after, in 1956, Herman Denham [28] suggested that these species might play a role in physiological events

Table 4.1 Reactive oxygen species of interest in oxidative stress

Compound		Remarks
$O_2^{\bullet}$	Superoxide anion	One electron reduction state, formed in many autooxidation reactions (e.g. mitochondrial activity, flavonoproteins, redox cycling)
$HO_2^{\bullet}$	Perhydroxy radical	Protonated form of $O_2^{\bullet}$ more lipid-soluble
$RO^{\bullet}$	Alkoxy radical	Oxygen-centered organic (e.g. lipid) radical
$HO^{\bullet}$	Hydroxyl radical	Three-electron reduction state, formed by Fenton reaction, metal(iron)-catalyzed Haber-Weiss reaction; highly reactive
$ROO^{\bullet}$	Peroxy radical	Formally formed from organic (e.g.) hydroxy peroxide ROOH, by By hydrogen abstraction
H_2O_2		Two-electron reduction state, formed from $O_2^{\bullet}$ $HO_2^{\bullet}$ by dismutation or directly from O_2
ROOH		Organic hydroperoxide (e.g. lipid-, thymine-OOH)
$ONOO^-$		Peroxynitrite is an anion. It is an unstable "valence isomer" of nitrate, NO_3^-. It can be generated from hydrogen peroxide and nitrite or from superoxide anion and NO radical.
$^1Dg\ O_2$	(also $O^*)_2$	Singlet molecular oxygen, first excited state, 22 kcal/mol above Ground state (triplet) O_2^3
^3RO	(also RO^*)	Excited carbonyl

and, particularly, in the aging process [29]. Naturally, brain aging is also associated with a progressive imbalance between antioxidant defenses and intracellular concentrations of ROS. However, the aging process does not affect the CNS uniformly [44] and brain regions as well as types of neurons differ substantially in the amount of DNA damage they accumulate during aging [57]. It has been shown that oxidative conditions cause not only structural damage but also changes in the set points of redox-sensitive signaling processes including the insulin receptor signaling pathway. In the absence of insulin, the otherwise low insulin receptor signaling is strongly enhanced by oxidative conditions. Oxidative damage to mitochondrial DNA and the electron transport chain, perturbations in brain iron and calcium homeostasis and changes in plasma cysteine homeostasis may altogether represent both causes and consequences of increased oxidative stress [17]. The mitochondria are the powerhouse of the cells and as such they generate high levels ROS near the electron transport chain. It is believed that they are among the important players in the ageing process. Mutations in mitochondrial DNA are accumulated with age and are thought to lead to local age related impairment of cellular energy supply [41]. Viveros et al. [68] described a mouse strain that displays premature aging. The animals that explore the maze slowly show impaired neuromuscular vigor and coordination, decreased locomotor activity, increased level of emotionality/anxiety, decreased levels of brain biogenic amines as well as immunosenescence and decreased life span compared to their control counterparts. In view of the link between oxidative stress and the aging process, the redox state of peritoneal leukocytes has been studied and has shown oxidative stress. Moreover, diet supplementation with antioxidants has been shown to be an effective strategy for protection against early immune and behavioral decline, altered redox state of leukocytes and premature mortality in prematurely aging mice, which supports the validity of this model.This provides evidence for correlation between premature aging and oxidative stress. Aging of the mammalian brain is associated with a continuous decrease in the capacity to produce ATP by oxidative phosphorylation. The impairment of mitochondrial function is mainly due to diminished electron transfer by complexes I and IV, whereas inner membrane H^+ impermeability and F1-ATP synthase activity are only slightly affected. Dysfunctional mitochondria in aged rodents show decreased rates of respiration and of electron transfer, decreased membrane potential, increased content of the oxidation products of phospholipids and proteins and increased size and fragility. Conditions that increased mice median life span, such as moderate exercise, vitamin E supplementation, caloric restriction, and high spontaneous neurological activity also improved neurological performance and mitochondrial function in the aged brain. The diffusion of mitochondrial NO and H_2O_2 to the cytosol is decreased in the aged brain and may be a factor for reduced mitochondrial biogenesis [8]. Increased oxidative stress was also associated with brain injuries [63]. Unfortunately the success in using free-radical scavengers to counteract the deleterious effect of accumulated radicals has been limited especially in human clinical studies [64]. These results suggest that neutralization of ROS is not sufficient to alleviate the toxic effects of various brain traumas and other factors must be treated as well. High levels of degenerating microglia were found in the brains of aged humans and humans with neurodegenerative disorders. It was postulated that loss of microglia and microglial neuroprotective functions could, at least in part, account for aging-related neurodegeneration [66]. Microglia-derived inflammatory neurotoxins play a principal role in the pathogenesis of neurodegenerative disorders including Alzheimer's disease, Parkinson's disease, amyotrophic lateral sclerosis and HIV-associated dementia; chief among these is reactive oxygen species. The detrimental effects of oxidative stress in the brain and nervous system are primarily a result of the diminished capacity of the central nervous system to prevent ongoing oxidative damage. A spectrum of environmental cues, mitochondrial dysfunction, accumulation of aberrant misfolded proteins, inflammation and defects in protein clearance are known to evolve and form as a result of disease progression. These factors likely affect glial function serving to accelerate the rate of disease progression [55].

4.3 Conclusions

Every complex multicellular organism requires a high level of energy in order to maintain its basic biological processes. To meet these energy demands, cells developed the mechanism of oxidative phosphorylation, which requires high levels of oxygen. The fact that oxygen is dangerous to the very life forms for which it has become an essential component for energy production is referred to as the "Oxygen Paradox." To balance between the harmful and beneficial effects of oxygen consumption, the cells developed intricate mechanisms to maintain the proper cellular redox state. A homeostatic redox state is critically important for proper functioning of every cell in our body, especially for neurons, which are long-living, highly active postmitotic cells. The redox state is involved in every stage of every cell—from birth to death. The redox state affects many cellular processes, intracellular organelles, DNA integrity, proteins and lipids. It also affects cellular metabolism and the activation of various death programs. Thus, understanding the molecular mechanisms that control and maintain redox state homeostasis in various types′ of cells and especially in neuronal cells is critically important for designing new strategies to treat various types of human redox-associated diseases.

References

1. Alom-Ruiz SP, Anilkumar N, Shah AM (2008) Reactive oxygen species and endothelial activation. Antioxid Redox Signal 10:1089–1100.
2. Ames BN, Shigenaga MK (1992) Oxidants are a major contributor to aging. Ann NY Acad Sci 663:85–96.
3. Atamna H, Cheung I, Ames BN (2000) A method for detecting abasic sites in living cells: age-dependent changes in base excision repair. Proc Natl Acad Sci USA 97:686–691.
4. Babior BM (2000) The NADPH oxidase of endothelial cells. IUBMB Life 50:267–269.
5. Barzilai A, Yamamoto K (2004) DNA damage responses to oxidative stress. DNA Repair (Amst) 3:1109–1115.
6. Barzilai A, Biton S, Shiloh Y (2008) The role of the DNA damage response in neuronal development, organization and maintenance. DNA Repair (Amst) 7:1010–1027.
7. Beckman JS, Ischiropoulos H, Zhu L, van der Woerd M, Smith C, Chen J, Harrison J, Martin JC, Tsai M (1992) Kinetics of superoxide dismutase- and iron-catalyzed nitration of phenolics by peroxynitrite. Arch Biochem Biophys 298:438–445.
8. Boveris A, Navarro A (2008) Brain mitochondrial dysfunction in aging. IUBMB Life 60: 308–314.
9. Brozmanova J, Dudas A, Henriques JA (2001) Repair of oxidative DNA damage – an important factor reducing cancer risk. Minireview Neoplasma 48:85–93.
10. Cadet J, Bourdat AG, D'Ham C, Duarte V, Gasparutto D, Romieu A, Ravanat JL (2000) Oxidative base damage to DNA: specificity of base excision repair enzymes. Mutat Res 462:121–128.
11. Cao G, Prior RL (1998) Comparison of different analytical methods for assessing total antioxidant capacity of human serum. Clin Chem 44:1309–1315.
12. Chance B, Sies H, Boveris A (1979) Hydroperoxide metabolism in mammalian organs. Physiol Rev 59:527–605.
13. Conger AD, Fairchild LM (1952) Breakage of Chromosomes by Oxygen. Proc Natl Acad Sci USA 38:289–299.
14. Davies KJ (2000) Oxidative stress, antioxidant defenses, and damage removal, repair, and replacement systems. IUBMB Life 50:279–289.
15. Dianov GL, Souza-Pinto N, Nyaga SG, Thybo T, Stevnsner T, Bohr VA (2001) Base excision repair in nuclear and mitochondrial DNA. Prog Nucleic Acid Res Mol Biol 68: 285–297.
16. Dizdaroglu M, Jaruga P, Birincioglu M, Rodriguez H (2002) Free radical-induced damage to DNA: mechanisms and measurement. Free Radic Biol Med 32:1102–1115.

17. Droge W, Schipper HM (2007) Oxidative stress and aberrant signaling in aging and cognitive decline. Aging Cell 6:361–370.
18. Elliott RM, Astley SB, Southon S, Archer DB (2000) Measurement of cellular repair activities for oxidative DNA damage. Free Radic Biol Med 28:1438–1446.
19. Evans PH (1993) Free radicals in brain metabolism and pathology. Br Med Bull 49: 577–587.
20. Gilbert D, (ed.) (1981) Perspective on the history of oxygen and life. New York: Springer Verlag.
21. Girotti AW (1998) Lipid hydroperoxide generation, turnover, and effector action in biological systems. J Lipid Res 39:1529–1542.
22. Griffiths HR (2000) Antioxidants and protein oxidation. Free Radic Res 33 Suppl:S47–58.
23. Forman HJ, Boveris A (1982) Superoxide radical and hydrogen peroxide in mitochondria, in: WA Pryor (Ed.), Free Radicals in Biology, Vol. V, New York: Academic Press, pp. 65–90.
24. Halliwell B (1996) Vitamin C: antioxidant or prooxidant in vivo. Free Rad Res 25:439–454.
25. Halliwell B (1997) What nitrates tyrosine? Is nitrotyrosine specific as a biomarker of peroxynitrite formation in vivo? FEBS Lett 411:157–160.
26. Halliwell B GJ (1999) Free Radicals in Biology and Medicine, third edition. Edition. Midsomer Norton, Avon, third edition. Oxford University Press, Oxford.
27. Halliwell B, Gutteridge, JMC, (1989) Free Radicals in Biology and Medicine Oxford, Clarendon Press, Oxford University Press, Oxford.
28. Harman D (1956) Aging: a theory based on free radical and radiation chemistry. J Gerontol 11:298–300.
29. Harman D (1981) The aging process. Proc Natl Acad Sci USA 78:7124–7128.
30. Hogg N, Kalyanaraman B (1999) Nitric oxide and lipid peroxidation. Biochim Biophys Acta 1411:378–384.
31. Holmgren A (1989) Thioredoxin and glutaredoxin systems. J Biol Chem 264:13963–13966.
32. Hrbac JKR (2000) Biological redox activity: Its importance, method for its quantification and implication for health and disease. Drug Develop Res 50:516–527.
33. Imlay JA, Chin SM, Linn S (1988) Toxic DNA damage by hydrogen peroxide through the Fenton reaction in vivo and in vitro. Science 240:640–642.
34. Ischiropoulos H, Zhu L, Chen J, Tsai M, Martin JC, Smith CD, Beckman JS (1992) Peroxynitrite-mediated tyrosine nitration catalyzed by superoxide dismutase. Arch Biochem Biophys 298:431–437.
35. Kamata H, Hirata H (1999) Redox regulation of cellular signalling. Cell Signal 11:1–14.
36. King PA, Anderson VE, Edwards JO, Gustafson G, Plumb RC, Sugas JW (1992) A stable solid that generates hydroxyl radical upon dissolution in aqueous solution: reaction with proteins and nucleic acids. J Am Chem Soc 114:5430 –5432.
37. Kohen R (1999) Skin antioxidants: their role in aging and in oxidative stress–new approaches for their evaluation. Biomed Pharmacother 53:181–192.
38. Kohen R, Gati I (2000) Skin low molecular weight antioxidants and their role in aging and in oxidative stress. Toxicology 148:149–157.
39. Levine RL, Moskovitz J, Stadtman ER (2000) Oxidation of methionine in proteins: roles in antioxidant defense and cellular regulation. IUBMB Life 50:301–307.
40. Mates JM, Sanchez-Jimenez F (1999) Antioxidant enzymes and their implications in pathophysiologic processes. Front Biosci 4:D339–345.
41. Meissner C (2007) Mutations of mitochondrial DNA – cause or consequence of the ageing process? Z Gerontol Geriatr 40:325–333.
42. Michalik V, Spotheim Maurizot M, Charlier M (1995) Calculation of hydroxyl radical attack on different forms of DNA. J Biomol Struct Dyn 13:565–575.
43. Moreno JJ, Pryor WA (1992) Inactivation of alpha 1-proteinase inhibitor by peroxynitrite. Chem Res Toxicol 5:425–431.
44. Morrison JH, Hof PR (2002) Selective vulnerability of corticocortical and hippocampal circuits in aging and Alzheimer's disease. Prog Brain Res 136:467–486.

45. Nohl H, Hegner D (1978) Do mitochondria produce oxygen radicals in vivo? Eur J Biochem 82:563–567.
46. Offord E, van Poppel G, Tyrrell R (2000) Markers of oxidative damage and antioxidant protection: current status and relevance to disease. Free Radic Res 33 Suppl:S5-19.
47. Packer MA, Murphy MP (1994) Peroxynitrite causes calcium efflux from mitochondria which is prevented by Cyclosporin A. FEBS Lett 345:237–240.
48. Palmer RM, Ferrige AG, Moncada S (1987) Nitric oxide release accounts for the biological activity of endothelium-derived relaxing factor. Nature 327:524–526.
49. Porwol T, Ehleben W, Brand V, Acker H (2001) Tissue oxygen sensor function of NADPH oxidase isoforms, an unusual cytochrome aa3 and reactive oxygen species. Respir Physiol 128:331–348.
50. Prior RL, Cao G (1999) In vivo total antioxidant capacity: comparison of different analytical methods. Free Radic Biol Med 27:1173–1181.
51. Radi R, Beckman JS, Bush KM, Freeman BA (1991a) Peroxynitrite-induced membrane lipid peroxidation: the cytotoxic potential of superoxide and nitric oxide. Arch Biochem Biophys 288:481–487.
52. Radi R, Beckman JS, Bush KM, Freeman BA (1991b) Peroxynitrite oxidation of sulfhydryls. The cytotoxic potential of superoxide and nitric oxide. J Biol Chem 266:4244–4250.
53. Radi R, Rodriguez M, Castro L, Telleri R (1994) Inhibition of mitochondrial electron transport by peroxynitrite. Arch Biochem Biophys 308:89–95.
54. Reiter RJ (1995) Oxidative processes and antioxidative defense mechanisms in the aging brain. FASEB J 9:526–533.
55. Reynolds A, Laurie C, Mosley RL, Gendelman HE (2007) Oxidative stress and the pathogenesis of neurodegenerative disorders. Int Rev Neurobiol 82:297–325.
56. Roy S, Khanna S, Wallace WA, Lappalainen J, Rink C, Cardounel AJ, Zweier JL, Sen CK (2003) Characterization of perceived hyperoxia in isolated primary cardiac fibroblasts and in the reoxygenated heart. J Biol Chem 278:47129–47135.
57. Rutten BP, Korr H, Steinbusch HW, Schmitz C (2003) The aging brain: less neurons could be better. Mech Ageing Dev 124:349–355.
58. Samuni A, Aronovitch J, Godinger D, Chevion M, Czapski G (1983) On the cytotoxicity of vitamin C and metal ions. A site-specific Fenton mechanism. Eur J Biochem 137: 119–124.
59. Schafer FQ, Buettner GR (2001) Redox environment of the cell as viewed through the redox state of the glutathione disulfide/glutathione couple. Free Radic Biol Med 30:1191–1212.
60. Semenza GL (2001) HIF-1, O(2), and the 3 PHDs: how animal cells signal hypoxia to the nucleus. Cell 107:1–3.
61. Sen CK (2001) Antioxidant and redox regulation of cellular signaling: introduction. Med Sci Sports Exerc 33:368–370.
62. Shapiro M (1972) Redox balance in the body: An approach to quantification. J Surg Res 3:138–152.
63. Shohami E, Beit-Yannai E, Horowitz M, Kohen R (1997) Oxidative stress in closed-head injury: brain antioxidant capacity as an indicator of functional outcome. J Cereb Blood Flow Metab 17:1007–1019.
64. Slemmer JE, Shacka JJ, Sweeney MI, Weber JT (2008) Antioxidants and free radical scavengers for the treatment of stroke, traumatic brain injury and aging. Curr Med Chem 15:404–414.
65. Sohal RS (1997) Mitochondria generate superoxide anion radicals and hydrogen peroxide. FASEB J 11:1269–1270.
66. Streit WJ, Miller KR, Lopes KO, Njie E (2008) Microglial degeneration in the aging brain–bad news for neurons? Front Biosci 13:3423–3438.
67. Trachootham D, Lu W, Ogasawara MA, Nilsa RD, Huang P (2008) Redox regulation of cell survival. Antioxid Redox Signal 10:1343–1374.

68. Viveros MP, Arranz L, Hernanz A, Miquel J, De la Fuente M (2007) A model of premature aging in mice based on altered stress-related behavioral response and immunosenescence. Neuroimmunomodulation 14:157–162.
69. Watson B (1993) Evaluation of the concomitance of lipid peroxidation in experimental models of cerebral ischemia and stroke. Elsevier Science Publishers, B.V., Amsterdam.
70. Wells WW, Yang Y, Deits TL, Gan ZR (1993) Thioltransferases. Adv Enzymol Relat Areas Mol Biol 66:149–201.
71. Yamasaki I, Piette LH (1991) EPR spin trapping study on the oxidizing species formed in the reaction of the ferrous ion with hydrogen peroxide. J Am Chem Soc 113:7588–7593.
72. Haber F, Weiss, J, (1934) The catalytic decomposition of hydrogen peroxide by iron salts. Proc R Soc London 147:332–351.

Chapter 5
Cell Cycle Regulation and DNA Damage

Ryo Sakasai and Randal S. Tibbetts

Abstract Regulation of the cell cycle is perhaps the best-understood facet of the DNA damage. DNA damage signals generated downstream of the apical DNA damage signaling kinases ATM and ATR lead to transient checkpoints in the cell cycle at the G1/S and G2/M boundaries, as well as transient inhibition of DNA replication. The focus of this review is to highlight the current understanding and established principles of DNA damage-dependent cell cycle regulation, including the roles of protein phosphorylation/dephosphorylation, and regulated protein degradation. The DNA damage-cell cycle paradigm has broad implications for understanding genomic instability, cancer susceptibility, and neurodegenerative disease.

Keywords Cell cycle · Checkpoint · DNA damage · Mitosis · Cyclin · Cyclin-dependent kinase · Anaphase-promoting complex · Ubiquitin · Proteasome · Ataxia-telangiectasia · Cancer · ATM · ATR · CHEK2 · CHEK1 · CDC25 · Tumor suppressor · DNA replication · Genetic instability

5.1 Introduction

Earlier chapters have introduced the general principles of DNA damage signaling, including the detection of DNA lesions, the activation of the apical signaling kinases ATM and ATR, and the assembly of large protein complexes that facilitate DNA damage signal amplification. This chapter will focus on perhaps the best characterized endpoint of the DNA damage response: cell cycle checkpoint arrest. The regulation of the cell cycle by DNA damage is multifaceted and the literature complex; at the time of this writing, more than 6400 published articles contain the key word "cell cycle checkpoint". There are many nuances to cell cycle checkpoint regulation in mammals, including cell-type dependent outcomes (for example

R. Sakasai (✉)
Department of Pharmacology, University of Wisconsin School of Medicine and Public Health, Madison, WI 53706, USA
e-mail: sakasai.pbc@mri.tmd.ac.jp

K.K. Khanna, Y. Shiloh (eds.), *The DNA Damage Response: Implications on Cancer Formation and Treatment*, DOI 10.1007/978-90-481-2561-6_5,

embryonic stem cells lack G_1 checkpoint arrest pathways), and differential signaling in response to myriad different types of genotoxic agents. It is not possible to summarize all of these findings, many of them quite important, here. Nor will we dedicate sufficient space to cell cycle checkpoint regulation in budding and fission yeasts, systems where many critical discoveries have been made. Instead we will summarize general principles of cell cycle checkpoint regulation, the best characterized pathways, and future directions, including opportunities for therapeutic intervention in human disease. We apologize in advance to our colleagues whose work is not cited in this chapter due to space constraints.

5.2 Overview of the Cell Cycle

The eukaryotic cell cycle is a paradox of conceptual simplicity and biological complexity that is perhaps unsurpassed as a paradigm for modern biological interrogation. Its objective is simple: to produce two genetically identical daughter cells from one mother cell. This outcome is accomplished by the flawless coordination of cell growth, DNA replication, and cell division (cytokinesis) steps that are temporally linked by a biochemical oscillator. The cell cycle is highly responsive to changes in external and internal cellular cues and is universally deregulated in cancer, where the unbridling of the cell cycle from its normal restraints contributes to genetic instability and tumorigenesis, as outlined in other chapters of this book. In this chapter we will focus on cell cycle regulation in the special context of DNA damage, which engages conserved signaling pathways that induce transient cell cycle arrest at defined stages of the cell cycle termed checkpoints. The topic of cell cycle checkpoint regulation has risen to prominence owing to its relevance to the pathogenesis and treatment of human disease states, including cancer and neurodegeneration syndromes such as ataxia-telangiectasia (A-T). Work on the cell cycle and its interface with the DNA damage response is also historically significant, resulting in a joint Nobel Prize for Tim Hunt, Leland Hartwell, and Paul Nurse whose pioneering work with model organisms provided much of the foundation for our current understanding of cell cycle regulation see [1–3]. The reader is encouraged to consult a number of excellent reviews cited in this chapter that present more in-depth discussions on specific aspects of cell cycle regulation.

The overall process of cell division can be divided into two stages: interphase, which comprises growth and DNA replication phases, and "M" (*m*itotic) phase, during which intracellular macromolecules are equally divided between the daughter cells (Fig. 5.1). Interphase is categorized into three different stages: G_1 (Gap 1) phase, S (Synthesis) phase and G_2 (Gap 2) phase. G_1 is the longest phase between M phase and S phase during which cells synthesize protein, replicate organelles, and prepare for the onset of S phase. The end of G_1 phase is particularly critical; it is at this time that a series of rapid molecular events, driven by feed-forward amplification loops, render the cell permissive for DNA replication in S phase. During

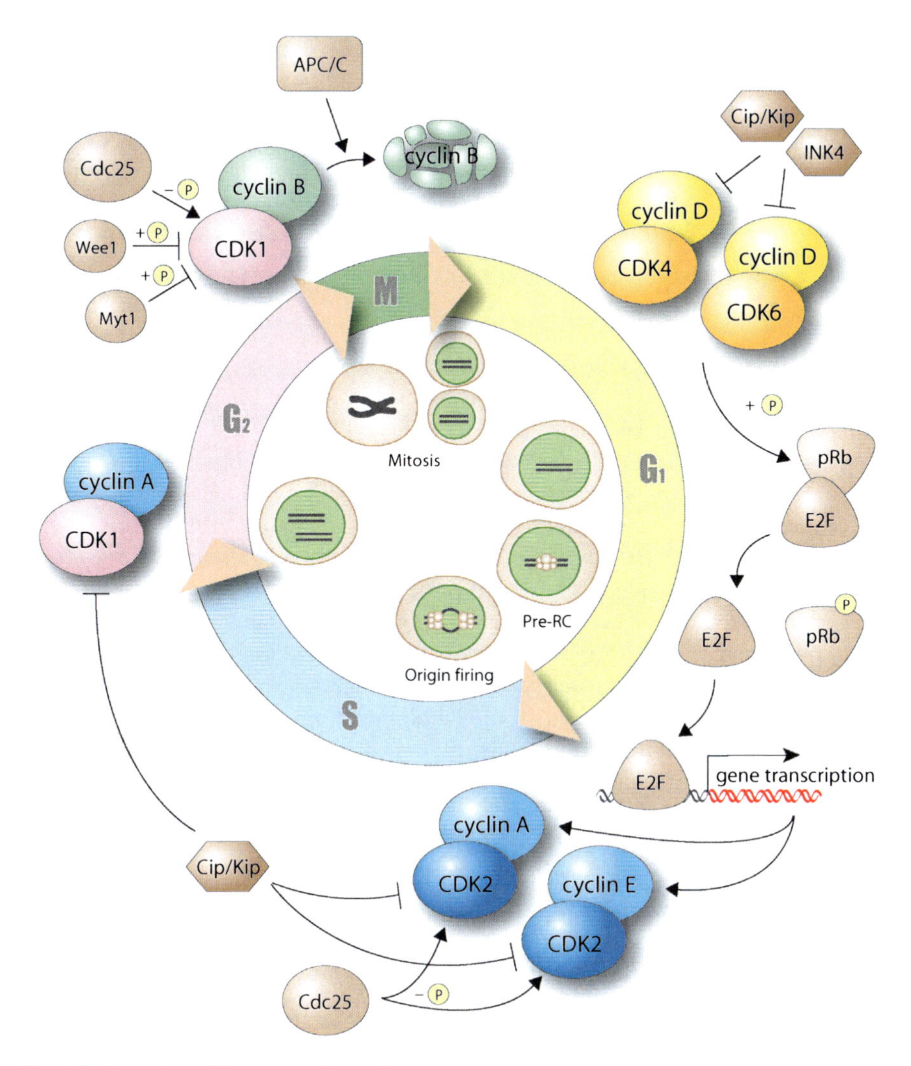

Fig. 5.1 Overview of the mammalian cell cycle

S phase, chromosomal DNA is replicated via hierarchical and temporally regulated activation of DNA replication origins. In mammals DNA replication takes approximately 6–10 hours to complete, depending on the cell type and, at any given time, dozens of chromosomal regions (replicons) are actively undergoing replication.

Completion of S phase leads to a variable G_2 phase delay prior to M-phase, a visually stunning process that was first observed more than 120 years ago by Walther Flemming—who introduced the terms "mitosis" and "chromatin" to describe the condensation and movement of dye-stained filaments that he observed upon microscopic examination of salamander gills [4]. M-phase in somatic cells is comprised of four steps: chromosome condensation (prophase), alignment of paired sister chromosomes on an equatorial plate (metaphase), chromosome segregation (anaphase), and chromosome decondensation/cytokinesis (telophase). The coupling of M phase to the successful completion of the prior S phase is often referred to as the S/M or DNA replication checkpoint. Failure of this checkpoint frequently results in cell death from catastrophic missegregation of chromosomes, but can also result in the production of genetically altered, karyotypically unstable cells that are potential tumor precursors.

5.2.1 Cyclins and Cyclin-Dependent Kinases

The molecular underpinnings of the cell cycle are now relatively well understood and several recurring themes—including protein phosphorylation/dephosphorylation, protein degradation, and feed-forward regulation—are central to ordered cell cycle dynamics. Of particular importance to this discussion are cyclin-dependent kinases (CDKs), which trigger cell cycle phase transitions through the phosphorylation of effector protein substrates. The activities of CDKs are tightly linked to obligatory regulatory subunits, the cyclins. The term "cyclin" was first coined by Hunt and colleagues, who discovered a protein species, cyclin B, whose expression level oscillated during synchronous cell divisions in fertilized sea urchin eggs [5]. Cyclin B, in complex with its cognate catalytic subunit, $p34^{CDC2/CDK1}$ (CDK1, cyclin-dependent kinase 1) constitutes maturation promoting factor (MPF, later named mitosis promoting factor), a biochemical activity with protein kinase function that stimulated G_2- to M-phase transition in Xenopus oocyte extracts [6]. MPF promotes the tell-tale signs of mitosis, including nuclear envelope breakdown, mitotic spindle formation and chromosome condensation, by catalyzing the phosphorylation of substrate proteins–numbering in the hundreds–that control these events. G_1- and S-phase transitions are similarly under the regulagory control of cyclin-CDKs. Commitment to enter S phase is promoted by one of several D-type cyclins in complex with CDK4 and CDK6 as well as cyclin E-CDK2. The initiation of DNA replication per se is triggered by cyclin E-CDK2 and cyclin A-CDK2 [7]. The specificity of cyclin-CDK interactions is determined via a cyclin box motif that provides a docking interface between the cyclin and its CDK partner. All told, there are 20 CDKs and 25 cyclin proteins encoded by the human genome, though not all of these proteins are likely to function as active cyclin-CDK complexes.

The activities of the CDKs are negatively regulated by cyclin-dependent kinase inhibitors (CKIs) a group of proteins that were identified on the basis of their high affinity association with CDKs. Whereas budding yeast expresses only a single CDKI, Sic1p, mammals encode two distinct families: the INK4 family including p15 (INK4B), p16 (INK4A) and p18 (INK4C), and Cip/Kip family comprised of p21 (WAF1/CIP1), p27 (KIP1) and p57 (KIP2) [8]. The INK4 family inhibits G_1 CDKs whereas the Cip/Kip family can inhibit all CDKs. If an automobile gas pedal analogy can be used to describe the cell cycle promoting activity of cyclins-CDKs, then the CKIs provide an opposing braking force that ensures the proper timing of a normal cell cycle. This CKI "brake" is also exploited by the cell cycle checkpoint apparatus to ensure appropriate cell cycle arrest in the face of genetic damage. Not unexpectedly then, CKIs function in a tumor suppressive capacity and mutations in CKIs, especially those belonging to the INK4 family, are associated with various human cancers, including melanoma and glioblastoma [9].

5.2.2 Control of Cyclin Stability

The waxing-waning pattern of cyclin protein accumulation is the most conspicuous biochemical correlate of the cell cycle oscillation, and proper control of cyclin B expression is critical for an orderly cell cycle to proceed. This is illustrated by the elegant study of Murray and Kirschner showing that microinjection of cyclin B mRNA is sufficient to induce mitosis in Xenopus oocytes [10]. Cyclin B expression is controlled by transcriptional and post-transcriptional mechanisms, but it is the post-translational regulation of cyclin B stability which imparts the rapid accumulation, disappearance, and reaccumulation of CDK activity in cycling cells. Cyclin B contains an amino-terminal destruction box of nine amino acids that targets the protein for ubiquitylation and proteasome-dependent degradation [10, 11]. Degradation of cyclin B is mediated by the anaphase promoting complex/cyclosome (APC/C) during the metaphase to anaphase transition. APC/C is a multimeric complex with ubiquitin E3 ligase activity that interacts with the destruction box and degrades cyclin B and many other protein targets in the cell, including proteins that mediate sister chromatid cohesion such as securin during the metaphase to anaphase transition [12–14]. Full activation of the APC/C toward its cyclin B substrate requires cyclin B-CDK1 dependent phosphorylation of the APC/C itself, so the APC/C in effect promotes its own inactivation as cells progress through anaphase [13, 15]. The mitotic function of the APC/C requires the incorporation of the CDC20 E3 ligase subunit; a functionally distinct APC/C complex containing an alternative E3 ligase, CDH1, functions in pre-RC assembly during S phase, and becomes active as cells progress into G_1 phase at the expense of CDC20, which is degraded by APC^{CDH1} [16, 17]. Intrinsic instability and tightly regulated ubiquitin-dependent degradation are general features of the cyclins. Cyclin A also posseses an amino-terminal destruction box and is degraded during prometaphase [18], whereas cyclin E is degraded by a phosphorylation-dependent mechanism in S phase [19, 20]. These degradation steps are essential for appropriate cell cycle phase transitions.

5.2.3 Post-Translational Regulation of CDK Activity

In addition to physical interactions with both positive (cyclins) and negative (CKIs) regulators, CDK activity is also subjected to bimodal regulation through direct post-translational modification, and this was first established using the cyclin B-CDK1 paradigm. Full activation of CDK1 requires phosphorylation by a complex comprised of Cdk7, cyclin H, and Mat1 referred to as CAK (CDK-activating kinase). CAK phosphorylates CDK1 on Thr-161, which lies within the kinase domain activation loop [21]. CAK also mediates CDK2 phosphorylation via Thr-160 phosphorylation, as well as the phosphorylation and activation of CDK4 and CDK6 [22]. On the other hand, CDK1 is negatively regulated via phosphorylation by the Wee1 and Myt1 protein kinases on Tyr-15 and Thr-14, which sterically blocks CDK1 access to peptide substrates [23, 24]. An equivalent to Tyr-15 is conserved in all human CDKs (CDK2, CDK4, CDK6) with documented cell cycle activity. Consequently, CDKs must be dephosphorylated on Tyr-15/Thr-14 to be activated. Dephosphorylation of these residues is carried out by the CDC25 (cell division cycle 25) family of dual specificity phosphatases, of which there are three closely related members (CDC25A, CDC25B, CDC25C) in mammals [25–28]. Whereas CDC25B and CDC25C dephosphorylate CDK1 and are primarily required for M-phase entry, Cdc25A is required for S-phase progression and G_2/M transition and dephosphorylates CDK2 and CDK1 [29]. As will be discussed below, CDC25 proteins represent important nodes for cell cycle checkpoint regulation.

5.2.4 Cell-Cycle Phase Transitions

5.2.4.1 G_1/S-Phase

All eukaryotic cells possess a commitment checkpoint which ensures that the all-or-none decision to replicate is made only under suitable environmental conditions. In yeast this commitment step is appropriately named "Start", whereas it is commonly referred to the "restriction point" in mammalian cells. Prior to the commitment step, the primary G_1 phase cyclins, the D-type cyclins, are maintained at low levels and the G_1 CDKs, CDK4 and CDK6, are maintained in a latent state through binding to INK4 proteins [30]. In response to mitogenic stimuli, such as serum growth factors, cyclin D1 expression is induced and CDK4 and/or CDK6 are phosphorylated by CAK. Active cyclin D-CDK4/CDK6 complexes then catalyze the multi-site phosphorylation of the retinoblastoma tumor suppressor (pRb) within a defined pocket region [31–33]. pRB fulfills a critical checkpoint function at the G_1/S boundary by restraining the activity of E2F family transcription factors, typified by E2F1, which regulate the expression of genes required for S phase entry and progression [34]. The phosphorylation of pRb liberates E2F1, which forms transcriptionally active heterodimers with one of several member of the DP family of transcriptional regulators. The activation of E2F1-DP-dependent genes including cyclin E, cyclin A, and DNA polymerase , represents an important commitment step that prepares the cell for the resource-intensive process of DNA replication.

The E2F1-dependent upregulation of cyclin E mRNA, combined with the concomitant stabilization of cyclin E protein, sets the stage for transition to a series of cyclin E-CDK2 catalyzed reactions, including further potentiation of pRb phosphorylation and E2F activation and the phosphorylation of the $p27^{KIP1}$ CKI, which is degraded ubiquitin-proteasome pathway [35]. Full activation of CDK2 requires its dephosphorylation on Thr-14/Tyr-15 by CDC25A; however, CDC25B and CDC25C probably also contribute [36]. Thus, activation of E2F1 ostensibly creates a positive feedback loop driving the rapid and irreversible commitment to S phase. The cyclin E-CDK2 signal is extinguished when cyclin E is extensively phosphorylated by CDK2 and other kinases, targeting it to the ubiquitin-proteasome pathway after S-phase entry [11]. Cyclin A replaces cyclin E (degraded at the end of G1) as the major partner for CDK2 [37, 38]. Cyclin A–CDK2 works in close cooperation with the DBF4-CDC7 kinase complex to stimulate the assembly and firing of licensed DNA replication origins [39, 40]. As outlined in an earlier chapter and reviewed elsewhere [41], licensed DNA replication origins assemble pre-replication centers (pre-RCs) upon the completion of mitosis when APC/C activity is high. Pre-RCs contain the heterohexameric MCM (minichromosome maintenance) DNA helicase which is recruited in ATP-dependent fashion by ORC (origin recognition complex), CDC6, and CDT1. Cyclin A/E-CDK2 associates with both licensed replication origins, and active DNA replication forks and phosphorylates numerous factors required for DNA replication, including the MCM proteins, which are also phosphorylated by Cdc7-Dbf4, DNA polymerase and PCNA [42–46]. The combined activities of CDK2 and CDC7-DBF4 promote the loading of Cdc45, MCM helicase activation, and DNA priming activity required for initiation of replication. The inactivation of APC/C via CDK-dependent phosphorylation prevents re-licensing of DNA replication origins and restricts DNA synthesis to once and only once per cell cycle [47]. Importantly, these early events in DNA replication origin firing are targeted by S-phase checkpoint pathways outlined below.

5.2.4.2 G_2/M Transition

Entry into and out of mitosis revolves around the tightly regulated catalytic activity of cyclin B-CDK1. Cyclin B levels are maintained at low levels in G_1 phase through mid-S-phase as a consequence of APC/C activity. In addition, during an amino-terminal nuclear export sequence ensures that cyclin B is excluded from the nucleus until G_2 phase [48–50]. As cells complete S phase and enter G_2 phase, stabilized cyclin B is phosphorylated by polo-like-kinase 1 (PLK) within its NES, leading to accumulation of cyclin B in the nucleus [48, 51, 52]. Cyclin B-CDK1 becomes activated following phosphorylation of CDK1 on Thr-161 by CAK and dephosphorylation of CDK1 on Tyr-15 and Thr-14, which is initially carried out by CDC25B and later by CDC25A and CDC25C [36]. Furthermore, PLK1 stimulates cyclin B-CDK1 activation by phosphorylating Wee1—targeting it for degradation—and CDC25C, which stimulates its activity [53, 54]. Active cyclin B-CDK1 phosphorylates Cdc25C on amino-terminal sites, stimulating its activity and initiating a positive feedback loop that leads to rapid and complete cyclin B-CDK1 activation [55, 56]. Active cyclin B-CDK1 mediates chromosome condensation, nuclear membrane disassembly and mitotic spindle formation [57, 58]. The APC/C-dependent degradation of cyclin B, coupled with the rephosphorylation of CDK1 by Wee1/Myt1 serves as a trigger for the metaphase to anaphase transition and cytokinesis, effectively completing one revolution of the cell cycle [59]. Cyclin A also participates in early M-phase events, including chromosome condensation through an association with CDK1 [60], though the basis for differential substrate specificity of cyclin E-CDK2 and cyclin A-CDK2 complexes is unclear.

5.3 Cell Cycle Interfaces of the DNA Damage Response

Cell cycle checkpoints are unprogrammed pauses in the cell cycle that are induced in response to alterations in genome or mitotic spindle integrity. The pathways governing cell cycle checkpoint arrest have been intensively studied for decades; however, like many signaling paradigms, the past decade has seen an exponential increase in our level of understanding of this process at the biochemical level. The reader has been introduced to the early signaling events initiated by different types of DNA damage, including agents that perturb DNA replication and the highly deleterious DNA double-strand break (DSB) (see Chapter 1 and Chapter 2). Thus, we will not belabor the principles underlying the early steps of DNA damage signaling, including the activation of ATM/ATR or recruitment of DNA damage response mediator proteins, in this chapter. Instead, we will outline how normal cell cycle restraints are coopted by the genomic surveillance apparatus to enforce cell cycle arrest in response to genetic lesions.

5.3.1 G_1/S Checkpoint

The induction of DNA damage during the G_1 phase of the cell cycle imparts a delay to S phase initiation referred to as the G_1/S checkpoint, which is comprised of "rapid" and "delayed" phases that initiate and sustain checkpoint arrest, respectively. The rapid phase of the G_1/S checkpoint is a protein synthesis-independent response that acutely inhibits the activity of cyclin E/A-CDK2, thus preventing the early steps required for the firing of DNA replication forks. The delayed G1/S checkpoint response is a protein synthesis dependent process that relies on the upregulation of CKIs, including $p21^{Waf1/Cip1}$ to inhibit CDK activity. This delayed pathway is critically dependent on the p53 tumor suppressor. Both pathways are dependent on the activities of the ATM and ATR protein kinases, which are activated in response to distinct and partially overlapping types of DNA damage. As we have seen from earlier chapters ATM controls the response to the DNA double-strand break (DSB), whereas ATR controls the response to DNA replication stress that perturbs DNA replication fork progression. However, these functional distinctions between ATM and ATR are not absolute and, in many cases, both kinases participate in checkpoint control pathways in response to a given genotoxic insult. In addition, the DNA-dependent protein kinase, DNA-PK, possesses overlapping substrate specificity with ATM and ATR and variably contributes to checkpoint signaling in mammalian cells [60].

5.3.1.1 Rapid G_1/S Checkpoint Arrest

An IR-induced genotoxic insult during the G_1 phase leads to the activation of ATM pathway within seconds to minutes followed by rapid downregulation of cyclin A/E-CDK2 catalytic activity. Inhibition of CDK2 occurs as a consequence of inactivation of CDC25A, which, as we have seen, mediates the activation of CDK2 through dephosphorylation of Tyr-15 and Thr-14. The inactivation of CDC25A, as well as CDC25B and CDC25C in response to DNA damage is carried out in large part by the checkpoint transducer kinases, CHK1 and CHK2. Though structurally unrelated, CHK1 and CHK2 fulfill partially overlapping functions in G_1/S, intra S-phase, and G_2/M checkpoint control downstream of ATR and ATM, respectively. CHK2 is orthologous to *S. cerevisiae* Rad53p and contains a phosphorylation site-rich amino-terminal domain adjacent to a FHA (forkhead-associated) domain that binds to phosphorylated Thr residues [61]. The *CHEK2* gene encoding CHK2 has received notoriety as a low-penetrance breast cancer allele and its mutation has also been linked to the Li-Fraumeni cancer susceptibility syndrome and multiorgan cancer [62].

CHK2 is rapidly phosphorylated by ATM in IR-damaged cells on Thr-68, which stimulates an intermolecular interaction between the phosphorylated amino terminus of one CHK2 molecule with the FHA domain of a second CHK2 protein [63–65]. CHK2 dimerization leads to the transautophosphorylation of the T loop within the carboxyl-terminal kinase domain, followed by the release of catalytically-active CHK2 monomers [62, 66]. Other protein kinases, including PLK1 and PLK3 are also known to contribute to amino-terminal CHK2 phosphorylation and activation [67, 68].

CDC25A is believed to be a critical cellular target of CHK2 within the rapid G_1/S checkpoint pathway. CHK2 phosphorylates CDC25A on three residues (Ser-123, Ser-178 and Ser-292), and also CHK1 phosphorylates Ser-76 in addition to these phosphorylation sites. Amino-terminal phosphorylation of CDC25A promotes its recognition by the E3 ubiquitin ligase $SCF^{\beta\text{-TrCP}}$, ubiquitylation, and subsequent proteasomal degradation [69–71]. The downregulation of CDC25A protein in G_1 phase results in the accumulation of Tyr-15/Thr-14-phosphorylated CDK2, failure to phosphorylate pRb and failure to accumulate active E2F1 transcriptional complexes required for the expression of S-phase genes. The assembly of functional pre-RCs (pre-replication complexes) at DNA replication origins during late G_1 phase is also acutely inhibited as a consequence of CDK2 inactivation [7]. The second critical regulator of origin firing, the CDC7-DBF4 kinase, is also the target of ATM/ATR-dependent checkpoint regulation [72], though its mechanism of inhibition in response to DNA damage is not well understood.

Although activation of the ATM-CHK2-CDC25A pathway is well documented, CHK2 itself is not absolutely required for rapid G_1/S checkpoint arrest [73–76]. The G_1/S checkpoint represents an interconnected collection of partially redundant pathways and the absence of CHK2 is likely compensated by the ATR-CHK1 arm of the S-phase checkpoint control network, which will be elaborated below. The near-normal checkpoint response of CHK2-deficient cells suggests that its critical tumor suppressive functions may lie outside cell cycle control. Instead, accumulating evidence suggests that pro-apoptotic functions of CHK2 may mediate tumor suppression [77].

5.3.1.2 Delayed G_1/S Checkpoint Arrest and the p53 Tumor Suppressor

The protein synthesis dependent G_1/S checkpoint arrest pathway manifests as a sustained accumulation of cells in G_1 phase hours after exposure to DNA damage and is critically dependent on the p53 tumor suppressor transcription factor. The p53 transcriptional network comprises genes that regulate cell cycle, apoptosis, and cell migration amongst many cellular processes. The pathways regulating p53 activity are perhaps more elaborate than those for any other protein encoded by the human genome and the reader is encouraged to consult the numerous reviews for a more adequate discussion of this important topic [78]. Although the expression and activity of this critical protein are upregulated in response to virtually all forms of cellular stress, DNA damage was one of the first and best characterized activators of the p53 pathway [79]. The transcriptional activity and stability of p53 are normally restrained by the MDM2 (murine double-minute) E3 ubiquitin ligase that targets it for ubiquitin-dependent degradation [80]. A second, closely related protein, dubbed MDMX (also termed MDM4) lacks E3 ligase activity, but antagonizes p53 via direct binding to its transactivation domain [81]. In response to DNA damage, p53 undergoes a complex and hierarchical series of post-translational modifications, including the ATM/ATR/DNA-PK-dependent phosphorylation of Ser-15, the phosphorylation of Ser-20 by CHK2 and possibly other kinases, and the acetylation of multiple Lys residues within the carboxyl-terminal tetramerization and central DNA-binding domains by the CBP/p300 and MYST family acetyl transferase proteins [82–86]. While the functional consequences of individual post-translational modifications in p53 activation are highly complex and context specific, the end result of this multifaceted regulation is the dissociation of MDM2 and the derepression of p53 transcriptional activity and stability [87]. The phosphorylation of MDM2 by ATM also contributes to p53 activation in part, by inducing MDM2 self-ubiquitinylation, as well as ubiquitinylation and degradation of MDMX [88, 89]. The ubiquitinylation and degradation of MDM2 and MDMX are negatively regulated by the Herpes virus-associated ubiquitin-specific protease (HAUSP). Dissociation of MDM2-HAUSP complexes in response to DNA damage leads to ubiquitinlylation and degradation of MDM2/MDMX and induction of p53 [90]. Finally, there are additional complexities controlling MDM2-p53 ubiquitinylation and stability that likely fine tune the activity and expression level of p53 during the DNA damage response [80].

An important end result of p53 induction by DNA damage is the transcriptional activation of the *CDKN1A* gene encoding the $p21^{Cip1}$ [91] p53 binds to the p21 promoter and is required for the induction of p21 mRNA and protein in response to DNA damage [92]. p21 elicts G_1-phase cell cycle arrest through inhibition of both CDK2 and CDK4-containing complexes [93, 94]. Cells genetically deficient for p21 recapitulate the G_1/S checkpoint defect observed in p53-deficient and the requirement for the translation of p21 mRNA imparts protein synthesis dependence to the delayed G_1/S checkpoint [95, 96]. In addition, p21 can directly inhibit DNA synthesis *in vitro* through binding to the DNA polymerase accessory factor, PCNA (proliferating cell nuclear antigen), suggesting a role in the intra-S-phase checkpoint outlined below [97, 98]. Other p53 target genes, including BTG2 and GADD45 also contribute to G_1/S checkpoint maintenance [99, 100]. The multi-faceted nature of p53-dependent checkpoint control has now been extended to include microRNAs (miRNA). p53 directly activates the miR-34 promoter and miR-34-dependent suppression of cell cycle genes such as Cyclin E may contribute to maintenance of G_1/S cell cycle arrest [101–103]. In sum, p53 contributes to G_1/S checkpoint arrest through multiple, transcription dependent pathways. In addition, as described below the p53 pathway is also required for sustenance of the G_2/M checkpoint in response to genotoxic stress.

5.3.2 S-Phase DNA Damage Checkpoints

The occurrence of DNA damage in S phase represents a particularly challenging and potentially deleterious insult to the cell. DNA replication forks encountering a DSB or other fork-obstructing lesion stall and can undergo a process termed replication fork collapse whereby the replication complex dissociates from the fork. Collapsed replication forks are subject to a number of DNA transactions, including homologous recombination and are susceptible to conversion into cytotoxic DSBs. Therefore, to mitigate the deleterious effects of DNA damage, the S-phase checkpoint must accomplish several objectives. These include: (i) delay the firing of uninitiated DNA replication origins; (ii) protect DNA replication forks that have stalled; and (iii) delay mitosis until S-phase damage has been repaired and DNA replication has been completed (S/M checkpoint). Intra-S-phase checkpoints employ overlapping ATR- and ATM-dependent signaling elements, with the precise pathways of S-phase checkpoint arrest being dictated by the type of offending lesion. As we have seen, the ATM-CHK2 pathway plays a dominant role in the signaling and activation of checkpoints in response to IR and radiomimetic drugs in G_1 and this pathway also contributes to the suppression of origin firing in S-phase cells. The ATR-CHK1 pathway, on the other hand, plays a supportive role in IR-induced checkpoint arrest, but becomes the preeminent pathway required for checkpoint arrest in response to stalled DNA replication.

5.3.2.1 ATM-Dependent Intra-S-Phase Checkpoint

The IR-induced intra-S-phase checkpoint culminates in the inhibition of S-phase cyclin-CDKs and can be viewed as a special example of the rapid G_1/S checkpoint. A defect in the IR-induced intra-S-phase DNA damage checkpoint is responsible for a hallmark abnormality of ATM-deficient cells: radioresistant DNA synthesis (RDS). Whereas wild-type cells exhibit a reduction in DNA synthesis (as measured by the incorporation of ^{3}H-thymidine label) following an acute IR exposure, ATM-deficient cells continue to synthesize DNA unabated [104]. RDS is also a property of NBS1-deficient cells from individuals with Njimegen Breakage Syndrome, and this phenotype provided early evidence that ATM and NBS1 function in a common pathway [105]. A simplified view of the intra-S-phase checkpoint pathway initiated by IR-induced DSBs is illustrated in Fig. 5.2. The ATM-CHK2-CDC25A pathway is an important signaling axis in this pathway, and its activation leads to inhibition of cyclin E/A-CDK2 and suppression of DNA replication origin firing [69]. DNA damage mediator proteins, including NBS1, FANCD2, SMC1, and BRCA1 are also critical for the response and, in several instances, ATM-dependent phosphorylation of mediator proteins stimulates pathway activation [106]. For example, ATM-dependent phosphorylation of NBS1 on Ser-343 is required for full intra-S-phase checkpoint activation, though this appears to be independent of the CHK2-CDC25A arm of the pathway [107–109]. Other critical phosphorylation targets for ATM in the S-phase checkpoint include SMC1, which is phosphorylated on Ser-957, and BRCA1, which is phosphorylated on multiple sites by ATM and ATR [110, 111]. Mutation of ATM phosphorylation sites in either SMC1 or BRCA1 attenuates S-phase checkpoint activation in response to IR [110, 112], though the underlying

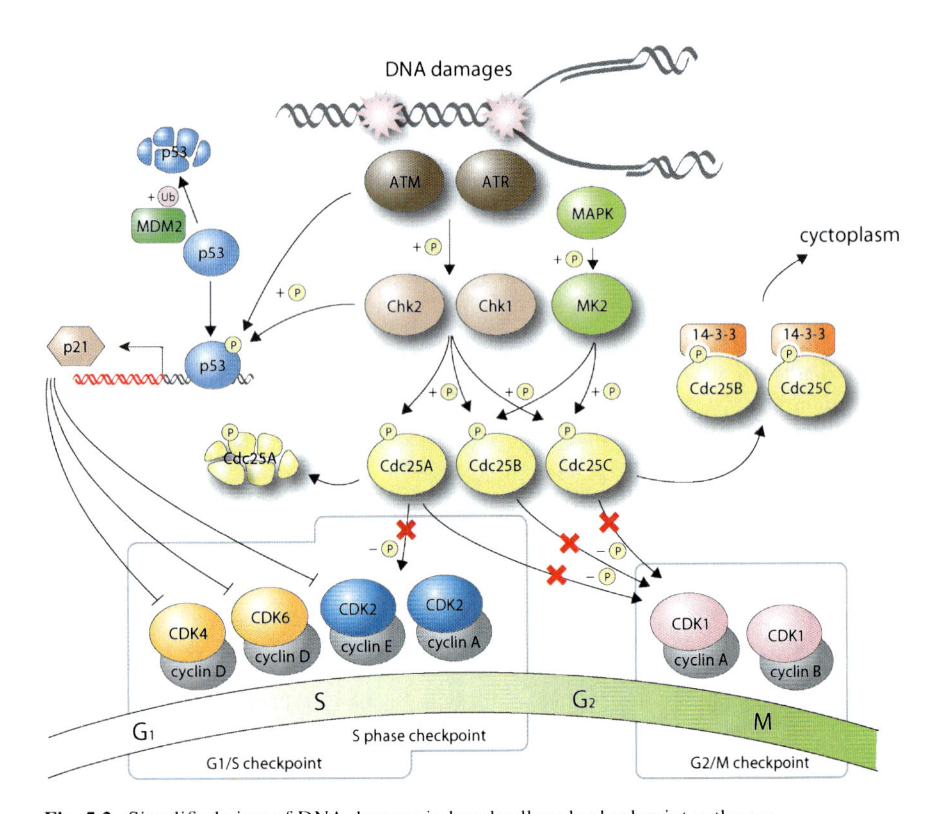

Fig. 5.2 Simplified view of DNA damage-induced cell cycle checkpoint pathways

biochemical mechanisms are not clear. Activation of the ATM-dependent S-phase checkpoint has also been recapitulated in vitro using *Xenopus laevis* oocyte extracts. Here, the addition of double-stranded DNA fragments is sufficient to inhibit functional pre-RC formation and DNA replication in an ATM-dependent manner. The suppression of DNA replication in this system is linked to the inhibition of cyclin E-CDK2 and resulting failure to assemble functional pre-RCs [113].

5.3.2.2 ATR-Dependent S-Phase Checkpoint Arrest

The ATR-CHK1 pathway plays a critical role in DNA transactions during S-phase, as evidenced by the fact that homozygous mutations in either ATR or CHK1 are associated with severe DNA replication abnormalities and embryonic lethality [114, 115]. The ATR-CHK1 pathway plays a supporting role in the IR-induced intra-S-phase checkpoint, and overexpression of ATR can partially suppress the RDS phenotype of A-T fibroblasts [116]. The activation of ATR by IR is temporally delayed in comparison to ATM activation and requires the DNA end-processing activity of the MRN complex and the DNA repair factor CtIP (BRCA1 C-terminus-interacting protein) [117, 118], which together mediate DNA end resection that is required for formation of ssDNA, ATR-ATRIP recruitment, and CHK1 phosphorylation. Activated CHK1 contributes to IR-induced S-phase checkpoint arrest through phosphorylation of CDC25A, CDC25B, and CDC25C, as outlined above.

In addition to a supportive role in IR-induced S-phase checkpoint arrest, the ATR-CHK1 axis fulfills critical functions in the initiation of S-phase arrest following exposure to agents that restrict DNA replication, including UV light, which induces bulky pyrmidine dimers and 6–4 photoproducts and hydroxyurea (HU), which inhibits ribonucleotide reductase and depletes cellular dNTP pools. Physiologic stimuli such as hypoxia also elicit ATR-CHK1-dependent checkpoint responses in mammalian cells [119]. As described in earlier chapters, the full-activation of ATR toward CHK1 in response to these agents requires several factors, including RPA, ATRIP, TopBP1 (Topoisomerase II binding protein 1), and claspin [120–125]. Central to the ATR activation model is its recruitment to RPA-coated ssDNA proximal to stalled DNA replication forks [125]. A series of hierarchical recruitment steps involving Rad17-RFC dependent loading of the Rad9-Rad1-Hus1 checkpoint sliding clamp and the Rad9-Rad1-Hus1-dependent recruitment of TopBP1 leads to TopBP1-dependent activation of colocalized ATR-ATRIP. CHK1 is recruited into close proximity of the active ATR complexes via Claspin. The increase in the local concentrations of ATR and CHK1 is presumably critical for rapid CHK1 phosphorylation and activation. ATR phosphorylates CHK1 on Ser-317 and Ser-345 [126–128], which stimulates CHK1 kinase activity through a poorly understood mechanism. Recent studies suggest that Claspin is not absolutely required for CHK1 phosphorylation and may in fact mediate CHK1-independent functions [129].

There are several ramifications of ATR-CHK1 pathway activation in S phase that appear to be partially DNA damage-type dependent. Using the Xenopus system, Costanzo et al. showed that Cdc-Dbf4 activity, and Cdc7-Dbf4-dependent loading of Cdc45 at Pre-RCs, is inhibited in an ATR-dependent fashion in response to etoposide, a topoisomerase poison that induces DSBs [72]. Somewhat surprisingly, inhibition of cyclin E-CDK2, which represents a biochemically defined endpoint of the ATM pathway in Xenopus, was not observed under these conditions. These results have led to the model whereby ATM and ATR inhibit pre-RC assembly independently via suppression of cyclin E-CDK2 and Cdc7-Dbf4, respectively. A second CDC7 regulator, DRF4, is also required for the S-phase checkpoint-dependent unloading of CDC45 and may function downstream of ATM/ATR [130]. Studies using UV-irradiated mammalian cells largely support the notion that ATR-dependent S-phase checkpoint arrest occurs via inhibition of Cdc7-Dbf4 [131]. Finally, in addition to being a target of the S-phase checkpoint, Cdc7 also contributes to CHK1 activation in response to DNA replication stress [132].

An interesting twist to S-phase checkpoint activation paradigms has been provided by studies linking the activation of DNA damage signaling to proteins circadian rhythm regulators [133]. The circadian rhythm is a transcriptional network-based oscillator that couples organismal physiology to photoperiod [134]. Cryptochrome photoreceptors respond receive photoperiod input are related to DNA photolyases, indicating an evolutionary link between circadian apparatus and the repair of UV light-induced DNA damage [135]. Circadian rhythm and genomic surveillance pathways may share other components, including Timeless (TIM), which was first identified as a circadian mutant in Drosophila [136]. Subsequently, TIM and its interacting protein, TIPIN, have been identified as components of the metazoan DNA replication fork [137, 138]. TIM is orthologous to *S. cerevisiae* Tof1, which is a DNA replication fork-associated protein [139]. TIPIN is required for CHK1 phosphorylation in response to DNA replication stress, and Tipin and TIM are implicated in the UV-induced S-phase checkpoint activation [137, 140–142]. The checkpoint functions of Tipin extend to IR-induced S-phase and G_2/M phase checkpoint activation [137]. Tipin-deficient Xenopus extracts fail to load Claspin in response to DNA replication inhibition. S-phase checkpoint defect of Tipin-deficient cells may be due to defective Claspin recruitment to chromatin [143]. The circadian regulator PER1 has also been implicated in the cell cycle checkpoint responses to DNA damage through interaction with ATM and CHK2, and PER1 is thought to mediate tumor suppression through its induction of DNA damage-response genes [144, 145]. The emerging evidence linking DNA damage and circadian signals to the cell cycle leads to interesting theoretical considerations. Circadian control of the cell cycle is hypothesized to restrict the bulk of DNA synthetic activity to the dark-light interface of the circadian oscillation, which potentially limits the UV exposure of actively replicating genomes, the so-called "flight from mutagenic light" [146].

5.3.2.3 S/M Checkpoint

Whereas S-phase DNA damage checkpoints function to delay restrain DNA replication origin firing in the presence of DNA damage, the S/M checkpoint monitors the status of DNA replication, preventing entry into mitosis prior to the completion of genome duplication. The attempted execution of mitosis in the face of uncompleted DNA replication leads to chromosome missegregation and fragmentation, often referred to as mitotic catastrophe or premature chromatin condensation (PCC) [147, 148]. The S/M checkpoint is activated by the same agents, including HU, that inhibit the firing of DNA replication origins. Activation of the S/M checkpoint in budding and fission yeasts requires the ATR orthologs Mec1 (*S. cerevisiae*) and Rad3 (*S. pombe*) and their respective effector kinases, Rad53 and Cds1 [149, 150]. Depletion of ATR from Xenopus extracts, or genetic disruption of CHK1 in chicken DT40 B cells, similarly uncouples S-phase completion from the execution of mitotic events [151–153]. The S/M checkpoint failure of ATR or CHK1 depleted extracts is attributable to defective CDC25 phosphorylation [154].

The S/M checkpoint circuitry may be more complicated in mammals. Genetic deficiency of CHK1 leads to the execution of mitotic events in some cellular contexts [155]. Furthermore, as predicted from the models outlined above, fibroblasts rendered conditionally deficient for ATR exhibit a profound defect in the downregulation of cyclin B-CDK1 activity following treatment with the DNA polymerase inhibitor, aphidicolin [147]. Nevertheless, these cells were restrained from prematurely entering mitosis in the presence of aphidicolin, indicating the existence of a distinct ATR-independent S/M checkpoint that does not target the canonical cyclin B-CDK2 complex [156]. The discordancy of S/M checkpoint status between ATR- and CHK1-deficient is not explained but could reflect cell-type dependent effects or possibly, ATR-independent signaling inputs to CHK1.

5.3.3 G_2/M Checkpoint

The G_2/M checkpoint represents a temporary G_2 delay that suppresses mitotic entry following an overt DNA-damaging insult. It is a highly sensitive response; in budding yeast the introduction of a single persistent DSB is sufficient to activate it [157, 158, 159]. Whereas the G_1/S checkpoint is frequently compromised in tumor cells via mutations in p53, Rb and their intertwined pathways, the G_2/M checkpoint remains robust. G_2/M checkpoint arrest is induced through a series of intertwined and partially redundant pathways that exploit all the regulatory themes discussed in Section 5.1: inhibitory CDK1 phosphorylation, CKI induction, and protein degradation. The checkpoint can be (somewhat arbitrarily) divided into initiation (protein synthesis independent) and maintenance (protein synthesis dependent) phases, which will be considered separately.

5.3.3.1 Initiation of G_2/M Arrest

The journey to G_2/M checkpoint arrest begins with DNA damage detection and activation of the apical DNA damage-signaling kinases ATM and ATR and, of course, ends with inactivation of CDK1. The contributions of ATM and ATR to G_2/M checkpoint activation depend on which phase of the cell cycle the initial genotoxic insult was sustained. For example, the exposure of an asynchronous culture of mammalian cells to a moderate dose of IR (1-2 Gy) results in a rapid decline in mitotic fraction that can be conveniently assayed by quantifying levels of histone H3 phosphorylation on Ser-10. This rapid decline in mitotic fraction is attributable to the arrest of cells that were in G_2 phase at the time of irradiation, is largely dose-independent, and requires ATM [160]. Although ATM-dependent G_2/M arrest has been generally modeled to occur through the canonical ATM-CHK2-CDC25A,B,C pathway, there does not appear to be an absolute requirement for CHK2 in acute IR-induced G_2/M arrest [73]. Other ATM-dependent pathways are therefore likely to play an important role. One strong candidate is the protein phosphatase, PP1. ATM indirectly stimulates PP1 phosphatase activity by phosphorylating its inhibitor protein, I-2 [161]. One consequence of ATM-dependent PP1 activation is the dephosphorylation and inactivation of Aurora kinase. Aurora is implicated as a major kinase of Ser-10 on Histone H3 (H3), the phosphorylation of which is critical for chromosome condensation during mitotic prophase [161]. Thus, ATM-dependent activation of PP1 and subsequent inactivation of Aurora may explain the rapid downregulation of phospho-H3 that is a strong correlate of IR-induced G_2/M arrest.

Whereas ATM is required for acute G_2/M checkpoint activation when cells are irradiated in G_2, Cells that are in S phase at the time of irradiation accumulate at the subsequent G_2/M boundary for a prolonged period of time, and early studies indicated that ATR was required for this aspect of G_2/M checkpoint activation [116, 147, 160]. Much as it is required for S-phase checkpoint activation, CHK1 plays a critical role in G_2/M checkpoint activation in response to a variety of genotoxic stimuli. The mitotic CDK1 phosphatases CDC25B and CDC25C are important targets of CHK1 in the G_2/M checkpoint. The first identified substrate for CHK1, CDC25C, is phosphorylated by CHK1 on Ser-216, creating a binding site for the 14-3-3ϵ protein. 14-3-3-bound CDC25C is exported from the nucleus, leading to the accumulation of inactive, Tyr-15/Thr-14-phosphorylated CDK1 and the suppression of mitotic entry [162–164]. The CDC25C regulation scheme is significantly more complicated than initial models suggested. Using a Xenopus oocyte model system, Margolis and colleagues demonstrated that PP2A-dependent dephosphorylation of the single CDC25 homolog on an amino acid upstream of the 14-3-3 binding site was required for 14-3-3 binding and nuclear export [165]. Analogous, coupled dephosphorylation and phosphorylation events appear to control the activity of mammalian CDC25C [166]. In addition, CDC25A is also targeted by 14-3-3 proteins after phosphorylation by CHK1 on Ser-78 and Thr-507 and this contributes to G_2/M checkpoint activation [167].

APC/C^{Cdh1} also plays a critical role in the initiation of IR-induced G_2/M checkpoint arrest and PLK1 is one if it important targets. Recall that PLK1 promotes mitotic progression through inactivating phosphorylation of the CDK1 inhibitory kinase, WEE1. Its position as a negative regulator of WEE1 might suggest that PLK1 is a target of the G_2/M checkpoint and, in fact, it is. PLK1 is degraded in an APC^{Cdh1}-dependent manner when cells are γ-irradiated in G_2 phase, leading to reactivation of WEE1 and inhibitory phosphorylation of CDK1 [168]. The activation of APC/C^{Cdh1} is mediated by the protein phosphatase, CDC14, which interacts with Cdh1 and is thought to reverse CDK-dependent inhibitory phosphorylation events [168].

5.3.3.2 Stress-Activated Kinases and G_2/M Delay

A third, pathway leading to G_2/M and intra-S-phase arrest is dependent on the activity of the stress-activated protein kinase (SAPK), p38, which belongs to the broader family of mitogen activated protein kinases (MAPKs). p38 and the functionally related Jun N-terminal kinase (JNK) are activated by a wide range of stresses, including protein synthesis inhibitors, osmotic shock, glycosylation inhibitors and DNA damaging agents such as UV light and IR [169, 170]. p38 and JNK function downstream of MAPK kinases (MAP2Ks) and their upstream kinases the MAP2K kinases (MAP3Ks) [169]. p38 has been implicated in UV-induced G_2/M arrest through phosphorylation of CDC25B1 on Ser-309, which recruits 14-3-3 proteins [171]. Later studies revealed that the p38-dependent substrate MAPK-associated protein kinase 2 (MK2) is an intermediary in p38-dependent checkpoint activation [172]. MK2 directly phosphorylates CDC25B2 on Ser-323 (analogous to Ser-309 in CDC25B1), promoting its binding to 14-3-3 and nuclear export in intact cells. MK2-deficient cells continue to replicate their DNA and entire mitosis following exposure to UV light, indicating that MK2 is required for both the intra-S-phase and G_2/M phase DNA damage checkpoints in response to this stimulus [172].

Upstream of p38, a family of MAP3Ks termed the TAO (thousand-and-one) kinases is activated by ATM in response to IR. TAO kinases are required for IR-induced p38 activation and, as would be predicted from the above model, defects in TAO kinase function are associated with a defect in the IR-induced G_2/M checkpoint [173]. Thus, the TAO-p38-MK2 pathway represents a third arm of ATM/ATR-dependent G_2/M and intra-S-phase checkpoint pathways. Interference with this pathway was shown to cause mitotic catastrophe and DNA damage hypersensitivity in p53-deficient tumor, but not in tumor cells expressing wild-type p53 [174]. These findings point toward MK2 as a possible therapeutic target in p53-deficient cancer cells and illustrate the important role of p53 in maintaining G_2/M checkpoint arrest (elaborated below).

The observation that both CDC25B and CDC25C are targets of checkpoint regulation by CHK1, CHK2, MK2, and other kinases, supports the notion that these proteins are important nodes for checkpoint-arrest pathways. Surprisingly, the combined genetic knockout of the *CDC25B* and *CDC25C* genes did not yield a discernible cell cycle phenotype in mice [175]. On the other hand, it has been reported that Cdc25A is required for normal G_2/M checkpoint response [176, 177]. Cdc25A can dephosphorylate CDK1 in addition to CDK2, and is phosphorylated by Chk1 leading to its degradation by the ubiquitin-proteasome pathway in response to DNA damage [71]. Thus, although this issue remains to be fully resolved, the degradation of CDC25A may represent the most important event controlling CDC25-dependent cell cycle events in mammalian cells exposed to DNA damage.

5.3.3.3 Transcription-Dependent G_2/M Checkpoint Pathways

As we have seen, p53 plays an important role in G_1/S phase checkpoint maintenance in response to DNA damage, and it plays an equally important role in G_2/M checkpoint maintenance [178]. A definitive G_2/M checkpoint function for p53 was first established using HCT116 colon carcinoma cells in which *TP53* gene loci had been disrupted through homologous recombination [179]. p53-deficient HCT116 cells exhibited normal downregulation of mitotic fraction several hours after DNA damage exposure, but failed to sustain G_2/M checkpoint arrest for longer periods of time [179]. Premature G_2/M checkpoint release was also observed in $p21^{Cip1}$-deficient HCT116 cells, indicating that the p53-dependent induction of $p21^{Cip1}$, and resulting inhibition of CDK1, is responsible for sustained G_2/M arrest in this cellular context [179]. Other DNA damage-inducible p53 target genes, including 14-3-3 σ and GADD45 (Growth arrest and DNA damage induced protein of 45 kDa), also contribute to the maintenance of G_2/M checkpoint arrest [180, 181]. 14-3-3σ enforcess IR-dependent G_2/M checkpoint arrest through cytoplasmic sequestration of cyclin B-CDK1 [182]. GADD45 binds to CDK1, dissociates cyclin B-CDK1 complexes, and is required for the UV-induced G_2/M checkpoint arrest, but not IR-induced checkpoint activation [181, 183]. It is unclear why UV and IR-induced G_2/M checkpoints exhibit differential requirements for GADD45 induction. Finally, the p53-dependent repression of cell cycle genes, including *CYCLINB*, *CDK1* and *CDC25C*, further represses available Cyclin B-CDK1 kinase activity [178, 184–186].

While p53 has received the most attention, other transcriptional regulators play important roles in checkpoint control. Among these, NF-kB is a strong candidate G_2/M checkpoint regulator in the DNA damage response. NF-kB is activated via ATM and ATR-dependent pathways in response to DNA damage and its induction correlates with protracted G_2/M arrest, p21 induction and cell survival [187–190]. Other transcription factors with well-established cell cycle promoting functions, including CREB, SP1, and E2F1 are also direct phosphorylation targets of ATM/ATR [191–194]. However, the roles of these factors in DNA damage-dependent cell cycle regulation are not yet characterized.

5.3.3.4 Recovery from G_2/M Checkpoint Arrest

G_2/M-arrested cells surviving genotoxic insult eventually re-enter the cell cycle. Time-dependent extinction of the DNA damage signal and consequent diminution of ATM/ATR activity is intuitively linked to reduced checkpoint signaling. However, although the extent of G_2/M arrest is roughly correlated to the dose of incipient DNA damage, studies using Artemis-deficient cells that are defective for DSB repair through non-homologous end joining (NHEJ) have shown that complete DNA repair is not required cell cycle reentry [195]. Cell cycle reentry in mammalian cells resembles the DNA damage adaptation phenomenon studied in yeast, and this process may have implications for oncogenesis following genotoxic insults [196]. At the molecular level, regulated degradation of Claspin has emerged as an important event in release from G_2/M checkpoint arrest. During G_2-phase, Claspin is phosphorylated by PLK1, leading to its ubiquitin dependent degradation by the $SCF^{\beta Trcp}$ E3 ubiquitin ligase [197–199]. The reduction in Claspin leads to reduced CHK1 activity, derepression of CDC25A, B, and C, activating dephosphorylation of Cyclin B-CDK1, and G_2/M progression. As we have seen, the APC^{Cdh1}-dependent controlled degradation of PLK1 is also important for DNA damage-induced G_2/M checkpoint arrest, and PLK1 is degraded at anaphase during a normal cell cycle [200]. In G_2-arrested cells, resumption of the cell cycle and entry in mitosis requires the reactivation of PLK1. Activation of PLK1 during a normal G_2 phase and after checkpoint delay is achieved through phosphorylation of its activation loop by Aurora kinase [201]. In addition, Claspin is targeted for degradation during G_0/G_1 phase by the APC/C^{Cdh1} E3 ligase ensuring low CHK1 activity and favorable conditions for S-phase initiation [168].

5.4 The DDR and Cell Cycle Latency: The Special Case of Neurons

The investigation of mammalian cell cycle checkpoint pathways has, of course, been carried out predominantly in actively cycling cells and tissues and often within the general context of understanding cancer cell biology. However, the same cell cycle checkpoint pathways that initiate growth arrest in actively dividing cells may be important for maintaining terminal growth arrest in post-mitotic tissues. Of particular relevance to this discussion is the neuron—a cell type often disproportionately affected by mutations in DDR genes [202]. The progressive neurodegenerative phenotypes of A-T and NBS patients may be partially attributable to the chronic accumulation of unrepaired DSBs, which can compromise gene expression and lead to oxidative stress-induced macromolecule damage [203]. However, it is becoming clear that loss of cell cycle control is also an important determinant of neurodegeneration in the A-T paradigm [204]. Evidence supporting cell cycle reentry in A-T came in the form of immunohistochemical staining of post-mortem brain tissue, which showed aberrantly high levels of proliferation markers, including PCNA and cyclin A and B [204]. Purkinje neurons in ATM-null mice also exhibit cell cycle reactivation markers, and the proliferation and differentiation of neural precursors is defective in these animals [204, 205]. Direct demonstration of a cell cycle-suppressing function in neurons was provided using a Drosophila model, in which neuron-directed RNAi against ATM stimulated neurons to synthesize DNA, attempt mitosis, and undergo programmed cell death [206]. Remarkably, mutation of the single Drosophila CDC25 ortholog, *string* prevented cell cycle reentry and restored neuron viability in this model of ATM deficiency. These findings indicate that, under some circumstances, ATM-dependent checkpoints are required to maintain cell cycle latency in terminally differentiated cells, which retain the potential to reenter the cell cycle when active checkpoint signaling pathways are disrupted. Cell cycle reactivation of neurons is also observed in other human neurodegenerative disorders such as Alzheimer's Disease and Parkinson's Disease. Although it remains to be seen whether attempts at cell cycle reentry in these conditions are caused by inactivation of checkpoint pathways, CDK1 activation in several neuron contexts leads to apoptotic cell death [207].

5.4.1 Concluding Remarks: Exploiting Checkpoint Defects Therapeutically

A reasonable conclusion derived from the above narrative is that the DDR-cell cycle paradigm has reached a point of intimidating complexity, and no doubt this complexity will continue to increase as finer biochemical details of the relevant signaling networks emerge. There are, however, important real-life ramifications and applications of this new knowledge, particularly in the universe of cancer chemotherapy. Specifically, the ubiquitous deregulation of the cell cycle and common loss of G_1/S checkpoint function in cancer provides a potential therapeutic window of opportunity. It has been posited that G_1/S checkpoint-defective tumor cells are overly reliant on the G_2/M checkpoint and may therefore be more susceptible to partial inhibition of G_2/M checkpoint control pathways. Among numerous recent examples supporting this general idea, p53-deficient cell lines and tumor xenografts are more susceptible to DNA damage-induced cell death when MK2 is silenced through RNAi [174]. This finding will no doubt stimulate interest in the development and testing of MK2 inhibitors that may show anti-tumor properties when combined with cytotoxic therapies. In fact, the development of inhibitors of other protein kinases with checkpoint function, including ATM and CHK1 is further along. The ATM-specific inhibitor KU-55933 sensitizes cancer cell lines to IR and various DSB-inducing agents and other ATM inhibitors are at various stages of testing or development [208, 209, 210]. Owing to its essential function in the G_2/M checkpoint, CHK1 is also an attractive chemotherapeutic target. ATP-competitive inhibitors of CHK1, including AZD7762 (AstraZeneca) and PF-00477736 (Pfizer) sensitize cancer cells and tumor xenografts to anti-cancer DNA-damaging agents [211, 212]. Both drugs are being tested in clinical trials individually and in combination with gemcitabine for efficacy against solid tumors [213].

The link between cell cycle reentry and neurodegeneration in A-T may provide another venue for cell cycle-targeted therapy. The suppression of neurodegeneration in ATM-deficient Drosophila by monoallelic inactivation of CDC25 suggests that pharmacologic inhibition of CDC25 or CDKs may slow motor function decline in A-T [206]. A number of CDC25 inhibitors have been described through the years, including quinonoids such as menadione, which have been pursued for potential anti-cancer properties. The electrophilicity and redox cycling properties of these compounds and associated toxicities in preclinical studies diminished enthusiasm for their clinical utility [214]. Nevertheless, next generation CDC25 inhibitors merit close scrutiny for potential application in the A-T setting. It is a reasonable expectation that, as our understanding of DNA damage signaling and cell cycle regulation becomes more complete, new targets for therapeutic intervention in cancer and A-T will be discovered.

References

1. Nurse, P. M. (2002) *Biosci Rep* 22, 487–499
2. Hunt, T. (2002) *Biosci Rep* 22, 465–486
3. Hartwell, L. H. (2004) *Biosci Rep* 24, 523–544
4. Paweletz, N. (2001) *Nat Rev* 2, 72–75
5. Evans, T., Rosenthal, E. T., Youngblom, J., Distel, D., and Hunt, T. (1983) *Cell* 33, 389–396
6. Masui, Y., and Markert, C. L. (1971) *J Exp Zool* 177, 129–145
7. Woo, R. A., and Poon, R. Y. (2003) *Cell cycle (Georgetown, Tex)* 2, 316–324
8. Sherr, C. J., and Roberts, J. M. (1999) *Genes Dev* 13, 1501–1512
9. Kim, W. Y., and Sharpless, N. E. (2006) *Cell* 127, 265–275
10. Murray, A. W., and Kirschner, M. W. (1989) *Nature* 339, 275–280
11. Glotzer, M., Murray, A. W., and Kirschner, M. W. (1991) *Nature* 349, 132–138
12. Yamano, H., Gannon, J., Mahbubani, H., and Hunt, T. (2004) *Molecular cell* 13, 137–147
13. Peters, J. M. (2006) *Nat Rev* 7, 644–656
14. King, R. W., Peters, J. M., Tugendreich, S., Rolfe, M., Hieter, P., and Kirschner, M. W. (1995) *Cell* 81, 279–288
15. Listovsky, T., Zor, A., Laronne, A., and Brandeis, M. (2000) *Exp Cell Res* 255, 184–191
16. Fang, G., Yu, H., and Kirschner, M. W. (1998) *Mol Cell* 2, 163–171
17. Prinz, S., Hwang, E. S., Visintin, R., and Amon, A. (1998) *Curr Biol* 8, 750–760
18. Geley, S., Kramer, E., Gieffers, C., Gannon, J., Peters, J. M., and Hunt, T. (2001) *J Cell Biol* 153, 137–148
19. Strohmaier, H., Spruck, C. H., Kaiser, P., Won, K. A., Sangfelt, O., and Reed, S. I. (2001) *Nature* 413, 316–322
20. Ye, X., Nalepa, G., Welcker, M., Kessler, B. M., Spooner, E., Qin, J., Elledge, S. J., Clurman, B. E., and Harper, J. W. (2004) *J Biol Chem* 279, 50110–50119
21. Desai, D., Wessling, H. C., Fisher, R. P., and Morgan, D. O. (1995) *Mol Cellular Biol* 15, 345–350
22. Lolli, G., and Johnson, L. N. (2005) *Cell cycle (Georgetown, Tex)* 4, 572–577
23. Parker, L. L., and Piwnica-Worms, H. (1992) *Science (New York, NY)* 257, 1955–1957
24. Mueller, P. R., Coleman, T. R., Kumagai, A., and Dunphy, W. G. (1995) *Science (New York, NY)* 270, 86–90
25. Galaktionov, K., and Beach, D. (1991) *Cell* 67, 1181–1194
26. Sadhu, K., Reed, S. I., Richardson, H., and Russell, P. (1990) *Proc Nat Acad Sci USA* 87, 5139–5143
27. Nagata, A., Igarashi, M., Jinno, S., Suto, K., and Okayama, H. (1991) *New Biol* 3, 959–968
28. Rudolph, J. (2007) *Nat Rev Cancer* 7, 202–211
29. Karlsson-Rosenthal, C., and Millar, J. B. (2006) *Trends Cell Biol* 16, 285–292
30. Sharpless, N. E. (2005) *Mutat Res* 576, 22–38
31. Buchkovich, K., Duffy, L. A., and Harlow, E. (1989) *Cell* 58, 1097–1105
32. Kato, J., Matsushime, H., Hiebert, S. W., Ewen, M. E., and Sherr, C. J. (1993) *Genes Dev* 7, 331–342
33. Fisher, R. P., and Morgan, D. O. (1994) *Cell* 78, 713–724
34. Polager, S., and Ginsberg, D. (2008) *Trends Cell Biol* 18, 528–535
35. Vlach, J., Hennecke, S., and Amati, B. (1997) *EMBO J* 16, 5334–5344
36. Boutros, R., Dozier, C., and Ducommun, B. (2006) *Curr Opin Cell Biol* 18, 185–191
37. Girard, F., Strausfeld, U., Fernandez, A., and Lamb, N. J. (1991) *Cell* 67, 1169–1179
38. Walker, D. H., and Maller, J. L. (1991) *Nature* 354, 314–317
39. Dowell, S. J., Romanowski, P., and Diffley, J. F. (1994) *Science (New York, NY)* 265, 1243–1246
40. Jiang, W., McDonald, D., Hope, T. J., and Hunter, T. (1999) *EMBO J* 18, 5703–5713
41. Diffley, J. F. (2004) *Curr Biol* 14, R778–786
42. Nasheuer, H. P., Moore, A., Wahl, A. F., and Wang, T. S. (1991) *J Biol Chem* 266, 7893–7903

43. Voitenleitner, C., Fanning, E., and Nasheuer, H. P. (1997) *Oncogene* 14, 1611–1615
44. Sclafani, R. A. (2000) *J Cell Sci* 113 (Pt 12), 2111–2117
45. Masai, H., and Arai, K. (2002) *J Cell Physiol* 190, 287–296
46. Montagnoli, A., Valsasina, B., Brotherton, D., Troiani, S., Rainoldi, S., Tenca, P., Molinari, A., and Santocanale, C. (2006) *J Biol Chem* 281, 10281–10290
47. Mailand, N., and Diffley, J. F. (2005) *Cell* 122, 915–926
48. Yang, J., Bardes, E. S., Moore, J. D., Brennan, J., Powers, M. A., and Kornbluth, S. (1998) *Genes Dev* 12, 2131–2143
49. Hagting, A., Karlsson, C., Clute, P., Jackman, M., and Pines, J. (1998) *EMBO J* 17, 4127–4138
50. Toyoshima, F., Moriguchi, T., Wada, A., Fukuda, M., and Nishida, E. (1998) *EMBO J* 17, 2728–2735
51. Toyoshima-Morimoto, F., Taniguchi, E., Shinya, N., Iwamatsu, A., and Nishida, E. (2001) *Nature* 410, 215–220
52. Yuan, J., Eckerdt, F., Bereiter-Hahn, J., Kurunci-Csacsko, E., Kaufmann, M., and Strebhardt, K. (2002) *Oncogene* 21, 8282–8292
53. Watanabe, N., Arai, H., Nishihara, Y., Taniguchi, M., Hunter, T., and Osada, H. (2004) *Proc Natl Acad Sci USA* 101, 4419–4424
54. Guardavaccaro, D., and Pagano, M. (2006) *Mol Cell* 22, 1–4
55. Bulavin, D. V., Higashimoto, Y., Demidenko, Z. N., Meek, S., Graves, P., Phillips, C., Zhao, H., Moody, S. A., Appella, E., Piwnica-Worms, H., and Fornace, A. J., Jr. (2003) *Nat Cell Biol* 5, 545–551
56. Strausfeld, U., Fernandez, A., Capony, J. P., Girard, F., Lautredou, N., Derancourt, J., Labbe, J. C., and Lamb, N. J. (1994) *J Biol Chem* 269, 5989–6000
57. Lamb, N. J., Fernandez, A., Watrin, A., Labbe, J. C., and Cavadore, J. C. (1990) *Cell* 60, 151–165
58. Blangy, A., Lane, H. A., d'Herin, P., Harper, M., Kress, M., and Nigg, E. A. (1995) *Cell* 83, 1159–1169
59. Wolf, F., Sigl, R., and Geley, S. (2007) *Cell cycle (Georgetown, Tex)* 6, 1408–1411
60. Furuno, N., den Elzen, N., and Pines, J. (1999) *J Cell Biol* 147, 295–306
61. Durocher, D., Henckel, J., Fersht, A. R., and Jackson, S. P. (1999) *Mol Cell* 4, 387–394
62. Antoni, L., Sodha, N., Collins, I., and Garrett, M. D. (2007) *Nat Rev Cancer* 7, 925–936
63. Matsuoka, S., Huang, M., and Elledge, S. J. (1998) *Science (New York, NY)* 282, 1893–1897
64. Xu, X., Tsvetkov, L. M., and Stern, D. F. (2002) *Mol Cell Biol* 22, 4419–4432
65. Ahn, J. Y., Li, X., Davis, H. L., and Canman, C. E. (2002) *J Biol Chem* 277, 19389–19395
66. Ahn, J., Urist, M., and Prives, C. (2004) *DNA Repair (Amst)* 3, 1039–1047
67. Tsvetkov, L., Xu, X., Li, J., and Stern, D. F. (2003) *J Biol Chem* 278, 8468–8475
68. Bahassi el, M., Myer, D. L., McKenney, R. J., Hennigan, R. F., and Stambrook, P. J. (2006) *Mutat Res* 596, 166–176
69. Falck, J., Mailand, N., Syljuasen, R. G., Bartek, J., and Lukas, J. (2001) *Nature* 410, 842–847
70. Sorensen, C. S., Syljuasen, R. G., Falck, J., Schroeder, T., Ronnstrand, L., Khanna, K. K., Zhou, B. B., Bartek, J., and Lukas, J. (2003) *Cancer Cell* 3, 247–258
71. Jin, J., Shirogane, T., Xu, L., Nalepa, G., Qin, J., Elledge, S. J., and Harper, J. W. (2003) *Genes Dev* 17, 3062–3074
72. Costanzo, V., Shechter, D., Lupardus, P. J., Cimprich, K. A., Gottesman, M., and Gautier, J. (2003) *Mol Cell* 11, 203–213
73. Takai, H., Naka, K., Okada, Y., Watanabe, M., Harada, N., Saito, S., anderson, C. W., Appella, E., Nakanishi, M., Suzuki, H., Nagashima, K., Sawa, H., Ikeda, K., and Motoyama, N. (2002) *EMBO J* 21, 5195–5205
74. Hirao, A., Cheung, A., Duncan, G., Girard, P. M., Elia, A. J., Wakeham, A., Okada, H., Sarkissian, T., Wong, J. A., Sakai, T., De Stanchina, E., Bristow, R. G., Suda, T., Lowe, S. W., Jeggo, P. A., Elledge, S. J., and Mak, T. W. (2002) *Mol Cell Biol* 22, 6521–6532
75. Jallepalli, P. V., Lengauer, C., Vogelstein, B., and Bunz, F. (2003) *J Biol Chem* 278, 20475–20479

76. Jack, M. T., Woo, R. A., Hirao, A., Cheung, A., Mak, T. W., and Lee, P. W. (2002) *Proc Natl Acad Sci USA* 99, 9825–9829
77. Stracker, T. H., Couto, S. S., Cordon-Cardo, C., Matos, T., and Petrini, J. H. (2008) *Mol Cell* 31, 21–32
78. Laptenko, O., and Prives, C. (2006) *Cell Death Differ* 13, 951–961
79. Kastan, M. B., Onyekwere, O., Sidransky, D., Vogelstein, B., and Craig, R. W. (1991) *Cancer Res* 51, 6304–6311
80. Brooks, C. L., and Gu, W. (2006) *Mol Cell* 21, 307–315
81. Shvarts, A., Steegenga, W. T., Riteco, N., van Laar, T., Dekker, P., Bazuine, M., van Ham, R. C., van der Houven van Oordt, W., Hateboer, G., van der Eb, A. J., and Jochemsen, A. G. (1996) *EMBO J* 15, 5349–5357
82. Canman, C. E., Lim, D. S., Cimprich, K. A., Taya, Y., Tamai, K., Sakaguchi, K., Appella, E., Kastan, M. B., and Siliciano, J. D. (1998) *Science (New York, NY)* 281, 1677–1679
83. Tibbetts, R. S., Brumbaugh, K. M., Williams, J. M., Sarkaria, J. N., Cliby, W. A., Shieh, S. Y., Taya, Y., Prives, C., and Abraham, R. T. (1999) *Genes Dev* 13, 152–157
84. Chehab, N. H., Malikzay, A., Appel, M., and Halazonetis, T. D. (2000) *Genes Dev* 14, 278–288
85. Gu, W., and Roeder, R. G. (1997) *Cell* 90, 595–606
86. Sykes, S. M., Mellert, H. S., Holbert, M. A., Li, K., Marmorstein, R., Lane, W. S., and McMahon, S. B. (2006) *Mol Cell* 24, 841–851
87. Toledo, F., and Wahl, G. M. (2006) *Nat Rev Cancer* 6, 909–923
88. Maya, R., Balass, M., Kim, S. T., Shkedy, D., Leal, J. F., Shifman, O., Moas, M., Buschmann, T., Ronai, Z., Shiloh, Y., Kastan, M. B., Katzir, E., and Oren, M. (2001) *Genes Dev* 15, 1067–1077
89. Pereg, Y., Shkedy, D., de Graaf, P., Meulmeester, E., Edelson-Averbukh, M., Salek, M., Biton, S., Teunisse, A. F., Lehmann, W. D., Jochemsen, A. G., and Shiloh, Y. (2005) *Proc Natl Acad Sci USA* 102, 5056–5061
90. Meulmeester, E., Maurice, M. M., Boutell, C., Teunisse, A. F., Ovaa, H., Abraham, T. E., Dirks, R. W., and Jochemsen, A. G. (2005) *Mol Cell* 18, 565–576
91. el-Deiry, W. S., Harper, J. W., O'Connor, P. M., Velculescu, V. E., Canman, C. E., Jackman, J., Pietenpol, J. A., Burrell, M., Hill, D. E., Wang, Y., et al. (1994) *Cancer Res* 54, 1169–1174
92. Macleod, K. F., Sherry, N., Hannon, G., Beach, D., Tokino, T., Kinzler, K., Vogelstein, B., and Jacks, T. (1995) *Genes Dev* 9, 935–944
93. Harper, J. W., Adami, G. R., Wei, N., Keyomarsi, K., and Elledge, S. J. (1993) *Cell* 75, 805–816
94. Harper, J. W., Elledge, S. J., Keyomarsi, K., Dynlacht, B., Tsai, L. H., Zhang, P., Dobrowolski, S., Bai, C., Connell-Crowley, L., Swindell, E., and et al. (1995) *Mol Biol Cell* 6, 387–400
95. Waldman, T., Kinzler, K. W., and Vogelstein, B. (1995) *Cancer Res* 55, 5187–5190
96. Deng, C., Zhang, P., Harper, J. W., Elledge, S. J., and Leder, P. (1995) *Cell* 82, 675–684
97. Li, R., Waga, S., Hannon, G. J., Beach, D., and Stillman, B. (1994) *Nature* 371, 534–537
98. Waga, S., Hannon, G. J., Beach, D., and Stillman, B. (1994) *Nature* 369, 574–578
99. Guardavaccaro, D., Corrente, G., Covone, F., Micheli, L., D'Agnano, I., Starace, G., Caruso, M., and Tirone, F. (2000) *Mol Cell Biol* 20, 1797–1815
100. Kastan, M. B., Zhan, Q., el-Deiry, W. S., Carrier, F., Jacks, T., Walsh, W. V., Plunkett, B. S., Vogelstein, B., and Fornace, A. J., Jr. (1992) *Cell* 71, 587–597
101. Bommer, G. T., Gerin, I., Feng, Y., Kaczorowski, A. J., Kuick, R., Love, R. E., Zhai, Y., Giordano, T. J., Qin, Z. S., Moore, B. B., MacDougald, O. A., Cho, K. R., and Fearon, E. R. (2007) *Curr Biol* 17, 1298–1307
102. Chang, T. C., Wentzel, E. A., Kent, O. A., Ramachandran, K., Mullendore, M., Lee, K. H., Feldmann, G., Yamakuchi, M., Ferlito, M., Lowenstein, C. J., Arking, D. E., Beer, M. A., Maitra, A., and Mendell, J. T. (2007) *Mol Cell* 26, 745–752

103. He, L., He, X., Lim, L. P., de Stanchina, E., Xuan, Z., Liang, Y., Xue, W., Zender, L., Magnus, J., Ridzon, D., Jackson, A. L., Linsley, P. S., Chen, C., Lowe, S. W., Cleary, M. A., and Hannon, G. J. (2007) *Nature* 447, 1130–1134
104. Painter, R. B., and Young, B. R. (1980) *Proc Natl Acad Sci USA* 77, 7315–7317
105. Girard, P. M., Foray, N., Stumm, M., Waugh, A., Riballo, E., Maser, R. S., Phillips, W. P., Petrini, J., Arlett, C. F., and Jeggo, P. A. (2000) *Cancer Res* 60, 4881–4888
106. Lavin, M. F., and Kozlov, S. (2007) *Cell Cycle (Georgetown, Tex)* 6, 931–942
107. Lim, D. S., Kim, S. T., Xu, B., Maser, R. S., Lin, J., Petrini, J. H., and Kastan, M. B. (2000) *Nature* 404, 613–617
108. Zhao, S., Weng, Y. C., Yuan, S. S., Lin, Y. T., Hsu, H. C., Lin, S. C., Gerbino, E., Song, M. H., Zdzienicka, M. Z., Gatti, R. A., Shay, J. W., Ziv, Y., Shiloh, Y., and Lee, E. Y. (2000) *Nature* 405, 473–477
109. Falck, J., Petrini, J. H., Williams, B. R., Lukas, J., and Bartek, J. (2002) *Nat Genet* 30, 290–294
110. Kim, S. T., Xu, B., and Kastan, M. B. (2002) *Genes Dev* 16, 560–570
111. Cortez, D., Wang, Y., Qin, J., and Elledge, S. J. (1999) *Science (New York, NY)* 286, 1162–1166
112. Xu, B., Kim, S., and Kastan, M. B. (2001) *Mol Cell Biol* 21, 3445–3450
113. Costanzo, V., Robertson, K., Ying, C. Y., Kim, E., Avvedimento, E., Gottesman, M., Grieco, D., and Gautier, J. (2000) *Mol Cell* 6, 649–659
114. de Klein, A., Muijtjens, M., van Os, R., Verhoeven, Y., Smit, B., Carr, A. M., Lehmann, A. R., and Hoeijmakers, J. H. (2000) *Curr Biol* 10, 479–482
115. Liu, Q., Guntuku, S., Cui, X. S., Matsuoka, S., Cortez, D., Tamai, K., Luo, G., Carattini-Rivera, S., DeMayo, F., Bradley, A., Donehower, L. A., and Elledge, S. J. (2000) *Genes Dev* 14, 1448–1459
116. Cliby, W. A., Roberts, C. J., Cimprich, K. A., Stringer, C. M., Lamb, J. R., Schreiber, S. L., and Friend, S. H. (1998) *EMBO J* 17, 159–169
117. Jazayeri, A., Falck, J., Lukas, C., Bartek, J., Smith, G. C., Lukas, J., and Jackson, S. P. (2006) *Nat Cell Biol* 8, 37–45
118. Sartori, A. A., Lukas, C., Coates, J., Mistrik, M., Fu, S., Bartek, J., Baer, R., Lukas, J., and Jackson, S. P. (2007) *Nature* 450, 509–514
119. Hammond, E. M., and Giaccia, A. J. (2004) *DNA Repair (Amst)* 3, 1117–1122
120. Makiniemi, M., Hillukkala, T., Tuusa, J., Reini, K., Vaara, M., Huang, D., Pospiech, H., Majuri, I., Westerling, T., Makela, T. P., and Syvaoja, J. E. (2001) *J Biol Chem* 276, 30399–30406
121. Yamane, K., Wu, X., and Chen, J. (2002) *Mol Cell Biol* 22, 555–566
122. Kumagai, A., Lee, J., Yoo, H. Y., and Dunphy, W. G. (2006) *Cell* 124, 943–955
123. Yan, S., Lindsay, H. D., and Michael, W. M. (2006) *J Cell Biol* 173, 181–186
124. Liu, S., Bekker-Jensen, S., Mailand, N., Lukas, C., Bartek, J., and Lukas, J. (2006) *Mol Cell Biol* 26, 6056–6064
125. Zou, L., and Elledge, S. J. (2003) *Science (New York, NY)* 300, 1542–1548
126. Abraham, R. T. (2001) *Genes Devel* 15, 2177–2196
127. Stiff, T., Walker, S. A., Cerosaletti, K., Goodarzi, A. A., Petermann, E., Concannon, P., O'Driscoll, M., and Jeggo, P. A. (2006) *EMBO J* 25, 5775–5782
128. Kumagai, A., and Dunphy, W. G. (2000) *Mol Cell* 6, 839–849
129. Petermann, E., Helleday, T., and Caldecott, K. W. (2008) *Mol Biol Cell* 19, 2373–2378
130. Yanow, S. K., Gold, D. A., Yoo, H. Y., and Dunphy, W. G. (2003) *J Biol Chem* 278, 41083–41092
131. Heffernan, T. P., Unsal-Kacmaz, K., Heinloth, A. N., Simpson, D. A., Paules, R. S., Sancar, A., Cordeiro-Stone, M., and Kaufmann, W. K. (2007) *J Biol Chem* 282, 9458–9468
132. Kim, J. M., Kakusho, N., Yamada, M., Kanoh, Y., Takemoto, N., and Masai, H. (2008) *Oncogene* 27, 3475–3482
133. Kondratov, R. V., and Antoch, M. P. (2007) *Trends Cell Biol* 17, 311–317

134. Takahashi, J. S., Hong, H. K., Ko, C. H., and McDearmon, E. L. (2008) *Nat Rev Genet* 9, 764–775
135. Lin, C., and Todo, T. (2005) *Genome Biol* 6, 220
136. Sehgal, A., Price, J. L., Man, B., and Young, M. W. (1994) *Science (New York, NY)* 263, 1603–1606
137. Chou, D. M., and Elledge, S. J. (2006) *Proc Natl Acad Sci USA* 103, 18143–18147
138. Gotter, A. L., Suppa, C., and Emanuel, B. S. (2007) *J Mol Biol* 366, 36–52
139. Katou, Y., Kanoh, Y., Bando, M., Noguchi, H., Tanaka, H., Ashikari, T., Sugimoto, K., and Shirahige, K. (2003) *Nature* 424, 1078–1083
140. Unsal-Kacmaz, K., Mullen, T. E., Kaufmann, W. K., and Sancar, A. (2005) *Mol Cell Biol* 25, 3109–3116
141. Yoshizawa-Sugata, N., and Masai, H. (2007) *J Biol Chem* 282, 2729–2740
142. Unsal-Kacmaz, K., Chastain, P. D., Qu, P. P., Minoo, P., Cordeiro-Stone, M., Sancar, A., and Kaufmann, W. K. (2007) *Mol Cell Biol* 27, 3131–3142
143. Errico, A., Costanzo, V., and Hunt, T. (2007) *Proc Natl Acad Sci USA* 104, 14929–14934
144. Gery, S., Komatsu, N., Baldjyan, L., Yu, A., Koo, D., and Koeffler, H. P. (2006) *Mol Cell* 22, 375–382
145. Fu, L., Pelicano, H., Liu, J., Huang, P., and Lee, C. (2002) *Cell* 111, 41–50
146. Allada, R., and Meissner, R. A. (2005) *Mol Cell Biochem* 274, 141–149
147. Brown, E. J., and Baltimore, D. (2003) *Genes Dev* 17, 615–628
148. Nghiem, P., Park, P. K., Kim, Y., Vaziri, C., and Schreiber, S. L. (2001) *Proc Natl Acad Sci USA* 98, 9092–9097
149. al-Khodairy, F., and Carr, A. M. (1992) *EMBO J* 11, 1343–1350
150. Weinert, T. A., Kiser, G. L., and Hartwell, L. H. (1994) *Genes Dev* 8, 652–665
151. Hekmat-Nejad, M., You, Z., Yee, M. C., Newport, J. W., and Cimprich, K. A. (2000) *Curr Biol* 10, 1565–1573
152. Zachos, G., Rainey, M. D., and Gillespie, D. A. (2005) *Mol Cell Biol* 25, 563–574
153. Guo, Z., Kumagai, A., Wang, S. X., and Dunphy, W. G. (2000) *Genes Dev* 14, 2745–2756
154. Kumagai, A., Guo, Z., Emami, K. H., Wang, S. X., and Dunphy, W. G. (1998) *J Cell Biol* 142, 1559–1569
155. Lam, M. H., Liu, Q., Elledge, S. J., and Rosen, J. M. (2004) *Cancer Cell* 6, 45–59
156. Brown, E. J. (2003) *Cell Cycle (Georgetown, Tex)*2, 188–189
157. Sandell, L. L., and Zakian, V. A. (1993) *Cell* 75, 729–739
158. Toczyski, D. P., Galgoczy, D. J., and Hartwell, L. H. (1997) *Cell* 90, 1097–1106
159. Lee, S. E., Moore, J. K., Holmes, A., Umezu, K., Kolodner, R. D., and Haber, J. E. (1998) *Cell* 94, 399–409
160. Xu, B., Kim, S. T., Lim, D. S., and Kastan, M. B. (2002) *Mol Cell Biol* 22, 1049–1059
161. Tang, X., Hui, Z. G., Cui, X. L., Garg, R., Kastan, M. B., and Xu, B. (2008) *Mol Cell Biol* 28, 2559–2566
162. Peng, C. Y., Graves, P. R., Thoma, R. S., Wu, Z., Shaw, A. S., and Piwnica-Worms, H. (1997) *Science (New York, NY)* 277, 1501–1505
163. Sanchez, Y., Wong, C., Thoma, R. S., Richman, R., Wu, Z., Piwnica-Worms, H., and Elledge, S. J. (1997) *Science (New York, NY)* 277, 1497–1501
164. Lopez-Girona, A., Furnari, B., Mondesert, O., and Russell, P. (1999) *Nature* 397, 172–175
165. Margolis, S. S., Perry, J. A., Forester, C. M., Nutt, L. K., Guo, Y., Jardim, M. J., Thomenius, M. J., Freel, C. D., Darbandi, R., Ahn, J. H., Arroyo, J. D., Wang, X. F., Shenolikar, S., Nairn, A. C., Dunphy, W. G., Hahn, W. C., Virshup, D. M., and Kornbluth, S. (2006) *Cell* 127, 759–773
166. Forester, C. M., Maddox, J., Louis, J. V., Goris, J., and Virshup, D. M. (2007) *Proc Natl Acad Sci USA* 104, 19867–19872
167. Chen, M. S., Ryan, C. E., and Piwnica-Worms, H. (2003) *Mol Cell Biol* 23, 7488–7497
168. Bassermann, F., Frescas, D., Guardavaccaro, D., Busino, L., Peschiaroli, A., and Pagano, M. (2008) *Cell* 134, 256–267

169. Ravingerova, T., Barancik, M., and Strniskova, M. (2003) *Mol Cell Biochem* 247, 127–138
170. Mitra, A. K., Singh, R. K., and Krishna, M. (2007) *Mol Cell Biochem* 294, 65–72
171. Bulavin, D. V., Higashimoto, Y., Popoff, I. J., Gaarde, W. A., Basrur, V., Potapova, O., Appella, E., and Fornace, A. J., Jr. (2001) *Nature* 411, 102–107
172. Manke, I. A., Nguyen, A., Lim, D., Stewart, M. Q., Elia, A. E., and Yaffe, M. B. (2005) *Mol Cell* 17, 37–48
173. Raman, M., Earnest, S., Zhang, K., Zhao, Y., and Cobb, M. H. (2007) *EMBO J* 26, 2005–2014
174. Reinhardt, H. C., Aslanian, A. S., Lees, J. A., and Yaffe, M. B. (2007) *Cancer Cell* 11, 175–189
175. Ferguson, A. M., White, L. S., Donovan, P. J., and Piwnica-Worms, H. (2005) *Mol Cell Biol* 25, 2853–2860
176. Mailand, N., Podtelejnikov, A. V., Groth, A., Mann, M., Bartek, J., and Lukas, J. (2002) *EMBO J* 21, 5911–5920
177. Zhao, H., Watkins, J. L., and Piwnica-Worms, H. (2002) *Proc Natl Acad Sci USA* 99, 14795–14800
178. Taylor, W. R., and Stark, G. R. (2001) *Oncogene* 20, 1803–1815
179. Bunz, F., Dutriaux, A., Lengauer, C., Waldman, T., Zhou, S., Brown, J. P., Sedivy, J. M., Kinzler, K. W., and Vogelstein, B. (1998) *Science (New York, NY)* 282, 1497–1501
180. Hermeking, H., Lengauer, C., Polyak, K., He, T. C., Zhang, L., Thiagalingam, S., Kinzler, K. W., and Vogelstein, B. (1997) *Mol Cell* 1, 3–11
181. Wang, X. W., Zhan, Q., Coursen, J. D., Khan, M. A., Kontny, H. U., Yu, L., Hollander, M. C., O'Connor, P. M., Fornace, A. J., Jr., and Harris, C. C. (1999) *Proc Natl Acad Sci USA* 96, 3706–3711
182. Chan, T. A., Hermeking, H., Lengauer, C., Kinzler, K. W., and Vogelstein, B. (1999) *Nature* 401, 616–620
183. Zhan, Q., Antinore, M. J., Wang, X. W., Carrier, F., Smith, M. L., Harris, C. C., and Fornace, A. J., Jr. (1999) *Oncogene* 18, 2892–2900
184. St Clair, S., Giono, L., Varmeh-Ziaie, S., Resnick-Silverman, L., Liu, W. J., Padi, A., Dastidar, J., DaCosta, A., Mattia, M., and Manfredi, J. J. (2004) *Mol Cell* 16, 725–736
185. Imbriano, C., Gurtner, A., Cocchiarella, F., Di Agostino, S., Basile, V., Gostissa, M., Dobbelstein, M., Del Sal, G., Piaggio, G., and Mantovani, R. (2005) *Mol Cell Biol* 25, 3737–3751
186. St Clair, S., and Manfredi, J. J. (2006) *Cell Cycle (Georgetown, Tex)* 5, 709–713
187. Piret, B., Schoonbroodt, S., and Piette, J. (1999) *Oncogene* 18, 2261–2271
188. Rocha, S., Garrett, M. D., Campbell, K. J., Schumm, K., and Perkins, N. D. (2005) *EMBO J* 24, 1157–1169
189. Wuerzberger-Davis, S. M., Chang, P. Y., Berchtold, C., and Miyamoto, S. (2005) *Mol Cancer Res* 3, 345–353
190. Wu, Z. H., Shi, Y., Tibbetts, R. S., and Miyamoto, S. (2006) *Science (New York, NY)* 311, 1141–1146
191. Shi, Y., Venkataraman, S. L., Dodson, G. E., Mabb, A. M., LeBlanc, S., and Tibbetts, R. S. (2004) *Proc Natl Acad Sci USA* 101, 5898–5903
192. Olofsson, B. A., Kelly, C. M., Kim, J., Hornsby, S. M., and Azizkhan-Clifford, J. (2007) *Mol Cancer Res* 5, 1319–1330
193. Iwahori, S., Yasui, Y., Kudoh, A., Sato, Y., Nakayama, S., Murata, T., Isomura, H., and Tsurumi, T. (2008) *Cell Signal* 20, 1795–1803
194. Lin, W. C., Lin, F. T., and Nevins, J. R. (2001) *Genes Devel* 15, 1833–1844
195. Lobrich, M., and Jeggo, P. A. (2007) *Nat Rev Cancer* 7, 861–869
196. Syljuasen, R. G. (2007) *Oncogene* 26, 5833–5839
197. Mailand, N., Bekker-Jensen, S., Bartek, J., and Lukas, J. (2006) *Mol Cell* 23, 307–318
198. Mamely, I., van Vugt, M. A., Smits, V. A., Semple, J. I., Lemmens, B., Perrakis, A., Medema, R. H., and Freire, R. (2006) *Curr Biol* 16, 1950–1955

199. Peschiaroli, A., Dorrello, N. V., Guardavaccaro, D., Venere, M., Halazonetis, T., Sherman, N. E., and Pagano, M. (2006) *Mol Cell* 23, 319–329
200. Lindon, C., and Pines, J. (2004) *J Cell Biol* 164, 233–241
201. Macurek, L., Lindqvist, A., Lim, D., Lampson, M. A., Klompmaker, R., Freire, R., Clouin, C., Taylor, S. S., Yaffe, M. B., and Medema, R. H. (2008) *Nature* 455, 119–123
202. Rass, U., Ahel, I., and West, S. C. (2007) *Cell* 130, 991–1004
203. Biton, S., Barzilai, A., and Shiloh, Y. (2008) *DNA Repair (Amst)* 7, 1028–1038
204. Yang, Y., and Herrup, K. (2005) *J Neurosci* 25, 2522–2529
205. Allen, D. M., van Praag, H., Ray, J., Weaver, Z., Winrow, C. J., Carter, T. A., Braquet, R., Harrington, E., Ried, T., Brown, K. D., Gage, F. H., and Barlow, C. (2001) *Genes Dev* 15, 554–566
206. Rimkus, S. A., Katzenberger, R. J., Trinh, A. T., Dodson, G. E., Tibbetts, R. S., and Wassarman, D. A. (2008) *Genes Dev* 22, 1205–1220
207. Becker, E. B., and Bonni, A. (2004) *Prog Neurobiol* 72, 1–25
208. Hollick, J. J., Rigoreau, L. J., Cano-Soumillac, C., Cockcroft, X., Curtin, N. J., Frigerio, M., Golding, B. T., Guiard, S., Hardcastle, I. R., Hickson, I., Hummersone, M. G., Menear, K. A., Martin, N. M., Matthews, I., Newell, D. R., Ord, R., Richardson, C. J., Smith, G. C., and Griffin, R. J. (2007) *J Med Chem* 50, 1958–1972
209. Hickson, I., Zhao, Y., Richardson, C. J., Green, S. J., Martin, N. M., Orr, A. I., Reaper, P. M., Jackson, S. P., Curtin, N. J., and Smith, G. C. (2004) *Cancer Res* 64, 9152–9159
210. Rainey, M. D., Charlton, M. E., Stanton, R. V., and Kastan, M. B. (2008) *Cancer Res* 68, 7466–7474
211. Blasina, A., Hallin, J., Chen, E., Arango, M. E., Kraynov, E., Register, J., Grant, S., Ninkovic, S., Chen, P., Nichols, T., O'Connor, P., and Anderes, K. (2008) *Mol Cancer Ther* 7, 2394–2404
212. Zabludoff, S. D., Deng, C., Grondine, M. R., Sheehy, A. M., Ashwell, S., Caleb, B. L., Green, S., Haye, H. R., Horn, C. L., Janetka, J. W., Liu, D., Mouchet, E., Ready, S., Rosenthal, J. L., Queva, C., Schwartz, G. K., Taylor, K. J., Tse, A. N., Walker, G. E., and White, A. M. (2008) *Mol Cancer Ther* 7, 2955–2966
213. Ashwell, S., Janetka, J. W., and Zabludoff, S. (2008) *Expert Opin Investig Drugs* 17, 1331–1340
214. Contour-Galcera, M. O., Sidhu, A., Prevost, G., Bigg, D., and Ducommun, B. (2007) *Pharmacol Ther* 115, 1–12

Chapter 6
Chromatin Modifications Involved in the DNA Damage Response to Double Strand Breaks

Julia Pagan, Emma Bolderson, Mathew Jones, and Kum Kum Khanna

Abstract In eukaryotes, genomic DNA is tightly compacted into a protein-DNA complex known as chromatin. This dense structure presents a barrier to DNA-dependent processes including transcription, replication and DNA repair. The repressive structure of chromatin is overcome by ATP-dependent chromatin remodelling complexes and chromatin-modifying enzymes. There is now ample evidence that DNA double-strand breaks (DSBs) elicit various histone modifications (such as acetylation, deacetylation, and phosphorylation) that function combinatorially to control the dynamic structure of the chromatin microenvironment. The role of these mechanisms during transcription and replication has been well studied, while the research into their impact on regulation of DNA damage response is rapidly gaining momentum. How chromatin structure is remodeled in response to DNA damage and how such alterations influence DSB repair are currently significant questions. This review will summarise the major chromatin modifications and chromatin remodelling complexes implicated in the DNA damage response to DSBs.

Keywords Chromatin · DNA damage · DNA repair · Histone modifications · Chromatin remodelling

6.1 Chromatin Structure

Within the nucleus of all eukaryotic cells, DNA is packaged by positively charged histone proteins (rich in lysine and arginine residues) which bind to the negatively charged phosphate backbone of DNA. The DNA bound histone complex is known as chromatin. The primary repeating unit of chromatin is the nucleosome, which comprises 147 base pairs wrapped around the core histone octamer. The histone octamer is composed of a heterotetramer of histones H3 and H4 surrounded by two heterodimers of histones H2A and H2B. The resulting nucleosomal array is stabilised

K.K. Khanna (✉)
Signal Transduction Laboratory, The Queensland Institute of Medical Research, Brisbane, Queensland 4006, Australia
e-mail: kumkum.khanna@qimr.edu.au

K.K. Khanna, Y. Shiloh (eds.), *The DNA Damage Response: Implications on Cancer Formation and Treatment*, DOI 10.1007/978-90-481-2561-6_6,

by binding of a linker histone, H1, to the core of each nucleosome, which locks the DNA at the entry and exit points from the nucleosome and plays an important role during chromatin condensation. Variation in this module can be introduced by the choice of histone variants which are assembled in specialised chromatin regions (reviewed in [55]).

Each histone within the nucleosome contains two important domains, a globular domain involved in histone-histone and histone-DNA interactions and the amino and carboxyl terminal tails which are targets of post-translational modifications. These modifications include phosphorylation on serine (S) and threonine (T) residues, acetylation, ubiquitination and sumoylation on lysine (K) residues and methylation of lysine (K) and arginine residues (R).

Chromatin remodeling during DNA damage detection and repair is dependent upon two broad groups of enzymes. Firstly, proteins that are classed as histone-modifying enzymes add a range of chemical residues to covalently modify chromatin. The histone-modifiying enzymes include histone acetyltransferases (HATs), histone deacetylases (HDACs), methyltransferases, demethylases, kinases, ubiquitin ligases and others. Secondly, ATP-dependent chromatin-remodellers alter the histone organisation and arrangement of nucleosomes. Histone modifications can directly affect chromatin structure by disrupting the charge of histone residues important for higher order chromatin structures. Alternatively, histone modifications can serve as recognition platforms for trans-acting remodelling factors [56]. These remodelling factors utilise ATP to elicit changes in the structure and function of the chromatin fiber [7, 56, 102].

6.2 Overview of DSB Repair Pathways

The purpose of this review is to illustrate the importance of chromatin context in the regulation of DSB repair and DSB response. Therefore, a brief description of DSB repair mechanisms is in order, a topic discussed more fully in chapter 8 by Williamson et al.

There are two principle methods of DNA DSB repair; non-homologous end-joining (NHEJ) [66] and homology directed repair (HDR) [36]. The principal DNA DSB repair mechanism in G_1 is NHEJ, during which DNA ends undergo minimal processing and are ligated directly together. In NHEJ the DSB is bound by the Ku70-Ku80 heterodimer and DNA-dependent kinase (DNA-PK). The two ends of the DNA are then ligated by the Lig4-XRCC4 complex. In contrast to NHEJ, HDR requires significant sequence homology and utilises a sister chromatid/chromosome as a template to repair DSBs. Therefore HDR functions in late S-G_2 phase. The Mre11-Rad50-Nbs1 (MRN) complex is implicated in sensing DNA DSBs in this method of repair and leads to resection of the DNA surrounding the DSB, generating single-stranded DNA (ssDNA). The ssDNA regions are bound to by the HDR machinery inducing the formation of Rad51 nucleoprotein filaments, which stimulate strand invasion, Holliday junction formation, branch migration and ligation of the DNA ends. Finally the heteroduplex is resolved, resulting in homozygosity in the damaged region [36].

The dynamic restructuring of chromatin surrounding the lesion is critically important to allow the accumulation of repair proteins to the lesion [57]. This remodelling occurs throughout large segments of chromatin flanking the damage site, allowing the HDR and NHEJ machinery access to the damaged DNA [43, 97, 105]. The "Access-Repair-Restore" model describes a process by which the DNA lesion is made accessible to the DNA damage repair machinery, facilitating DNA repair and eventually leading to the restoration of high-order chromatin organisation [98]. Modifications of histone tails and remodelling of chromatin by remodelling factors provide access of repair factors to the sites of DNA damage. Once the appropriate repair machinery has gained access and repair is completed, the nucleosomes are cleared of modifications and the chromatin environment is restored in order to cease checkpoint signalling.

6.3 Histone Modifications Associated with DNA Damage Repair

The covalent modifications of specific residues in histones by phosphorylation, acetylation and methylation have all been linked to DNA damage repair. These modifications alter the charge of specific residues thereby affecting histone-histone and histone-DNA interactions and can act as a signal for the recruitment of repair proteins to sites of DNA breaks.

6.3.1 Phosphorylation of H2AX

The most prominent chromatin modification after DSB induction is the phosphorylation of H2AX on its C-terminal tail at a conserved SQE motif, which plays a primary role in DNA damage repair by facilitating the access of repair proteins to sites of DNA breaks [86, 87]. The role of H2AX phosphorylation has been reviewed extensively (for more thorough reviews see [28, 52, 89]. The analogous phosphorylation occurs on yeast H2A at residue 129 in H2A (referred to from here on as H2AX for both yeast and mammals), also 4 amino acids from the carboxy terminus [22]. The phosphorylation occurs on the C-terminal tail domain which projects out toward the front of the nucleosome and contacts the linker DNA entering and exiting the nucleosome [85]. H2AX phosphorylation at the conserved SQE motif can be performed by phosphatidylinositol 3 kinase like kinases, including ATM, ATR or DNA-PK [11, 101, 116]. Detailed mapping of H2AX phosphorylation (γ-H2AX) has revealed that phosphorylation starts in the areas immediately surrounding the DSB, and extends up to 2 Mb in higher eukaryotes [86]. In yeast, γ-H2AX is detected 50 kb from the break. Chromatin immunoprecipitation (ChIP) analysis of γ-H2AX around a single inducible DSB site revealed that the highest level of γ-H2AX is present 3–5 kb from the break and was undetectable in the immediate proximity of the DSB [96, 109].

Currently it is believed that H2AX phosphorylation stabilises the interaction of DSB response proteins, such as 53BP1 [117], BRCA1 and Nbs1 at the repair site [12, 13, 83] and acts as an assembly platform to facilitate the accumulation of DNA damage response proteins onto damaged chromatin. The γ-H2AX phospho-epitope directly binds to the BRCT repeat of MDC1/NFBD1 [62, 103]. This complex formation regulates H2AX phosphorylation and is required accordingly for recruitment of DSB response proteins to flanking chromatin [5, 67, 68] and for normal radioresistance [103]. Cells lacking H2AX are able to undergo checkpoint activation and cell cycle arrest [13], but are unable to maintain arrest in the presence of persistent damage [26, 77]. This defective G_2/M checkpoint response is likely to be caused by the impaired accumulation of checkpoint factors such as MDC1, 53BP1 at DSBs, which serve as an amplification step at low levels of DNA damage.

H2AX phosphorylation has also been shown to recruit cohesin subunits [109], as well as chromatin remodelling factors like the NuA4 complex and Ino80 complex [21, 72, 112]. Together these studies demonstrate that phosphorylated H2AX, γ-H2AX, functions to recruit repair and chromatin remodelling proteins. However, these remodelling factors appear at DSBs much later than the very rapid phosphorylation of H2A, suggesting existence of additional mechanisms that dictate the recruitment of the chromatin remodelling proteins. Furthermore, attempts to mimic the role of γ-H2AX in yeast by substituting the phospho-serine residue with a phospho-mimicking residues in H2A, have failed to establish any role for this modification in actual chromatin rearrangement [29]. In addition, H2A phosphorylation and nucleosome eviction controlled by the Ino80 complex has been shown to occur independently [108]. Cells with inactivated Ino80, SWR1-C and NuA4 manifest greater sensitivity to DNA damage than H2A-S129A mutant, further suggesting that chromatin remodelling complexes function in a γ-H2AX independent manner in initial stages following DNA damage. Importantly, the chromatin remodelling factors themselves are required to allow H2AX phosphorylation since the inactivation of RSC (remodels the structure of chromatin) or SWI/SNF complex in mammalian cells causes a defect in H2AX phosphorylation [81].

H2AX-deficient mice resemble mice with chromosomal instability disorders in that they display radiosensitivity, male specific infertility, small size, chromosome instability and reduced levels of secondary immunoglobulin isotypes [13]. H2AX deficiency also modifies tumor susceptibility in mice [4, 13]. Loss of a single H2AX allele compromised genome integrity and enhanced cancer susceptibility in the absence of p53 [12]. The genomic instability and radiosensitivity in these heterozygous mice was rescued by restoring the null allele with wild type H2AX but not with alanine or glutamic acid phosphorylation sites mutants [12].

Studies have suggested that phosphorylation of H2AX contributes to both DSB-repair pathways (NHEJ and HDR). Substitution of the H2A S129 with Alanine in yeast led to increased sensitivity to DNA damage and a slight defect in NHEJ repair [22]. H2AX deficient cells exhibit impaired sister chromatid exchange and DNA-end joining [30, 122]. H2AX is required for efficient resolution of DSBs during class switch recombination [13]. H2AX has also been proposed to function to tether DNA ends together, thereby reducing the chances of illegitimate recombination events [3]. Although the precise mechanism is unclear, it is thought that H2AX might increase the local concentration of repair factors near the lesion [27].

Once DSB repair is completed, γ-H2AX must be converted to unphosphorylated H2A to restore the un-damaged chromatin structure. It is likely that the H2AX phosphorylation signal is removed by the combined action of histone exhange, possible through the FACT histone chaperone complex [37], and phosphatases. The Pph3 phosphatase enzyme regulates the dephosphorylation of H2AX in yeast following its removal from the chromatin [48]. Interestingly it appears that Pph3 is only able to dephosphosphorylate H2AX after it has been removed from chromatin. In mammalian cells, the phosphatases PP2A [18], PP4 [19, 76] and PP2Cgamma [51] have been shown to contribute to dephosphorylation of H2AX. The inability to dephosphorylate H2AX by inactivation of PP2A leads to defective DNA repair and failure to inactivate checkpoints [48]. In *Drosophila*, phosphorylated H2Av (the H2AX homologue) is removed by a mechanism involving acetylation by the Tip60/p400 complex and exchange with unmodified H2Av [60]. H2AX is normally protected from dephosphorylation through an interaction with its binding partner MDC1 [103].

6.3.2 Additional Histone Phosphorylation Events

In *Saccharomyces cerevisiae*, two additional residues on H2A (Ser-122 and Thr-126) are phosphorylated in vivo after DNA damage [35, 71]. Serine 122 mutants (threonine 119 in higher eukaryotes) are more hypersensitive to DNA damaging agents than the serine 129 mutants [35]. Phosphorylation of serine 122 may function in the same manner as γ-H2AX and act as a scaffolding platform for repair proteins or alter the chromatin structure to facilitate DNA repair. In support of a repair "histone code", Moore et al. found that distinct modification patterns were involved in different repair pathways [71].

In addition to H2A, other histones also undergo phosphorylation after DNA damage. Histone H2B is phosphorylated on serine 14 following DSB induction [25]. H2B serine 14 foci formation does require phosphorylation of H2AX, perhaps suggesting that γ-H2AX may be required to retain the kinase that is responsible for H2B phosphorylation at close proximity to the chromatin.

Histone H4 is also phosphorylated in response to DSBs at serine 1 by casein kinase II (CKII) [17, 110]. Serine 1 phosphorylation is enhanced in chromatin surrounding DSBs and was shown to play a role in NHEJ repair using a plasmid-based assay [17]. Histone H4 serine 1 phosphorylation has been implicated in restoration of chromatin following DNA repair. In a support of a role for serine 1 phosphorylation following DNA repair, the kinase responsible for this phosphorylation event, CKII binds to the Sin3-Rpd3 HDAC complex to facilitate late repair events. Secondly, the NuA4 Histone acetyl transferase (HAT) complex (Tip60 in human cells) required for DNA repair is inhibited by phosphorylation of H4 at serine 1

[8, 110]. Histone H4 serine 1 phosphorylation occurs after NuA4 action during double-strand break repair. The inhibition of NuA4 by H4 serine 1 phosphorylation is thought to re-establish chromatin modifications by preventing re-acetylation and turning off the DNA-damage signaling [110].

Recently, it has been discovered that Chk1 kinase constitutively phosphorylates H3 on Thr11, a novel chromatin modification correlated with transcriptional activation of *cyclin B* and *Cdk1* genes. Following DNA damage, Chk1 dissociates from chromatin, rapidly reducing phosphorylation of H3Thr11, and consequently reducing the expression of Cyclin B and Cdk1, through decreased binding of the GCN5 acetylase. H3Thr11 phosphorylation by Chk1 is considered to regulate DNA-damage-induced transcriptional repression [94].

6.4 Methylation of Histones

In addition to protein phosphorylation other covalent histone modifications also function during DNA repair. Histone H3 methylation at Lysine-79 (H3K79me) is recognised by the Tudor domain of the DNA repair protein 53BP1 [41]. Methylation at Lysine-79 is not induced after damage; instead the methylated residue becomes accessible after DNA damage due to damage-dependent distortion of the chromatin structure. This is an attractive hypothesis because it provides the DNA damage sensors with a mechanism, which allows rapid sensing of DNA damage.

Cells deficient in Dot1, the histone methyl-transferase responsible for H3K79 methylation are unable to form 53BP1 foci after damage [120]. In budding yeast, this process is conserved with Rad9, the 53BP1 equivalent, also recruited to methylated H3K79 [32]. In contrast, the fission yeast orthologue of 53BP1, Crb2, binds to methylated H4K20 residues [34, 88]. Mutations of Set9, which is required for H4K20me, cause hypersensitivity to different forms of DNA damage. Interestingly, 53BP1 in mammalian cells also binds to H4K20me directly [10]. The crystal structure of the 53BP1-H4K20me shows that 53BP1 contains a 5 amino-acid "cage" that specifically interacts with dimethyl residues, and not trimethylated residues [10], explaining the methylation state specific recognition of the histone by the Tudor domain. A recent report implicates dimethylated K20 of H4 in the recruitment of 53BP1 for XRCC4-dependent NHEJ, independent of H2AX. In contrast, the HDR response is dependent on the interaction of MDC1 with γ-H2AX, suggesting that distinct post-translational modifications on histones might contribute to the cells decision to undergo NHEJ or HDR [121].

6.5 Ubiquitination of Histones

Several recent papers have demonstrated the importance of the ubiquitin modification in the DNA damage response. Ubiquitin is a conserved 76 amino acid protein, which is conjugated to lysine residues in the process of ubiquitination. The ubiquitin tag can function as a signal for protein degradation in many instances, but is also known to regulate protein stability, function and/or the localization of a wide number of proteins. Both H2A and H2AX are polyubiquitinated in response to ionising radiation, catalysed by RNF8 RING finger ubiquitin ligase [40, 70] together with the E2 Ubc13 enzyme [54, 114]. RNF8 is rapidly recruited to sites of DNA damage in an MDC1-dependent manner through its functional FHA domain and RNF8 is required to recruit Ubc13. Cells lacking Ubc13 are unable to form ubiquitin-dependent foci in response to IR [124]. Rap80 is the mediator protein, which targets proteins to DSBs through its "ubiquitin-interaction motif", which recognizes polyubiquitin chains generated by RNF8 (Fig. 6.1). Through its "Abraxas Interaction Region" RAP80 recruits the Abraxas complex consisting of BRCA1, Abraxas and Brcc36 [115, 99, 49]. Thus, RNF8 is critically required for recruitment of RAP80, BRCA1, Abraxas, BRCC36 and 53BP1 to irradiation induced foci (Fig. 6.1) and is dispensable for the accumulation of early acting proteins such as γH2AX and MDC1 into foci. The depletion of ubiquitin itself results in a reduction in recruitment of 53BP1 and BRCA1 to foci [54, 70, 114]. Overall it appears that the RNF8 catalyzed ubiquitin modification does not lead to protein degradation because the primary ubiquitin linkage implicated in the Ubc13 mediated pathways is that of lys-63 linkage, rather than the lys-48, the canonical degradative signal. An interesting possibility is that RNF8 executed histone ubiquitination maintains genomic integrity by promoting the restructuring of chromatin at DSBs to facilitate the concentration of late acting repair factors including, p53BP1 and BRCA1 near the lesion (Fig. 6.1).

6.6 Histone Acetylation and Deacetylation

Many studies have also implicated histone acetylation in DNA repair, in essence by facilitating the access of repair proteins to the damaged region. The acetylation of lysine residues on histone proteins neutralises their basic charge and weakens histone-DNA binding. The affect of histone acetylation on chromatin structure is best demonstrated by its inhibitory effect on higher order chromatin structure [95]. In contrast, the hypoacetylation of histones decreases DNA accessibility and supports higher order chromatin structures [24].

In a repair context, dynamic changes in histone acetylation have been shown to correlate with homologous recombination [106]. The ability to regulate histone acetylation is essential for viability following homologous recombination. In yeast, mutations of modified lysine residues K5, K8, K12 and K16 in H4 led to increased sensitivity to DNA-damaging agents, suggesting that acetylation is necessary for DNA repair [8].

The turnover of histone acetylation is regulated by balancing the activity of acetyltransferases (HATs) and histone deacetylases (HDACs). In yeast, specific HATs including Hsa1 [8] Gcn [90] and Hat1 [84], are recruited to regions flanking DSBs. The recruitment of HATs following DSB induction correlates with the onset

Table 6.1 Histone modifications and their function

Histone	S. cerevisae Amino acid	Modification	Enzyme	Mammals Amino acid	Modification	Enzyme	Functions Regulated
H2A/X	S129	Phosphorylation	Mec1 Tel1	S139	Phosphorylation	ATM ATR DNA-PK	Condensation, Repair
H2B	S10	Phosphorylation	Ste20	S14	Phosphorylation	ND	ND
	K123	Ubiquitylation	Rad6 Bre1	K120	Ubiquitylation	RAD6	Checkpoint activation
H3	K56	Acetyation	Gcn5	ND	ND	ND	Repair
	K79	Methylation	Dot1	K79	Methylation	DOT1L	Repair
H4	S1	Phosphorylation	CK2	ND	ND	ND	Repair
	K5	Acetylation	Esa1	K5	Acetylation	TIP60	Repair
	K8	Acetylation	Esa1	K8	Acetylation	TIP60	Repair
	K12	Acetylation	Esa1	K12	Acetylation	TIP60	Repair
	K16	Acetylation	Esa1	K16	Acetylation	TIP60	Repair
	ND	ND	ND	K20	Di- or trimethylation	PR-SET7	Repair

ND, not determined
Adapted from Downs et al. [23]

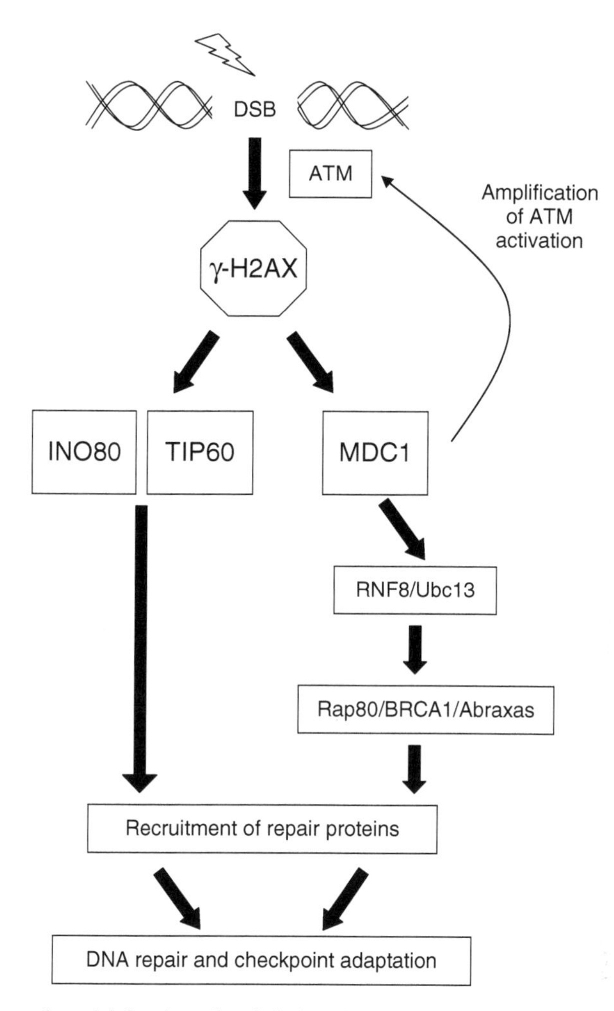

Fig. 6.1 Proposed model for the role of distinct histone modifications in response to DSBs. Following the acquisition of DNA DSBs in mammalian cells, ATM kinase is activated and histone H2AX is phosphorylated on S139. This phosphorylated residue directly binds the mediator protein MDC1, which is required for recruitment and stabilization of DSB proteins at the site of the lesion. MDC1 binds the RNF8/Ubc13 ubiquitin ligase responsible for the ubiquitination of H2A/H2AX. This ubiquitination event recruits the RAP80, via its ubiquitin-interaction motif. RAP80 is required for relocalization of the BRCA1/ABRAXAS complex to the site of DNA damage. Based on the delayed recruitment of 53BP1 to lesions, it has been proposed that MDC1 promotes the restructuring of chromatin to allow exposure of methylated-histones, which recruit 53BP1. Phosphorylated H2AX also binds directly the chromatin remodelling factors including INO80 and Tip60. These remodelling factors play crucial roles in both HDR and NHEJ pathway of DSB repair by recruiting repair proteins, facilitating the rejoining of DNA ends by ligases (NHEJ) and involvement in processes such as end resection and strand exchange for HDR

of acetylation on histone H3 K9, K14, K18, K23 and K27 residues and histone H4 K5, K8, K12 and K16 residues [106]. Deletion of four acetylated lysine residues in the N terminal tail of H4 in yeast results in hypersensitivity to DNA damaging agents. Indeed in yeast it has been observed that repair can be directed towards a replication-coupled repair pathway by acetylation of a single lysine residue on histone H4. In contrast, the acetylation of more than one lysine was required for repair of the DSB by NHEJ [8].

The mammalian Tip60 acetylation complex which acetylates H2A and H4 is recruited to DSBs where it is critically required for the accumulation of repair proteins at DNA lesions, demonstrating the need for actetylation in subsequent chromatin relaxation required for DSB repair. Furthermore, the relaxation of chromatin via salt treatments alleviates defective repair protein recruitment induced by lack of acetylation [75]. The involvement of histone acetylation and HAT complexes in DNA repair demonstrates that cells use the same basic mechanism to regulate different cellular processes, such as transcription and DNA repair, where the inhibitory structure of chromatin needs to be overcome.

The increased acetylation begins to subside two hours after DSB induction and this correlates with the arrival of HDACs including Sir2 [106], Rpd3 [110] and Hst1[106]. In mammalian cells, HDAC4 has been shown to colocalise in foci with 53BP1, which gradually disappeared in repair-proficient cells but persisted in repair-deficient cell lines, suggesting that resolution of HDAC4 foci is linked to repair [47a]. In yeast, the Sirtuin deacetylases participate in DNA repair [58]. Furthermore, it has been shown that histone deacetylation (at Lys16H4) is required for DNA repair [44]. The removal of acetylation from histones may result in a chromatin structure that enhances the likelihood of assembling an efficient repair complex, possibly by tethering the ends of the broken DNA together.

Histone acetylation can also directly recruit repair proteins. Acetylation was the first chromatin modification shown to directly bind a protein module [20]. Acetyl lysine residues bind specifically to bromodomain containing proteins. Bromodomains are highly conserved protein modules found in many chromatin associated proteins and nearly all known HATs [45]. The acetyl-lysine specificity targets bromodomain containing proteins to acetylated chromatin regions to facilitate chromatin remodeling [118]. Proteins containing bromodomains such as Esa1 and its mammalian homologue Tip60 may recognise specific combinations of acetylated histone residues and therefore be recruited to sites of DSBs [123].

In order for the highly ordered state of the chromatin to be restored following DNA repair, nucleosomes have to be reassembled in all newly replicated regions of the DNA. Chromatin reassembly has been shown to be dependent upon the acetylation of histone H3 on lysine 56 via the HAT activity of Rtt109 [16, 64]. The histone chaperone protein, Asf1, facilitates this event. Cells lacking ASF1 undergo normal resection of the DSB, including disassembly of chromatin and normal repair, however these cells are seriously defective in their ability to reassemble chromatin and ability to re-enter the cell cycle. A similar phenotype is observed when Rtt109 is deleted, demonstrating that H3K56 acetylation is required for chromatin assembly and checkpoint recovery after DNA repair. The reversal of lysine 56 acetylation is catalysed in budding yeast by the Hst3 and Hst4 (Class III) HDACs, which interestingly are down-regulated in response to DNA damage [69].

Inhibition of histone deacetylases leads to radiosensitivity in yeast [77a] and human cells (reviewed in [14]), although the mechanism is unclear. The inhibition of the repair of radiation-induced DNA damage may be involved [14] or transcription regulation of target genes might contribute to HDACi inhibitor induced radiosensitivity. Alternatively, the HDAC inhibitors might induce radiosensitivity through the increased accessibility of acetylated DNA resulting in susceptibility of chromatin to acquire DNA lesions. It is clear that the level of global chromatin compaction can influence the extent of DNA damage [74].

6.7 Recruitment of Chromatin Remodelling Factors

In addition to the histone-modifying proteins, there is a second class of proteins, the ATP-dependent chromatin-remodellers that influence the structure of chromatin (reviewed in [79]). These proteins disrupt histone-DNA interactions using energy derived from ATP hydrolysis to alter nucleosome structure. The action of chromatin-remodellers in vivo is not completely understood, however like the histone modifiers this group of proteins are suggested to facilitate the access to the chromatin of proteins involved in repair, transcription and replication.

DSB repair has been shown to be dependent upon several distinct ATP-dependent chromatin-remodelling complexes. Much of the work on these factors has been carried out on yeast mutants, using an assay in which a single DSB can be induced at the MAT (yeast mating type) locus via galactose-induced expression of the HO endonuclease [63]. If there are HMRa or HMRα donor sequences present at the site of DSB, the lesion will be repaired by HDR, if they are absent NHEJ will repair the DSB. This assay is routinely used to study DSB repair as the break can be formed in nearly 100% of cells. This method in combination with chromatin immunoprecipitation allows extensive examination of the recruitment of DNA repair proteins and chromatin remodelling factors.

6.7.1 SWI/SNF

SWI/SNF, a multi-subunit complex is a well-established member of the group of ATP-dependent chromatin remodelling complexes. The activity of the SWI/SNF complex has been shown to be required for DNA DSB repair, particularly in homologous recombination [15]. In yeast, the SWI/SNF complex contains eleven subunits including the catalytic subunit swi2/snf2. Mutations that disrupt the catalytic activity of SWI/SNF complex lead to hypersensitivity to DNA damaging agents in both human and yeast cells [6, 15]. In addition, the SWI/SNF complex is recruited to an HO-endonuclease induced break in budding yeast *S. cerevisiae* [15]. Interestingly, SWI/SNF displays different recruitment kinetics than other chromatin remodellers, arguing against redundancy between chromatin remodellers. In vitro the chromatin remodelling properties of SWI/SNF include altering interactions between DNA and histones to move and remodel nucleosomes and transfer histone dimers, but its role in vivo remains unclear.

The SWI/SNF complex is required to repair HO-induced DSBs at the MATa locus, strongly implicating it in the HDR pathway. Consistent with this, the SWI/SNF complex interacts with the HMLα HDR donor locus and MATα recipient locus and is required for strand invasion to occur during repair of HO-induced breaks [15]. It has also been suggested that it may recruit the HDR proteins Rad51/52 to sites of DSBs [15]. The precise role of SWI/SNF in HDR is unknown, but it is thought that it may remodel compact chromatin during strand invasion thereby facilitating interactions between repair proteins and DNA [113]. SWI/SNF has also been shown to play a role in VDJ recombination by stimulating RAG1/RAG2 cleavage in vitro and associating with loci undergoing rearrangement in vivo [61, 82]. Inactivation of SWI/SNF in mammalian cells leads to defective H2AX phosphorylation, foci formation and DNA repair, suggesting that the chromatin remodelling activity of the complex may have a substantial role in DNA repair [81].

6.7.2 The INO80 Remodelling Complex

The INO80 chromatin remodelling complex is a large, abundant, conserved multi-subunit complex with roles in transcription and DNA repair (reviewed in [78]). This complex is comprised of at least fifteen principal components including INO80 [90], actin-related proteins (Arp4, Arp5, Arp8, Nhp10) and two putative helicases, Rvb1 and Rvb2, RuvB like proteins, which are homologs of bacterial RuvB, the Holliday Junction resolvase [107]. The human INO80 complex also contains five additional, unique subunits [47]. The INO80 complex displays various catalytic activities including 3'–5' helicase activity, nucleosome sliding and DNA-dependent ATPase activity, however the actual remodelling activity of the complex remains unclear [90].

Evidence from several studies has implicated INO80 in DNA repair. The deletion of ATPase activity of INO80, Arp5 or Arp8 results in the loss of catalytic activity of the complex, which sensitises cells to DNA damaging agents [90, 91, 112]. In addition, many of the subunits from this remodelling complex are also recruited to ~1.6 kb region in the vicinity of HO-endonuclease induced DSBs at the *MAT* locus dependent on the presence of phosphorylated H2A [21, 72, 112]. Accordingly, the recruitment of INO80 to sites of DSBs is greatly reduced in strains that lack Mec1 (ATR homolog) and Tel1 (ATM homolog) kinases, as well as strains carrying a mutant of H2A that cannot be phosphorylated. Consistent with this, mammalian INO80 has been shown to bind γ-H2AX [72].

The INO80 complex is implicated in both NHEJ and HDR DSB repair pathways. Mutants deficient in the INO80 components *Arp5Δ* and *arp8Δ* are hypersensitive to DSBs when the HRR pathway is defective, suggesting they are required for the correct function of the NHEJ pathway [112]. INO80 has also been implicated in the HDR pathway in several model systems, though its precise role remains elusive [31,

108, 119]. In support for a role in the HDR pathway, INO80 has been implicated in the recruitment of repair factors, Rad51 and Rad52, which are required for the formation of the nucleo-protein filaments during strand invasion [108]. It is possible that this recruitment is dependent on INO80's demonstrated role in histone eviction. Defective nucleosome depletion around a DSB has been observed in INO80 deficient yeast strains [108] and mammalian systems [57]. In light of the above, it is proposed that the INO80 complex eliminates histones from around the DSB, allowing DNA resection and facilitating HDR. INO80 has also been implicated in evicting γ-H2AX allowing it to be dephosphorylated by the Pph3 phosphatase and thereby terminating the DNA damage signal [79, 108]. It is unclear if a histone chaperone is involved in the nucleosome disassembly activity of INO80, but one potential contender is ASF1 which has been previously assigned a role in DNA repair [73]. It should be noted however that in a study by Papamichos-Chronakis and colleagues the INO80 complex was found to play only a minor, if any, role in HDR and NHEJ repair [80].

INO80 has also been implicated in the generation of ssDNA via involvement in 3'–5' DNA resection that occurs prior to strand invasion during HDR, whereby mutants in INO80 are unable to recruit the Mre11 nuclease and show a lack of ssDNA formation [111]. The INO80 complex has been suggested to regulate checkpoint adaptation [80], although at present it is unclear whether this is due to the role in repair resolution or an additional function.

6.7.3 Remodels the Structure of Chromatin (RSC)

The RSC complex is comprised of fifteen components and like SWI/SNF is implicated in chromatin remodelling to facilitate DNA repair. RSC is essential for viability in yeast [6]. RSC mutants without catalytic activity are hypersensitive to DNA damaging agents, like SWI/SNF mutants and exhibit defective DNA repair via both NHEJ and HDR pathways [15, 93]. Like other DNA repair proteins RSC is rapidly recruited to a HO-induced DSB in an Mre11-dependent manner [93]. In turn, RSC is required for the recruitment of Mre11 to DSBs [92]. Once at the DSB the RSC together with the Mre11/Rad50/Xrs2 (MRX) complex is postulated to remodel chromatin and maintain H2AX phosphorylation. In addition, it has been shown that components of the RSC complex interact with Tel1/ATM and Mec1/ATR following DNA damage and are required for their recruitment to sites of DSBs and subsequent H2AX phosphorylation [65]. Yeast mutants in the RSC component RSC2 display a defect in nucleosome displacement following DSB induction, suggesting that DSB-induced nucleosome remodelling is dependent upon the RSC complex [65].

In addition to this early role in DNA repair the RSC is suggested to have a role in the later events of HDR, possibly during strand invasion as RSC mutants are defective in the ligation of DNA ends before the completion of HDR [15]. RSC interacts with the NHEJ protein Ku80, which suggests that it may also have a role in

NHEJ. In support of a role in NHEJ, depletion of Sth1, the ATPase subunit of RSC, reduces phosphorylation of H2A and significantly reduces DNA end-joining [92]. Notably, RSC may load cohesin, onto chromosomes to keep chromosome arms in close proximity during repair [38, 39]. Therefore RSC may have a role at multiple steps during the detection and repair of DSB.

6.7.4 Tip60/p400 and the NuA4 Complex

In higher eukaryotes the NuA4 complex consists of two types of chromatin remodellers, an ATP-dependent chromatin remodeller (p400) and an enzyme with HAT activity (Tip60). In yeast the NuA4 and SWR1 complexes appear to be the functional homologues of the mammalian NuA4 complex and contain the orthologues of p400 (Swr1) and Tip60 (Esa1).

Tip60 was first implicated in DNA repair when exogenous expression of a HAT-defective Tip60 mutant led to the accumulation of DSBs following IR-induced DNA damage [42]. Mammalian TIP60 has both acetylation and ATPase activities and contains a number of subunits homologous to subunits of both the SWR1 and NuA4 complex in yeast. Tip60 is likely to be involved in DNA-damage sensing, signalling and repair through several independent mechanisms. Several members of the NuA4 complex, including Arp4p, Epl1p, Eaf1p and Esa1p in yeast are recruited to HO-induced DSBs [8, 21]. In agreement with this, Tip60 also binds to the chromatin surrounding DSBs in mammalian cells [75].

The NuA4 subunit Arp4 is required for recruitment of the complex to DSBs and functions by recognising S129 phosphorylation of H2A in yeast [21]. The HAT component of the yeast NuA4 complex, Esa1 performs the earliest known acetylation event, following a DSB on the N-terminal tail of histone H4 [8, 21, 110]. This occurs within 1 h of induction of a HO-induced DSB. The mammalian orthologue of Esa1, Tip60 has also been shown to acetylate histone H4 which is suggested to facilitate the loading of various repair proteins, including Brca1, 53bp1 and Rad51 onto DNA [75]. It should be noted however that the recruitment of some DNA repair proteins, including Nbs1 and Mdc1 is not dependent upon the acetylation status of chromatin.

Besides the role in the recruitment of DNA repair factors to the site of DSBs, Tip60 is also required to regulate the phosphorylation of H2AX in order to switch off DNA damage signalling. The chromatin remodelling proteins Rvb1 and Rvb2 are members of various remodeling complexes including INO80 and NuA4. The HAT activity of TIP60 and dephosphorylation of γ-H2AX has recently been shown to be regulated by Rvb1 and Rvb2 [46]. In support of a role for Tip60 in regulating the removal of H2AX phosphorylation, depletion of Tip60 from human cells leads to the accumulation of γ-H2AX on chromatin [46]. In addition to studies in human cells, Tip60 has been shown to exchange phosphorylated H2Av for unmodified H2Av in Drosophila [59]. Together, these studies suggest that Tip60 is required for the removal and dephosphorylation of H2AX and subsequently the attenuation of the DNA damage signal, restoring order to the chromatin. Interestingly ATM is acetylated by Tip60 and this acetylation event seems to be essential for the activation of this DNA damage kinase [104], supporting the hypothesis that the activity of Tip60 might also be coupled to sensing and signalling of DNA damage response.

6.8 Recent Advances in the Chromatin-Repair Field

Recent work has demonstrated that the modulation of chromatin structure is required to activate DNA damage response pathways. Treatment of cells with agents that alter the chromatin structure leads to activation of the DNA damage response pathway in the absence of DSBs [2]. Consistent with this, the DNA damage response can be activated in the absence of exogenous DNA damage by the tethering of DNA damage response proteins to chromatin, demonstrating the importance of chromatin as a scaffold in the activation and amplification of the DNA damage response. This was recently shown in elegant studies by two separate groups, in mouse cells [100] and yeast cells [9]. Both groups incorporated an integrated array of Lac operators (LacO), the binding site for the Lac Repressor, into the genome of respective model system. Fusion of DSB response proteins to the Lac Repressor tethered these proteins to the chromatin in the Lac array thereby mimicking the accumulation of repair and signaling proteins in the absence of any lesion. Together, these studies suggest that the DNA damage response is not absolutely dependent upon DNA damage. In yeast, the simultaneous tethering of 9-1-1 complex and Ddc2 to the LacO array is required for checkpoint activation. In mammalian cells, the accumulation of any one of MDC1, Mre11, Nbs1 or ATM is sufficient to achieve checkpoint activation. This work extends the seminal discovery by Bakkenist and Kastan [2], as discussed in the chapter by Daniel Durocher, that the ATM kinase can be activated by agents that disrupt global chromatin structure in the absence of any discernible DNA damage [2, 53]. Whether ATM directly senses the disturbance in chromatin structure or requires an unidentified DSB sensor to transmit the signal is thus far unclear.

New studies have revealed details about how the DNA damage response signals to key chromatin regulators. New molecular players involved in regulating the transient increase in global chromatin accessibility following induction of DSBs include the Kap1 corepressor, which is phosphorylated by ATM on S824 [125]. Knocking down the expression of Kap1 or overexpression of a Kap1 mutant that mimics constitutive phosphorylation of Kap1 leads to chromatin relaxation, while expression of mutant Kap1 that cannot be phosphorylated abrogates DNA damage-induced chromatin relaxation. ATM-mediated phosphorylation of Kap1 may also be involved in increasing the accessibility of heterochromatin to repair proteins [33]. Notably, it has been shown that DSBs induced dissociation of HP1-beta from heterochromatin is required for H2AX phosphorylation surrounding the DNA lesion, uncovering further molecular details regarding the accessibility of heterochromatin to repair factors [1]. Further studies have shown that new players like the nucleosome-binding protein, HMGN1, modulate the interaction of ATM with chromatin before and after

DSB induction, thereby optimising its activation [50]and thus revealing that the DNA damage response is influenced by chromatin structure. The presumed hypothesis is that chromatin modulation is an integral part of the DNA damage response and the relaxation of chromatin facilitates surveillance of the genome and access of DNA damage sensors and repair proteins to DSBs. Future studies clarifying the role of these proteins, and other chromatin remodelling factors, will be helpful in deciphering the epigenetic mechanisms that control genome stability and the DNA damage response.

6.9 Conclusions

It is clear that chromatin restructuring in response to DNA damage is essential for initiating, propagating and terminating DSB repair and may even precede DNA end resection. The same chromatin remodelling machineries and histone modifications that are involved in transcription are also exploited for effective DNA repair. Amongst the different histone modifications phosphorylation of all four histones as well as the variant H2AX play an important role in DNA repair by facilitating the access of repair proteins to DSB sites. The challenge is to better understand the regulation of the factors that remodel chromatin during the DNA damage response. Further studies are needed to understand the full nature of chromatin changes and histone modifications induced by DNA damage and how these modifications act in a combinatorial fashion as a histone code to regulate DNA damage response, including initiation, propagation and termination of repair response or involvement in the checkpoint adaptation. The array of histone modifications could signal within the cell cycle stage and chromatin context for the appropriate functional outcome. A more thorough knowledge of the histone code at DSBs will be key to determine whether distinct modifications and chromatin remodelling machines favour specific repair pathways.

References

1. Ayoub, N., A.D. Jeyasekharan, J.A. Bernal, and A.R. Venkitaraman. 2008. HP1-beta mobilization promotes chromatin changes that initiate the DNA damage response. *Nature*. 453:682–6.
2. Bakkenist, C.J., and M.B. Kastan. 2003. DNA damage activates ATM through intermolecular autophosphorylation and dimer dissociation. *Nature*. 421:499–506.
3. Bassing, C.H., and F.W. Alt. 2004. H2AX may function as an anchor to hold broken chromosomal DNA ends in close proximity. *Cell Cycle*. 3:149–53.
4. Bassing, C.H., K.F. Chua, J. Sekiguchi, H. Suh, S.R. Whitlow, J.C. Fleming, B.C. Monroe, D.N. Ciccone, C. Yan, K. Vlasakova, D.M. Livingston, D.O. Ferguson, R. Scully, and F.W. Alt. 2002. Increased ionizing radiation sensitivity and genomic instability in the absence of histone H2AX. *Proc Natl Acad Sci USA*. 99:8173–8.
5. Bekker-Jensen, S., C. Lukas, R. Kitagawa, F. Melander, M.B. Kastan, J. Bartek, and J. Lukas. 2006. Spatial organization of the mammalian genome surveillance machinery in response to DNA strand breaks. *J Cell Biol*. 173:195–206.

6. Bennett, C.B., L.K. Lewis, G. Karthikeyan, K.S. Lobachev, Y.H. Jin, J.F. Sterling, J.R. Snipe, and M.A. Resnick. 2001. Genes required for ionizing radiation resistance in yeast. *Nat Genet*. 29:426–34.
7. Bernstein, B.E., A. Meissner, and E.S. Lander. 2007. The mammalian epigenome. *Cell*. 128:669–81.
8. Bird, A.W., D.Y. Yu, M.G. Pray-Grant, Q. Qiu, K.E. Harmon, P.C. Megee, P.A. Grant, M.M. Smith, and M.F. Christman. 2002. Acetylation of histone H4 by Esa1 is required for DNA double-strand break repair. *Nature*. 419:411–5.
9. Bonilla, C.Y., J.A. Melo, and D.P. Toczyski. 2008. Colocalization of sensors is sufficient to activate the DNA damage checkpoint in the absence of damage. *Mol Cell*. 30:267–76.
10. Botuyan, M.V., J. Lee, I.M. Ward, J.E. Kim, J.R. Thompson, J. Chen, and G. Mer. 2006. Structural basis for the methylation state-specific recognition of histone H4-K20 by 53BP1 and Crb2 in DNA repair. *Cell*. 127:1361–73.
11. Burma, S., B.P. Chen, M. Murphy, A. Kurimasa, and D.J. Chen. 2001. ATM phosphorylates histone H2AX in response to DNA double-strand breaks. *J Biol Chem*. 276:42462–7.
12. Celeste, A., O. Fernandez-Capetillo, M.J. Kruhlak, D.R. Pilch, D.W. Staudt, A. Lee, R.F. Bonner, W.M. Bonner, and A. Nussenzweig. 2003. Histone H2AX phosphorylation is dispensable for the initial recognition of DNA breaks. *Nat Cell Biol*. 5:675–9.
13. Celeste, A., S. Petersen, P.J. Romanienko, O. Fernandez-Capetillo, H.T. Chen, O.A. Sedelnikova, B. Reina-San-Martin, V. Coppola, E. Meffre, M.J. Difilippantonio, C. Redon, D.R. Pilch, A. Olaru, M. Eckhaus, R.D. Camerini-Otero, L. Tessarollo, F. Livak, K. Manova, W.M. Bonner, M.C. Nussenzweig, and A. Nussenzweig. 2002. Genomic instability in mice lacking histone H2AX. *Science*. 296:922–7.
14. Cerna, D., K. Camphausen, and P.J. Tofilon. 2006. Histone deacetylation as a target for radiosensitization. *Curr Top Dev Biol*. 73:173–204.
15. Chai, B., J. Huang, B.R. Cairns, and B.C. Laurent. 2005. Distinct roles for the RSC and Swi/Snf ATP-dependent chromatin remodelers in DNA double-strand break repair. *Genes Dev*. 19:1656–61.
16. Chen, C.C., J.J. Carson, J. Feser, B. Tamburini, S. Zabaronick, J. Linger, and J.K. Tyler. 2008. Acetylated lysine 56 on histone H3 drives chromatin assembly after repair and signals for the completion of repair. *Cell*. 134:231–43.
17. Cheung, W.L., F.B. Turner, T. Krishnamoorthy, B. Wolner, S.H. Ahn, M. Foley, J.A. Dorsey, C.L. Peterson, S.L. Berger, and C.D. Allis. 2005. Phosphorylation of histone H4 serine 1 during DNA damage requires casein kinase II in S. cerevisiae. *Curr Biol*. 15:656–60.
18. Chowdhury, D., M.C. Keogh, H. Ishii, C.L. Peterson, S. Buratowski, and J. Lieberman. 2005. gamma-H2AX dephosphorylation by protein phosphatase 2A facilitates DNA double-strand break repair. *Mol Cell*. 20:801–9.
19. Chowdhury, D., X. Xu, X. Zhong, F. Ahmed, J. Zhong, J. Liao, D.M. Dykxhoorn, D.M. Weinstock, G.P. Pfeifer, and J. Lieberman. 2008. A PP4-phosphatase complex dephosphorylates gamma-H2AX generated during DNA replication. *Mol Cell*. 31:33–46.
20. Dhalluin, C., J.E. Carlson, L. Zeng, C. He, A.K. Aggarwal, and M.M. Zhou. 1999. Structure and ligand of a histone acetyltransferase bromodomain. *Nature*. 399:491–6.
21. Downs, J.A., S. Allard, O. Jobin-Robitaille, A. Javaheri, A. Auger, N. Bouchard, S.J. Kron, S.P. Jackson, and J. Cote. 2004. Binding of chromatin-modifying activities to phosphorylated histone H2A at DNA damage sites. *Mol Cell*. 16:979–90.
22. Downs, J.A., N.F. Lowndes, and S.P. Jackson. 2000. A role for Saccharomyces cerevisiae histone H2A in DNA repair. *Nature*. 408:1001–4.
23. Downs, J.A., M.C. Nussenzweig and A. Nussenzweig. 2007. Chromatin dynamics and the preservation of genetic information. Nature 47:951–958.
24. Eberharter, A., and P.B. Becker. 2002. Histone acetylation: a switch between repressive and permissive chromatin. Second in review series on chromatin dynamics. *EMBO Rep*. 3:224–9.
25. Fernandez-Capetillo, O., C.D. Allis, and A. Nussenzweig. 2004a. Phosphorylation of histone H2B at DNA double-strand breaks. *J Exp Med*. 199:1671–7.

26. Fernandez-Capetillo, O., H.T. Chen, A. Celeste, I. Ward, P.J. Romanienko, J.C. Morales, K. Naka, Z. Xia, R.D. Camerini-Otero, N. Motoyama, P.B. Carpenter, W.M. Bonner, J. Chen, and A. Nussenzweig. 2002. DNA damage-induced G2-M checkpoint activation by histone H2AX and 53BP1. *Nat Cell Biol.* 4:993–7.
27. Fernandez-Capetillo, O., A. Lee, M. Nussenzweig, and A. Nussenzweig. 2004b. H2AX: the histone guardian of the genome. *DNA Repair (Amst).* 3:959–67.
28. Fillingham, J., M.C. Keogh, and N.J. Krogan. 2006. GammaH2AX and its role in DNA double-strand break repair. *Biochem Cell Biol.* 84:568–77.
29. Fink, M., D. Imholz, and F. Thoma. 2007. Contribution of the serine 129 of histone H2A to chromatin structure. *Mol Cell Biol.* 27:3589–600.
30. Franco, S., M. Gostissa, S. Zha, D.B. Lombard, M.M. Murphy, A.A. Zarrin, C. Yan, S. Tepsuporn, J.C. Morales, M.M. Adams, Z. Lou, C.H. Bassing, J.P. Manis, J. Chen, P.B. Carpenter, and F.W. Alt. 2006. H2AX prevents DNA breaks from progressing to chromosome breaks and translocations. *Mol Cell.* 21:201–14.
31. Fritsch, O., G. Benvenuto, C. Bowler, J. Molinier, and B. Hohn. 2004. The INO80 protein controls homologous recombination in Arabidopsis thaliana. *Mol Cell.* 16:479–85.
32. Giannattasio, M., F. Lazzaro, P. Plevani, and M. Muzi-Falconi. 2005. The DNA damage checkpoint response requires histone H2B ubiquitination by Rad6-Bre1 and H3 methylation by Dot1. *J Biol Chem.* 280:9879–86.
33. Goodarzi, A.A., A.T. Noon, D. Deckbar, Y. Ziv, Y. Shiloh, M. Lobrich, and P.A. Jeggo. 2008. ATM signaling facilitates repair of DNA double-strand breaks associated with heterochromatin. *Mol Cell.* 31:167–77.
34. Greeson, N.T., R. Sengupta, A.R. Arida, T. Jenuwein, and S.L. Sanders. 2008. Di-methyl H4 lysine 20 targets the checkpoint protein CRB2 to sites of DNA damage. *J Biol Chem* 283(48):33168–74.
35. Harvey, A.C., S.P. Jackson, and J.A. Downs. 2005. Saccharomyces cerevisiae histone H2A Ser122 facilitates DNA repair. *Genetics.* 170:543–53.
36. Helleday, T., J. Lo, D.C. van Gent, and B.P. Engelward. 2007. DNA double-strand break repair: from mechanistic understanding to cancer treatment. *DNA Repair (Amst).* 6:923–35.
37. Heo, K., H. Kim, S.H. Choi, J. Choi, K. Kim, J. Gu, M.R. Lieber, A.S. Yang, and W. An. 2008. FACT-mediated exchange of histone variant H2AX regulated by phosphorylation of H2AX and ADP-ribosylation of Spt16. *Mol Cell.* 30:86–97.
38. Huang, J., J.M. Hsu, and B.C. Laurent. 2004. The RSC nucleosome-remodeling complex is required for Cohesin′s association with chromosome arms. *Mol Cell.* 13:739–50.
39. Huang, J., and B.C. Laurent. 2004. A Role for the RSC chromatin remodeler in regulating cohesion of sister chromatid arms. *Cell Cycle.* 3:973–5.
40. Huen, M.S., R. Grant, I. Manke, K. Minn, X. Yu, M.B. Yaffe, and J. Chen. 2007. RNF8 transduces the DNA-damage signal via histone ubiquitylation and checkpoint protein assembly. *Cell.* 131:901–14.
41. Huyen, Y., O. Zgheib, R.A. Ditullio, Jr., V.G. Gorgoulis, P. Zacharatos, T.J. Petty, E.A. Sheston, H.S. Mellert, E.S. Stavridi, and T.D. Halazonetis. 2004. Methylated lysine 79 of histone H3 targets 53BP1 to DNA double-strand breaks. *Nature.* 432:406–11.
42. Ikura, T., V.V. Ogryzko, M. Grigoriev, R. Groisman, J. Wang, M. Horikoshi, R. Scully, J. Qin, and Y. Nakatani. 2000. Involvement of the TIP60 histone acetylase complex in DNA repair and apoptosis. *Cell.* 102:463–73.
43. Jaberaboansari, A., G.B. Nelson, J.L. Roti Roti, and K.T. Wheeler. 1988. Postirradiation alterations of neuronal chromatin structure. *Radiat Res.* 114:94–104.
44. Jazayeri, A., A.D. McAinsh and S.P. Jackson. 2004. Saccharomyces cerevisiae Sin3p facilitates DNA double-strand break repair. *Proc Natl Acad Sci USA* 101:1644–9.
45. Jeanmougin, F., J.M. Wurtz, B. Le Douarin, P. Chambon, and R. Losson. 1997. The bromodomain revisited. *Trends Biochem Sci.* 22:151–3.
46. Jha, S., E. Shibata, and A. Dutta. 2008. Human Rvb1/Tip49 is required for the histone acetyltransferase activity of Tip60/NuA4 and for the downregulation of phosphorylation on H2AX after DNA damage. *Mol Cell Biol.* 28:2690–700.

47. Jin, J., Y. Cai, T. Yao, A.J. Gottschalk, L. Florens, S.K. Swanson, J.L. Gutierrez, M.K. Coleman, J.L. Workman, A. Mushegian, M.P. Washburn, R.C. Conaway, and J.W. Conaway. 2005. A mammalian chromatin remodeling complex with similarities to the yeast INO80 complex. *J Biol Chem.* 280:41207–12.
47a. Kao, G.D., W.G. McKenna, M.G. Guenther, R.J. Muschel, M.A. Lazar, and T.J. Yen. 2003. Histone deacetylase 4 interacts with 53BP1 to mediate the DNA damage response. *J Cell Biol.* 160(7):1017–27.
48. Keogh, M.C., J.A. Kim, M. Downey, J. Fillingham, D. Chowdhury, J.C. Harrison, M. Onishi, N. Datta, S. Galicia, A. Emili, J. Lieberman, X. Shen, S. Buratowski, J.E. Haber, D. Durocher, J.F. Greenblatt, and N.J. Krogan. 2006. A phosphatase complex that dephosphorylates gammaH2AX regulates DNA damage checkpoint recovery. *Nature.* 439:497–501.
49. Kim, H., J. Chen, and X. Yu. 2007. Ubiquitin-binding protein RAP80 mediates BRCA1-dependent DNA damage response. *Science* 316:1202–5.
50. Kim, Y.C., G. Gerlitz, T. Furusawa, F. Catez, A. Nussenzweig, K.S. Oh, K.H. Kraemer, Y. Shiloh and M. Bustin. 2009. Activation of ATM depends on chromatin interactions occurring before induction of DNA damage. *Nat Cell Biol.* 11: 92–6.
51. Kimura, H., N. Takizawa, E. Allemand, T. Hori, F.J. Iborra, N. Nozaki, M. Muraki, M. Hagiwara, A.R. Krainer, T. Fukagawa, and K. Okawa. 2006. A novel histone exchange factor, protein phosphatase 2Cgamma, mediates the exchange and dephosphorylation of H2A-H2B. *J Cell Biol.* 175:389–400.
52. Kinner, A., W. Wu, C. Staudt, and G. Iliakis. 2008. Gamma-H2AX in recognition and signaling of DNA double-strand breaks in the context of chromatin. *Nucleic Acids Res.* 36:5678–94.
53. Kitagawa, R., C.J. Bakkenist, P.J. McKinnon, and M.B. Kastan. 2004. Phosphorylation of SMC1 is a critical downstream event in the ATM-NBS1-BRCA1 pathway. *Genes Dev.* 18:1423–38.
54. Kolas, N.K., J.R. Chapman, S. Nakada, J. Ylanko, R. Chahwan, F.D. Sweeney, S. Panier, M. Mendez, J. Wildenhain, T.M. Thomson, L. Pelletier, S.P. Jackson, and D. Durocher. 2007. Orchestration of the DNA-damage response by the RNF8 ubiquitin ligase. *Science.* 318:1637–40.
55. Kornberg, R.D., and Y. Lorch. 1999. Twenty-five years of the nucleosome, fundamental particle of the eukaryote chromosome. *Cell.* 98:285–94.
56. Kouzarides, T. 2007. Chromatin modifications and their function. *Cell.* 128:693–705.
57. Kruhlak, M.J., A. Celeste, G. Dellaire, O. Fernandez-Capetillo, W.G. Muller, J.G. McNally, D.P. Bazett-Jones, and A. Nussenzweig. 2006. Changes in chromatin structure and mobility in living cells at sites of DNA double-strand breaks. *J Cell Biol.* 172:823–34.
58. Kruszewski, M., and I. Szumiel. 2005. Sirtuins (histone deacetylases III) in the cellular response to DNA damage–facts and hypotheses. *DNA Repair (Amst).* 4:1306–13.
59. Kusch, T., L. Florens, W. Macdonald, S. Swanson, R. Glaser, J.r. Yates, S. Abmayr, M. Washburn, and J. Workman. 2004a. Acetylation by Tip60 is required for selective histone variant exchange at DNA lesions. *Science.* 306:2084–7.
60. Kusch, T., L. Florens, W.H. Macdonald, S.K. Swanson, R.L. Glaser, J.R. Yates, 3rd, S.M. Abmayr, M.P. Washburn, and J.L. Workman. 2004b. Acetylation by Tip60 is required for selective histone variant exchange at DNA lesions. *Science.* 306:2084–7.
61. Kwon, J., K.B. Morshead, J.R. Guyon, R.E. Kingston, and M.A. Oettinger. 2000. Histone acetylation and hSWI/SNF remodeling act in concert to stimulate V(D)J cleavage of nucleosomal DNA. *Mol Cell.* 6:1037–48.
62. Lee, M.S., R.A. Edwards, G.L. Thede, and J.N. Glover. 2005. Structure of the BRCT repeat domain of MDC1 and its specificity for the free COOH-terminal end of the gamma-H2AX histone tail. *J Biol Chem.* 280:32053–6.
63. Lee, S.E., J.K. Moore, A. Holmes, K. Umezu, R.D. Kolodner, and J.E. Haber. 1998. Saccharomyces Ku70, mre11/rad50 and RPA proteins regulate adaptation to G2/M arrest after DNA damage. *Cell.* 94:399–409.

64. Li, Q., H. Zhou, H. Wurtele, B. Davies, B. Horazdovsky, A. Verreault, and Z. Zhang. 2008. Acetylation of histone H3 lysine 56 regulates replication-coupled nucleosome assembly. *Cell*. 134:244–55.
65. Liang, B., J. Qiu, K. Ratnakumar, and B.C. Laurent. 2007. RSC functions as an early double-strand-break sensor in the cell's response to DNA damage. *Curr Biol*. 17:1432–7.
66. Lieber, M.R. 2008. The mechanism of human nonhomologous DNA end joining. *J Biol Chem*. 283:1–5.
67. Lou, Z., K. Minter-Dykhouse, S. Franco, M. Gostissa, M.A. Rivera, A. Celeste, J.P. Manis, J. van Deursen, A. Nussenzweig, T.T. Paull, F.W. Alt, and J. Chen. 2006. MDC1 maintains genomic stability by participating in the amplification of ATM-dependent DNA damage signals. *Mol Cell*. 21:187–200.
68. Lukas, C., F. Melander, M. Stucki, J. Falck, S. Bekker-Jensen, M. Goldberg, Y. Lerenthal, S.P. Jackson, J. Bartek, and J. Lukas. 2004. Mdc1 couples DNA double-strand break recognition by Nbs1 with its H2AX-dependent chromatin retention. *Embo J*. 23: 2674–83.
69. Maas, N.L., K.M. Miller, L.G. DeFazio, and D.P. Toczyski. 2006. Cell cycle and checkpoint regulation of histone H3 K56 acetylation by Hst3 and Hst4. *Mol Cell*. 23:109–19.
70. Mailand, N., S. Bekker-Jensen, H. Faustrup, F. Melander, J. Bartek, C. Lukas, and J. Lukas. 2007. RNF8 ubiquitylates histones at DNA double-strand breaks and promotes assembly of repair proteins. *Cell*. 131:887–900.
71. Moore, J.D., O. Yazgan, Y. Ataian, and J.E. Krebs. 2007. Diverse roles for histone H2A modifications in DNA damage response pathways in yeast. *Genetics*. 176:15–25.
72. Morrison, A.J., J. Highland, N.J. Krogan, A. Arbel-Eden, J.F. Greenblatt, J.E. Haber, and X. Shen. 2004. INO80 and gamma-H2AX interaction links ATP-dependent chromatin remodeling to DNA damage repair. *Cell*. 119:767–75.
73. Mousson, F., F. Ochsenbein, and C. Mann. 2007. The histone chaperone Asf1 at the crossroads of chromatin and DNA checkpoint pathways. *Chromosoma*. 116:79–93.
74. Murga, M., I. Jaco, Y. Fan, R. Soria, B. Martinez-Pastor, M. Cuadrado, S.M. Yang, M.A. Blasco, A.I. Skoultchi, and O. Fernandez-Capetillo. 2007. Global chromatin compaction limits the strength of the DNA damage response. *J Cell Biol*. 178:1101–8.
75. Murr, R., J.I. Loizou, Y.G. Yang, C. Cuenin, H. Li, Z.Q. Wang, and Z. Herceg. 2006. Histone acetylation by Trrap-Tip60 modulates loading of repair proteins and repair of DNA double-strand breaks. *Nat Cell Biol*. 8:91–9.
76. Nakada, S., G.I. Chen, A.C. Gingras, and D. Durocher. 2008. PP4 is a gammaH2AX phosphatase required for recovery from the DNA damage checkpoint. *EMBO Rep*. 9: 1019–26.
77. Nakamura, T.M., L.L. Du, C. Redon, and P. Russell. 2004. Histone H2A phosphorylation controls Crb2 recruitment at DNA breaks, maintains checkpoint arrest, and influences DNA repair in fission yeast. *Mol Cell Biol*. 24:6215–30.
77a. Nicolas, E., T. Yamada, H.P. Cam, P.C. Fitzgerald, R. Kobayashi, and S.I. Grewal. 2007. Nat Struct Mol Biol. 14(5):372–80.
78. Osley, M.A., and X. Shen. 2006. Altering nucleosomes during DNA double-strand break repair in yeast. *Trends Genet*. 22:671–7.
79. Osley, M.A., T. Tsukuda, and J.A. Nickoloff. 2007. ATP-dependent chromatin remodeling factors and DNA damage repair. *Mutat Res*. 618:65–80.
80. Papamichos-Chronakis, M., J.E. Krebs, and C.L. Peterson. 2006. Interplay between Ino80 and Swr1 chromatin remodeling enzymes regulates cell cycle checkpoint adaptation in response to DNA damage. *Genes Dev*. 20:2437–49.
81. Park, J.H., E.J. Park, H.S. Lee, S.J. Kim, S.K. Hur, A.N. Imbalzano, and J. Kwon. 2006. Mammalian SWI/SNF complexes facilitate DNA double-strand break repair by promoting gamma-H2AX induction. *Embo J*. 25:3986–97.
82. Patenge, N., S.K. Elkin, and M.A. Oettinger. 2004. ATP-dependent remodeling by SWI/SNF and ISWI proteins stimulates V(D)J cleavage of 5 S arrays. *J Biol Chem*. 279:35360–7.

83. Paull, T.T., E.P. Rogakou, V. Yamazaki, C.U. Kirchgessner, M. Gellert, and W.M. Bonner. 2000. A critical role for histone H2AX in recruitment of repair factors to nuclear foci after DNA damage. *Curr Biol.* 10:886–95.
84. Qin, S., and M.R. Parthun. 2006. Recruitment of the type B histone acetyltransferase Hat1p to chromatin is linked to DNA double-strand breaks. *Mol Cell Biol.* 26:3649–58.
85. Redon, C., D. Pilch, E. Rogakou, O. Sedelnikova, K. Newrock, and W. Bonner. 2002. Histone H2A variants H2AX and H2AZ. *Curr Opin Genet Dev.* 12:162–9.
86. Rogakou, E.P., C. Boon, C. Redon, and W.M. Bonner. 1999. Megabase chromatin domains involved in DNA double-strand breaks in vivo. *J Cell Biol.* 146:905–16.
87. Rogakou, E.P., D.R. Pilch, A.H. Orr, V.S. Ivanova, and W.M. Bonner. 1998. DNA double-stranded breaks induce histone H2AX phosphorylation on serine 139. *J Biol Chem.* 273:5858–68.
88. Sanders, S.L., M. Portoso, J. Mata, J. Bahler, R.C. Allshire, and T. Kouzarides. 2004. Methylation of histone H4 lysine 20 controls recruitment of Crb2 to sites of DNA damage. *Cell.* 119:603–14.
89. Sedelnikova, O.A., D.R. Pilch, C. Redon, and W.M. Bonner. 2003. Histone H2AX in DNA damage and repair. *Cancer Biol Ther.* 2:233–5.
90. Shen, X., G. Mizuguchi, A. Hamiche, and C. Wu. 2000. A chromatin remodelling complex involved in transcription and DNA processing. *Nature.* 406:541–4.
91. Shen, X., R. Ranallo, E. Choi, and C. Wu. 2003. Involvement of actin-related proteins in ATP-dependent chromatin remodeling. *Mol Cell.* 12:147–55.
92. Shim, E.Y., S.J. Hong, J.H. Oum, Y. Yanez, Y. Zhang, and S.E. Lee. 2007. RSC mobilizes nucleosomes to improve accessibility of repair machinery to the damaged chromatin. *Mol Cell Biol.* 27:1602–13.
93. Shim, E.Y., J.L. Ma, J.H. Oum, Y. Yanez, and S.E. Lee. 2005. The yeast chromatin remodeler RSC complex facilitates end joining repair of DNA double-strand breaks. *Mol Cell Biol.* 25:3934–44.
94. Shimada, M., H. Niida, D.H. Zineldeen, H. Tagami, M. Tanaka, H. Saito, and M. Nakanishi. 2008. Chk1 is a histone H3 threonine 11 kinase that regulates DNA damage-induced transcriptional repression. *Cell.* 132:221–32.
95. Shogren-Knaak, M., H. Ishii, J.M. Sun, M.J. Pazin, J.R. Davie, and C.L. Peterson. 2006. Histone H4-K16 acetylation controls chromatin structure and protein interactions. *Science.* 311:844–7.
96. Shroff, R., A. Arbel-Eden, D. Pilch, G. Ira, W.M. Bonner, J.H. Petrini, J.E. Haber, and M. Lichten. 2004. Distribution and dynamics of chromatin modification induced by a defined DNA double-strand break. *Curr Biol.* 14:1703–11.
97. Sidik, K., and M.J. Smerdon. 1990. Nucleosome rearrangement in human cells following short patch repair of DNA damaged by bleomycin. *Biochemistry.* 29:7501–11.
98. Smerdon, M.J. 1991. DNA repair and the role of chromatin structure. *Curr Opin Cell Biol.* 3:422–8.
99. Sobhian, B., G. Shao, D.R. Lilli, A.C. Culhane, L.A. Moreau, B. Xia, D.M. Livingston, and R.A. Greenberg. 2007. RAP80 targets BRCA1 to specific ubiquitin structures at DNA damage sites. *Science* 316:1198–202.
100. Soutoglou, E., and T. Misteli. 2008. Activation of the cellular DNA damage response in the absence of DNA lesions. *Science.* 320:1507–10.
101. Stiff, T., M. O'Driscoll, N. Rief, K. Iwabuchi, M. Lobrich, and P.A. Jeggo. 2004. ATM and DNA-PK function redundantly to phosphorylate H2AX after exposure to ionizing radiation. *Cancer Res.* 64:2390–6.
102. Strahl, B.D., and C.D. Allis. 2000. The language of covalent histone modifications. *Nature.* 403:41–5.
103. Stucki, M., J.A. Clapperton, D. Mohammad, M.B. Yaffe, S.J. Smerdon, and S.P. Jackson. 2005. MDC1 directly binds phosphorylated histone H2AX to regulate cellular responses to DNA double-strand breaks. *Cell.* 123:1213–26.

104. Sun, Y., X. Jiang, S. Chen, N. Fernandes, and B.D. Price. 2005. A role for the Tip60 histone acetyltransferase in the acetylation and activation of ATM. *Proc Natl Acad Sci USA*. 102:13182–7.
105. Takahashi, K., and I. Kaneko. 1985. Changes in nuclease sensitivity of mammalian cells after irradiation with 60Co gamma-rays. *Int J Radiat Biol Relat Stud Phys Chem Med*. 48:389–95.
106. Tamburini, B.A., and J.K. Tyler. 2005. Localized histone acetylation and deacetylation triggered by the homologous recombination pathway of double-strand DNA repair. *Mol Cell Biol*. 25:4903–13.
107. Tsaneva, I.R., G. Illing, R.G. Lloyd and S.C. West. 1992. Purification and properties of the RuvA + RuvB proteins *Escherichia coli*. *Mol Gen Genet*. 235:1–10.
108. Tsukuda, T., A.B. Fleming, J.A. Nickoloff, and M.A. Osley. 2005. Chromatin remodelling at a DNA double-strand break site in Saccharomyces cerevisiae. *Nature*. 438: 379–83.
109. Unal, E., A. Arbel-Eden, U. Sattler, R. Shroff, M. Lichten, J.E. Haber, and D. Koshland. 2004. DNA damage response pathway uses histone modification to assemble a double-strand break-specific cohesin domain. *Mol Cell*. 16:991–1002.
110. Utley, R.T., N. Lacoste, O. Jobin-Robitaille, S. Allard, and J. Cote. 2005. Regulation of NuA4 histone acetyltransferase activity in transcription and DNA repair by phosphorylation of histone H4. *Mol Cell Biol*. 25:8179–90.
111. van Attikum, H., O. Fritsch, and S.M. Gasser. 2007. Distinct roles for SWR1 and INO80 chromatin remodeling complexes at chromosomal double-strand breaks. *Embo J*. 26: 4113–25.
112. van Attikum, H., O. Fritsch, B. Hohn, and S.M. Gasser. 2004. Recruitment of the INO80 complex by H2A phosphorylation links ATP-dependent chromatin remodeling with DNA double-strand break repair. *Cell*. 119:777–88.
113. Varga-Weisz, P.D., and P.B. Becker. 2006. Regulation of higher-order chromatin structures by nucleosome-remodelling factors. *Curr Opin Genet Dev*. 16:151–6.
114. Wang, B., and S.J. Elledge. 2007. Ubc13/Rnf8 ubiquitin ligases control foci formation of the Rap80/Abraxas/Brca1/Brcc36 complex in response to DNA damage. *Proc Natl Acad Sci USA*. 104:20759–63.
115. Wang, B., S. Matsuoka, B.A. Ballif, D. Zhang, A. Smogorzewska, S.P. Gygi, and S.J. Elledge. 2007. Abraxas and RAP80 form a BRCA1 protein complex required for the DNA damage response. *Science* 316:1194–8.
116. Ward, I.M., and J. Chen. 2001. Histone H2AX is phosphorylated in an ATR-dependent manner in response to replicational stress. *J Biol Chem*. 276:47759–62.
117. Ward, I.M., K. Minn, K.G. Jorda, and J. Chen. 2003. Accumulation of checkpoint protein 53BP1 at DNA breaks involves its binding to phosphorylated histone H2AX. *J Biol Chem*. 278:19579–82.
118. Winston, F., and C.D. Allis. 1999. The bromodomain: a chromatin-targeting module? *Nat Struct Biol*. 6:601–4.
119. Wu, S., Y. Shi, P. Mulligan, F. Gay, J. Landry, H. Liu, J. Lu, H.H. Qi, W. Wang, J.A. Nickoloff, C. Wu, and Y. Shi. 2007. A YY1-INO80 complex regulates genomic stability through homologous recombination-based repair. *Nat Struct Mol Biol*. 14:1165–72.
120. Wysocki, R., A. Javaheri, S. Allard, F. Sha, J. Cote, and S.J. Kron. 2005. Role of Dot1-dependent histone H3 methylation in G1 and S phase DNA damage checkpoint functions of Rad9. *Mol Cell Biol*. 25:8430–43.
121. Xie, A., A. Hartlerode, M. Stucki, S. Odate, N. Puget, A. Kwok, G. Nagaraju, C. Yan, F.W. Alt, J. Chen, S.P. Jackson, and R. Scully. 2007. Distinct roles of chromatin-associated proteins MDC1 and 53BP1 in mammalian double-strand break repair. *Mol Cell*. 28: 1045–57.
122. Xie, A., N. Puget, I. Shim, S. Odate, I. Jarzyna, C.H. Bassing, F.W. Alt, and R. Scully. 2004. Control of sister chromatid recombination by histone H2AX. *Mol Cell*. 16: 1017–25.

123. Zeng, L., Q. Zhang, G. Gerona-Navarro, N. Moshkina, and M.M. Zhou. 2008. Structural basis of site-specific histone recognition by the bromodomains of human coactivators PCAF and CBP/p300. *Structure*. 16:643–52.
124. Zhao, G.Y., E. Sonoda, L.J. Barber, H. Oka, Y. Murakawa, K. Yamada, T. Ikura, X. Wang, M. Kobayashi, K. Yamamoto, S.J. Boulton, and S. Takeda. 2007. A critical role for the ubiquitin-conjugating enzyme Ubc13 in initiating homologous recombination. *Mol Cell*. 25:663–75.
125. Ziv, Y., D. Bielopolski, Y. Galanty, C. Lukas, Y. Taya, D.C. Schultz, J. Lukas, S. Bekker-Jensen, J. Bartek, and Y. Shiloh. 2006. Chromatin relaxation in response to DNA double-strand breaks is modulated by a novel ATM- and KAP-1 dependent pathway. *Nat Cell Biol*. 8:870–6.

Chapter 7
Telomere Metabolism and DNA Damage Response

Tej K. Pandita

Abstract Genomic stability is maintained by telomeres, the end terminal DNA protein structures that protect chromosomes from fusion or degradation. Telomeres are essential for the proper maintenance of chromosomes and may play a role in aging, cancer and several other diseases. Shortening or loss of telomeric DNA repeats or altered telomere chromatin structure is correlated with telomere dysfunction such as chromosome end-to-end associations/telomere fusions that could lead to gene amplification and genomic instability. The DNA structure at the end of telomeres is distinguished from DNA double strand breaks (DSBs) in order to avoid nonhomologous end-joining (NHEJ), which requires unique, higher order nucleoprotein structure. Telomeres are attached to the nuclear matrix and have a specific chromatin structure. Whether this special structure is maintained by specific chromatin changes is yet to be thoroughly investigated. Altered telomere chromatin structure has been linked to defective DNA damage response (DDR), and eukaryotic cells utilize the DDR mechanisms of proficient DNA repair and cell cycle checkpoints in order to maintain genomic stability. Studies of the DNA damage response has lead to the identification of sensors and transducers which constitute a hierarchical signaling paradigm for the transduction of the initial damage signal to numerous downstream effectors, some of which have a role in both genomic stability and telomere metabolism. This review will summarize the factors involved in telomere maintenance and the influence of such factors on the DNA damage response.

Keywords Telomere structure · DDR · Chromatin modifications · Histone code · ATM

T.K. Pandita (✉)
Department of Radiation Oncology, Washington University School of Medicine, 4511 Forest Park Ave., St Louis, MO 63108, USA
e-mail: pandita@wustl.edu

K.K. Khanna, Y. Shiloh (eds.), *The DNA Damage Response: Implications on Cancer Formation and Treatment*, DOI 10.1007/978-90-481-2561-6_7,

7.1 Telomeres

Telomeres are specialized nucleoprotein structures that protect the chromosomes from end-to-end fusion, degradation, and inappropriate recombination. They are structurally and functionally complex. Telomeres consist of an array of simple DNA repeats at the extreme end of the chromosome, with a more complex array of repeats immediately adjacent to it. In most eukaryotes, telomeric DNA is composed of G-rich repeated sequences that are synthesized by the telomerase ribonucleoprotein complex whose integral RNA component, the telomerase RNA or TERC RNA, contains the sequences that act as a template for the synthesis of these repeats. Because DNA polymerase I cannot copy the extreme end of a DNA strand in the absence of alternative replicating mechanism, telomeres get shorter with each cell division. When telomeres become critically short, cell cycle arrest or cell death occurs. Based on this assumption, it is believed that telomere length may act as a molecular clock that regulates the life span of a cell. Telomerase activity counteracts the continuous telomere shortening caused by cell replication. However, in human telomerase activity is not observed in most somatic cells, or occurs only at very low levels. In contrast, germ cells, stem cells and their immediate progeny, activated T cells, monocytes, and notably most cancer cells express telomerase activity, but only in germ cells and most of cancer cells is telomerase activity sufficient to prevent telomere shortening.

Human telomeres consist of thousands of base pairs of TTAGGG repeats along with several specially associated proteins. The TTAGGG repeat array of most human telomeres ranges in size from 5 to 15 kb. However, some human telomeres seem to contain 20–30 kb of TTAGGG repeats. These repeats are oriented such that the G+T-rich strand extends toward the 3' ends of each chromosome and are of variable length. The termini of telomeres carry single-stranded TTAGGG repeats, called as G-strand overhangs, that appear to be present in all cells irrespective of the presence of telomerase. The average size of G-rich overhangs is 130–210 bases in length. The disruption of a G-overhang disrupts telomere function. Electron microscope analysis of telomeres has revealed that the chromosome ends form a higher order structure called the "T" loop (Fig. 7.1). Cells where the chromosomes were known to have long telomeres contained larger "T" loops than those that had short telomeres. Such loops may protect telomeres by physically stitching the potentially vulnerable single-stranded G-strand terminus back into the double stranded telomere sequence, several kilobases internal to the terminus.

The telomeres of human somatic cells range greatly in length, depending on the type of tissue and the person's age. Several causes could lead to such variation e.g., the telomeric array at each chromosome end is heterogeneous or physical maps of subtelomeric DNA are variable and/ or differential loss of telomeric repeats during development and cellular proliferation. Most of the eukaryotic telomeric DNA is organized in tightly packed nucleosomes which are separated by 10–20 bp of linker DNA [1]. Several specific proteins contribute to the telomeric structure; however, the exact organization of telomere-associated proteins is still

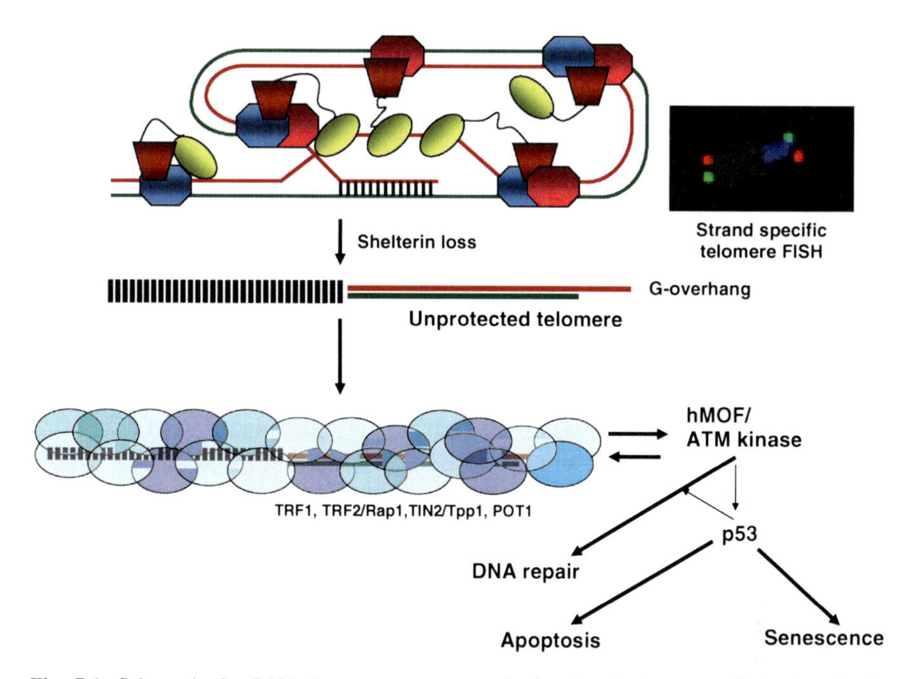

Fig. 7.1 Schematic for DNA damage response at dysfunctional telomeres. Shelterin subunits (TRF1, TRF2/Rap1, TIN2/TPP1, and POT1 protect telomeres. Loss of shelterin subunits results in loss of telomere protection. The exact mechanism is not known but it can happen with or without overt change in the structure of the DNA (the latter is depicted). Upon loss of protection, the telomere is then associated with DNA damage response factors. Telomere damage is associated with ATM activation (in absence of ATM, ATR kinase is activated), which can lead to DNA double-strand break (DSB) repair along with p53-dependent G1/S arrest or induction of either apoptosis or senescence. [Reproduced from: Misri S, Pandita S, Kumar R and Pandita TK. Telomeres, histone code and DNA damage. Cytogenetics Genome Res. 122:297–307 (2008).]

unclear. A complex formed by TRF1 (telomere repeat binding factor 1), TRF2, TIN2 (telomere-associated protein), Rap1 (replication associated protein 1), TPP1 (telomere protein-protein 1), and POT1 (protection of telomere 1), the major telomere-specific proteins that associate with telomeres to protect chromosome ends has been designated as Shelterin (Fig. 7.1) [2]. Three shelterin subunits, TRF1, TRF2, and POT1 directly recognize $(TTAGGG)_n$ repeats. They are interconnected by three additional shelterin proteins, TIN2, TPP1 , and Rap1, forming a complex that allows cells to distinguish telomeres from sites of DNA damage. Shelterin is a dynamic structural component of the telomere and is emerging as a protein complex with DNA remodeling activity that acts together with several associated DNA repair factors to change the structure of telomeric DNA. Although telomere interacting proteins TRF2 and POT1, which are part of shelterin complex, have been linked independently to repression of the activation of two DNA damage response pathways at chromosome ends, it has been a matter of great debate whether such proteins have any role in sensing global DNA damage or play any critical role in DNA DSB repair. Certainly, depletion of both proteins from an in vitro DNA end-joining assay had no effect on ligation, thus it is an open question whether these proteins have any specific role in DNA DSB repair in vivo. However, these structural proteins may have a specific role in staging the proteins that are required for the normal stability of telomeres.

Current evidence suggests that TRF2 and POT1 are the proteins primarily responsible for protection of telomeres. TRF2 stabilizes the telomeres by protecting the single-stranded DNA overhang from degradation or by remodeling the telomeres into the t-loop structure [3]. Invasion of the single-stranded overhang into the double-stranded telomeric tract forms the t-loop structure. In vitro, t-loop assembly involves the binding of TRF2, near the 3′ telomeric overhang [4]. TRF2 also possesses topoisomerase activity that likely assists in promoting strand invasion within the t-loop junction. Depletion of TRF2 results in cleavage of the telomeric G-overhang by ERCC3/XPF1, and telomeric end-to-end fusion by the NHEJ pathway [5–7]. Dominant-negative mutation of TRF2 effectively strips TRF2 and its interacting factors off the telomeres and causes ATM- and p53-dependent apoptosis, and cellular senescence [8, 9]. Disruption of TRF2 function also induces telomere dysfunctional induced foci that represent co-localization of telomeres with several DNA DSB response and repair factors including 53BP1, γ-H2AX, MDC1, Rad17, Rif1, the MRE11/NBS1/RAD50 complex and the activated form of ATM (ATM-ser1981) [2, 10, 11].

The telomeric single-stranded overhangs have been implicated as a critical component of the telomeric structure that is required for proper telomeric function [12, 13]. The current evidence suggests that all eukaryotes use a single-stranded DNA-binding protein to cap the telomeres [14]. Telomeres are protected by a capping protein(s) that binds to the single-stranded DNA commonly found at the ends of chromosomes. The erosion of the single-stranded overhangs immediately cause telomere dysfunction. Yet, while the telomeric double-stranded TTAGGG repeats persisted, the telomeres failed to protect chromosome ends. The loss of telomeric single-stranded overhangs consequently induces cellular responses and chromosomal instability. The *Oxytricha* telomeric overhang is tightly bound by a single-stranded DNA-binding protein, TEBP (telomere end-binding protein) composed of α and ß subunits [15]. TEBP forms an extensive interface along the overhang and buries the telomeric end in a deep hydrophobic pocket to provide an effective way to hide chromosome ends from DNA repair enzymes [16]. Telomeres of budding yeast are capped by a sequence-specific single-stranded DNA-binding protein, Cdc13. Unlike TEBP (telomere end-binding protein), Cdc13 is a recruitment factor that brings Stn1p to the telomere [17]. Stn1p binds a second capping protein, Ten1p [18], to protect the telomere.

POT1, an evolutionarily conserved shelterin component, is the best-known candidate for a human functional homologue of TEBP and Cdc13. It shares weak sequence similarity with the N-terminal DNA-binding domain of TEBP from ciliated protozoa. Crystal structures show that the POT1 protein adopts an oligonucleotide-oligosaccharide-binding fold with two loops that protrude to form a clamp for single-stranded DNA binding [19]. POT1 protein reportedly functions to protect telomeric 3′ overhangs, inhibit anaphase bridge formation, maintain cell viability and proliferation, regulate telomerase and telomere length, define the recessed 5′ end of telomeres, and enhance the ability of WRN and BLM helicases to unwind telomeric DNA [20–25].

Mice have two POT1 proteins (POT1a and POT1b), which are encoded by two distinct genes that likely originated from a recent gene duplication event [26]. Mouse POT1a and POT1b proteins seem to have overlapping and distinct functions in telomere protection and DNA damage signaling [26], providing functional diversity and complexity for telomere biology in mice. In contrast, humans have only one *POT1* gene. In addition to the full-length POT1 protein (also termed variant v1), at least four NH_2-terminally or COOH-terminally truncated isoforms (termed v2, v3, v4, and v5) are generated from the human *POT1* gene due to alternative RNA splicing [27]. The variant forms of human POT1 protein have different biological activities on telomeres, suggesting a different mode of complexity in telomere regulation in humans from that in mice. The functional analyses of the POT1 variants show that neither the single-stranded telomere-binding ability of the NH_2-terminal oligonucleotide-binding (OB) folds nor the telomerase-dependent telomere elongation activity mediated by the COOH-terminal TPP1-interacting domain is telomere protective by itself [28]. Importantly, a COOH-terminally truncated variant (v5), which consists of the NH_2-terminal OB folds and the central region of unknown function, is found to protect telomeres and prevent cellular senescence as efficiently as v1. These data reveal mechanistic and functional differences between v1 and v5: (*a*) v1, but not v5, functions through the maintenance of telomeric 3′ overhangs; (*b*) p53 is indispensable to v5 knockdown-induced senescence; and (*c*) v5 functions at only a fraction of telomeres to prevent DNA damage signaling. Furthermore, v5 is preferentially expressed in mismatch repair (MMR)-deficient cells and tumor tissues, suggesting its role in chromosome stability is associated with MMR deficiency.

POT1 and TRF2 are the telomeric proteins associated with the maintenance of the correct DNA configuration of the telomeric single-stranded overhangs. In a telomere-capping model, t-loops can provide cells with an architectural solution to the telomere protection problem in mammals. The exact structure at the base of the t-loops is not known, but it is clear that there is a short segment of single-stranded DNA, likely representing a D-loop of TTAGGG repeats that are displaced by the invading single-stranded overhang (Fig. 7.1). The current model suggests that telomeric single-stranded binding proteins and single-stranded DNA overhangs are required for t-loop formation [3].

Whereas the role played by telomeric proteins in telomere function and regulation has been widely investigated, little is known about the contribution of nucleosomes to the protection of chromosome ends. Wu and de Lange [29] have reported that telomeres lacking TRF2, POT1a, and POT1b have no nucleosome eviction in telomeres suggesting that such proteins do not have any chromatin modifying functions. However, it is not yet clear whether depletion of TRF2, POT1a, and POT1b influences higher order chromatin structure or telomere interactions with the nuclear matrix. It will be interesting to see whether the absence of TRF2, POT1a, and POT1b influences the telomere-specific histone modifications.

7.2 Telomere Dysfunction

Telomeres shorten as a function of aging in cells derived from normal human blood, skin and colonic mucosa. This shortening of telomeres result in a loss of telomere function. Such a loss could result in telomeric association (non-covalent interactions) or fusion (covalent interactions between the DNA of two telomeres), which could either involve single chromatid or double chromatids and can occur between sister chromatids or between chromatids of different chromosomes. Telomere associations or fusions can be seen at metaphase, and with the use of premature chromosome condensation technology, they have been observed in the G1 and G2 phase of the cell cycle [30, 31]. Occasionally, telomeric associations or fusions could lead to a bridge formation at anaphase and thus to genomic instability, gene amplification and tumorigenesis.

There are several factors that can lead to telomere dysfunction. For example telomere dysfunction can be induced by cellular stress, replicative exhaustion, or chemical treatment, which can lead to the inhibition of cell proliferation [32]. Telomere dysfunction has also been reported in Werner's Syndrome [33]. Dysfunctional telomeres can also be a substrate for nonhomologous end joining, leading to chromosome end-to-end fusions, anaphase bridge formation, and global genome instability [34–36]. The structure and function of telomeres are regulated by a multiprotein complex containing TRF1 and TRF2 that bind specifically to double-stranded telomeric DNA and POT1, which binds single-stranded G-rich telomeric sequence [2]. These three proteins are part of a six-subunit telomeric-specific complex termed "shelterin" [2], which includes three proteins, TIN2, TPP1 and RAP1. Depletion of telomeric proteins cause a loss of telomeric overhangs, apoptosis, senescence, and chromosome abnormalities [8, 9, 20, 37, 38]. Dyskeratosis congenita (DC) a human disease has been directly linked to an impairment of telomere maintenance [39]. Mutations in three different genes have been identified in patients with DC – DKC1, TERC, and TERT. The products of these genes, dyskerin encoded by DKC1, the RNA component of telomerase, TERC, and the catalytic component of telomerase, TERT, form the catalytically active telomerase. It is thought that telomere length rather than impaired telomerase activity is responsible for disease in patients with DC.

7.3 DNA Damage Foci at Dysfunctional Telomeres

Telomere dysfunction by inhibition of TRF2 induces a DNA damage response in mammalian cells which is associated with factors, such as 53BP1, γ-H2AX, Rad17, ATM, and Mre11 [40]. The domain of telomere-associated DNA damage factors are known as a Telomere Dysfunction-Induced Focus (TIF). The accumulation of 53BP1 on uncapped telomeres was reduced in the presence of the PI3 kinase inhibitors caffeine and wortmannin, which affect ATM, ATR, and DNA-PK. By contrast, Mre11 TIFs were resistant to caffeine, consistent with previous findings on the Mre11 response to ionizing radiation. A-T cells had a diminished 53BP1 TIF response, indicating that the ATM kinase is a major transducer of this pathway. However, in the absence of ATM, TRF2 inhibition still induced TIFs and senescence, pointing to a second ATM-independent pathway. It is becoming evident that the cellular response to telomere dysfunction is governed by proteins that also control the DNA damage response. Induction of TIFs through TRF2 inhibition provides an opportunity to study the DNA damage response in defined regions of chromosomes with physically marked lesions.

7.4 Factors Common in DNA Damage Response and Telomere Metabolism

Perusal of literature reveals that there is increasing evidence that most DNA damage response proteins are directly or indirectly involved in telomere maintenance (Figs. 7.1 and 7.2). Cells defective in DNA DSB repair proteins including ATM, Ku, DNA-PKcs, RAD51D, and the MRN (MRE11/RAD50/NBS1) complex show faulty telomere metabolism [41]. Telomere behavior is abnormal in *ATM* deficient cells, both somatic as well as germ cells [36, 42–50], but the precise mechanism by which ATM regulates telomere structure and function is not known. Recent studies suggest that ATM interacts with various chromatin modifying factors that could influence telomere function (Fig. 7.2) [51–53].

Chromatin, the physiological packaging structure of histones and DNA, is gaining appreciation as a factor maintaining genomic stability and as a relevant target of signaling pathways [54, 55]. Chromatin structure is an important determinant of protein-DNA interactions, with consequences for DNA metabolism and transcription control [56–59]. After DNA damage, eukaryotic cells orchestrate a complex array of responses essential for cell survival [60], coordinating checkpoint-mediated cell cycle arrest and the repair of damaged DNA. Histone modifications by phosphorylation, acetylation, or methylation have all been linked to DNA damage repair [61]. Phosphorylation of histone H2A in fission yeast (referred to as H2AX phosphorylation) or the H2A variant H2AX in mammals also has a role in checkpoint control [62, 63]. Recent studies have indicated that histone modifications in response to DNA damage are influenced by nonhistone chromosomal proteins. We will discuss in detail the role of nonhistone chromosomal proteins in telomere maintenance and DDR.

Chromatin changes frequently occur at both local and global levels and specifically telomeres acquire a chromatin structure that ensures their proper function [64]. Telomere chromatin also contain nucleosomes, the basic unit of chromatin, that is composed of an octamer of histones, two subunits of each H3, H4, H2A and H2B, and DNA that wraps around the histone core [65]. The N-terminal tails of the histones protrude from the nucleosome core and are subjected to different post-translational modifications. The types of histone post-translational modifications and the degree of DNA methylation determine the specific chromatin structure

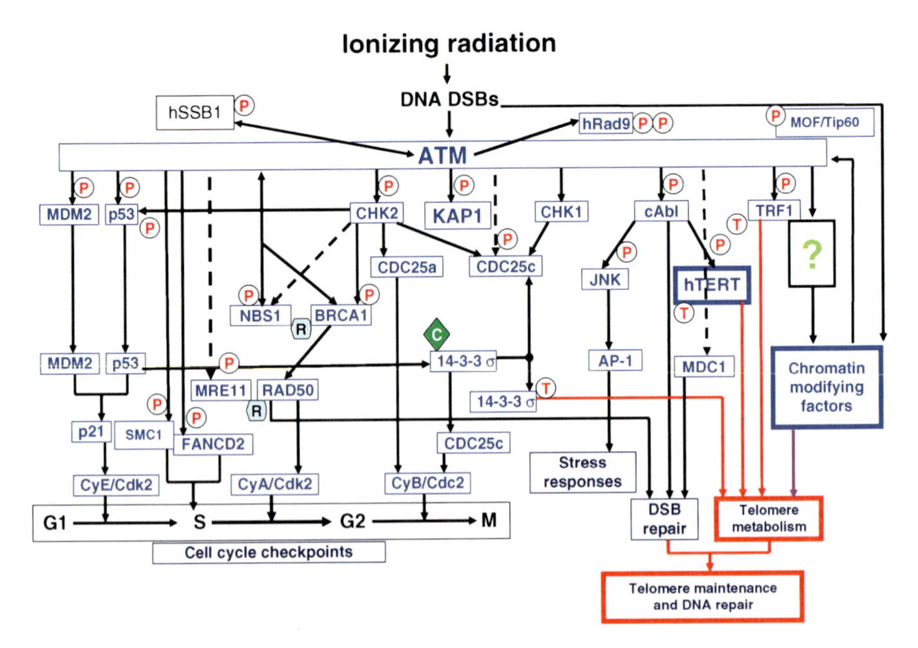

Fig. 7.2 Role of ATM in genomic stability and telomere metabolism. ATM kinase activity increases immediately after exposure to ionizing radiation (IR). ATM mediates the early stages of the rapid induction of several signaling pathways, which include regulation of the cell-cycle checkpoint controls, activation of the DNA-DSB (double-strand break) repair pathway, activation of stress responses, and maintenance of telomeres. "P" with white solid arrows indicates reported phosphorylation events; dashed arrows represent possible signaling steps and do not imply direct interaction between proteins; "C" indicates sequestering in cytoplasm; "R" indicates repair complexes; and "T" indicates a role for the protein in telomere metabolism. ATM influences telomere function through direct interaction with TRF1. In addition, c-Abl phosphorylates hTERT and negatively regulates telomerase activity. The other downstream effector of ATM in telomere metabolism is 14-3-3σ. Abbreviations: AP-1: Apetala1 transcription factor; BRCA1: breast cancer susceptibility gene product 1; cAbl: Abelson protein tyrosine kinase; cyclin B; CDC25: cell division cycle 25; Cdk2: cyclin-dependent kinase; CHK: checkpoint kinase; CyA, CyB, CyE: cyclinA, cyclin B, cyclin E; FANCD2: Fanconi anaemia protein; hSSB1: human single strand DNA binding protein; hTERT: human catalytic unit of telomerase; JNK: Jun N-terminal protein kinase; KAP1: co-repressor of KRAB associated protein; MDC1: mediator of damage checkpoint 1; MRE11: meiotic recombination 11 gene product; MDM2: mouse double minute 2 (p53-binding protein); NBS1: Nijmegen breakage syndrome 1 protein (p95); RAD50: a radiation-damage-repair-associated protein; SMC1: structural maintenance of chromosome 1; TRF1: telomere-repeat-finding factor 1; MOF and Tip60 are members of the MYST family and have histone acetyltranferase activity, which has been implicated in ATM function. [Reproduced from: Misri S, Pandita S, Kumar R and Pandita TK. Telomeres, histone code and DNA damage. Cytogenetics Genome Res. 122:297–307 (2008).]

[64, 66–68]. Though mammalian telomere ends do not exist as DNA DSBs due to the terminal T-loop structures (Fig. 7.1), DNA repair proteins have been shown to interact with telomeres. Whether these interactions are dependent upon any specific histone modification is under investigation. Recent investigations have begun to elucidate the chromatin changes that occur upon DNA damage [69–77].

The swiftness of the DNA DSB response is facilitated by hierarchical signaling networks that orchestrate chromatin structural changes, cell cycle checkpoints, and multiple enzymatic activities to repair the broken DNA ends. Sensors and transducers signal to numerous downstream cellular effectors which function primarily by substrate posttranslational modifications including phosphorylation, acetylation, methylation, and ubiquitylation. In particular, the past several years have provided important insight into the specific modifications to histones to control DNA damage detection, signaling, and repair. We will describe below the recently identified common factors involved both in DDR as well as telomere metabolism.

7.5 ATM

The ATM protein belongs to a growing family of PIKK kinases and functions as an intrinsic part of the cell cycle machinery that surveys genomic integrity, cell cycle progression, and processing of DNA damage [36, 50]. ATM protein kinase is primarily activated in response to DNA DSBs caused by IR or radiomimetic drugs, and also detects DSBs during meiosis or mitosis, or breaks consequent to damage by free radicals (Fig. 7.2) [78]. It shows similarity to several yeast and mammalian proteins involved in meiotic recombination and cell cycle progression, namely, the products of *MEC1* in the budding yeast *S. cerevisiae* and *RAD3* in the fission yeast *S. pombe* [79, 80] and the Tor proteins found in yeasts and mammals [81, 82]. Cells deficient in ATM have been shown to have a high frequency of spontaneous chromosomal aberrations, high rates of intrachromosomal recombination and error-prone recombination [36, 50]. Such cells have higher initial and residual chromosomal aberrations in the G1- and G2-phases after IR exposure as determined by premature chromosome condensation [30, 83, 84]. Cells defective in ATM function also display higher frequencies of chromosomal aberrations seen at metaphase after IR exposure [84]. Mice mutated in the *Atm* gene display similar pleiotropic defects [85–87].

ATM is observed at the sites of DNA damage, where it is autophosphorylated and is dissociated from its nonactive dimeric form to the active monomeric form [88, 89]. Lee and Paul reported that ATM stimulation appeared to be primarily through an increase in substrate recruitment by ATM [90]. From these findings, we propose that the Mre11-complex binds to ATM, inducing conformational changes that facilitate an increase in the affinity of ATM towards its substrates.

Detection and signaling of DNA damage are mediated through downstream targets of ATM (Fig. 7.2) [50]. ATM directly phosphorylates p53 and interacts with many molecules directly involved in DSB repair (Fig. 7.2), as well as in cell signaling. Some of these molecular targets include: Atr, c-Abl, Chk-1, Chk-2, Rpa, Brca1, Brca2, NF-kB/IkB alpha, β-adaptin, hSSB1 and autophosphorylation of ATM itself. Thus, ATM is a "hierarchical kinase," capable of initiating many pathways simultaneously [50]. After recruitment to sites of DNA damage, ATM phosphorylates a number of substrates, including Chk1 and Chk2, which in turn target other proteins to induce cell-cycle arrest and facilitate DNA repair (Fig. 7.2). Although ATM is known to be a central transducer of DNA damage signals, recent studies have demonstrated that ATM stabilizes chromosomal V(D)J recombination DSB intermediates, facilitates DNA end joining, and prevents broken ends from participating in chromosomal deletions, inversions, and translocations [91, 92]. Furthermore, it has been reported that ATM mediated checkpoints block the persistence and transmission of un-repaired DSBs in developing lymphocytes [93]. Recently we have reported that ATM phosphorylates single-strand DNA (ssDNA)-binding proteins (SSBs) hSSB1 in response to DNA double-strand breaks (DSBs) [94]. This phosphorylation event is required for DNA damage-induced stabilization of hSSB1. Upon induction of DNA damage, hSSB1 accumulates in the nucleus and co-localizes with other known repair proteins. In contrast to RPA, hSSB1 does not localize to replication foci in S-phase cells and hSSB1 deficiency does not influence S-phase progression. Depletion of hSSB1 abrogates the cellular response to DSBs, including activation of ATM and phosphorylation of ATM targets after exposure to ionizing radiation.

DNA damage checkpoint pathways in *S. cerevisiae* are governed by the ATM homolog Tel1 and the Mre11-complex [95]. In mitotic cells, the Tel1-Mre11 complex pathway triggers Rad53 activation and its interaction with Rad9, whereas in meiosis it activates Rad9 and the Rad53 paralog Mre4/Mek1. Activation of the Tel1-Mre11 complex checkpoint appears to depend upon the Mre11-complex as a damage sensor and, at least in meiotic cells, to depend on unprocessed DNA DSBs. The DSB repair functions of the Mre11-complex are enhanced by the pathway, suggesting that the complex both initiates and is regulated by the Tel1-dependent DSB signal. These findings suggest that the Mre11-complex has a role in the meiotic recombination as well [95].

Cells derived from individuals with ataxia-telangiectasia display genomic instability and defective DNA repair [30, 36, 50, 83]. The first report describing the association of ATM with DNA damage repair and telomere metabolism was described in 1995 [42]. Subsequently, the role of ATM in telomere function became clear by comparing the meiotic prophase 1 of spermatocytes from mice with and without ATM, revealing that ATM inactivation results in aberrant telomere clustering during meiotic prophase [36, 44, 46–48, 50]. These studies prompted several groups to determine whether ATM is physically associated with telomeres. Chan et al. suggested that a primary function of the ATM-family kinases in telomere maintenance is to act on the telomere itself rather than to activate the enzymatic activity of telomerase [96]. Recent studies have shown that cells expressing mutated telomerase (containing mutated template-bearing RNA) have activated ATM as detected by foci formation. However, transient depletion of ATM resulted in the lack of telomere fusions [97], further suggesting the role of ATM in modulating chromatin status to provide a template for DNA ligation. ATM is associated with chromatin [46, 98], and a fraction of ATM is detected in nuclear aggregates [99]. The striking correlation between the appearance of retained ATM and of phosphorylated histone H2AX (γ-H2AX) and the rapid association of a fraction of ATM with γ-H2AX foci are consistent with a major role of ATM in the early detection of DSBs and subsequent induction of cellular responses [99]. It has been reported that the ATM protein interacts with telomere-repeat-binding factor 1 (TRF1; also known as Pin2) [100, 101], and that lack of TRF2 results in apoptosis in an ATM-dependent fashion [8]. These reports further suggest a role for ATM in telomere maintenance. Interestingly, it has been shown that ATM function influences the fraction of telomeres anchored to the nuclear matrix [102]. Subsequent studies revealed that the central DNA damage response protein ATM is physically associated with telomeres [103, 104]. These results are not surprising as ATM is a protein kinase that interacts with several substrates and is implicated in mitogenic signal transduction, chromosome condensation, meiotic recombination, cell-cycle control, and telomere maintenance (Fig. 7.2) [36, 50]. Besides ATM, several of its effectors influence both DDR as well as telomere stability as discussed below.

7.6 MDC1

Mediator of DNA damage checkpoint 1(MDC1) plays a role in the detection and repair of human and mouse telomeres rendered dysfunctional through inhibition of TRF2 [10]. Consistent with its role in promoting DNA damage foci, MDC1 knock-down affected the formation of telomere dysfunction-induced foci (TIFs), abrogates the accumulation of phosphorylated ATM, 53BP1, Nbs1, and to a lesser extent, γ-H2AX. In addition to this effect on TIFs, the rate of NHEJ of dysfunctional telomeres has been reported to significantly decrease when MDC1 itself or its recruitment to chromatin was inhibited. MDC1 appeared to promote a step in the NHEJ pathway after the removal of the 3′ telomeric overhang. The acceleration of NHEJ was unlikely to be due to increased presence of 53BP1 and Mre11 in TIFs, since knock-down of neither factor affected telomere fusions [10]. These studies suggested that the binding of MDC1 to γ-H2AX directly affects NHEJ.

7.7 c-Abl

The activation of nuclear c-Abl tyrosine kinase by ionizing radiation requires a functional ATM protein [105, 106], and c-Abl in turn associates with the catalytic unit of telomerase (hTERT) in human cell lines through binding of its Src-homology 3 (SH3) domain to a region (amino acids 308–316) of hTERT [107]. Ionizing radiation induces c-Abl-dependent tyrosine phosphorylation of hTERT, which thereby inhibits hTERT activity and negatively regulates telomere length [107].

7.8 Mammalian Rad9

The mammalian homologue of *Schizosaccharomyces pombe Rad9* was isolated in 1996 [108], and the gene product in human cells is phosphorylated by ATM in response to DNA damage [109]. Together with hRad1 and hHus1, hRad9 forms a nuclear complex that resembles PCNA (proliferating cell nuclear antigen) and is believed to sense DNA damage [110, 111]. However, the cellular colocalization of hRad9 with the phosphorylated form of histone H2AX (γ-H2AX) after DNA damage is independent of ATM function [112], raising the possibility that hRad9 may also have a direct DNA damage-sensing function. Recent studies have revealed that inactivation of mammalian Rad9 increases chromosome end-to-end associations and the frequency of telomere loss [113]. This telomere instability correlated with enhanced S- and G2-phase-specific cell killing, delayed kinetics of γ-H2AX focus appearance and disappearance, and reduced chromosomal repair after IR exposure, suggesting that Rad9 plays a role in cell cycle phase-specific DNA damage repair [113]. Furthermore, mammalian Rad9 interacted with Rad51, and inactivation of mammalian Rad9 also resulted in decreased homologous recombinational repair, which occurs predominantly in the S and G2 phases of the cell cycle [113]. Together, these findings provide evidence for a role of mammalian Rad9 in telomere metabolism and DDR in order to maintain genomic stability.

7.9 DNA-PK

The DNA-dependent protein kinase (DNA-PK) is a multicomponent complex consisting of the DNA-PK catalytic subunit (DNA-PKcs) and the Ku heterodimer (Ku80 and Ku70). The DNA-dependent protein kinase catalytic subunit (DNA-PKcs) is critical for DNA repair via the NHEJ pathway and is mutated in SCID mice. Interestingly, in SCID mice, the spermatogonia are radiosensitive [114]. Spermatogonial apoptosis occurs faster in irradiated DNA-PKcs-deficient SCID testis compared to their wild-type counterparts. However, p53 induction is unaffected in SCID cells, suggesting that in spermatogonia DNA-PK functions in DNA damage repair rather than p53 accumulation. Spontaneous apoptosis of spermatocytes occurred in the SCID testis, despite the fact that these cells lack the Ku proteins. This suggests that DNA-PKcs functions independently of the Ku proteins to promote DNA repair in these cells. The majority of these apoptotic spermatocytes are found at stage IV of the seminiferous epithelium where a meiotic checkpoint has been suggested to exist. DSBs are less accurately repaired in SCID spermatocytes that then fail to pass the meiotic checkpoint. Thus the role for DNA-PKcs during the meiotic prophase differs from that in mitotic cells since it is not influenced by IR and is independent of the Ku heterodimer.

Gilley and coworkers [115] reported that a deficiency in DNA-PKcs severely disrupts telomere capping function but does not affect telomere length. Furthermore, telomerase activity is not affected by a DNA-PKcs deficiency. DNA-PKcs-deficient mouse embryonic fibroblasts also exhibited elevated levels of chromosome fragments and breaks, which correlated with increased telomere fusions. Thus DNA-PKcs is a telomere maintenance protein which functions specifically in telomere capping and not in telomere length control.

7.10 Ku

Ku is a heterodimer of Ku70 and Ku80, which is crucial for NHEJ repair [116–118]. It binds site specifically to particular DNA sequences [119–121], functions in site-specific recombination of V(D)J gene segments [122], and plays an important role at the telomere [123–127]. During the repair of DNA DSBs, Ku binds non-specifically to DNA ends with high affinity [128–131]. However, telomeric ends are capped or bound by specific telomere proteins that serve to conceal and disguise the telomeric DNA end, thereby preventing telomere end fusion events as well as preventing cellular DNA damage signaling. Several groups have found that Ku is in close proximity to the mammalian telomere and that loss of Ku, in virally transformed Ku-deficient mouse cell lines, resulted in telomere fusion events [125, 132]. Ku forms a high-affinity protein/protein interaction with TRF1 to localize to internal regions of telomeric DNA and thus Ku provides an essential telomere capping function in primary mammalian cells to prevent telomere fusions. Therefore, Ku functions in a different way at the telomere than during DNA DSB repair by the NHEJ pathway.

7.11 MRN

The components of MRN complex (Mre11, Rad50 and Nbs1/Xrs1) play a critical role in DNA DSB repair in eukaryotes, and in yeast all three MRN components bind telomeres [133–135]. The role of MRN complex in telomere maintenance parallels its function in processing DNA DSBs during repair. The MRN complex together with ATM promotes activation of the ATR kinase in G2-phase post IR exposure. During homologous recombination, the key step in the activation mechanism is a Mre11-dependent resection of DSBs which leads to the formation of a replication proteins A (RPA)-coated ssDNA [136]. A similar situation is observed at yeast telomeres, where Mre11 is implicated in telomere resection and loading of Mec1p during late S phase [137]. These observations indicate that the telomeric function of MRe11 might reflect the role of Mre11 in recognizing and processing of DNA DSB.

7.12 14-3-3σ

14-3-3σ has been implicated in the G2 checkpoint [138]. Its association with different kinases in the cytosol and on the nuclear membrane might contribute to kinase activation during intracellular signaling, and the protein appears to sequester the mitotic initiation complex Cdc2/cyclinB1 in the cytoplasm after DNA damage [139]. Cells with both copies of the gene inactivated displayed frequent loss of telomeric repeat sequences, enhanced frequencies of chromosome end-to-end associations, and terminal nonreciprocal translocations [140]. These phenotypes correlated with a reduction in the amount of G-strand overhangs at the telomeres and an altered nuclear matrix association of telomeres in these cells. Since the p53-mediated G1 checkpoint is operative in *14-3-3σ* $^{-/-}$ (gene symbol: *SFN*) cells, the observed chromosomal aberrations occurred preferentially in G2 after exposure to IR, corroborating the role of the 14-3-3σ protein in G2–M progression. Dhar et al. [140] also reported that, even in untreated cycling cells, occasional chromosomal breaks or telomere-telomere associations trigger a G2 checkpoint arrest followed by repair of these aberrant chromosome structures before entering M phase. Since *14-3-3σ* $^{-/-}$ cells are defective in maintaining G2 arrest, they enter M phase without repair of the aberrant chromosome structures and undergo cell death during mitosis. Together, these studies provide evidence for the correlation between a dysfunctional G2–M checkpoint control, genomic instability, and loss of telomeres in human cells mediated by a downstream effector of ATM (Fig. 7.2).

7.13 Heterochromatin Protein 1 (HP1)

In eukaryotes, there are two major types of chromatin: heterochromatin and euchromatin [141]. Heterochromatin corresponds to the relatively gene-poor, late-replicating, repetitious sequences found near centromeric and telomeric locations. In contrast, euchromatin contains single copy sequences, including the majority of genes, and replicates relatively early in the cell cycle. Euchromatin and heterochromatin are further distinguished by specific histone tail modifications. In addition, euchromatin and heterochromatin show differences in non-histone chromosomal protein constituents. One of the best-studied examples of a nonhistone chromatin protein is heterochromatin protein 1 (HP1) initially identified in *Drosophila* and named for its predominant localization to centromeric heterochromatin [141].

In mammals, there are three HP1 isoforms termed HP1α, HP1β (M31, CBX1) and HP1γ (M32) [142–144]. All HP1 isoforms share a high degree of sequence similarity although they localize to distinct chromosomal territories; HP1α and HP1β are usually found at sites of constitutive heterochromatin, while HP1γ has a more euchromatic distribution [145, 146]. The best studied isoform, HP1β, is a small 26 kDa protein with two sequence-related domains, the N-terminal chromodomain (CD) and the C-terminal chromoshadow domain (CSD), which are separated by a "hinge" region [reviewed in [68]].

Telomeres are associated with the nuclear matrix and are thought to be heterochromatic. In human cells the overexpression of HP1 isoforms $HP1^{Hs\alpha}$ or $HP1^{Hs\beta}$ results in reduction of the G-strand overhangs and overall telomere sizes [147]. Cells overexpressing $HP1^{Hs\alpha}$ or $HP1^{Hs\beta}$ also display a higher frequency of chromosome end-to-end associations, spontaneous chromosomal damage and reduced chromosome damage repair than the parental cells [147].

Recent studies have revealed that DNA DSB formation is followed by ATM dependent chromatin relaxation [148]. ATM's effector in this pathway is KRAB-associated protein (KAP-1, also known as TIF1β, KRIP-1 or TRIM28) [148]. KAP-1 is phosphorylated in an ATM-dependent manner on Ser 824 exclusively at

DNA damage sites. Knocking down or mimicking constitutive phosphorylation of KAP-1 leads to chromatin relaxation [148] while deletion of the KAP-1 phosphorylation site leads to loss of DSB-induced chromatin decondensation and renders the cells hypersensitive to DSB-inducing agents. These results suggest that chromatin relaxation is a critical regulatory point in the DNA-damage response pathway. KAP-1 also recruits HP1 proteins to form small HP1-containing heterochromatin domains that repress gene activity [149, 150] and expression of mutant HP1β results in abrogation of damage-induced H2AX phosphorylation [151]. Further studies have shown that DNA damage influences the chromatin associated movement of HP1β, supporting the relationship between HP1β and DDR [151]. HP1β colocalizes with methylated lysine 20 of histone H4 [H4K20me3] which is also a binding motif for 53BP1 [152]. These studies further support the view that HP1β influences both telomere metabolism as well as DDR.

7.14 Chromatin Modification in Response to DNA DSBs

Eukaryotic genomes are packaged into chromatin comprised of DNA wrapped around positively charged histone proteins. Nucleosomes are the basic unit of chromatin consisting of 146 bp of DNA, wrapped around an octamer containing two copies each of core histones H2A, H2B, H3, and H4. Nucleosomes are further packaged and linked together with histone H1. Chromatin creates a natural barrier against access to DNA during transcription, damage repair, and recombination. Cells have developed various mechanisms by which chromatin structure can be manipulated to regulate access to DNA. These include (a) ATP-dependent chromatin remodeling, (b) incorporation of histone variants into nucleosomes, and (c) covalent histone modifications [153]. Chromatin remodeling by multi-subunit complexes utilizes the energy from ATP hydrolysis to affect histone-DNA interactions that results in sliding nucleosomes on the DNA molecule and thus regulating access to specific sequences. Histone variants possess biophysical properties distinct from those of canonical core histones, and their substitution into nucleosomes can bring about alterations into the higher order chromatin structure. In addition, core histone modification itself can alter chromatin structure. For example acetylation of histone H4 at lysine 16 (H4K16ac) disrupts higher-order chromatin structure and changes the functional interactions between chromatin-associated proteins [154]. Covalent modifications of histones can alter the charge of specific residues, affecting both histone-histone and histone-DNA interactions, and can act as signals for binding by the proteins involved in DNA DSB repair.

There is an increasing body of evidence about the role of histone modifications in the DDR [50, 147, 155, 156] and DSB repair [71, 72, 157]. There is also a strong correlation between defective DSB repair, genomic instability, and telomere dysfunction, and further investigation into this area would determine whether telomere stability is based on the same paradigms [31, 147, 158]. Recent studies suggest that chromatin structure has an important role in initiating, propagating and terminating the DNA DSB repair process [159] and the chromatin response in general may precede DNA end resection [160] (Fig. 7.2). Among the different histone modifications, phosphorylation of all 4 histones as well as the variant H2AX plays a primary role in DNA damage response by facilitating access of repair proteins to DNA breaks.

7.15 DSB Signaling and Checkpoint Activation

In response to DNA DSBs, a complex network of cell-cycle checkpoint proteins is activated, resulting in cell-cycle arrest at all three DNA damage cell-cycle checkpoints (G1-S, intra-S and G2-M). The signaling molecules that orchestrate the DNA damage cell-cycle checkpoints are the phosphatidylinositol-3 kinase-related (PIKK) class of protein kinases; ATM, ATR, and DNA-PK. Survival after DNA damage of cells with impaired ATM or ATR function is compromised, and they are defective in initiating DNA damage-induced cell-cycle arrest. During cell-cycle arrest a DSB is repaired by either HR or NHEJ. Both the pathways have been linked with the stability of the telomeres as well as genomic DNA.

7.16 Conclusions and Future Prospects

All organisms respond to interruptions in genomic stability by launching the DDR and maintaining telomere stability. DNA damage response can be considered a signaling transduction pathway where the DNA damage is detected by "sensors" that trigger the activation of a transduction system composed of protein kinases and a series of adaptor proteins, some of which are known to interact with telomeres (Fig. 7.2). This kinase cascade amplifies the initial DNA damage signal and triggers the activation of effector proteins (Fig. 7.2) [161]. The end result is a delay in cell cycle progression until the damage has been repaired or the inability to properly repair DNA damage is detected, inducing either apoptosis or senescence. Two pathways for the repair of DSBs, NHEJ and HR, have evolved in eukaryotes [161]. These pathways, like processes such as transcription and replication, act on DNA of DSBs or telomeres that are embedded in chromatin. A growing body of evidence supports a role for chromatin-modifying and remodeling activities and their proper signaling in DSB repair and telomere maintenance by these pathways. Further evidence for a role of chromatin structure in DNA damage repair comes from studies showing that treatment of cells with agents that induce changes in chromatin structure lead to activation of DNA damage response pathways and enhanced telomerase activity in the absence of DSBs [88, 162, 163]. This indicates that chromatin modulation is an integral component of DDR as well as telomere stability. The connection between chromatin structure and DNA repair has been made [164], but only limited studies have clearly demonstrated the role of specific chromatin modifications (histone code) in DNA repair processes and telomere maintenance. Although the relationship between telomeres and DDR is highly dynamic, their functionality may be intertwined by specific histone codes, which need to be identified. Deciphering novel epigenetic mechanisms like the histone code behind DDR and telomere stability will add to our understanding of genomic stability, which will be of enormous therapeutic value.

Acknowledgements The work in my laboratory is supported by National Institute of Health grants CA129537 and CA123232 (T.K.P). We thank the members of my laboratory Q. Yang and S. Scott for their help and thoughtful discussion.

References

1. Pisano S, Galati A, Cacchione S. Telomeric nucleosomes: Forgotten players at chromosome ends. Cell Mol Life Sci 2008;65:3553–3563.
2. de Lange T. Shelterin: the protein complex that shapes and safeguards human telomeres. Genes Dev 2005;19:2100–10.
3. Griffith JD, Comeau L, Rosenfield S, et al. Mammalian telomeres end in a large duplex loop. Cell 1999;97:503–14.
4. Stansel RM, de Lange T, Griffith JD. T-loop assembly in vitro involves binding of TRF2 near the 3′ telomeric overhang. EMBO J 2001;20:5532–40.
5. Celli GB, de Lange T. DNA processing is not required for ATM-mediated telomere damage response after TRF2 deletion. Nat Cell Biol 2005;7:712–8.
6. Zhu XD, Niedernhofer L, Kuster B, Mann M, Hoeijmakers JH, de Lange T. ERCC1/XPF removes the 3′ overhang from uncapped telomeres and represses formation of telomeric DNA-containing double minute chromosomes. Mol Cell 2003;12:1489–98.
7. Konishi A, de Lange T. Cell cycle control of telomere protection and NHEJ revealed by a ts mutation in the DNA-binding domain of TRF2. Genes Dev 2008;22:1221–30.
8. Karlseder J, Broccoli D, Dai Y, Hardy S, de Lange T. p53- and ATM-dependent apoptosis induced by telomeres lacking TRF2. Science 1999;283:1321–5.
9. Karlseder J, Smogorzewska A, de Lange T. Senescence induced by altered telomere state, not telomere loss. Science 2002;295:2446–9.
10. Dimitrova N, de Lange T. MDC1 accelerates nonhomologous end-joining of dysfunctional telomeres. Genes Dev 2006;20:3238–43.
11. De Lange T. Telomere-related genome instability in cancer. Cold Spring Harb Symp Quant Biol 2005;70:197–204.
12. Baumann P, Cech TR. Pot1, the putative telomere end-binding protein in fission yeast and humans. Science 2001;292:1171–5.
13. Blackburn EH. Switching and signaling at the telomere. Cell 2001;106:661–73.
14. de Lange T. Protection of mammalian telomeres. Oncogene 2002;21:532–40.
15. Gottschling DE, Cech TR. Chromatin structure of the molecular ends of Oxytricha macronuclear DNA: phased nucleosomes and a telomeric complex. Cell 1984;38:501–10.
16. Horvath MP, Schweiker VL, Bevilacqua JM, Ruggles JA, Schultz SC. Crystal structure of the Oxytricha nova telomere end binding protein complexed with single strand DNA. Cell 1998;95:963–74.
17. Pennock E, Buckley K, Lundblad V. Cdc13 delivers separate complexes to the telomere for end protection and replication. Cell 2001;104:387–96.
18. Grandin N, Damon C, Charbonneau M. Ten1 functions in telomere end protection and length regulation in association with Stn1 and Cdc13. EMBO J 2001;20:1173–83.
19. Lei M, Podell ER, Baumann P, Cech TR. DNA self-recognition in the structure of Pot1 bound to telomeric single-stranded DNA. Nature 2003;426:198–203.
20. Yang Q, Zheng YL, Harris CC. POT1 and TRF2 cooperate to maintain telomeric integrity. Mol Cell Biol 2005;25:1070–80.

21. Hockemeyer D, Sfeir AJ, Shay JW, Wright WE, de Lange T. POT1 protects telomeres from a transient DNA damage response and determines how human chromosomes end. Embo J 2005;24:2667–78.
22. Opresko PL, Mason PA, Podell ER, et al. POT1 stimulates RecQ helicases WRN and BLM to unwind telomeric DNA substrates. J Biol Chem 2005;280:32069–80.
23. Wang X, Liu L, Montagna C, Ried T, Deng CX. Haploinsufficiency of Parp1 accelerates Brca1-associated centrosome amplification, telomere shortening, genetic instability, apoptosis, and embryonic lethality. Cell Death Differ 2007;14:924–31.
24. Loayza D, De Lange T. POT1 as a terminal transducer of TRF1 telomere length control. Nature 2003;423:1013–8.
25. Veldman T, Etheridge KT, Counter CM. Loss of hPot1 function leads to telomere instability and a cut-like phenotype. Curr Biol 2004;14:2264–70.
26. Wu L, Multani AS, He H, et al. Pot1 deficiency initiates DNA damage checkpoint activation and aberrant homologous recombination at telomeres. Cell 2006;126:49–62.
27. Baumann P, Podell E, Cech TR. Human Pot1 (protection of telomeres) protein: cytolocalization, gene structure, and alternative splicing. Mol Cell Biol 2002;22:8079–87.
28. Yang Q, Zhang R, Horikawa I, et al. Functional diversity of human protection of telomeres 1 isoforms in telomere protection and cellular senescence. Cancer Res 2007;67:11677–86.
29. Wu P, de Lange T. No Overt Nucleosome Eviction at Deprotected Telomeres. Mol Cell Biol 2008.
30. Pandita TK, Hittelman WN. Initial chromosome damage but not DNA damage is greater in ataxia telangiectasia cells. Radiat Res 1992;130:94–103.
31. Sharma GG, Gupta A, Wang H, et al. hTERT associates with human telomeres and enhances genomic stability and DNA repair. Oncogene 2003;22:131–46.
32. Tahara H, Kusunoki M, Yamanaka Y, Matsumura S, Ide T. G-tail telomere HPA: simple measurement of human single-stranded telomeric overhangs. Nat Methods 2005;2:829–31.
33. Crabbe L, Verdun RE, Haggblom CI, Karlseder J. Defective telomere lagging strand synthesis in cells lacking WRN helicase activity. Science 2004;306:1951–3.
34. Maser RS, DePinho RA. Telomeres and the DNA damage response: why the fox is guarding the henhouse. DNA Repair (Amst) 2004;3:979–88.
35. Feldser DM, Hackett JA, Greider CW. Telomere dysfunction and the initiation of genome instability. Nat Rev Cancer 2003;3:623–7.
36. Pandita TK. ATM function and telomere stability. Oncogene 2002;21:611–8.
37. van Steensel B, Smogorzewska A, de Lange T. TRF2 protects human telomeres from end-to-end fusions. Cell 1998;92:401–13.
38. Karlseder J, Hoke K, Mirzoeva OK, et al. The Telomeric Protein TRF2 Binds the ATM Kinase and Can Inhibit the ATM-Dependent DNA Damage Response. PLoS Biol 2004;2:E240.
39. Gu B, Bessler M, Mason PJ. Dyskerin, telomerase and the DNA damage response. Cell Cycle 2009;8.
40. Takai H, Smogorzewska A, de Lange T. DNA damage foci at dysfunctional telomeres. Curr Biol 2003;13:1549–56.
41. Slijepcevic P. The role of DNA damage response proteins at telomeres – an "integrative" model. DNA Repair (Amst) 2006;5:1299–306.
42. Pandita TK, Pathak S, Geard CR. Chromosome end associations, telomeres and telomerase activity in ataxia telangiectasia cells. Cytogenet Cell Genet 1995;71:86–93.
43. Pandita TK, Hall EJ, Hei TK, et al. Chromosome end-to-end associations and telomerase activity during cancer progression in human cells after treatment with alpha-particles simulating radon progeny. Oncogene 1996;13:1423–30.
44. Pandita TK, Westphal CH, Anger M, et al. Atm inactivation results in aberrant telomere clustering during meiotic prophase. Mol Cell Biol 1999;19:5096–105.
45. Pandita TK, Dhar S. Influence of ATM function on interactions between telomeres and nuclear matrix. Radiat Res 2000;154:133–9.

46. Scherthan H, Jerratsch M, Dhar S, Wang YA, Goff SP, Pandita TK. Meiotic telomere distribution and Sertoli cell nuclear architecture are altered in Atm- and Atm-p53-deficient mice. Mol Cell Biol 2000;20:7773–83.
47. Pandita TK, Hunt CR, Sharma GG, Yang Q. Regulation of telomere movement by telomere chromatin structure. Cell Mol Life Sci 2007;64:131–8.
48. Pandita TK. The role of ATM in telomere structure and function. Radiat Res 2001;156: 642–7.
49. Pandita TK. Telomeres and Telomerase Encyclopedia of Cancer 2002;4:335–362.
50. Pandita TK. A multifaceted role for ATM in genome maintenance. Exp Rev Mol Med 2003;5:1–21.
51. Gupta A, Guerin-Peyrou TG, Sharma GG, et al. The mammalian ortholog of Drosophila MOF that acetylates histone H4 lysine 16 is essential for embryogenesis and oncogenesis. Mol Cell Biol 2008;28:397–409.
52. Gupta A, Sharma GG, Young CS, et al. Involvement of human MOF in ATM function. Mol Cell Biol 2005;25:5292–305.
53. Sun Y, Jiang X, Chen S, Fernandes N, Price BD. A role for the Tip60 histone acetyltransferase in the acetylation and activation of ATM. Proc Natl Acad Sci USA 2005;102: 13182–7.
54. Fischle W, Wang Y, Allis CD. Histone and chromatin cross-talk. Curr Opin Cell Biol 2003;15:172–83.
55. Legube G, Trouche D. Regulating histone acetyltransferases and deacetylases. EMBO Rep 2003;4:944–7.
56. Pruss D, Reeves R, Bushman FD, Wolffe AP. The influence of DNA and nucleosome structure on integration events directed by HIV integrase. J Biol Chem 1994;269: 25031–41.
57. Pruss D, Bushman FD, Wolffe AP. Human immunodeficiency virus integrase directs integration to sites of severe DNA distortion within the nucleosome core. Proc Natl Acad Sci USA 1994;91:5913–7.
58. Otten AD, Tapscott SJ. Triplet repeat expansion in myotonic dystrophy alters the adjacent chromatin structure. Proc Natl Acad Sci USA 1995;92:5465–9.
59. Wallrath LL, Lu Q, Granok H, Elgin SC. Architectural variations of inducible eukaryotic promoters: preset and remodeling chromatin structures. Bioessays 1994;16:165–70.
60. Hoeijmakers JH. DNA repair mechanisms. Maturitas 2001;38:17–22; discussion -3.
61. Fernandez-Capetillo O, Mahadevaiah SK, Celeste A, et al. H2AX is required for chromatin remodeling and inactivation of sex chromosomes in male mouse meiosis. Dev Cell 2003;4:497–508.
62. Fernandez-Capetillo O, Nussenzweig A. Linking histone deacetylation with the repair of DNA breaks. Proc Natl Acad Sci USA 2004;101:1427–8.
63. Nakamura TM, Du LL, Redon C, Russell P. Histone H2A phosphorylation controls Crb2 recruitment at DNA breaks, maintains checkpoint arrest, and influences DNA repair in fission yeast. Mol Cell Biol 2004;24:6215–30.
64. Sims RJ, 3rd, Nishioka K, Reinberg D. Histone lysine methylation: a signature for chromatin function. Trends Genet 2003;19:629–39.
65. Olins DE, Olins AL. Chromatin history: our view from the bridge. Nat Rev Mol Cell Biol 2003;4:809–14.
66. Workman JL, Kingston RE. Alteration of nucleosome structure as a mechanism of transcriptional regulation. Annu Rev Biochem 1998;67:545–79.
67. Strahl BD, Grant PA, Briggs SD, et al. Set2 is a nucleosomal histone H3-selective methyltransferase that mediates transcriptional repression. Mol Cell Biol 2002;22:1298–306.
68. Khorasanizadeh S. The nucleosome: from genomic organization to genomic regulation. Cell 2004;116:259–72.
69. Green CM, Almouzni G. When repair meets chromatin. First in series on chromatin dynamics. EMBO Rep 2002;3:28–33.

70. Vidanes GM, Bonilla CY, Toczyski DP. Complicated tails: histone modifications and the DNA damage response. Cell 2005;121:973–6.
71. Peterson CL, Cote J. Cellular machineries for chromosomal DNA repair. Genes Dev 2004;18:602–16.
72. van Attikum H, Gasser SM. The histone code at DNA breaks: a guide to repair? Nat Rev Mol Cell Biol 2005;6:757–65.
73. Lydall D, Whitehall S. Chromatin and the DNA damage response. DNA Repair (Amst) 2005;4:1195–207.
74. Wurtele H, Verreault A. Histone post-translational modifications and the response to DNA double-strand breaks. Curr Opin Cell Biol 2006;18:137–44.
75. Loizou JI, Murr R, Finkbeiner MG, Sawan C, Wang ZQ, Herceg Z. Epigenetic information in chromatin: the code of entry for DNA repair. Cell Cycle 2006;5:696–701.
76. Lund AH, van Lohuizen M. Epigenetics and cancer. Genes Dev 2004;18:2315–35.
77. Aucott R, Bullwinkel, J., Yu, Y., Shi, W., Billur, M., Brown, J.P., Menzel, U., Kioussis, D., Wang, G., Reisert, I., Weimer, J., Pandita, R.K., Sharma, G.G., Pandita, T.K., Fundele R, Singh P.B. HP1? is required for development of the cerebral neocortex and diaphragmatic neuromuscular junctions. J Cell Biol 2008;183:597–606.
78. Lavin MF, Kozlov S. ATM activation and DNA damage response. Cell Cycle 2007;6: 931–42.
79. Bentley NJ, Holtzman DA, Flaggs G, et al. The Schizosaccharomyces pombe rad3 checkpoint gene. Embo J 1996;15:6641–51.
80. Lydall D, Nikolsky Y, Bishop DK, Weinert T. A meiotic recombination checkpoint controlled by mitotic checkpoint genes. Nature 1996;383:840–3.
81. Keith CT, Schreiber SL. PIK-related kinases: DNA repair, recombination, and cell cycle checkpoints. Science 1995;270:50–1.
82. Savitsky K, Bar-Shira A, Gilad S, et al. A single ataxia telangiectasia gene with a product similar to PI-3 kinase. Science 1995;268:1749–53.
83. Pandita TK, Hittelman WN. The contribution of DNA and chromosome repair deficiencies to the radiosensitivity of ataxia-telangiectasia. Radiat Res 1992;131:214–23.
84. Morgan SE, Lovly C, Pandita TK, Shiloh Y, Kastan MB. Fragments of ATM which have dominant-negative or complementing activity. Mol Cell Biol 1997;17:2020–9.
85. Barlow C, Hirotsune S, Paylor R, et al. Atm-deficient mice: a paradigm of ataxia telangiectasia. Cell 1996;86:159–71.
86. Elson A, Wang Y, Daugherty CJ, et al. Pleiotropic defects in ataxia-telangiectasia protein-deficient mice. Proc Natl Acad Sci USA 1996;93:13084–9.
87. Xu Y, Ashley T, Brainerd EE, Bronson RT, Meyn MS, Baltimore D. Targeted disruption of ATM leads to growth retardation, chromosomal fragmentation during meiosis, immune defects, and thymic lymphoma. Genes Dev 1996;10:2411–22.
88. Bakkenist CJ, Kastan MB. DNA damage activates ATM through intermolecular autophosphorylation and dimer dissociation. Nature 2003;421:499–506.
89. Kozlov SV, Graham ME, Peng C, Chen P, Robinson PJ, Lavin MF. Involvement of novel autophosphorylation sites in ATM activation. EMBO J 2006;25:3504–14.
90. Lee JH, Paull TT. Direct activation of the ATM protein kinase by the Mre11/Rad50/Nbs1 complex. Science 2004;304:93–6.
91. Bredemeyer AL, Sharma GG, Huang CY, et al. ATM stabilizes DNA double-strand-break complexes during V(D)J recombination. Nature 2006;442:466–70.
92. Huang CY, Sharma GG, Walker LM, Bassing CH, Pandita TK, Sleckman BP. Defects in coding joint formation in vivo in developing ATM-deficient B and T lymphocytes. J Exp Med 2007;204:1371–81.
93. Callen E, Jankovic M, Difilippantonio S, et al. ATM prevents the persistence and propagation of chromosome breaks in lymphocytes. Cell 2007;130:63–75.
94. Richard DJ, Bolderson E, Cubeddu L, et al. Single-stranded DNA-binding protein hSSB1 is critical for genomic stability. Nature 2008;453:677–81.

95. Usui T, Ogawa H, Petrini JH. A DNA damage response pathway controlled by Tel1 and the Mre11 complex. Mol Cell 2001;7:1255–66.
96. Chan SW, Chang J, Prescott J, Blackburn EH. Altering telomere structure allows telomerase to act in yeast lacking ATM kinases. Curr Biol 2001;11:1240–50.
97. Stohr BA, Blackburn EH. ATM mediates cytotoxicity of a mutant telomerase RNA in human cancer cells. Cancer Res 2008;68:5309–17.
98. Gately DP, Hittle JC, Chan GK, Yen TJ. Characterization of ATM expression, localization, and associated DNA-dependent protein kinase activity. Mol Biol Cell 1998;9: 2361–74.
99. Andegeko Y, Moyal L, Mittelman L, Tsarfaty I, Shiloh Y, Rotman G. Nuclear retention of ATM at sites of DNA double strand breaks. J Biol Chem 2001;276:38224–30.
100. Kishi S, Lu KP. A critical role for Pin2/TRF1 in ATM-dependent regulation. Inhibition of Pin2/TRF1 function complements telomere shortening, radiosensitivity, and the G(2)/M checkpoint defect of ataxia-telangiectasia cells. J Biol Chem 2002;277:7420–9.
101. Kishi S, Zhou XZ, Ziv Y, et al. Telomeric protein Pin2/TRF1 as an important ATM target in response to double strand DNA breaks. J Biol Chem 2001;276:29282–91.
102. Smilenov LB, Dhar S, Pandita TK. Altered telomere nuclear matrix interactions and nucleosomal periodicity in ataxia telangiectasia cells before and after ionizing radiation treatment. Mol Cell Biol 1999;19:6963–71.
103. Takata H, Kanoh Y, Gunge N, Shirahige K, Matsuura A. Reciprocal association of the budding yeast ATM-related proteins Tel1 and Mec1 with telomeres in vivo. Mol Cell 2004;14:515–22.
104. Verdun RE, Crabbe L, Haggblom C, Karlseder J. Functional human telomeres are recognized as DNA damage in G2 of the cell cycle. Mol Cell 2005;20:551–61.
105. Baskaran R, Wood LD, Whitaker LL, et al. Ataxia telangiectasia mutant protein activates c-Abl tyrosine kinase in response to ionizing radiation. Nature 1997;387:516–9.
106. Shafman T, Khanna KK, Kedar P, et al. Interaction between ATM protein and c-Abl in response to DNA damage. Nature 1997;387:520–3.
107. Kharbanda S, Kumar V, Dhar S, et al. Regulation of the hTERT telomerase catalytic subunit by the c-Abl tyrosine kinase. Curr Biol 2000;10:568–75.
108. Lieberman HB, Hopkins KM, Nass M, Demetrick D, Davey S. A human homolog of the Schizosaccharomyces pombe rad9+ checkpoint control gene. Proc Natl Acad Sci USA 1996;93:13890–5.
109. Chen MJ, Lin YT, Lieberman HB, Chen G, Lee EY. ATM-dependent phosphorylation of human Rad9 is required for ionizing radiation-induced checkpoint activation. J Biol Chem 2001;276:16580–6.
110. St Onge RP, Udell CM, Casselman R, Davey S. The human G2 checkpoint control protein hRAD9 is a nuclear phosphoprotein that forms complexes with hRAD1 and hHUS1. Mol Biol Cell 1999;10:1985–95.
111. Venclovas C, Thelen MP. Structure-based predictions of Rad1, Rad9, Hus1 and Rad17 participation in sliding clamp and clamp-loading complexes. Nucleic Acids Res 2000;28: 2481–93.
112. Greer DA, Besley BD, Kennedy KB, Davey S. hRad9 rapidly binds DNA containing double-strand breaks and is required for damage-dependent topoisomerase II beta binding protein 1 focus formation. Cancer Res 2003;63:4829–35.
113. Pandita RK, Sharma GG, Laszlo A, et al. Mammalian rad9 plays a role in telomere stability, s- and g2-phase-specific cell survival, and homologous recombinational repair. Mol Cell Biol 2006;26:1850–64.
114. Hamer G, Roepers-Gajadien HL, van Duyn-Goedhart A, et al. DNA double-strand breaks and gamma-H2AX signaling in the testis. Biol Reprod 2003;68:628–34.
115. Gilley D, Tanaka H, Hande MP, et al. DNA-PKcs is critical for telomere capping. Proc Natl Acad Sci USA 2001;98:15084–8.
116. Taccioli GE, Gottlieb TM, Blunt T, et al. Ku80: product of the XRCC5 gene and its role in DNA repair and V(D)J recombination. Science 1994;265:1442–5.

117. Critchlow SE, Jackson SP. DNA end-joining: from yeast to man. Trends Biochem Sci 1998;23:394–8.
118. Kanaar R, Hoeijmakers JH, van Gent DC. Molecular mechanisms of DNA double strand break repair. Trends Cell Biol 1998;8:483–9.
119. Giffin W, Torrance H, Rodda DJ, Prefontaine GG, Pope L, Hache RJ. Sequence-specific DNA binding by Ku autoantigen and its effects on transcription. Nature 1996;380: 265–8.
120. Ludwig DL, Chen F, Peterson SR, Nussenzweig A, Li GC, Chen DJ. Ku80 gene expression is Sp1-dependent and sensitive to CpG methylation within a novel cis element. Gene 1997;199:181–94.
121. Galande S, Kohwi-Shigematsu T. Poly(ADP-ribose) polymerase and Ku autoantigen form a complex and synergistically bind to matrix attachment sequences. J Biol Chem 1999;274:20521–8.
122. Nussenzweig A, Chen C, da Costa Soares V, et al. Requirement for Ku80 in growth and immunoglobulin V(D)J recombination. Nature 1996;382:551–5.
123. Boulton SJ, Jackson SP. Identification of a Saccharomyces cerevisiae Ku80 homologue: roles in DNA double strand break rejoining and in telomeric maintenance. Nucleic Acids Res 1996;24:4639–48.
124. Porter SE, Greenwell PW, Ritchie KB, Petes TD. The DNA-binding protein Hdf1p (a putative Ku homologue) is required for maintaining normal telomere length in Saccharomyces cerevisiae. Nucleic Acids Res 1996;24:582–5.
125. Hsu HL, Gilley D, Blackburn EH, Chen DJ. Ku is associated with the telomere in mammals. Proc Natl Acad Sci USA 1999;96:12454–8.
126. Gravel S, Larrivee M, Labrecque P, Wellinger RJ. Yeast Ku as a regulator of chromosomal DNA end structure. Science 1998;280:741–4.
127. Gasser SM. A sense of the end. Science 2000;288:1377–9.
128. Mimori T, Hardin JA, Steitz JA. Characterization of the DNA-binding protein antigen Ku recognized by autoantibodies from patients with rheumatic disorders. J Biol Chem 1986;261:2274–8.
129. Paillard S, Strauss F. Analysis of the mechanism of interaction of simian Ku protein with DNA. Nucleic Acids Res 1991;19:5619–24.
130. Cary RB, Peterson SR, Wang J, Bear DG, Bradbury EM, Chen DJ. DNA looping by Ku and the DNA-dependent protein kinase. Proc Natl Acad Sci USA 1997;94:4267–72.
131. Dynan WS, Yoo S. Interaction of Ku protein and DNA-dependent protein kinase catalytic subunit with nucleic acids. Nucleic Acids Res 1998;26:1551–9.
132. Bailey SM, Meyne J, Chen DJ, et al. DNA double-strand break repair proteins are required to cap the ends of mammalian chromosomes. Proc Natl Acad Sci U S A 1999;96:14899–904.
133. Lombard DB, Guarente L. Nijmegen breakage syndrome disease protein and MRE11 at PML nuclear bodies and meiotic telomeres. Cancer Res 2000;60:2331–4.
134. Nakamura TM, Moser BA, Russell P. Telomere binding of checkpoint sensor and DNA repair proteins contributes to maintenance of functional fission yeast telomeres. Genetics 2002;161:1437–52.
135. Takata H, Tanaka Y, Matsuura A. Late S phase-specific recruitment of Mre11 complex triggers hierarchical assembly of telomere replication proteins in Saccharomyces cerevisiae. Mol Cell 2005;17:573–83.
136. Jazayeri A, Falck J, Lukas C, et al. ATM- and cell cycle-dependent regulation of ATR in response to DNA double-strand breaks. Nat Cell Biol 2006;8:37–45.
137. Larrivee M, LeBel C, Wellinger RJ. The generation of proper constitutive G-tails on yeast telomeres is dependent on the MRX complex. Genes Dev 2004;18:1391–6.
138. Chan TA, Hermeking H, Lengauer C, Kinzler KW, Vogelstein B. 14-3-3Sigma is required to prevent mitotic catastrophe after DNA damage. Nature 1999;401:616–20.
139. Xing H, Kornfeld K, Muslin AJ. The protein kinase KSR interacts with 14-3-3 protein and Raf. Curr Biol 1997;7:294–300.

140. Dhar S, Squire JA, Hande MP, Wellinger RJ, Pandita TK. Inactivation of 14-3-3sigma influences telomere behavior and ionizing radiation-induced chromosomal instability. Mol Cell Biol 2000;20:7764–72.
141. Li Y, Kirschmann DA, Wallrath LL. Does heterochromatin protein 1 always follow code? Proc Natl Acad Sci USA 2002;99 Suppl 4:16462–9.
142. Singh PB, Miller JR, Pearce J, et al. A sequence motif found in a Drosophila heterochromatin protein is conserved in animals and plants. Nucleic Acids Res 1991;19:789–94.
143. Jones DO, Cowell IG, Singh PB. Mammalian chromodomain proteins: their role in genome organisation and expression. Bioessays 2000;22:124–37.
144. Eissenberg JC, Elgin SC. The HP1 protein family: getting a grip on chromatin. Curr Opin Genet Dev 2000;10:204–10.
145. Wreggett KA, Hill F, James PS, Hutchings A, Butcher GW, Singh PB. A mammalian homologue of Drosophila heterochromatin protein 1 (HP1) is a component of constitutive heterochromatin. Cytogenet Cell Genet 1994;66:99–103.
146. Saunders WS, Chue C, Goebl M, et al. Molecular cloning of a human homologue of Drosophila heterochromatin protein HP1 using anti-centromere autoantibodies with anti-chromo specificity. J Cell Sci 1993 ;104 (Pt 2):573–82.
147. Sharma GG, Hwang KK, Pandita RK, et al. Human heterochromatin protein 1 isoforms HP1(Hsalpha) and HP1(Hsbeta) interfere with hTERT-telomere interactions and correlate with changes in cell growth and response to ionizing radiation. Mol Cell Biol 2003;23: 8363–76.
148. Ziv Y, Bielopolski D, Galanty Y, et al. Chromatin relaxation in response to DNA double-strand breaks is modulated by a novel ATM- and KAP-1 dependent pathway. Nat Cell Biol 2006;8:870–6.
149. Lechner MS, Levitan I, Dressler GR. PTIP, a novel BRCT domain-containing protein interacts with Pax2 and is associated with active chromatin. Nucleic Acids Res 2000;28: 2741–51.
150. Schultz DC, Ayyanathan K, Negorev D, Maul GG, Rauscher FJ, 3rd. SETDB1: a novel KAP-1-associated histone H3, lysine 9-specific methyltransferase that contributes to HP1-mediated silencing of euchromatic genes by KRAB zinc-finger proteins. Genes Dev 2002;16:919–32.
151. Ayoub N, Jeyasekharan AD, Bernal JA, Venkitaraman AR. HP1-beta mobilization promotes chromatin changes that initiate the DNA damage response. Nature 2008;453:682–6.
152. Aucott R, Bullwinkel J, Yu Y, et al. HP1-{beta} is required for development of the cerebral neocortex and neuromuscular junctions. J Cell Biol 2008;183:597–606.
153. Vaquero A, Loyola A, Reinberg D. The constantly changing face of chromatin. Sci Aging Knowledge Environ 2003;2003:RE4.
154. Shogren-Knaak M, Ishii H, Sun JM, Pazin MJ, Davie JR, Peterson CL. Histone H4-K16 acetylation controls chromatin structure and protein interactions. Science 2006;311: 844–7.
155. Gupta A, Sharma, G.G., Young, C.S.H., Agarwal, M.,Smith, E.R., Paull, T.T., Lucchesi, J.C., Khanna, K.K., Ludwig, T., and Pandita, T.K.,. Involvement of human MOF in ATM function. Mol Cell Biol 2005;25:5292–305.
156. Kusch T, Florens L, Macdonald WH, et al. Acetylation by Tip60 is required for selective histone variant exchange at DNA lesions. Science 2004;306:2084–7.
157. Thiriet C, Hayes JJ. Chromatin in need of a fix: phosphorylation of H2AX connects chromatin to DNA repair. Mol Cell 2005;18:617–22.
158. Sharma GG, Hall EJ, Dhar S, Gupta A, Rao PH, Pandita TK. Telomere stability correlates with longevity of human beings exposed to ionizing radiations. Oncol Rep 2003;10:1733–6.
159. Downs JA, Nussenzweig MC, Nussenzweig A. Chromatin dynamics and the preservation of genetic information. Nature 2007;447:951–8.
160. Mailand N, Bekker-Jensen S, Faustrup H, et al. RNF8 ubiquitylates histones at DNA double-strand breaks and promotes assembly of repair proteins. Cell 2007;131:887–900.

161. Scott SP, Pandita TK. The cellular control of DNA double-strand breaks. J Cell Biochem 2006;99:1463–1475.
162. Hunt CR, Pandita RK, Laszlo A, et al. Hyperthermia activates a subset of ataxia-telangiectasia mutated effectors independent of DNA strand breaks and heat shock protein 70 status. Cancer Res 2007;67:3010–7.
163. Agarwal M, Pandita S, Hunt CR, et al. Inhibition of telomerase activity enhances hyperthermia-mediated radiosensitization. Cancer Res 2008;68:3370–8.
164. Pandita TK, Richardson, C. Chromatin remodling finds its place in the DNA double-strand break response. Nucleic Acids Res 2009 (online).
165. Misri S, Pandita S, Kumar R and Pandita TK. Telomeres, histone code and DNA damage. Cytogenetics Genome Res. 2008;122:297–307.

Chapter 8
DNA Double Strand Break Repair: Mechanisms and Therapeutic Potential

Laura M. Williamson, Chris T. Williamson, and Susan P. Lees-Miller

Abstract DNA double strand breaks are widely considered to be the most cytotoxic form of DNA damage. They can be induced by exposure to exogenous agents such as ionizing radiation and topoisomerase poisons and can also occur naturally due to replication fork collapse and other cellular processes. DNA double strand breaks are also produced in the immune system as a consequence of V(D)J and class switch recombination. Here, we will review the major pathways for the detection and repair of DNA double strand breaks in human cells, namely non-homologous end joining and homologous recombination/homology-directed repair. We will also discuss how a better understanding of these pathways could lead to new, more targeted approaches for cancer therapy.

Keywords Alternative-NHEJ · DSB repair · Homologous recombination · Non-homologous end joining · Poly ADP ribose polymerase-1 · Radiosensitizers · Synthetic lethality

Abbreviations

A-T	Ataxia-Telangiectasia
ATM	ataxia-telangiectasia mutated
ATR	ATM- and Rad3-related
BRCA	Breast and ovarian cancer susceptibility allele
CSR	class switch recombination
ds	double-stranded
DNA-PK	DNA-dependent protein kinase
DNA-PKcs	catalytic subunit of the DNA-dependent protein kinase
DSB	DNA double strand break
DSBR	DNA double strand break repair

S.P. Lees-Miller (✉)
Department of Biochemistry and Molecular Biology, Southern Alberta Cancer Research Institute, University of Calgary, 3330 Hospital Drive NW, Calgary, AB T2N 4N1, Canada
e-mail: leesmill@ucalgary.ca

The first two authors contributed equally to this work.

K.K. Khanna, Y. Shiloh (eds.), *The DNA Damage Response: Implications on Cancer Formation and Treatment*, DOI 10.1007/978-90-481-2561-6_8,

FANCD1	Fanconi Anemia complementary group D1
FA	Fanconi Anemia
HDR	homology directed repair
IR	ionizing radiation
NBS	Nijmegen Breakage Syndrome
NHEJ	non-homologous end joining
PARP-1	poly ADP ribose polymerase-1
PIKK	phosphatidylinositol 3-kinase-like protein kinase
PNKP	polynucleotide kinase/phosphatase
RS-SCID	Radiosensitive severe combined immunodeficiency
SNP	single nucleotide polymorphism
ss	single-stranded
SSB	DNA single strand break
XLF	XRCC4-like factor

8.1 Introduction

DNA double strand breaks (DSBs) are widely considered to be the most cytotoxic form of DNA damage [52]. Indeed, a single DSB can be lethal to the cell [12]. DSBs can be introduced by exposure to exogenous agents, such as ionizing radiation (IR), radiomimetics (e.g. bleomycin and neocarzinostatin) and topoisomerase poisons (e.g. doxorubicin, etoposide and camptothecin), as well as endogenous agents (e.g. reactive oxygen species). DSBs can also arise as a result of collapsed replication forks (reviewed in [44, 85, 111, 115]) and during the processes of V(D)J recombination and class switch recombination (CSR), which are required for the development and maturation of the vertebrate immune system (reviewed in [8, 63, 65, 135]). Also, telomeres, which resemble DSBs, must be protected to prevent initiation of an inappropriate DNA damage response (reviewed in [32, 86]). Understanding how DSBs are detected and repaired is of critical importance as unrepaired DSBs can lead to cell death, while incorrectly repaired DSBs have the potential to produce chromosomal translocations and genomic instability, which can contribute to malignant transformation [59, 67].

In human cells there are two major pathways for the repair of DSBs, namely non-homologous end joining (NHEJ) and homologous recombination, also termed homology directed repair (HDR). In mammalian cells, the majority of IR-induced DSBs are repaired by NHEJ. NHEJ functions throughout the cell cycle, and is the major pathway for the repair of DSBs that arise during G0 and G1 [123]. In addition, NHEJ is responsible for the repair of programmed DSBs generated during V(D)J recombination [8, 63]. In contrast, HDR functions only in late S and G2 when an intact sister chromatid is available to act as a template for repair. One of the main functions of HDR is the repair of endogenous DSBs that are produced when replication forks collapse, and the reader is referred to several excellent reviews on this topic [55, 80, 137]. In this review, we will discuss our current understanding of the processes by which mammalian cells detect and repair IR-induced DSBs. We will also highlight how a better understanding of these pathways could lead to new, more targeted therapies for the treatment of cancer.

8.2 Detection and Repair of IR-Induced DNA Damage

8.2.1 IR-Induced Forms of DNA Damage

IR induces multiple forms of DNA damage, including damage to the bases and the ribose sugar groups as well as cleavage of the phosphate backbone to produce DNA single strand breaks (SSBs). DSBs are produced when two SSBs occur a short distance apart on opposite DNA strands. IR-induced DSBs are rarely simple blunt ended breaks with ligatable ends, but instead often contain multiple, complex or clustered lesions and overhanging ends that may contain non-ligatable ends such as 5′- hydroxyl, 3′-phosphate or 3′-phosphoglycolate groups which must be removed before the DNA ends can be rejoined [44].

8.2.2 The Major DSB Repair Pathways in Mammalian Cells

8.2.2.1 Non-Homologous End Joining (NHEJ)

The basic steps in NHEJ are (1) detection of the DSB and tethering of the DNA ends, (2) processing of the DNA termini to remove non-ligatable end groups and (3) religation of the processed DNA ends. Since NHEJ does not require a DNA template for repair, it is widely regarded as being error prone. Six core proteins are known to be required for NHEJ in human cells: the Ku70/80 heterodimer (Ku), the catalytic subunit of the DNA-dependent protein kinase (DNA-PKcs), XRCC4, DNA ligase IV, Artemis and XRCC4-like factor (XLF, also called Cernunnos) (reviewed in [81, 91, 146]). Deletion or inactivation of any of these core NHEJ proteins confers radiation sensitivity as well as defects in DSB repair and V(D)J recombination (reviewed in [111]). In addition to the core NHEJ proteins, repair of IR-induced DSBs likely involves additional processing factors such as polynucleotide kinase/phosphatase (PNKP), members of the DNA polymerase X family and perhaps other proteins (reviewed in [91]).

The first step in NHEJ is detection of the DSB by Ku. Ku is conserved throughout evolution, and is required for NHEJ in plants [16], budding and fission yeast [36, 119] and even some bacteria [133]. Accordingly, Ku is regarded as a core component of the NHEJ pathway [81, 82, 91, 146]. Ku binds with high affinity to the ends of double-stranded (ds) DNA in vitro, with little or no DNA sequence specificity (reviewed in [35, 91]). In vivo, the Ku heterodimer is recruited to sites of DNA damage rapidly and independently of other NHEJ or DSB repair proteins (Fig. 8.1) [69]. Once bound to the DSB, Ku acts as a scaffold to which other NHEJ proteins are recruited. Cells lacking Ku are radiosensitive and Ku-deficient mice are immunodeficient due to defective V(D)J recombination, and also exhibit growth

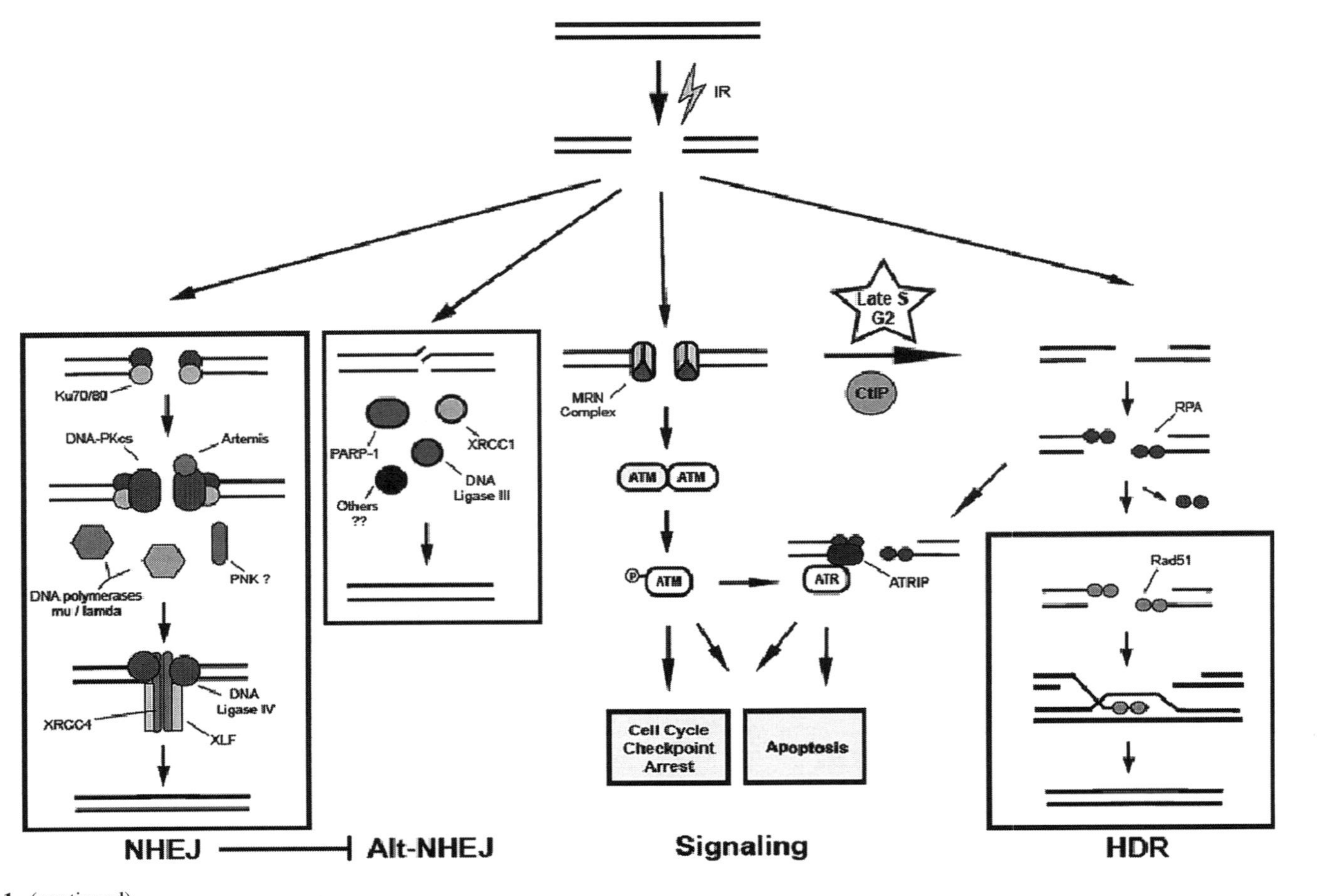

Fig. 8.1 (continued)

and telomere defects [50, 51, 106]. Interestingly, Ku-deficient human cells are non-viable, possibly due to defects in telomere function, suggesting possible differences in Ku function in mouse and human cells [42].

One of the most well characterized binding partners of Ku is DNA-PKcs. DNA-PKcs is a member of the phosphatidylinositol-3 kinase-like family of serine-threonine protein kinases (PIKKs), and like other PIKK family members, Ataxia-Telangiectasia Mutated (ATM) and ATM-, Rad3-related (ATR), DNA-PKcs phosphorylates its substrates primarily on serines or threonines that are followed by glutamines (SQ/TQ motifs) (reviewed in [1, 71, 91, 95]). Alone, DNA-PKcs has very weak protein kinase activity which is stimulated 5–10 fold by its interaction with DNA-bound Ku to form the DNA-dependent protein kinase complex, DNA-PK. DNA-PKcs null cells are radiosensitive [77] and have defective DSB repair [17, 34, 121]. Although there are no human syndromes associated with DNA-PKcs deficiency, a human patient with radiosensitive severe combined immunodeficiency (RS-SCID) resulting from a missense mutation in DNA-PKcs was recently identified [140]. Significantly, cells in which the protein kinase activity of DNA-PKcs is inactivated, whether by mutation [68, 70] or by small molecule inhibitors [27, 108, 113, 122, 156], are highly sensitive to IR, indicating that the protein kinase activity of DNA-PKcs is critical for NHEJ.

Unlike ATM and ATR, which both phosphorylate multiple proteins in response to DNA damage in vivo [28, 71-73], relatively few physiologically relevant substrates have been identified for DNA-PK. Indeed, although DNA-PK phosphorylates many proteins in vitro, including many of the core NHEJ proteins (Ku70, Ku80, XRCC4, DNA ligase IV, Artemis and XLF), there is little evidence that DNA-PK-dependent phosphorylation of any of these proteins is required for DSB repair in vivo (reviewed in [91]). To date, the best candidate for a physiologically relevant substrate of DNA-PKcs is DNA-PKcs itself. DNA-PKcs is highly phosphorylated both in vitro and in vivo and IR-induced phosphorylation of DNA-PKcs in vivo is largely DNA-PK-dependent (reviewed in [91, 95, 146]). Moreover, autophosphorylation of DNA-PKcs is required for the dissociation of DNA-PKcs from Ku in vitro [24] and from sites of DNA damage in vivo [139]. These findings have led to a model for the initial stages of NHEJ in which DNA-PKcs is first recruited to DNA-bound Ku to form

Fig. 8.1 DNA double strand break repair pathways in human cells. IR-induced DSBs in human cells can be detected and repaired by one of two main pathways: non-homologous end joining (NHEJ) or homology directed repair (HDR). In addition, an alternative end-joining pathway (Alt-NHEJ), which utilizes regions of microhomology between the broken ends to mediate repair has been identified in NHEJ-deficient cells. The Alt-NHEJ pathway is inhibited by NHEJ. NHEJ functions throughout the cell cycle whereas HDR is active only in late S and G2 phases of the cell cycle. The MRN complex detects DSBs in all stages of the cell cycle and is required for activation of ATM throughout the cell cycle and for activation of ATR and initiation of HDR in late S and G2. The activity of the MRN complex is regulated by CtIP, which is expressed only after DNA replication, thus linking initiation of HDR and activation of ATR to S and G2 phases of the cell cycle. See text for details

the DNA-PK complex which tethers the DNA ends together protecting them from inappropriate end processing. Once assembled at the ends of the DSB, DNA-PKcs undergoes autophosphorylation in *trans*, leading to dissociation of DNA-PKcs from the DNA ends, which facilitates access of downstream factors to the ends of the DSB [91, 95, 96]. Together, these studies suggest that the main function of DNA-PKcs is to protect DNA ends and to regulate the access of other proteins to the DSB, which in turn regulates pathway progression (reviewed in [91, 95, 97, 145, 146]).

Another role for DNA-PKcs is to recruit the end-processing factor Artemis to DSBs [89]. Artemis has 5′-3′ exonuclease activity and in the presence of DNA-PKcs and ATP displays endonuclease activity towards DNA hairpins and single stranded (ss) DNA flaps [90]. The nuclease activity of Artemis is required for opening DNA hairpins formed at coding joints during V(D)J recombination and mutations in Artemis are associated with radiation-sensitive severe combined immunodeficiency (RS-SCID) [103] (Table 8.1). The DNA-PKcs/Artemis complex can also remove 3′-phosphoglycolate groups from some DNA ends suggesting an additional role in end-processing [116]. However, Artemis-deficient cells are only modestly radiosensitive and have a mild DNA repair defect, suggesting that Artemis is required for the processing of only a subset of IR-induced DSBs in vivo [121]. It is also likely that other DNA end processing enzymes such as PNKP and DNA polymerases are involved in NHEJ. PNKP, which interacts with XRCC4, can convert non-ligatable 3’-phosphate and 5’-hydroxyl groups to 3’-hydroxyl and 5’-phosphate groups, respectively, while DNA polymerases mu and lambda may play a gap filling role during NHEJ (reviewed in [91]).

Once the DNA ends have been processed, they are ligated by the XRCC4-DNA ligase IV complex (X4-L4). XRCC4 and DNA ligase IV are required for both NHEJ

Table 8.1 Human syndromes associated with defects in DSB repair pathways

Protein	Syndrome	DSB Response Pathway Affected	Cancer Predisposition	Immune Defects	Reference
Artemis	RS-SCID	NHEJ	+	+	[102]
ATM	A-T	DSBR Signaling	+	+	[128]
ATR	ATR-Seckel	DSBR Signaling	–	–	[112]
BRCA2 (FANCD1)	FA (D1 group)	HDR / FA pathway	+	+	[61]
Ligase IV	Ligase IV syndrome	NHEJ	+	+	[109]
Mre11	ATLD	HDR/DSBR signaling	–	+	[136]
Nbs1	NBS	HDR/DSBR signaling	+	+	[125]
XLF/ Cernnunos	RS-CID with Microcephaly	NHEJ	–	+	[21]

References refer to the first paper identifying the protein responsible for the syndrome. For a more detailed review of phenotypes associated with defects in DSB repair proteins please refer to the following reviews: [30, 33, 72, 110, 111, 120, 143].

and V(D)J recombination and mice lacking either protein die in utero due to massive neuronal apoptosis [7, 43]. Hypomorphic mutations in DNA ligase IV are associated with ligase IV syndrome, which is characterized by neuronal degeneration and extreme radiosensitivity [111] (Table 8.1). The recently identified NHEJ factor XLF interacts with XRCC4 [3, 21] and Ku [153] and stimulates the activity of the XRCC4-DNA ligase IV complex towards non-compatible DNA ends [49, 88], suggesting that it plays a regulatory role in NHEJ. Consistent with its role in NHEJ, patients with defects in XLF exhibit immunodeficiency and radiosensitivity [21] (Table 8.1).

8.2.2.2 Alternative Non-Homologous End Joining (Alt-NHEJ)

As described above, NHEJ is the major pathway for the repair of IR-induced DSBs in human cells. It is predicted to occur with a minimal requirement for DNA microhomology and relatively little processing of the DNA surrounding the ends. There is increasing evidence that cells also contain an alternative end-joining pathway (Alt-NHEJ) that ligates DNA ends in the absence of the classical NHEJ factors (Ku, DNA-PKcs, XRCC4 and DNA ligase IV). This alternative end-joining pathway may involve the SSB and base excision repair proteins poly-ADP ribose polymerase-1 (PARP-1), XRCC1 and DNA ligase III (Fig. 8.1) [5, 141, 142]. In the absence of classical NHEJ proteins, PARP-1 is thought to bind DNA ends and stimulate synapsis prior to ligation by the XRCC1/ligase III complex. Ku inhibits this process by blocking access of PARP-1 to DNA ends, suggesting that the presence of a functional classical-NHEJ pathway inhibits the Alt-NHEJ pathway, and therefore, that Alt-NHEJ may only function in NHEJ-deficient cells [142]. Additional evidence for the presence of an alternative end joining pathway comes from studies of CSR in mammalian cells that lack XRCC4, DNA ligase IV or Ku70/80 and exhibit robust NHEJ-independent, highly error prone end joining [65, 107, 152].

8.2.3 Homology Directed Repair (HDR)

The second major pathway for the repair of DSBs in human cells is HDR. HDR mediates the repair of DSBs induced by IR and collapsed replication forks as well as inter-strand cross-links and is active only during late S and G2 phases of the cell cycle. The study of HDR-defective cell lines, as well as experiments using recombination reporter assays in yeast and higher eukaryotes, has led to the development of several models for HDR that require a sister chromatid to act as a DNA template for repair, and are therefore considered to be a relatively error free repair pathways (reviewed in [55, 80, 126, 137]). Other more error prone variations of the HDR pathway, such as single-strand annealing have also been described and are discussed elsewhere [55, 80, 126, 137].

The initial step in HDR is detection of the DSB by the Mre11, Rad50, Nbs1 (MRN) complex, which is followed by 5′-3′ resection to produce a long ssDNA 3′ extension (reviewed in [55, 80, 126, 137]) (Fig. 8.1). In the next step, termed strand invasion, the 3′-ssDNA strand acquires the ability to invade the homologous sequence to form an intermediate called a D loop (Fig. 8.1). For this to occur, the ssDNA must be bound by Rad51 to form a presynaptic filament. Although Rad51 readily binds ssDNA to form a nucleoprotein filament in vitro [144], in vivo, formation of the filament is highly regulated. The first step in filament formation involves binding of the ssDNA binding protein RPA to the long 3′-ssDNA extension, which prevents premature strand invasion. Multiple proteins, including Rad52, BRCA2 and the Rad51 paralogues (Rad51B, Rad51C, Rad51D, XRCC2 and XRCC3), are subsequently involved in the replacement of RPA by Rad51, and stabilization of the filament (reviewed in [126, 137], see also Chapter 18). Subsequently, in steps that are not well characterized at a molecular level, the 3′-DNA end is extended by a DNA polymerase, likely pol eta, and a second DSB end is captured by annealing to the extended D loop. This leads to the formation of one or more crossed strand structures termed Holliday junctions. Finally, the Holliday junctions are resolved to produce either crossover or non-crossover products depending on which HDR subpathway is utilized (see [55, 80, 126, 137] for details). Proteins that have been implicated in these later steps of HDR include Rad54, WRN, BLM, p53 BLAP75, hMSH2-hMSH6, XPF, ERCC1, DNA polymerases delta and epsilon, PCNA and DNA ligase 1 (reviewed in [55]).

The MRN complex plays an important role in multiple aspects of the DNA damage response (reviewed in [149]). Mre11 exhibits both 3′-5′ exonuclease and ssDNA endonuclease activities and its nuclease activity is required for HDR [22]. Interestingly, the ssDNA endonuclease activity of Mre11, but not its exonuclease activity, is required for strand resection [148]. In human cells, MRN interacts with CtIP (C-terminal region of adenovirus E1A-binding protein [CtBP]-interacting protein), which has been shown to stimulate the endonuclease activity of Mre11 and to promote MRN-dependent resection [127]. The functional homologues of CtIP, Sporulation in the Absence of Spo Eleven (Sae2) and CtIP-related protein 1 (Ctp1) probably play similar roles in *S. cerevisiae* and *S. pombe*, respectively [78, 83, 138]. However, it should be noted that apart from a conserved C-terminal region, CtIP, Sae2 and Ctp1 have very limited amino acid sequence similarity, which could indicate different mechanisms of regulation between yeast and mammalian cells. In yeast, the MRX (Mre11, Rad50, Xrs2) complex together with Sae2 first trims the ends of DNA, making them suitable for further processing. This step requires the RecQ helicase, Sgs1, together with an exonuclease, possibly exonuclease 1 and/or Dna2 [99, 157]. Interestingly, CtIP is also required for Alt-NHEJ in extracts from mammalian cells, suggesting that it may play a role in end resection in other DSB repair pathways [11]. Another important function of CtIP/Sae2/Ctp1 is to couple initiation of HDR with cell cycle phase. CtIP levels are low in G1 and increase in S and G2, thus linking CtIP-mediated stimulation of MRN-dependent resection to S and G2 phases of the cell cycle [83, 138]. The end-resecting abilities of CtIP/Sae2/Ctp1 are also regulated by CDK-dependent phosphorylation allowing for further coupling of HDR to cell cycle stage [25, 62, 154]. ATM phosphorylates CtIP and regulates its association with BRCA1 [79], an interaction that is required for ssDNA formation at DSB ends suggesting that this interaction regulates the resection-promoting ability of CtIP [25]. CtIP is also targeted for ubiquitination which is catalyzed by its interacting partner BRCA1, thereby, providing another potential level of regulation of HDR [155].

Another important function of the MRN complex is in activation of ATM, the key signaling kinase involved in IR-induced checkpoint arrest (reviewed in [72, 75] and Chapter 10). The MRN complex is required for the activation of ATM in vitro and in vivo [72, 75, 130], however, this function does not require the nuclease activity of Mre11 [22]. In contrast, the nuclease activity of Mre11 is required for IR-induced activation of ATR in S and G2 [22, 64]. ATM is also required for the activation of ATR in response to IR [64] and also may regulate several aspects of HDR [132]. Indeed, ATM-deficient chicken DT40 cells have reduced rates of HDR [101].

8.2.4 DSB Repair Pathway Choice

As discussed, human cells contain multiple pathways that can detect and repair IR-induced DSBs. Although recent studies have revealed the importance of CtIP in coupling the initiation of HDR and the activation of ATR to late S and G2, many other aspects of how the various DSB repair pathways are coordinated are still poorly understood (reviewed in [18, 132]). For example, initiation of NHEJ and MRN-dependent activation of ATM both occur in G0 and G1 [64, 75, 123] but what determines whether a given DSB is detected by Ku (to initiate NHEJ) or MRN (to activate ATM and subsequently ATR) is not known. Indeed, laser microbeam irradiation experiments reveal that Ku and MRN are recruited to the same sites of DNA damage independently throughout the cell cycle but whether they compete for binding to the same DSB or bind to different DSBs is not known [69]. It is also possible that chromatin structure in the vicinity of the break will have a major effect on DSB repair and pathway choice ([47], see also Chapter 6) and this is an area that warrants further investigation.

8.3 The Therapeutic Potential of DSB Repair Pathways

8.3.1 DSB Repair Pathways as Predictors of Radiation Response and Treatment Outcome

As discussed above, cells lacking any of the core NHEJ components are hypersensitive to IR and other DSB-inducing agents. Moreover, multiple studies have shown that knock down or inactivation of NHEJ proteins leads to radiosensitivity in an experimental setting (reviewed in [113]). Although HDR-deficient cells are only modestly radiation sensitive, they are sensitive to several classes of chemotherapeutics including topoisomerase poisons and DNA cross-linking agents (reviewed in [84, 117]). Therefore, the level of expression and/or the activity of DSB repair proteins in malignant cells could influence response to radiation and chemotherapy in cancer patients. Proteins that function in DSB repair pathways could therefore serve as important biomarkers and predictors of treatment response. Indeed, low levels of expression of Ku have been linked to increased radiosensitivity and/or improved survival in a variety of solid tumours [54, 76, 151]. Moreover, high DNA-PKcs levels correlate with poor prognosis in B-cell chronic lymphocytic leukemia (B-CLL) [150]. Similarly, altered levels of the HDR protein Rad51 may predict HDR activity and response to radiation therapy [14, 56, 100]. Single nucleotide polymorphisms (SNPs) in DNA repair genes may also have potential as biomarkers of clinical response. For example, SNPs in DNA ligase IV have been linked to radiation toxicity in prostate cancer patients [31], while SNPs in Ku70, Ku80 and XRCC4 have been linked to genomic instability and overall risk of breast cancer [45, 147].

8.3.2 Small Molecule Inhibitors of DSB Repair Pathways

The development of small molecules that target key proteins in DSB repair pathways is emerging as an exciting new avenue for possible therapeutic intervention. As described above, the protein kinase activity of DNA-PKcs is required for NHEJ [68, 70]. Early studies showed that PIKK inhibitors such as wortmannin inhibit DSB repair and radiosensitize cells in vitro [27, 122], leading to the suggestion that small molecule inhibitors of DNA-PKcs kinase activity could have clinical potential as radiosensitizing agents [113]. Given the similarities in amino acid sequence between the kinase domain of the PIKK family members and the catalytic subunit of PI3K itself, several first generation small molecule inhibitors of DNA-PK were derived from the PI3K inhibitor LY294002 (reviewed in [108]) and shown to have activity against DNA-PK in vitro [4, 48, 53, 60, 74]. Subsequently, more selective inhibitors of DNA-PK kinase activity, such as NU7441, were developed. NU7441 inhibits DNA-PK kinase activity with nanomolar potency [74], and induces radiosensitivity and reduced repair of IR-induced DSBs in human cell lines [156]. Another small molecule inhibitor of DNA-PK, IC87361, produced radiosensitization in cell lines and a murine xenograft tumor model [131]. The DNA-PK inhibitor vanillin also increases the sensitivity of cells to cisplatin and mitomycin C [29, 38]. The combination of potency and specificity of these compounds makes them potentially clinically useful as radio- or chemo-sensitizers (reviewed in [108]). However, caveats are associated with this approach. For example, most currently available DNA-PK inhibitors are ATP analogues, and given the fact that human cells contain over 500 protein kinases [92], the possibility exists for less than optimum specificity in vivo. Indeed, a widely used ATM kinase inhibitor KU55933 [57] and the newly identified ATM inhibitor CP466722 [118], which were both identified in ATM kinase activity assays, both inhibit phosphorylation of the tyrosine kinase, Src [118]. Also, given that DNA-PKcs is widely expressed in both normal and cancer cells, the therapeutic window for radiosensitization using DNA-PK inhibitors could be quite small.

It is also possible that other proteins in the NHEJ pathway could be targets for therapeutic intervention. Indeed, elegant studies from the Tomkinson group have identified a series of inhibitors of human DNA ligases, including NHEJ-specific DNA ligase IV. These inhibitors inhibit DNA rejoining in vitro and sensitize cell lines to DNA damaging agents [26]. Another potential therapeutic target is PNKP, which has been implicated in SSB repair, base excision repair and NHEJ. Reduced expression of PNKP in human cells induces radio- and chemo-sensitivity, suggesting that inhibitors of PNKPs kinase and/or phosphatase activities may have therapeutic potential (reviewed in [13]). Other potential ways to target NHEJ or other DSB repair pathways could involve identification of inhibitors that disrupt critical protein-protein or protein-DNA interactions.

The HDR pathway is also a target for small molecule inhibitors with therapeutic potential. A small molecule named Mirin has been shown to inhibit the activation of ATM following DNA damage by blocking the interaction between ATM and Mre11. This compound also severely disrupted HDR and the damage-induced G2/M checkpoint in mammalian cells [37]. In addition, a small molecule histone deacetylase inhibitor PCI-24781 decreases expression of Rad51, inhibiting HDR in vitro and in vivo [2]. Small molecule inhibitors of DSB repair pathways may also be useful in targeting tumour stem cells (also called tumour initiating cells) [9, 10, 23, 58] since it has been suggested that cancer stem cells are resistant to chemotherapy and radiation therapy due to enhanced DSB response pathways and DNA repair capacity [6, 15, 114].

8.3.3 Synthetic Lethality

Recently, a more targeted approach to cancer therapy has emerged which capitalizes on the fact that tumour cells which have defects in one DSB repair pathway become more reliant on alternative, compensatory repair pathways. Inactivation of the compensatory pathway can then lead to cell death in the absence of additional genotoxic stress. This approach, termed synthetic lethality, has been widely used in genetics to describe cell death resulting from disruption of multiple compensatory pathways in a single cell, and has now been applied to human cells containing defects in DSB repair pathways [66, 87, 93].

In a series of landmark studies, human cells containing mutations in BRCA1 or BRCA2 were shown to be highly sensitive to inhibition of PARP-1 [20, 41]. As discussed above, PARP-1 is involved in base excision repair and DNA SSB repair pathways. It functions by adding poly-ADP ribose (PAR) groups to various acceptor proteins, including histones, which is thought to induce relaxation of chromatin structure surrounding the break site, allowing better access of repair proteins [129]. Inhibition of PARP-1 is thought to promote the accumulation of SSBs, which are subsequently converted to DSBs during DNA replication and then repaired by the HDR pathway. Since BRCA1 or BRCA2 defective cells are compromised in their ability to repair DSBs via HDR [104, 105, 134], inhibition of PARP-1 has been proposed to lead to accumulation of DSBs, which results in cell death in the absence of additional genotoxic stressors. These findings have led to the development of PARP-1 inhibitors for BRCA1- and BRCA2-deficient breast and ovarian tumors [40, 93,

98, 124]. Indeed, PARP-1 inhibitors are currently in phase 1 and 2 clinical trials as monotherapy for breast and ovarian cancer [87].

The synthetic lethality approach may also be applicable to other tumour types since siRNA knockdown of other HDR proteins such as Rad51, Rad54 and RPA1 also results in sensitivity to PARP-1 inhibitors [94]. Similarly, human cells deficient in other proteins involved in NHEJ (Ku80 and DNA-PKcs) and DNA damage signaling (ATM and Nbs1) are also sensitive to PARP-1 inhibition [19, 46, 94]. Moreover, PARP-1 deficient cells are sensitive to small molecule inhibitors of both ATM and DNA-PK [19]. However, although the synthetic lethality approach is appealing, it is not without its limitations. For example, BRCA2-deficient cell lines have recently been shown to become resistant to PARP inhibition due to secondary mutations in BRCA2 that resulted in a restored ability to undergo HDR [39].

8.4 Summary

In summary, DNA DSBs can be caused by a variety of exogenous insults as well as through endogenous processes, and, if not repaired, can lead to genomic instability and/or cell death. In human cells DSBs are repaired by NHEJ or HDR, however, increasing evidence suggests that alternative end joining pathways may also be involved, at least in NHEJ-deficient cells. How these various pathways are regulated and coordinated in the cell to promote efficient repair is an area of active investigation. As our understanding of the mechanisms by which human cells detect and repair DSBs increases, potential ways to manipulate these pathways for therapeutic gain have begun to emerge. The recent development of small molecule inhibitors that directly target the NHEJ and HDR pathways have the potential to lead to new approaches to sensitize tumour cells to chemo and/or radiation therapy. Similarly, the synthetic lethality approach, which exploits genetic defects in specific tumour types, may lead to more targeted approaches to cancer therapy. One of the most exciting developments in this regard is the use of PARP-1 inhibitors to target DSB repair defective tumours. As our understanding of the cellular mechanisms for the detection and repair of DSBs improves, additional avenues for new therapies for cancer treatment are likely to emerge.

Acknowledgements Thanks to Drs E. Kurz, T. Beattie, J. Cobb and G. Bebb for helpful comments. LMW and CTW are supported by graduate studentships from the Alberta Heritage Foundation for Medical Research (AHFMR) (LMW) and Alberta Health Services (LMW and CTW). SPLM is a scientist of the AHFMR and holds the Engineered Air Chair in Cancer Research.

References

1. Abraham, R. T. 2004. PI 3-kinase related kinases: 'big' players in stress-induced signaling pathways. DNA Repair (Amst) 3:883–7.
2. Adimoolam, S., M. Sirisawad, J. Chen, P. Thiemann, J. M. Ford, and J. J. Buggy. 2007. HDAC inhibitor PCI-24781 decreases RAD51 expression and inhibits homologous recombination. Proc Natl Acad Sci USA 104:19482–7.

3. Ahnesorg, P., P. Smith, and S. P. Jackson. 2006. XLF interacts with the XRCC4-DNA ligase IV complex to promote DNA nonhomologous end-joining. Cell 124:301–13.
4. Aristegui, S. R., M. D. El-Murr, B. T. Golding, R. J. Griffin, and I. R. Hardcastle. 2006. Judicious application of allyl protecting groups for the synthesis of 2-morpholin-4-yl-4-oxo-4H-chromen-8-yl triflate, a key precursor of DNA-dependent protein kinase inhibitors. Org Lett 8:5927–9.
5. Audebert, M., B. Salles, and P. Calsou. 2004. Involvement of poly(ADP-ribose) polymerase-1 and XRCC1/DNA ligase III in an alternative route for DNA double-strand breaks rejoining. J Biol Chem 279:55117–26.
6. Bao, S., Q. Wu, R. E. McLendon, Y. Hao, Q. Shi, A. B. Hjelmeland, M. W. Dewhirst, D. D. Bigner, and J. N. Rich. 2006. Glioma stem cells promote radioresistance by preferential activation of the DNA damage response. Nature 444:756–60.
7. Barnes, D. E., G. Stamp, I. Rosewell, A. Denzel, and T. Lindahl. 1998. Targeted disruption of the gene encoding DNA ligase IV leads to lethality in embryonic mice. Curr Biol 8:1395–8.
8. Bassing, C. H., W. Swat, and F. W. Alt. 2002. The mechanism and regulation of chromosomal V(D)J recombination. Cell 109 Suppl:S45–55.
9. Baumann, M., W. Dubois, and H. D. Suit. 1990. Response of human squamous cell carcinoma xenografts of different sizes to irradiation: relationship of clonogenic cells, cellular radiation sensitivity in vivo, and tumor rescuing units. Radiat Res 123:325–30.
10. Baumann, M., M. Krause, and R. Hill. 2008. Exploring the role of cancer stem cells in radioresistance. Nat Rev Cancer 8:545–54.
11. Bennardo, N., A. Cheng, N. Huang, and J. M. Stark. 2008. Alternative-NHEJ is a mechanistically distinct pathway of mammalian chromosome break repair. PLoS Genet 4:e1000110.
12. Bennett, C. B., A. L. Lewis, K. K. Baldwin, and M. A. Resnick. 1993. Lethality induced by a single site-specific double-strand break in a dispensable yeast plasmid. Proc Natl Acad Sci USA 90:5613–7.
13. Bernstein, N. K., F. Karimi-Busheri, A. Rasouli-Nia, R. Mani, G. Dianov, J. N. Glover, and M. Weinfeld. 2008. Polynucleotide kinase as a potential target for enhancing cytotoxicity by ionizing radiation and topoisomerase I inhibitors. Anticancer Agents Med Chem 8:358–67.
14. Bindra, R. S., P. J. Schaffer, A. Meng, J. Woo, K. Maseide, M. E. Roth, P. Lizardi, D. W. Hedley, R. G. Bristow, and P. M. Glazer. 2004. Down-regulation of Rad51 and decreased homologous recombination in hypoxic cancer cells. Mol Cell Biol 24:8504–18.
15. Blazek, E. R., J. L. Foutch, and G. Maki. 2007. Daoy medulloblastoma cells that express CD133 are radioresistant relative to CD133- cells, and the CD133+ sector is enlarged by hypoxia. Int J Radiat Oncol Biol Phys 67:1–5.
16. Bleuyard, J. Y., M. E. Gallego, and C. I. White. 2006. Recent advances in understanding of the DNA double-strand break repair machinery of plants. DNA Repair (Amst) 5:1–12.
17. Blunt, T., N. J. Finnie, G. E. Taccioli, G. C. Smith, J. Demengeot, T. M. Gottlieb, R. Mizuta, A. J. Varghese, F. W. Alt, P. A. Jeggo, and et al. 1995. Defective DNA-dependent protein kinase activity is linked to V(D)J recombination and DNA repair defects associated with the murine scid mutation. Cell 80:813–23.
18. Branzei, D., and M. Foiani. 2008. Regulation of DNA repair throughout the cell cycle. Nat Rev Mol Cell Biol 9:297–308.
19. Bryant, H. E., and T. Helleday. 2006. Inhibition of poly (ADP-ribose) polymerase activates ATM which is required for subsequent homologous recombination repair. Nucleic Acids Res 34:1685–91.
20. Bryant, H. E., N. Schultz, H. D. Thomas, K. M. Parker, D. Flower, E. Lopez, S. Kyle, M. Meuth, N. J. Curtin, and T. Helleday. 2005. Specific killing of BRCA2-deficient tumours with inhibitors of poly(ADP-ribose) polymerase. Nature 434:913–7.
21. Buck, D., L. Malivert, R. de Chasseval, A. Barraud, M. C. Fondaneche, O. Sanal, A. Plebani, J. L. Stephan, M. Hufnagel, F. le Deist, A. Fischer, A. Durandy, J. P. de Villartay, and P.

Revy. 2006. Cernunnos, a novel nonhomologous end-joining factor, is mutated in human immunodeficiency with microcephaly. Cell 124:287–99.
22. Buis, J., Y. Wu, Y. Deng, J. Leddon, G. Westfield, M. Eckersdorff, J. M. Sekiguchi, S. Chang, and D. O. Ferguson. 2008. Mre11 nuclease activity has essential roles in DNA repair and genomic stability distinct from ATM activation. Cell 135:85–96.
23. Chalmers, A. J. 2007. Radioresistant glioma stem cells–therapeutic obstacle or promising target? DNA Repair (Amst) 6:1391–4.
24. Chan, D. W., and S. P. Lees-Miller. 1996. The DNA-dependent protein kinase is inactivated by autophosphorylation of the catalytic subunit. J Biol Chem 271:8936–41.
25. Chen, L., C. J. Nievera, A. Y. Lee, and X. Wu. 2008. Cell cycle-dependent complex formation of BRCA1.CtIP.MRN is important for DNA double-strand break repair. J Biol Chem 283:7713–20.
26. Chen, X., S. Zhong, X. Zhu, B. Dziegielewska, T. Ellenberger, G. M. Wilson, A. D. MacKerell, Jr., and A. E. Tomkinson. 2008. Rational design of human DNA ligase inhibitors that target cellular DNA replication and repair. Cancer Res 68:3169–77.
27. Chernikova, S. B., R. L. Wells, and M. M. Elkind. 1999. Wortmannin sensitizes mammalian cells to radiation by inhibiting the DNA-dependent protein kinase-mediated rejoining of double-strand breaks. Radiat Res 151:159–66.
28. Cimprich, K. A., and D. Cortez. 2008. ATR: an essential regulator of genome integrity. Nat Rev Mol Cell Biol 9:616–27.
29. Collis, S. J., T. L. DeWeese, P. A. Jeggo, and A. R. Parker. 2005. The life and death of DNA-PK. Oncogene 24:949–61.
30. D'Amours, D., and S. P. Jackson. 2002. The Mre11 complex: at the crossroads of DNA repair and checkpoint signalling. Nat Rev Mol Cell Biol 3:317–27.
31. Damaraju, S., D. Murray, J. Dufour, D. Carandang, S. Myrehaug, G. Fallone, C. Field, R. Greiner, J. Hanson, C. E. Cass, and M. Parliament. 2006. Association of DNA repair and steroid metabolism gene polymorphisms with clinical late toxicity in patients treated with conformal radiotherapy for prostate cancer. Clin Cancer Res 12:2545–54.
32. de Lange, T. 2005. Shelterin: the protein complex that shapes and safeguards human telomeres. Genes Dev 19:2100–10.
33. Demuth, I., and M. Digweed. 2007. The clinical manifestation of a defective response to DNA double-strand breaks as exemplified by Nijmegen breakage syndrome. Oncogene 26:7792–8.
34. DiBiase, S. J., Z. C. Zeng, R. Chen, T. Hyslop, W. J. Curran, Jr., and G. Iliakis. 2000. DNA-dependent protein kinase stimulates an independently active, nonhomologous, end-joining apparatus. Cancer Res 60:1245–53.
35. Downs, J. A., and S. P. Jackson. 2004. A means to a DNA end: the many roles of Ku. Nat Rev Mol Cell Biol 5:367–78.
36. Dudasova, Z., A. Dudas, and M. Chovanec. 2004. Non-homologous end-joining factors of Saccharomyces cerevisiae. FEMS Microbiol Rev 28:581–601.
37. Dupre, A., L. Boyer-Chatenet, R. M. Sattler, A. P. Modi, J. H. Lee, M. L. Nicolette, L. Kopelovich, M. Jasin, R. Baer, T. T. Paull, and J. Gautier. 2008. A forward chemical genetic screen reveals an inhibitor of the Mre11-Rad50-Nbs1 complex. Nat Chem Biol 4:119–25.
38. Durant, S., and P. Karran. 2003. Vanillins – a novel family of DNA-PK inhibitors. Nucleic Acids Res 31:5501–12.
39. Edwards, S. L., R. Brough, C. J. Lord, R. Natrajan, R. Vatcheva, D. A. Levine, J. Boyd, J. S. Reis-Filho, and A. Ashworth. 2008. Resistance to therapy caused by intragenic deletion in BRCA2. Nature 451:1111–5.
40. Evers, B., R. Drost, E. Schut, M. de Bruin, E. van der Burg, P. W. Derksen, H. Holstege, X. Liu, E. van Drunen, H. B. Beverloo, G. C. Smith, N. M. Martin, A. Lau, M. J. O'Connor, and J. Jonkers. 2008. Selective inhibition of BRCA2-deficient mammary tumor cell growth by AZD2281 and cisplatin. Clin Cancer Res 14:3916–25.
41. Farmer, H., N. McCabe, C. J. Lord, A. N. Tutt, D. A. Johnson, T. B. Richardson, M. Santarosa, K. J. Dillon, I. Hickson, C. Knights, N. M. Martin, S. P. Jackson, G. C. Smith, and

A. Ashworth. 2005. Targeting the DNA repair defect in BRCA mutant cells as a therapeutic strategy. Nature 434:917–21.
42. Fattah, K. R., B. L. Ruis, and E. A. Hendrickson. 2008. Mutations to Ku reveal differences in human somatic cell lines. DNA Repair (Amst) 7:762–74.
43. Frank, K. M., N. E. Sharpless, Y. Gao, J. M. Sekiguchi, D. O. Ferguson, C. Zhu, J. P. Manis, J. Horner, R. A. DePinho, and F. W. Alt. 2000. DNA ligase IV deficiency in mice leads to defective neurogenesis and embryonic lethality via the p53 pathway. Mol Cell 5:993–1002.
44. Friedberg, E. C., G. C. Walker, W. Siede, R. D. Wood, R. A. Schultz, and T. Ellenberger. 2005. DNA Repair and Mutagenesis, 2 ed. ASM Press, Washington, D.C.
45. Fu, Y. P., J. C. Yu, T. C. Cheng, M. A. Lou, G. C. Hsu, C. Y. Wu, S. T. Chen, H. S. Wu, P. E. Wu, and C. Y. Shen. 2003. Breast cancer risk associated with genotypic polymorphism of the nonhomologous end-joining genes: a multigenic study on cancer susceptibility. Cancer Res 63:2440–2446.
46. Gaymes, T. J., S. Shall, F. Farzaneh, and G. J. Mufti. 2008. Chromosomal instability syndromes are sensitive to poly ADP ribose polymerase inhibitors. Haematologica.
47. Goodarzi, A. A., A. T. Noon, D. Deckbar, Y. Ziv, Y. Shiloh, M. Lobrich, and P. A. Jeggo. 2008. ATM signaling facilitates repair of DNA double-strand breaks associated with heterochromatin. Mol Cell 31:167–77.
48. Griffin, R. J., G. Fontana, B. T. Golding, S. Guiard, I. R. Hardcastle, J. J. Leahy, N. Martin, C. Richardson, L. Rigoreau, M. Stockley, and G. C. Smith. 2005. Selective benzopyranone and pyrimido[2,1-a]isoquinolin-4-one inhibitors of DNA-dependent protein kinase: synthesis, structure-activity studies, and radiosensitization of a human tumor cell line in vitro. J Med Chem 48:569–85.
49. Gu, J., H. Lu, A. G. Tsai, K. Schwarz, and M. R. Lieber. 2007. Single-stranded DNA ligation and XLF-stimulated incompatible DNA end ligation by the XRCC4-DNA ligase IV complex: influence of terminal DNA sequence. Nucleic Acids Res 35:5755–62.
50. Gu, Y., S. Jin, Y. Gao, D. T. Weaver, and F. W. Alt. 1997. Ku70-deficient embryonic stem cells have increased ionizing radiosensitivity, defective DNA end-binding activity, and inability to support V(D)J recombination. Proc Natl Acad Sci USA 94:8076–81.
51. Gu, Y., K. J. Seidl, G. A. Rathbun, C. Zhu, J. P. Manis, N. van der Stoep, L. Davidson, H. L. Cheng, J. M. Sekiguchi, K. Frank, P. Stanhope-Baker, M. S. Schlissel, D. B. Roth, and F. W. Alt. 1997. Growth retardation and leaky SCID phenotype of Ku70-deficient mice. Immunity 7:653–65.
52. Hall, E. J., and A. J. Giaccia. 2006. Radiobiology for the Radiobiologist, 6th ed. Lippincott Williams & Wilkins, Philadelphia.
53. Hardcastle, I. R., X. Cockcroft, N. J. Curtin, M. D. El-Murr, J. J. Leahy, M. Stockley, B. T. Golding, L. Rigoreau, C. Richardson, G. C. Smith, and R. J. Griffin. 2005. Discovery of Potent Chromen-4-one Inhibitors of the DNA-Dependent Protein Kinase (DNA-PK) Using a Small-Molecule Library Approach. J Med Chem 48:7829–7846.
54. Harima, Y., S. Sawada, Y. Miyazaki, K. Kin, H. Ishihara, M. Imamura, M. Sougawa, N. Shikata, and T. Ohnishi. 2003. Expression of Ku80 in cervical cancer correlates with response to radiotherapy and survival. Am J Clin Oncol 26:e80–5.
55. Helleday, T., J. Lo, D. C. van Gent, and B. P. Engelward. 2007. DNA double-strand break repair: from mechanistic understanding to cancer treatment. DNA Repair (Amst) 6:923–35.
56. Henning, W., and H. W. Sturzbecher. 2003. Homologous recombination and cell cycle checkpoints: Rad51 in tumour progression and therapy resistance. Toxicology 193:91–109.
57. Hickson, I., Y. Zhao, C. J. Richardson, S. J. Green, N. M. Martin, A. I. Orr, P. M. Reaper, S. P. Jackson, N. J. Curtin, and G. C. Smith. 2004. Identification and characterization of a novel and specific inhibitor of the ataxia-telangiectasia mutated kinase ATM. Cancer Res 64:9152–9.
58. Hill, R. P., and L. Milas. 1989. The proportion of stem cells in murine tumors. Int J Radiat Oncol Biol Phys 16:513–8.
59. Hoeijmakers, J. H. 2001. Genome maintenance mechanisms for preventing cancer. Nature 411:366–74.

60. Hollick, J. J., L. J. Rigoreau, C. Cano-Soumillac, X. Cockcroft, N. J. Curtin, M. Frigerio, B. T. Golding, S. Guiard, I. R. Hardcastle, I. Hickson, M. G. Hummersone, K. A. Menear, N. M. Martin, I. Matthews, D. R. Newell, R. Ord, C. J. Richardson, G. C. Smith, and R. J. Griffin. 2007. Pyranone, thiopyranone, and pyridone inhibitors of phosphatidylinositol 3-kinase related kinases. Structure-activity relationships for DNA-dependent protein kinase inhibition, and identification of the first potent and selective inhibitor of the ataxia telangiectasia mutated kinase. J Med Chem 50:1958–72.
61. Howlett, N. G., T. Taniguchi, S. Olson, B. Cox, Q. Waisfisz, C. De Die-Smulders, N. Persky, M. Grompe, H. Joenje, G. Pals, H. Ikeda, E. A. Fox, and A. D. D'Andrea. 2002. Biallelic inactivation of BRCA2 in Fanconi anemia. Science 297:606–9.
62. Huertas, P., F. Cortes-Ledesma, A. A. Sartori, A. Aguilera, and S. P. Jackson. 2008. CDK targets Sae2 to control DNA-end resection and homologous recombination. Nature 455: 689–92.
63. Jankovic, M., A. Nussenzweig, and M. C. Nussenzweig. 2007. Antigen receptor diversification and chromosome translocations. Nat Immunol 8:801–8.
64. Jazayeri, A., J. Falck, C. Lukas, J. Bartek, G. C. Smith, J. Lukas, and S. P. Jackson. 2006. ATM- and cell cycle-dependent regulation of ATR in response to DNA double-strand breaks. Nat Cell Biol 8:37–45.
65. Jolly, C. J., A. J. Cook, and J. P. Manis. 2008. Fixing DNA breaks during class switch recombination. J Exp Med 205:509–13.
66. Kaelin, W. G., Jr. 2005. The concept of synthetic lethality in the context of anticancer therapy. Nat Rev Cancer 5:689–98.
67. Kastan, M. B., and J. Bartek. 2004. Cell-cycle checkpoints and cancer. Nature 432:316–23.
68. Kienker, L. J., E. K. Shin, and K. Meek. 2000. Both V(D)J recombination and radioresistance require DNA-PK kinase activity, though minimal levels suffice for V(D)J recombination. Nucleic Acids Res 28:2752–61.
69. Kim, J. S., T. B. Krasieva, H. Kurumizaka, D. J. Chen, A. M. Taylor, and K. Yokomori. 2005. Independent and sequential recruitment of NHEJ and HR factors to DNA damage sites in mammalian cells. J Cell Biol 170:341–7.
70. Kurimasa, A., S. Kumano, N. V. Boubnov, M. D. Story, C. S. Tung, S. R. Peterson, and D. J. Chen. 1999. Requirement for the kinase activity of human DNA-dependent protein kinase catalytic subunit in DNA strand break rejoining. Mol Cell Biol 19:3877–84.
71. Kurz, E. U., and S. P. Lees-Miller. 2004. DNA damage-induced activation of ATM and ATM-dependent signaling pathways. DNA Repair (Amst) 3:889–900.
72. Lavin, M. F. 2008. Ataxia-telangiectasia: from a rare disorder to a paradigm for cell signalling and cancer. Nat Rev Mol Cell Biol 9:759–69.
73. Lavin, M. F., and S. Kozlov. 2007. ATM activation and DNA damage response. Cell Cycle 6:931–42.
74. Leahy, J. J., B. T. Golding, R. J. Griffin, I. R. Hardcastle, C. Richardson, L. Rigoreau, and G. C. Smith. 2004. Identification of a highly potent and selective DNA-dependent protein kinase (DNA-PK) inhibitor (NU7441) by screening of chromenone libraries. Bioorg Med Chem Lett 14:6083–7.
75. Lee, J. H., and T. T. Paull. 2007. Activation and regulation of ATM kinase activity in response to DNA double-strand breaks. Oncogene 26:7741–8.
76. Lee, S. W., K. J. Cho, J. H. Park, S. Y. Kim, S. Y. Nam, B. J. Lee, S. B. Kim, S. H. Choi, J. H. Kim, S. D. Ahn, S. S. Shin, E. K. Choi, and E. Yu. 2005. Expressions of Ku70 and DNA-PKcs as prognostic indicators of local control in nasopharyngeal carcinoma. Int J Radiat Oncol Biol Phys 62:1451–7.
77. Lees-Miller, S. P., R. Godbout, D. W. Chan, M. Weinfeld, R. S. Day, 3rd, G. M. Barron, and J. Allalunis-Turner. 1995. Absence of p350 subunit of DNA-activated protein kinase from a radiosensitive human cell line. Science 267:1183–5.
78. Lengsfeld, B. M., A. J. Rattray, V. Bhaskara, R. Ghirlando, and T. T. Paull. 2007. Sae2 is an endonuclease that processes hairpin DNA cooperatively with the Mre11/Rad50/Xrs2 complex. Mol Cell 28:638–51.

79. Li, S., N. S. Ting, L. Zheng, P. L. Chen, Y. Ziv, Y. Shiloh, E. Y. Lee, and W. H. Lee. 2000. Functional link of BRCA1 and ataxia telangiectasia gene product in DNA damage response. Nature 406:210–5.
80. Li, X., and W. D. Heyer. 2008. Homologous recombination in DNA repair and DNA damage tolerance. Cell Res 18:99–113.
81. Lieber, M. R. 2008. The mechanism of human nonhomologous DNA end joining. J Biol Chem 283:1–5.
82. Lieber, M. R., H. Lu, J. Gu, and K. Schwarz. 2008. Flexibility in the order of action and in the enzymology of the nuclease, polymerases, and ligase of vertebrate non-homologous DNA end joining: relevance to cancer, aging, and the immune system. Cell Res 18:125–33.
83. Limbo, O., C. Chahwan, Y. Yamada, R. A. de Bruin, C. Wittenberg, and P. Russell. 2007. Ctp1 is a cell-cycle-regulated protein that functions with Mre11 complex to control double-strand break repair by homologous recombination. Mol Cell 28:134–46.
84. Liu, S. K., P. L. Olive, and R. G. Bristow. 2008. Biomarkers for DNA DSB inhibitors and radiotherapy clinical trials. Cancer Metastasis Rev 27:445–58.
85. Lobrich, M., and P. A. Jeggo. 2007. The impact of a negligent G2/M checkpoint on genomic instability and cancer induction. Nat Rev Cancer 7:861–9.
86. Longhese, M. P. 2008. DNA damage response at functional and dysfunctional telomeres. Genes Dev 22:125–40.
87. Lord, C. J., and A. Ashworth. 2008. Targeted therapy for cancer using PARP inhibitors. Curr Opin Pharmacol 8(4):363–369.
88. Lu, H., U. Pannicke, K. Schwarz, and M. R. Lieber. 2007. Length-dependent binding of human XLF to DNA and stimulation of XRCC4.DNA ligase IV activity. J Biol Chem 282:11155–62.
89. Ma, Y., U. Pannicke, K. Schwarz, and M. R. Lieber. 2002. Hairpin opening and overhang processing by an Artemis/DNA-dependent protein kinase complex in nonhomologous end joining and V(D)J recombination. Cell 108:781–94.
90. Ma, Y., K. Schwarz, and M. R. Lieber. 2005. The Artemis:DNA-PKcs endonuclease cleaves DNA loops, flaps, and gaps. DNA Repair (Amst) 4:845–51.
91. Mahaney, B. L., K. Meek, and S. P. Lees Miller. 2009 Repair of ionizing radiation induced DNA double strand breaks by non-homologous end-joining. Biochem. J. 417:639–50.
92. Manning, G., D. B. Whyte, R. Martinez, T. Hunter, and S. Sudarsanam. 2002. The protein kinase complement of the human genome. Science 298:1912–34.
93. Martin, S. A., C. J. Lord, and A. Ashworth. 2008. DNA repair deficiency as a therapeutic target in cancer. Curr Opin Genet Dev 18:80–6.
94. McCabe, N., N. C. Turner, C. J. Lord, K. Kluzek, A. Bialkowska, S. Swift, S. Giavara, M. J. O'Connor, A. N. Tutt, M. Z. Zdzienicka, G. C. Smith, and A. Ashworth. 2006. Deficiency in the repair of DNA damage by homologous recombination and sensitivity to poly(ADP-ribose) polymerase inhibition. Cancer Res 66:8109–15.
95. Meek, K., V. Dang, and S. P. Lees Miller. 2008. DNA-PK: the means to justify the ends? Adv Immunol 99:33–58.
96. Meek, K., P. Douglas, X. Cui, Q. Ding, and S. P. Lees-Miller. 2007. trans Autophosphorylation at DNA-dependent protein kinase's two major autophosphorylation site clusters facilitates end processing but not end joining. Mol Cell Biol 27:3881–90.
97. Meek, K., S. Gupta, D. A. Ramsden, and S. P. Lees-Miller. 2004. The DNA-dependent protein kinase: the director at the end. Immunol Rev 200:132–41.
98. Menear, K. A., C. Adcock, R. Boulter, X. L. Cockcroft, L. Copsey, A. Cranston, K. J. Dillon, J. Drzewiecki, S. Garman, S. Gomez, H. Javaid, F. Kerrigan, C. Knights, A. Lau, V. M. Loh, Jr., I. T. Matthews, S. Moore, M. J. O'Connor, G. C. Smith, and N. M. Martin. 2008. 4-[3-(4-cyclopropanecarbonylpiperazine-1-carbonyl)-4-fluorobenzyl]-2H-phthalazin-1-one: a novel bioavailable inhibitor of poly(ADP-ribose) polymerase-1. J Med Chem 51:6581–91.

99. Mimitou, E. P., and L. S. Symington. 2008. Sae2, Exo1 and Sgs1 collaborate in DNA double-strand break processing. Nature 455:770–4.
100. Miyagawa, K. 2008. Clinical relevance of the homologous recombination machinery in cancer therapy. Cancer Sci 99:187–94.
101. Morrison, C., E. Sonoda, N. Takao, A. Shinohara, K. Yamamoto, and S. Takeda. 2000. The controlling role of ATM in homologous recombinational repair of DNA damage. Embo J 19:463–71.
102. Moshous, D., I. Callebaut, R. de Chasseval, B. Corneo, M. Cavazzana-Calvo, F. Le Deist, I. Tezcan, O. Sanal, Y. Bertrand, N. Philippe, A. Fischer, and J. P. de Villartay. 2001. Artemis, a novel DNA double-strand break repair/V(D)J recombination protein, is mutated in human severe combined immune deficiency. Cell 105:177–86.
103. Moshous, D., L. Li, R. Chasseval, N. Philippe, N. Jabado, M. J. Cowan, A. Fischer, and J. P. de Villartay. 2000. A new gene involved in DNA double-strand break repair and V(D)J recombination is located on human chromosome 10p. Hum Mol Genet 9: 583–8.
104. Moynahan, M. E., T. Y. Cui, and M. Jasin. 2001. Homology-directed dna repair, mitomycin-c resistance, and chromosome stability is restored with correction of a Brca1 mutation. Cancer Res 61:4842–50.
105. Moynahan, M. E., A. J. Pierce, and M. Jasin. 2001. BRCA2 is required for homology-directed repair of chromosomal breaks. Mol Cell 7:263–72.
106. Nussenzweig, A., C. Chen, V. da Costa Soares, M. Sanchez, K. Sokol, M. C. Nussenzweig, and G. C. Li. 1996. Requirement for Ku80 in growth and immunoglobulin V(D)J recombination. Nature 382:551–5.
107. Nussenzweig, A., and M. C. Nussenzweig. 2007. A backup DNA repair pathway moves to the forefront. Cell 131:223–5.
108. O'Connor, M. J., N. M. Martin, and G. C. Smith. 2007. Targeted cancer therapies based on the inhibition of DNA strand break repair. Oncogene 26:7816–24.
109. O'Driscoll, M., K. M. Cerosaletti, P. M. Girard, Y. Dai, M. Stumm, B. Kysela, B. Hirsch, A. Gennery, S. E. Palmer, J. Seidel, R. A. Gatti, R. Varon, M. A. Oettinger, H. Neitzel, P. A. Jeggo, and P. Concannon. 2001. DNA ligase IV mutations identified in patients exhibiting developmental delay and immunodeficiency. Mol Cell 8:1175–85.
110. O'Driscoll, M., A. R. Gennery, J. Seidel, P. Concannon, and P. A. Jeggo. 2004. An overview of three new disorders associated with genetic instability: LIG4 syndrome, RS-SCID and ATR-Seckel syndrome. DNA Repair (Amst) 3:1227–35.
111. O'Driscoll, M., and P. A. Jeggo. 2006. The role of double-strand break repair – insights from human genetics. Nat Rev Genet 7:45–54.
112. O'Driscoll, M., V. L. Ruiz-Perez, C. G. Woods, P. A. Jeggo, and J. A. Goodship. 2003. A splicing mutation affecting expression of ataxia-telangiectasia and Rad3-related protein (ATR) results in Seckel syndrome. Nat Genet 33:497–501.
113. Pastwa, E., and M. Malinowski. 2007. Non-homologous DNA end joining in anticancer therapy. Curr Cancer Drug Targets 7:243–50.
114. Phillips, T. M., W. H. McBride, and F. Pajonk. 2006. The response of CD24(-/low)/CD44+ breast cancer-initiating cells to radiation. J Natl Cancer Inst 98:1777–85.
115. Povirk, L. F. 2006. Biochemical mechanisms of chromosomal translocations resulting from DNA double-strand breaks. DNA Repair (Amst) 5:1199–212.
116. Povirk, L. F., T. Zhou, R. Zhou, M. J. Cowan, and S. M. Yannone. 2007. Processing of 3'-phosphoglycolate-terminated DNA double strand breaks by Artemis nuclease. J Biol Chem 282:3547–58.
117. Powell, S. N., and L. A. Kachnic. 2008. Therapeutic exploitation of tumor cell defects in homologous recombination. Anticancer Agents Med Chem 8:448–60.
118. Rainey, M. D., M. E. Charlton, R. V. Stanton, and M. B. Kastan. 2008. Transient inhibition of ATM kinase is sufficient to enhance cellular sensitivity to ionizing radiation. Cancer Res 68:7466–74.

119. Raji, H., and E. Hartsuiker. 2006. Double-strand break repair and homologous recombination in Schizosaccharomyces pombe. Yeast 23:963–76.
120. Revy, P., L. Malivert, and J. P. de Villartay. 2006. Cernunnos-XLF, a recently identified non-homologous end-joining factor required for the development of the immune system. Curr Opin Allergy Clin Immunol 6:416–20.
121. Riballo, E., M. Kuhne, N. Rief, A. Doherty, G. C. Smith, M. J. Recio, C. Reis, K. Dahm, A. Fricke, A. Krempler, A. R. Parker, S. P. Jackson, A. Gennery, P. A. Jeggo, and M. Lobrich. 2004. A pathway of double-strand break rejoining dependent upon ATM, Artemis, and proteins locating to gamma-H2AX foci. Mol Cell 16:715–24.
122. Rosenzweig, K. E., M. B. Youmell, S. T. Palayoor, and B. D. Price. 1997. Radiosensitization of human tumor cells by the phosphatidylinositol3-kinase inhibitors wortmannin and LY294002 correlates with inhibition of DNA-dependent protein kinase and prolonged G2-M delay. Clin Cancer Res 3:1149–56.
123. Rothkamm, K., I. Kruger, L. H. Thompson, and M. Lobrich. 2003. Pathways of DNA double-strand break repair during the mammalian cell cycle. Mol Cell Biol 23: 5706–15.
124. Rottenberg, S., J. E. Jaspers, A. Kersbergen, E. van der Burg, A. O. Nygren, S. A. Zander, P. W. Derksen, M. de Bruin, J. Zevenhoven, A. Lau, R. Boulter, A. Cranston, M. J. O'Connor, N. M. Martin, P. Borst, and J. Jonkers. 2008. High sensitivity of BRCA1-deficient mammary tumors to the PARP inhibitor AZD2281 alone and in combination with platinum drugs. Proc Natl Acad Sci USA 105:17079–17084.
125. Saar, K., K. H. Chrzanowska, M. Stumm, M. Jung, G. Nurnberg, T. F. Wienker, E. Seemanova, R. D. Wegner, A. Reis, and K. Sperling. 1997. The gene for the ataxia-telangiectasia variant, Nijmegen breakage syndrome, maps to a 1-cM interval on chromosome 8q21. Am J Hum Genet 60:605–10.
126. San Filippo, J., P. Sung, and H. Klein. 2008. Mechanism of eukaryotic homologous recombination. Annu Rev Biochem 77:229–57.
127. Sartori, A. A., C. Lukas, J. Coates, M. Mistrik, S. Fu, J. Bartek, R. Baer, J. Lukas, and S. P. Jackson. 2007. Human CtIP promotes DNA end resection. Nature 450:509–14.
128. Savitsky, K., A. Bar-Shira, S. Gilad, G. Rotman, Y. Ziv, L. Vanagaite, D. A. Tagle, S. Smith, T. Uziel, S. Sfez, and et al. 1995. A single ataxia telangiectasia gene with a product similar to PI-3 kinase. Science 268:1749–53.
129. Schreiber, V., F. Dantzer, J. C. Ame, and G. de Murcia. 2006. Poly(ADP-ribose): novel functions for an old molecule. Nat Rev Mol Cell Biol 7:517–28.
130. Shiloh, Y. 2006. The ATM-mediated DNA-damage response: taking shape. Trends Biochem Sci 31:402–410.
131. Shinohara, E. T., L. Geng, J. Tan, H. Chen, Y. Shir, E. Edwards, J. Halbrook, E. A. Kesicki, A. Kashishian, and D. E. Hallahan. 2005. DNA-dependent protein kinase is a molecular target for the development of noncytotoxic radiation-sensitizing drugs. Cancer Res 65: 4987–92.
132. Shrivastav, M., L. P. De Haro, and J. A. Nickoloff. 2008. Regulation of DNA double-strand break repair pathway choice. Cell Res 18:134–47.
133. Shuman, S., and M. S. Glickman. 2007. Bacterial DNA repair by non-homologous end joining. Nat Rev Microbiol 5:852–61.
134. Snouwaert, J. N., L. C. Gowen, A. M. Latour, A. R. Mohn, A. Xiao, L. DiBiase, and B. H. Koller. 1999. BRCA1 deficient embryonic stem cells display a decreased homologous recombination frequency and an increased frequency of non-homologous recombination that is corrected by expression of a brca1 transgene. Oncogene 18:7900–7.
135. Stavnezer, J., J. E. Guikema, and C. E. Schrader. 2008. Mechanism and regulation of class switch recombination. Annu Rev Immunol 26:261–92.
136. Stewart, G. S., R. S. Maser, T. Stankovic, D. A. Bressan, M. I. Kaplan, N. G. Jaspers, A. Raams, P. J. Byrd, J. H. Petrini, and A. M. Taylor. 1999. The DNA double-strand break repair gene hMRE11 is mutated in individuals with an ataxia-telangiectasia-like disorder. Cell 99:577–87.

137. Sung, P., and H. Klein. 2006. Mechanism of homologous recombination: mediators and helicases take on regulatory functions. Nat Rev Mol Cell Biol 7:739–50.
138. Takeda, S., K. Nakamura, Y. Taniguchi, and T. T. Paull. 2007. Ctp1/CtIP and the MRN complex collaborate in the initial steps of homologous recombination. Mol Cell 28:351–2.
139. Uematsu, N., E. Weterings, K. Yano, K. Morotomi-Yano, B. Jakob, G. Taucher-Scholz, P. O. Mari, D. C. van Gent, B. P. Chen, and D. J. Chen. 2007. Autophosphorylation of DNA-PKcs regulates its dynamics at DNA double-strand breaks. J Cell Biol 177:219–29.
140. van der Burg, M., H. Ijspeert, N. S. Verkaik, T. Turul, W. W. Wiegant, K. Morotomi-Yano, P. O. Mari, I. Tezcan, D. J. Chen, M. Z. Zdzienicka, J. J. van Dongen, and D. C. van Gent. 2009. A DNA-PKcs mutation in a radiosensitive T-B- SCID patient inhibits Artemis activation and nonhomologous end-joining. J Clin Invest 119:91–8.
141. Wang, H., B. Rosidi, R. Perrault, M. Wang, L. Zhang, F. Windhofer, and G. Iliakis. 2005. DNA ligase III as a candidate component of backup pathways of nonhomologous end joining. Cancer Res 65:4020–30.
142. Wang, M., W. Wu, B. Rosidi, L. Zhang, H. Wang, and G. Iliakis. 2006. PARP-1 and Ku compete for repair of DNA double strand breaks by distinct NHEJ pathways. Nucleic Acids Res 34:6170–82.
143. Wang, X., and A. D. D'Andrea. 2004. The interplay of Fanconi anemia proteins in the DNA damage response. DNA Repair (Amst) 3:1063–9.
144. West, S. C. 2003. Molecular views of recombination proteins and their control. Nat Rev Mol Cell Biol 4:435–45.
145. Weterings, E., and D. J. Chen. 2007. DNA-dependent protein kinase in nonhomologous end joining: a lock with multiple keys? J Cell Biol 179:183–6.
146. Weterings, E., and D. J. Chen. 2008. The endless tale of non-homologous end-joining. Cell Res 18:114–24.
147. Willems, P., K. Claes, A. Baeyens, V. Vandersickel, J. Werbrouck, K. De Ruyck, B. Poppe, R. Van den Broecke, A. Makar, E. Marras, G. Perletti, H. Thierens, and A. Vral. 2008. Polymorphisms in nonhomologous end-joining genes associated with breast cancer risk and chromosomal radiosensitivity. Genes Chromosomes Cancer 47:137–48.
148. Williams, R. S., G. Moncalian, J. S. Williams, Y. Yamada, O. Limbo, D. S. Shin, L. M. Groocock, D. Cahill, C. Hitomi, G. Guenther, D. Moiani, J. P. Carney, P. Russell, and J. A. Tainer. 2008. Mre11 dimers coordinate DNA end bridging and nuclease processing in double-strand-break repair. Cell 135:97–109.
149. Williams, R. S., J. S. Williams, and J. A. Tainer. 2007. Mre11-Rad50-Nbs1 is a keystone complex connecting DNA repair machinery, double-strand break signaling, and the chromatin template. Biochem Cell Biol 85:509–20.
150. Willmore, E., S. L. Elliott, T. Mainou-Fowler, G. P. Summerfield, G. H. Jackson, F. O'Neill, C. Lowe, A. Carter, R. Harris, A. R. Pettitt, C. Cano-Soumillac, R. J. Griffin, I. G. Cowell, C. A. Austin, and B. W. Durkacz. 2008. DNA-dependent protein kinase is a therapeutic target and an indicator of poor prognosis in B-cell chronic lymphocytic leukemia. Clin Cancer Res 14:3984–92.
151. Wilson, C. R., S. E. Davidson, G. P. Margison, S. P. Jackson, J. H. Hendry, and C. M. West. 2000. Expression of Ku70 correlates with survival in carcinoma of the cervix. Br J Cancer 83:1702–6.
152. Yan, C. T., C. Boboila, E. K. Souza, S. Franco, T. R. Hickernell, M. Murphy, S. Gumaste, M. Geyer, A. A. Zarrin, J. P. Manis, K. Rajewsky, and F. W. Alt. 2007. IgH class switching and translocations use a robust non-classical end-joining pathway. Nature 449:478–482.
153. Yano, K., K. Morotomi-Yano, S. Y. Wang, N. Uematsu, K. J. Lee, A. Asaithamby, E. Weterings, and D. J. Chen. 2008. Ku recruits XLF to DNA double-strand breaks. EMBO Rep 9:91–6.
154. Yu, X., and J. Chen. 2004. DNA damage-induced cell cycle checkpoint control requires CtIP, a phosphorylation-dependent binding partner of BRCA1 C-terminal domains. Mol Cell Biol 24:9478–86.

155. Yu, X., S. Fu, M. Lai, R. Baer, and J. Chen. 2006. BRCA1 ubiquitinates its phosphorylation-dependent binding partner CtIP. Genes Dev 20:1721–6.
156. Zhao, Y., H. D. Thomas, M. A. Batey, I. G. Cowell, C. J. Richardson, R. J. Griffin, A. H. Calvert, D. R. Newell, G. C. Smith, and N. J. Curtin. 2006. Preclinical evaluation of a potent novel DNA-dependent protein kinase inhibitor NU7441. Cancer Res 66:5354–62.
157. Zhu, Z., W. H. Chung, E. Y. Shim, S. E. Lee, and G. Ira. 2008. Sgs1 helicase and two nucleases Dna2 and Exo1 resect DNA double-strand break ends. Cell 134:981–94.

Chapter 9
DNA Base Excision Repair: A Recipe for Survival

Rabindra Roy and Sankar Mitra

Abstract Base excision repair (BER) is an evolutionarily conserved process for maintaining genomic integrity by eliminating several dozen altered (oxidized, deaminated or alkylated) or inappropriate bases that could be mutagenic. These base lesions are generated endogenously or induced by various genotoxicants, including reactive oxygen species (ROS), reactive nitrogen species (RNS) or alkylating agents. BER involves 4 or 5 steps starting with excision of the lesion base by distinct DNA glycosylases and then a common pathway to repair the resulting abasic (AP) site or its cleavage products. The same process is also used for repair of single-strand breaks generated directly by ROS and radiation. This pathway is usually initiated by an AP-endonuclease (APE) that generates a 3′ OH terminus at the damage site, followed by repair synthesis with a DNA polymerase and finally nick sealing by a DNA ligase. Multiple DNA glycosylases, far fewer than the substrate lesions, have usually broad and overlapping substrate range, and could serve as back-up enzymes of one another in vivo. Mammalian cells, surprisingly, encode only one major APE, APE1, unlike two APEs in lower organisms. In spite of the overall similarity, BER in the mammals is more complex than in *E. coli*. The mammalian glycosylases form complexes with downstream proteins to carry out efficient repair via distinct subpathways one of which, responsible for repair of strand breaks with 3′ phosphate termini generated either by the endonuclease VIII like (NEIL) family glycosylases or directly by ROS, requires the phosphatase activity of polynucleotide kinase (PNK) instead of APE1. Different complexes may utilize distinct DNA polymerases and ligases. Mammalian glycosylases have invariably nonconserved extensions at one of the termini, that are involved in modulating their activity, and also in providing the interaction interface for complex formation with other BER and non-BER proteins and in organelle targeting. Polymorphisms in NEILs and 8-oxoguanine-DNA glycosylase (OGG1) have been linked to various cancers, whereas reduced

---

R. Roy (✉)
Lombardi Comprehensive Cancer Center, Georgetown University, 20057-1468, Washington DC, USA
e-mail: rr228@georgetown.edu

K.K. Khanna, Y. Shiloh (eds.), *The DNA Damage Response: Implications on Cancer Formation and Treatment*, DOI 10.1007/978-90-481-2561-6_9,

activities of N-methylpurine-DNA glycosylase (MPG) and thymine-DNA glycosylase (TDG) have been linked to adenocarcinoma type of lung cancer. Mutation in MutY homolog (MYH) also has been linked to colorectal cancer of polyposis type. Thus, BER plays an essential role in maintaining genome integrity, disease prevention and survival.

Keywords Base damage · DNA glycosylase · Reactive oxygen species · Alkyl base adducts · AP-endonuclease

Abbreviations

APC	Adenomatous polyposis coli
AP	abasic
APE	AP-endonuclease
BER	base excision repair
EMS	ethyl methane sulfonate
ENU	ethylnitrosourea
FEN-1	5′flap endonuclease 1
HhH	helix-hairpin-helix
H2TH	helix-2 turn-helix
IR	ionizing radiation
LEC	Long Evans Cinnamon
LET	linear energy transfer
LP-BER	long-patch BER
MeCl	methyl chloride
MMS	methyl methanesulfonate
MNNG	N-methyl-N′-nitro-n-nitrosoguanidine
MNU	N-methyl nitrosourea
MPG	N-methylpurine-DNA glycosylase
mtDNA	mitochondrial DNA
MTIC	5-(3-methyltriazeno)-imidazole-4-carboxamide
MYH	MutY homolog
Nfo	endonuclease IV
Nth	endonuclease III
NTH1	human endonuclease III
OGG1	8-oxoguanine-DNA glycosylase
$OH^{\bullet}$	hydroxyl radical
$O_2^{\bullet -}$	superoxide radical anion
PCNA	proliferating cell nuclear antigen
PNK	polynucleotide kinase
RFC	replication factor-C
RNS	reactive nitrogen species (RNS)
ROS	reactive oxygen species
RPA	replication protein A
SAM	S-adenosylmethionine

SN-BER	single-nucleotide BER
TDG	thymine-DNA glycosylase
TMZ	Temozolomide
TRF2	telomerase repeat binding factor 2
UDG	uracil-DNA glycosylase
XP	Xeroderma pigmentosum

9.1 Introduction

The genome is inherently unstable due to spontaneous chemical reactions, and its fidelity is compromised due to rare but significant replication errors. Moreover, the genomes of all organisms are continuously exposed to a wide variety of insults, e.g., endogenous reactive oxygen species (ROS), and environmental genotoxic agents such as cigarette smoke, UV light, and ionizing radiation. Furthermore, chemotherapeutic drugs including alkylating agents are invariably genotoxic and induce damage in both healthy and tumor cell genomes. ROS, because of their continuous generation as by-products of respiration in the mitochondria [34] represent the predominant group of chemicals that damage genomes of aerobic organisms. ROS are also generated by ionizing radiation and during oxidative metabolism of xenobiotic agents by p450 s. ROS include $O_2^{\bullet -}$ (superoxide radical anion), $OH^{\bullet}$ (hydroxyl radical) and H_2O_2 (hydrogen peroxide). $O_2^{\bullet -}$ is also generated during inflammatory response by NADH oxidases [60]. ROS- induced cellular changes have been implicated in a multitude of diseases, ranging from cardiovascular dysfunction to arthritis and cancer, as well as in aging and age-related disorders. To counteract ROS-induced DNA damage, cells have evolved several defense mechanisms that act at different levels to prevent or repair such damage. The base excision repair (BER) pathway is responsible for repairing most endogenous base lesions and abnormal or inappropriate bases in the genome as well as for repairing similar lesions generated by several groups of environmental agents, or their metabolic intermediates. The BER pathway is also involved in repair of DNA single–strand breaks. Such breaks resulting from free radical reaction of deoxyribose residues in DNA invariably possess blocked termini, which have to be removed during the repair process. Contrary to the earlier concept that BER is a simple process requiring very few proteins, recent discoveries paint a far more complex picture of mammalian BER with several subpathways. This review has touched briefly on bacterial and yeast BER pathways, while focusing primarily on the mammalian BER system.

9.2 DNA Damage

DNA damage can be divided into two major classes: endogenous and environmental. The endogenous group includes mainly hydrolytic and oxidative DNA products. Oxygen radicals that are generated unavoidably from normal intracellular metabolic processes are highly reactive. The environmental reactants include physical and chemical agents that cause DNA damage, often extracellularly generated. All components of DNA (bases, sugars and phosphodiester linkages) could be affected by both endogenous and exogenous reactants, but the damage to the nitrogenous bases is critical, because these specify the genetic code.

9.2.1 Endogenous DNA Lesions

Endogenous DNA damage arises by several mechanisms. Nitrosative stress and changes in pH and temperature can cause deamination of exocyclic amino groups of cytosine, adenine, guanine, and 5-methylcytosine and result in the conversion of the affected bases to promutagenic uracil, hypoxanthine, xanthine, and thymine, respectively [105, 106].

Another important source of endogenous DNA damage is oxidative stress. Reactive oxygen species (ROS) and the radical products of lipid peroxidation are constantly generated as by-products of aerobic metabolism (e.g., leakage from the electron transport chain, associated with the reduction of oxygen to water during mitochondrial respiration). Cellular defense processes including phagocytosis and peroxisomal metabolism [7, 73], and exposure to various natural and synthetic agents also induce ROS production. In spite of multiple defenses against reactive oxygen and aldehyde products, the latter could still sneak into the nucleus and react with DNA bases generating promutagenic and toxic lesions, such as 8-oxoguanine, thymine glycol, 5-hydroxycytosine, ethenoadenine (exocyclic adduct), and malondialdehyde and 4-hydroxynonenal-induced propano adducts [114].

Although the alkylating agents are generally exogenous, at least one endogenous source of alkylating agents, is S-adenosylmethionine (SAM); SAM, a methyl group donor in enzymatic DNA methylation reactions in vivo, can nonenzymatically alkylate G at O-6 position [106, 111].

9.2.2 Exogenous Lesions

Environmental DNA damage can be caused by several agents including ionizing radiation (IR) or various chemical agents. Ionizing radiation causes damage directly by ionizing the DNA or indirectly by generating $OH^{\bullet}$ radical from water, the most active oxidant that randomly damages all cellular components, and induces a variety of DNA base lesions and strand breaks. Typically, 1-Gy low linear energy transfer (LET) radiation (X-rays or gamma rays) can generate 600–1,000 single-strand breaks, 16–40 double-stand breaks, $\sim$250 damaged thymine residues and many other oxidized bases in DNA [191].

9.2.2.1 Drugs and Other Alkylating Agents

Chemical agents comprise the other important class of environmental DNA damage. Mustard gas [di(2-chloro-ethyl)sulphide] used as a chemical weapon during World War I, and its derivatives have been subsequently used in chemotherapy of cancer. These like most other chemotherapeutic agents kill cells by damaging DNA. Alkylating agents directly or after metabolic activation act as electrophiles to react with many nucleophilic sites in DNA. In the early days of chemotherapy, simple monofunctional or bifunctional alkylating agents, such as methylnitrosourea (MNU), N-methyl-N′-nitro-n-nitrosoguanidine (MNNG), methyl methanesulfonate (MMS), ethylnitrosourea (ENU), and ethyl methane sulfonate (EMS) were widely used. later, they became clinically obsolete, because of severe toxicity, and induction of secondary cancers. However, they are still used as model compounds in laboratories for repair studies of alkylation DNA damage. In addition to these chemotherapeutic agents, methyl chloride (MeCl), an abundant alkylating environmental mutagen and carcinogen, directly methylates DNA [50, 100, 101]. Temozolomide (TMZ) is an orally administered alkylating chemotherapeutic agent, with facile transport to the cerebrospinal fluid. With near ideal pharmacokinetics, and 100% bioavailability (in less than 2 h after administration), TMZ is used to treat gliomas, melanomas, mesotheliomas, sarcomas, lymphomas, leukemias, and carcinomas of the colon and ovary. Prolonged treatment with mild to moderate toxicity is the preferred treatment regimen. TMZ does not require metabolic activation; rather, it spontaneously is converted into the active metabolite 5-(3-methyltriazeno)-imidazole-4-carboxamide (MTIC) at physiological pH, which generates the methylating methyldiazonium species. Due to these ideal qualities, TMZ is sometimes the preferred therapeutic agent over other drugs, such as dacarbazine and procarbazine, both of which require metabolic activation in the liver [53]. The DNA adducts induced by TMZ or other methylating agents include 3-methyladenine, 7-methylguanine, O^6-methylguanine and abasic (AP) sites. These adducts have distinct biological effects: 3-methyladenine accumulation is cytotoxic, whereas O^6-methylguanine may become cytotoxic due to futile repair cycles resulting in DNA strand breaks [74, 75, 163, 91]. 7-methylguanine is not cytotoxic, but is spontaneously depurinated or is cleaved by DNA glycosylases (e.g., *N*-methyl purine-DNA glycosylase (MPG)) generating AP sites, which are highly cytotoxic as well as mutagenic.

9.3 Base Excision Repair (BER): A Pathway for Repairing Inappropriate Bases and Single-Strand Breaks: Early Observations

The non-bulky oxidized, deaminated and alkylated bases and exocyclic ethenoadducts, as well as DNA single-strand breaks, generated by endogenous, exogenous and tumoricidal agents as described already, are generally repaired via the BER pathway, most versatile among all excision repair pathways. The basic mechanism of BER was first elucidated in *E. coli* (reviewed in [54]). Subsequent studies showed that the BER process is conserved in the eukaryotes including mammals. BER is distinct from the classical nucleotide excision repair (NER) pathway, which was discovered for UV lesions in 1963–1964. Lack of NER activity in

Xeroderma pigmentosum (XP) patients is responsible for UV sensitivity and skin cancer proneness in XP individuals [54]. Unlike NER involving multitude of proteins, BER in *E. coli* requires only four or five enzymes to carry out repair. Repair is initiated with removal of the damaged, abnormal or inappropriate base by a DNA glycosylase, resulting in an AP site. The first DNA glycosylase to be discovered by Demple and Linn and subsequently by Lindahl, was uracil DNA glycosylase (UDG) in *E. coli,*during the search for an enzyme that recognizes uracil [105, 35]. This was predicted in order to prevent mutagenesis because the unrepaired U in DNA is generated due to deamination of cytosine and G•U pair would be mutagenic when U pairs with adenine during DNA replication [105]. Subsequently, uracil excision activity was identified in other bacteria, yeast, plants and mammalian cells, and mitochondria-specific UDGs were also characterized [96]. The AP-site after base excision is repaired by successive cleavages of the DNA strand next to the AP sites by AP-endonuclease (APE). Like UDG, APE was also discovered in *E. coli,*however, first as an 3' exonuclease/DNA 3'phosphatase. Then an exonuclease/deoxyribosephosphohydrolase activity of DNA polymerases remove the sugar phosphate at the 5' terminus of the strand break generated by APE. The same polymerase could fill in the resulting single-strand gap to produce a nick. Finally, a DNA ligase seals the nick to restore the original DNA structure ([84, 19, 43, 44, 196, 78, 55, 125, 88]; Fig. 9.1). The steps in the basic BER pathway were established in a nonsequential fashion. These steps in the BER pathway are, however, conserved in organisms ranging from *E. coli* to the mammals.

9.3.1 Further Clarification of the Base Excision Step

DNA glycosylases specific for oxidized base lesions are bifunctional with intrinsic AP lyase activity that cleave the DNA strand at the AP site generated after base excision [127]. They were originally named "Type I endonuclease" in contrast to APEs which were considered as type II endonucleases. Thus, the names of endonuclease III (Nth) and endonuclease VIII (including the *E. coli* Fpg) were given based on their bifunctional cleaving activity, active site characteristics and also their tertiary structures (reviewed in [119]). Verly and colleagues subsequently established that these AP lyases are not true endonucleases but carry out βor βδ elimination reaction on the free deoxyribose residue leading to phosphodiester bond cleavage [8, 9]. The Nth family utilizes an internal Lys residue as the active site for β elimination reaction, generating a 3′ phospho α,β-unsaturated aldehyde (3′ PUA; formal name: 3′ phospho 4-hydroxylpentenal), at the strand break. In contrast, the members of the endonuclease VIII (nei) family catalyze βδ elimination at the AP site by utilizing N-terminal Pro as the nucleophile and remove the deoxyribose residue to produce 3′ phosphate terminus at the DNA strand break [200]. In both cases, the 5′ terminus retains the phosphate moiety. Figure 9.1 depicts the β and βδ elimination reactions. These two glycosylases families also have distinct structural features, namely, the presence of either an helix-hairpin-helix (HhH) in Nth or an helix-2

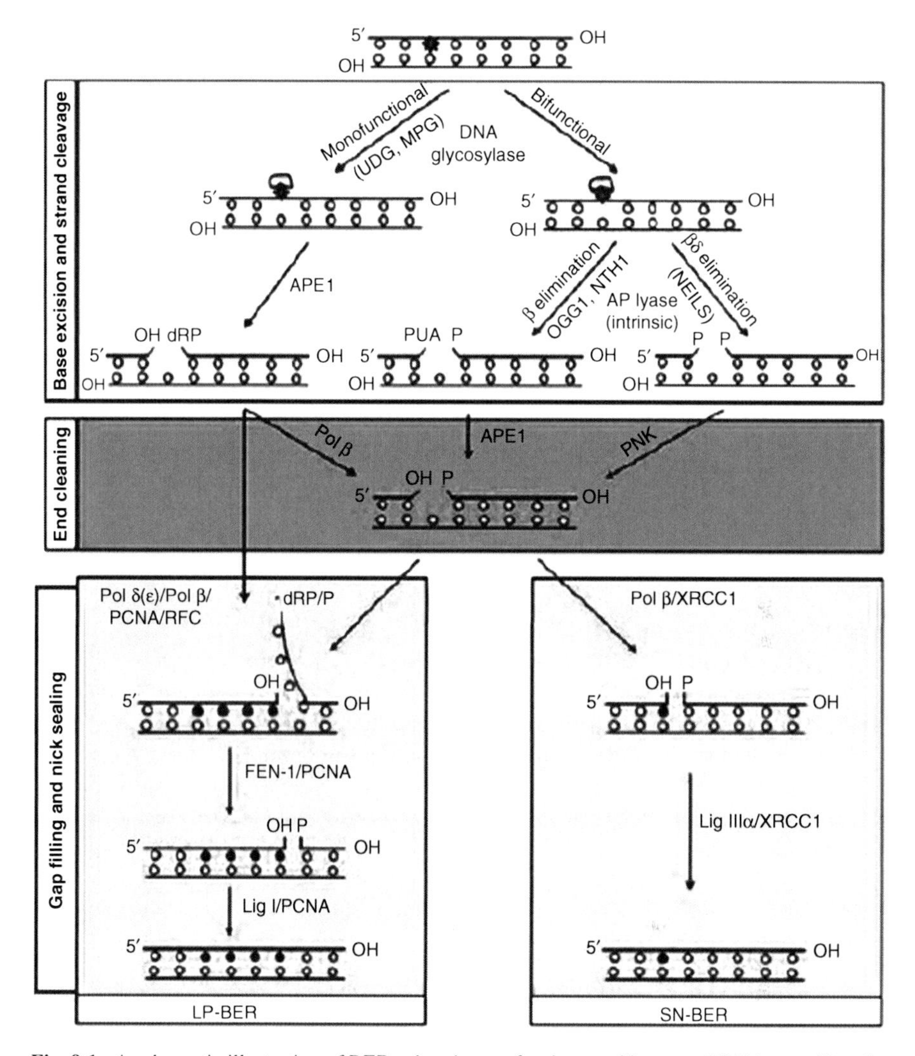

Fig. 9.1 A schematic illustration of BER subpathways for damaged bases and DNA strand breaks. The damaged base is represented as a star (∗). Divergent base excision steps converge to common steps for end processing, followed by gap filling/repair DNA synthesis (represented as *filled dots*) and nick sealing. Polβ could also be involved in LP-BER by collaborating with FEN-1. Other details are discussed in the text

turn-helix (H2TH) motif in the Nei/Fpg family [81]. Representative X-ray crystallographic structures have been elucidated for both families. Unlike in the case of APE cleavage of an AP-site generated by monofunctional glycosylases where the 5′ terminus is blocked with deoxyribose phosphate, the 3′ terminus of ROS-induced DNA strand breaks is blocked, and needs to be cleaned prior to subsequent repair synthesis.

9.4 Distinct Catalytic Mechanisms of Mono and Bifunctional DNA Glycosylases

The monofunctional DNA glycosylases typically use an activated water molecule as a nucleophile in attacking sugar C1′ of the target nucleotide to cleave the glycosidic bond, and excise the damaged base to generate an AP-site. In contrast, the bifunctional glycosylases/AP lyases often use ε-NH_2 of a lysine or the N-terminal proline as the active site nucleophile [119]. An intermediate reaction step for these enzymes is the formation of a transient Schiff base between the amino group and the C1′ of deoxyribose for both base excision and subsequent DNA strand cleavage. Usually, the base excision and lyase reaction occur in a concerted sequence. However, in some cases, e.g., for human 8-oxoguanine-DNA glycosylase (OGG1) and also to some extent for human endonuclease III (NTH1), the β-elimination reaction is relatively weak. Thus, intact AP sites are the major product after OGG1- and NTH1-catalyzed cleavage of 8-oxoguanine and thymine glycol from DNA, respectively [158, 189, 76, 107, 112].

9.5 A Common Mechanism for Substrate Recognition by Mono and Bifunctional DNA-Glycosylases

Despite differences in the specific residues used to recognize damaged bases, one unifying mechanism for BER initiation by mono or bifunctional DNA glycosylases is the extrahelical flipping of the damaged deoxynucleotide into a lesion-specific recognition pocket. All DNA glycosylases studied so far bind to the minor groove, kink DNA at the site of damage, disrupt the base pairing (present, although, in many cases the damaged bases are unpaired, e.g., εA:T), extrude the damaged nucleotide from the interior (major groove) of the DNA helix and place the target base into the catalytic pocket. Thus each DNA glycosylase is damage specific to the extent that only the substrate bases that could be accommodated to provide the necessary contacts and orientation for their excision. The initial recognition apparently exploits the deformability of the DNA at a base pair destabilized by the presence of a lesion [81, 17]. However, the fact that many of the glycosylases are somewhat promiscuous in their substrate preference suggests that the catalytic pocket is plastic in order to allow induced fit of diverse substrates.

9.6 Mechanism of Discrimination of Damaged from Normal Bases by DNA Glycosylases

It has been estimated that less than ten 8-oxoG residues per 10^6 guanine bases are present in the eukaryotic nuclear genome [58]. This low frequency and structural similarity of damaged vs. normal bases pose a formidable challenge for the DNA glycosylases to search through millions of normal base pairs in order to identify rare damage bases [58, 11, 10, 12, 18]. It appears that these enzymes' ability to bind non-target (non-specific) DNA is the key to their rapid target recognition. The law of mass action dictates that even proteins with modest non-specific binding constants should spend the majority of time bound to non-specific DNA under the in vivo condition of high DNA concentration [190]. "Facilitated diffusion" is a process which involves multistep target location of initial association of the glycosylases with nonspecific DNA leading to the target binding. It is generally accepted that rapid target search of these glycosylases in vivo is made possible through combination of one-dimensional diffusion along DNA strands and three-dimensional transfer among DNA segments [18, 154, 178, 146, 14, 64, 80].

9.7 Distinct Steps Following Base Excision by DNA Glycosylases: Repair of AP Sites and Single-Strand Interruption with Nonligatable Termini

DNA glycosylases excise damaged or inappropriate bases generating repair intermediates, namely, AP sites and strand breaks with blocked termini. Repair of these lesions shares the common steps with that of ROS-induced DNA single-strand breaks, and spontaneously generated AP sites. AP sites could be further oxidized to various products due to fragmentation of deoxyribose moiety and strand cleavage [67]. The major products include 2′-deoxyribonolactone, 2-deoxypentos-4-ulose and 3′ phosphoglycolate [39]. But, all oxidized base-specific glycosylases generate 3′ phospho α,β unsaturated aldehyde or 3′ phosphate respectively, and 5′ phosphate at the strand break. In addition to the AP sites, the single-strand breaks with 3′ blocking residues have to be repaired to maintain genomic integrity.

9.7.1 AP-Endonuclease (APE), a Ubiquitous Repair Protein with Dual Nucleolytic Activities

In view of continuous generation of AP sites and DNA strand breaks, all of which are more toxic than damaged bases themselves [65], it is not surprising that multiple enzymatic processes have evolved to repair these lesions among which APE is the most prominent and ubiquitous [65]. As already mentioned, APE was discovered in *E. coli*as a 3′ exonuclease/DNA 3′ phosphatase and was named exonuclease III (Xth; reviewed in [38]) and subsequently identified as AP site-specific endonuclease that generates 3′ OH terminus after cleaving the DNA strand 5′ to the abasic site. *E. coli*Xth is thus a highly processive 3′ exonuclease/phosphatase and produces 3′ OH terminus at strand breaks with 3′ blocks. Another AP-endonuclease, endonuclease IV (Nfo), was subsequently discovered in *E. coli* whose activities as AP site-specific endonuclease and 3′ end cleaning phosphodiesterase are very similar to that of Xth except that Nfo lacks the 3′ exonuclease activity and does not require Mg^{2+} for catalysis [36, 109]. *E. coli* Xth and Nfo are the prototypes of the two APE families that are conserved in bacteria and lower eukaryotes ranging from yeast to *C. elegans*. These enzymes have similar but not identical enzymatic functions and conserved structures.

Xth and Nfo type APEs, named APN2 and APN1, respectively, are expressed in yeast [153, 89]. In contrast to the situation in *E. coli*, APN1 is the major contributor of APE activity in this unicellular eukaryote, *S. cerevisiae,*whereas. the major APE activity in *S. pombe* is contributed by APN2 [160]. Notably, as in *E. coli*, deletion of both APN genes does not cause lethality although it increases sensitivity to alkylating and oxidizing agents. However, simultaneous inactivation of the nucleotide excision repair pathway is lethal as was observed in *E. coli* indicating that the AP sites may be repaired via the NER pathway as a back-up process in lower eukaryotes [62].

9.7.2 Mammalian Cells Express Only Xth type APE, APE1

Unlike in *E. coli*or yeast, there is no evidence for an Nfo ortholog in mammalian cells. Mammalian APE1 was characterized based on activity more than two decades ago [37]. Recent biophysical studies showed that the affinity of hAPE1 binding toward DNA increased as much as 6-fold after replacing a single adenine (equilibrium dissociation constant, K_D, 5.3 nm) with an AP site (K_D, 0.87 nm). The enzyme-substrate complex formation appears to be thermodynamically stabilized and favored by a large change in Gibbs free energy. Low activation energy, the enthalpy of activation and dipole-dipole interactions appear to play critical role in AP site binding by APE1 [4]. Subsequent to its cloning and extensive characterization of APE1, a second Xth ortholog, named APE2 was cloned [63, 187]. However, the recombinant APE2 has negligible APE-specific activity, which suggests its lack of function in BER although it possesses $3'$-$5'$exonuclease and $3'$phosphodiesterase activities [187, 21, Izumi T, unpublished data]. In any case, expression of APE2 in various mammalian cells does implicate it in unknown functions in vivo.

9.7.3 Additional APE's Identified in Mammals

Akira Yasui's group has recently identified a new protein with APE activity which they have named PALF because of some similarity with aprataxin and PNK [92]. PALF accumulates at DNA strand breaks in vivo, dependent on poly ADP-ribose polymerase (PARP), which directly interacts with PALF. Because PALF also interacts with Ku86, XRCC4 and DNA Lig IV, all involved in nonhomologous end joining of DNA double-strand breaks, PALF may have a broad function in repair of both DNA single- and double-strand breaks. The 'end cleaning' of termini at DNA strand breaks is a critical step in both BER and single-strand break repair, to prepare the substrate for repair synthesis, and hence provides a link between DNA glycosylases and DNA polymerases in BER.

9.7.4 Additional Complexities: Involvement of PNK in a BER Subpathway for Mammalian Cells

NTH1 and OGG1 generate 3′ PUA by β-elimination of the AP site, which is efficiently removed by APE1 in the next step. NEILs, on the other hand, have βδ lyase activity and thus generate 3′ phosphate, a poor substrate for APE1. However, mammals, unlike *E. coli*, express high level of polynucleotide kinase (PNK) with dual 5′ kinase/3′phosphatase activities [120]. PNK has already been implicated in repair of ROS-induced single-strand breaks [193]. In fact, PNK directly interacts with XRCC1 and is stimulated by the latter for repairing single-strand breaks [193]. We showed that NEIL-initiated BER in mammals unlike in *E. coli* does not involve APE1, but requires PNK [194, 31]. The presence of a NEIL-initiated, APE-independent BER subpathway in mammalian cells was further supported by the presence of PNK and not APE1 in the NEIL1 or NEIL2 immunocomplex isolated from cell extracts [194, 31]. We also showed that NEILs carry out SN-BER mediated by PNK, Polβ, Lig IIIa and XRCC1. Stable interaction of NEILs with the downstream repair proteins Polβ, Lig IIIa and XRCC1 led us to propose that these DNA glycosylases, as the first enzyme in the repair process, determine the specific BER sub-pathway [194, 31]. Thus for oxidized bases, DNA glycosylases/AP lyases that carry out either β elimination or βδ elimination determine the subsequent steps. APE1 is responsible for processing the β-elimination product while PNK is required for generating 3′-OH terminus from 3′-phosphate, a product of βδ elimination catalyzed by NEIL1 or NEIL2.Furthermore, AP sites and 3′ PUA generated by other DNA glycosylases can also be processed through a NEIL-PNK dependent pathway [68, 69, 132]. This alternative route of repair thus may provide important redundancy in mammalian BER, a critical safeguard against oxidative and spontaneous DNA damage [70].

9.8 Repair of Alkylated Bases by Monofunctional DNA Glycosylases and by MGMT, an Unusual Suicide Protein

As already mentioned, endogenous base lesions include methyl adducts which are likely to be generated by chemical reactions with S-adenosylmethionine [169]. The methylated base adducts could also be generated by reactions with exogenous methylating agents (reviewed in [128]). A multitude of methyl base adducts are produced some of which including O^6-methylguanine (O^6-methyl G) and O^4-methylthymine are highly mutagenic. These methylated adducts are repaired in all organisms via two pathways, one of which is direct reversal of the damage without repair synthesis. This was first discovered in *E. coli* as the regulator for adaptive response to alkylating agent and the protein was named Ada [167]. Adaptive response in *E. coli* is defined as the resistance of cells to a toxic and mutagenic dose of alkylating agents following exposure to a low nontoxic dose of similar agents [167]. This induced resistance is due to activation of several repair genes encoding Ada protein itself and several other proteins including AlkA, a glycosylase responsible for removing multiple alkylated bases (reviewed in [169]). We and Lindahl's group characterized the activity of Ada as a suicide repair protein that demethylates O^6-methyl G in DNA in a stiochiometric reaction and gets inactivated in the process [52, 142] and reviewed in [129]. This activity is ubiquitous which we named O^6- methyl G methyltransferase (MGMT). A second direct reversal enzyme named AlkB in *E. coli*and its mammalian orthologs named ABH 1–3 demethylates 1-methylA and 3-methylC in a complex reaction via oxidative process [169]. In contrast, other methyl adducts as well as hypoxanthine and xanthine, deamination products of A and G respectively, 1, N^6-ethenoadenine (εA), a lipid-peroxidation product, are repaired via the conventional BER pathway by a stable [166, 3] monofunctional N-methylpurine-DNA glycosylase (MPG, also named AAG and ANPG). This enzyme and its *E. coli*ortholog AlkA excise N-alkylpurines, primarily 3-methylA, 3-ethylG and 7-methylA. *E. coli*expresses another glycosylase Tag which specifically excises 3-ethyladenine. AlkA also excises O^2-methyladducts of C and T (reviewed in [124, 165]). A recent study showed that MPG is the predominant enzyme for the BER process in removing εA in mammalian genomes. Although, εA is a fairly a bulky adduct compared to other small BER substrate lesions, NER pathway is not involved in repair of this adduct [27]. The N-terminal tail in MPG plays a critical role in overcoming the product inhibition, which is achieved by reducing the differences of MPG binding affinity toward the substrate lesion and its reaction product, AP-sites and thus, is essential for the turnover [2, 3]. This is in contrast to the fact that the N-terminal extension of hNTH1, an oxidative DNA damage repair glycosylase, inhibited the enzyme turnover and thus inhibited hNTH1's activity significantly [107].

9.9 Distal Steps in BER

APEs generate 3' OH directly by cleaving AP sites or removing the 3' blocking groups generated by bifunctional DNA glycosylases or oxidative fragmentation of deoxyribose after base removal. In the former situation the phosphodeoxyribose moiety blocking the 5' terminus after DNA strand cleavage needs to be removed. *E. coli* DNA polymerase I have intrinsic 5' exonuclease activity. In mammalian cells, DNA polymerase β (polβ) in the nucleus and the replicative DNA polymerase γ (polγ) in the mitochondria possess intrinsic 5' deoxyribose-phosphate (dRp) lyase activity that would generate 5' phosphate terminus [116, 151, 156, 117, 110]. Thus for monofunctional DNA glycosylases, APE together with dRP lyase activity and for bifunctional DNA glycosylases, APE alone will produce a 1-nucleotide gap with 3' OH/5'P termini which could be then filled in by DNA Polβ or γ; sealing of the resulting nick to complete the repair process is catalyzed by DNA ligase IIIα in the nucleus and a variant form in the mitochondria [99].

9.10 Complexity of BER in Mammalian Cells: SN- vs. LP-BER

In the simple BER model, as described above, excision of the damaged base leaves a 1-nt gap whose repair has been named single-nucleotide repair (SN-BER), also called short-patch repair (SP-BER; [98]). In contrast, a repair patch size of 2–8 deoxynucleotides was observed initially in reconstituted BER systems and subsequently in vivo, which was named long-patch BER (LP-BER). During in vitro repair of an AP site analog, lacking the aldehyde group, Polβ could not remove the APE-generated 5′ blocking group via its dRP lyase activity [55, 118, 94]. Matsumoto's and Dogliotti's groups were the first to show that such a 5′ blocking group could be removed by the 5′-flap endonuclease 1 (FEN-1) whose normal function is to remove the 5′ RNA primers from Okazaki fragments during DNA replication. Subsequent steps in LP-BER are identical to that of DNA replication involving DNA polymerases δ/ε and DNA ligase I (Lig I). These enzymes including FEN-1 are recruited to the DNA template by proliferating cell nuclear antigen (PCNA), a sliding clamp, which enhances their activity [55, 118]. The choice of LP-BER vs. SN-BER may thus depend on, among other factors, the nature of the 5′-dRP terminal moiety. With an unaltered aldehyde group the 5′ dRP could be removed by Polβ to could carry out SN-BER [116, 151]. On the other hand, LP-BER is necessary when the AP sites are further oxidized by ROS and the resulting 5′ blocking groups after cleavage of the DNA strand by APE or the DNA strand is directly broken by ROS with a sugar fragment, (such as 2-deoxyribonolactone), could not be removed by Polβ 5′ AP lyase activity. In that case, repair occurs via the LP-BER subpathway in which FEN-1 excises the 5′-oxidized dRP-containing segment displaced as a flap during repair synthesis [104]. More recently, FEN-1 was found to have additional 5′ exonuclease and gap-specific endonuclease activities [172]. The role of FEN-1's exonuclease function in LP-BER however, is yet to be clarified.

Several other factors have been reported to be involved in LP-BER including DNA replication factor C (RFC) that acts as a sliding clamp loader in LP-BER as it does during replication [103]. It is likely that Polδ/ε can substitute for Polβ in repair synthesis and carry out LP-BER, because Polβ-null cells grow normally and are not hypersensitive to oxidative stress [140]. However, Polβ also interacts with FEN-1 and Lig I and could thus carry out LP-BER [108]. These distinct pathways are schematically shown in Fig. 9.1. Few other regulatory proteins (such as p21, Adenomatous polyposis coli (APC) and telomerase repeat binding factor 2 (TRF2)) may also be involved in LP-BER [87, 137, 134]. Such long-patch repair might be the main mode of BER of modified AP-sites in resting cells. This model may be consistent with some other reports [157, 79, 55], especially when repair is considered in proliferating cancer cells without challenge by carcinogens. However, FEN1 may be essential for long-patch BER irrespective of the DNA polymerase type. The selection of BER branches should be controlled by the rate-limiting step. To date, it has been suggested that the overall rate-limiting step in BER is the removal of dRp when BER is initiated by a monofunctional DNA glycosylase (e.g. UNG) [177], or the removal of 3′-blocking deoxyribose residue when a base damage is removed by a bifunctional DNA glycosylase (e.g. OGG1; [83, 158]). However, in separate studies, the joining of repair ends by a DNA ligase and XRCC1 and the ATP concentration was independently suggested to be a rate-limiting that determines the selection of the BER subpathway [181, 148, 141, 149].

Although in vitro experiments suggest that repair of base damage and AP sites could involve either SN- or LP-BER, the complete mechanism and relative importance of these competing processes in vivo are unknown. The BER mechanisms of εA and Hx have been intensively investigated in vitro, using cell-free extracts or purified proteins, but little information is available in vivo [55, 118, 94]. In fact, earlier in vitro studies and analysis of kinetic parameters have suggested a predominant role for short-patch repair, which questions in vivo occurrence of LP-BER [149, 175, 40]. However, one in vivo study reported the long-patch BER of 8-oxoG [168] suggesting that at least oxidized bases could be repaired via the LP-BER pathway.

More recently, we have explored NEIL1-mediated LP-BER based on the early observation that NEIL1 unlike other oxidized base-specific human DNA glycosylases is activated during the S-phase [68]. Subsequently we showed that NEIL1 and NEIL2, unlike OGG1 and NTH1, all with overlapping substrate specificity for oxidized bases, excise the base lesions from single-stranded DNA which is transiently formed during DNA replication and transcription [45]. We have since proposed that NEIL1 preferentially repairs damage in replicating genomes while NEIL2 may repair transcribed sequences [72]. In support of NEIL1's involvement in replication-associated repair (RAR), we have shown that NEIL1 physically and functionally interacts with DNA replication-associated proteins including PCNA, FEN-1 and DNA polymerase δ [46, 72]. Likely involvement with the replication proteins strongly suggests NEIL1's association in LP-BER [71].

9.11 Repair Interactome – A New Paradigm in BER

The broad view of BER in the last decade was that of a sequential process in which individual repair enzymes carried out isolated reactions independently of one another [84]. Based on the emerging evidence for the existence of the "protein interactome", we propose a new paradigm for BER that postulates collaboration of multiple proteins in coordinated fashion involving specific protein-protein interactions to enhance the efficiency of repair. It is now evident that cellular processes involving repair of both endogenous and induced genomic damage are essential for maintaining genomic integrity and homeostasis, which involve dynamic and complex interactions among a multitude of proteins in the cellular proteome in response to both endogenous cellular activities and environmental stress [5]. Such protein complexes are not unique to BER, and similar situations appear to exist for other DNA excision repair pathways as well. However, there is increasing evidence that the mammalian BER proteins are organized in temporally controlled complexes for cross-talks among themselves and with DNA, presumably for optimum efficiency of damaged base recognition and repair.

Formation of the multiprotein complexes for oxidative damage repair is enhanced by ROS in both nucleus and mitochondria [31, 33, 182]. It was previously shown that NEILs carry out SN-BER mediated by PNK, Polβ, Lig IIIα and XRCC1 by stably interacting with Polβ, Lig IIIα and XRCC1 [31, 194]. Other interactions such as that between OGG1 and XRCC1 or MPG and XRCC1 were reported earlier [115, 22]. It now appears that additional proteins including those of the DNA replication machinery are also involved in BER [126, 147]. Several DNA glycosylases interact with replication related-proteins, namely, PCNA and replication protein A (RPA). MYH and UNG interact with both PCNA and RPA [136, 144, 145]. We observed stable and functional association between NEIL1 and the WRN protein [32]. Although these associations may be needed merely for carrying out LP-BER, several observations, e.g., (i) co-localization of the UNG-PCNA-RPA complex at replication foci and (ii) S-phase-specific increase in the levels of UNG2, MYH and NEIL1 [20, 68, 71, 144], support the model that certain glycosylases are involved in replication-associated BER.

9.12 Coordination of Reaction Steps in the BER Pathway

A key unanswered question is whether there is coordinated handover among interacting partners in smaller sub-complexes or a larger complex is preformed to carry out efficient repair. The coordination in protein complex formation was originally compared to the passing of the baton, where the repair product of each enzyme in the BER pathway is "handed" over to the next enzyme in the pathway [197]. In general, the mammalian BER enzymes are slow in being released from reaction products. This could be envisioned as an advantage for the cell in preventing formation of unprotected DNA strand breaks. Notably, the 3'phosphodiesterase activity of hAPE1 is very weak and rate-limiting in the NTH-mediated BER pathway [83]. The 3′ α,β-unsaturated aldehyde at the single strand break created by hNTH1 needs to be repaired by hAPE to provide an appropriate substrate for the downstream BER enzymes, namely, a DNA polymerase and a ligase to complete the repair process. Otherwise, an imbalance in repair could lead to accumulation of DNA strand breaks. Therefore, minimization of unprotected AP sites and/or strand breaks, generated by hNTH1 should be beneficial to the cell. In fact, the low activity and product inhibition are not only the general features of DNA glycosylases but are also true for other mammalian BER enzymes. HAPE1 was also shown to be structurally optimized to retain the DNA product containing a cleaved AP site; moreover, substitution of residues that penetrate the DNA helix stimulated the endonuclease activity of APE [133]. In reality, each BER intermediate is in itself a type of damage, whose sequestration is required in order to prevent activation of cellular damage responses. While the activity of several mammalian glycosylases was shown to be enhanced by APE1 via repair coordination [197, 113, 76], recent study showed that Mg^{2+} could also act as a coordinator of MPG in repair of alkylated bases [1]. Mg^{2+} inhibited MPG activity by abrogating substrate binding. Notably, Mg^{2+} unlike for MPG, is an essential cofactor for the downstream repair enzymes, APE1, Polβ and DNA ligase. Thus, Mg^{2+} could act as a repair regulator by inhibiting DNA glycosylase, such as MPG, for efficient and balanced repair of damaged bases which may be less toxic and/or mutagenic than the subsequent product intermediates [1, 59, 155, 164, 173, 174, 140, 51, 161, 162].

9.13 Essentiality and Biological Consequences of BER Deficiency

9.13.1 Nonessentiality of Individual DNA Glycosylases in Mammals

The genes and cDNAs of mammalian OGG1 and NTH1 were cloned in late 1990 s, and NEIL1 and NEIL2 genes were identified in mammalian genome databases in 2002. The genomic cloning led to the generation of mouse mutants lacking MPG, UDG, TDG, OGG1, NTH1 and NEIL1 as well as MYH [48, 138, 28, 95, 123, 143, 199]. All mouse homozygous null mutants, except of TDG, are viable. Furthermore, no strong phenotype was observed in the mutant mice in regard to enhanced incidence of spontaneous cancer or accelerated aging. Although a small increase in cancer incidence was observed in OGG1-null mice after exposure to bromate and in MPG knock out mice when exposed to inflammatory agents [6, 122]. The lack of immediate effect was unexpected, particularly because the deficiency of other BER proteins, such as APE1 and DNA Polβ, results in embryonic lethality in mutant mice [61, 198]. Notably, TDG, unlike other DNA glycosylases, is essential for embryonic development possibly due to its other function in gene regulation [28]. There is compelling evidence that the oxidized base lesions are endogenously generated at a reasonably high frequency many of which are mutagenic and thus potentially carcinogenic. Repair of such mutagenic and carcinogenic lesions is essential, which is presumably carried out by other glycosylases in the glycosylase deficient mutants and that, unlike in NER or MMR, multiple glycosylases could provide back-up functions because of their overlapping substrate range. In contrast, deficiency in NER or MMR processes significantly enhances cancer susceptibility [131, 53, 54, 29]. We could argue that oxidative damage repair is too important to be left to a single repair enzyme. At the same time, as we have already pointed out, the glycosylases do have preferred substrates and it is also likely that the individual enzyme acts preferentially depending on the state of the genome, i.e., whether the cells are cycling or postmitotic and whether the damage is located in transcriptionally active vs. inactive sequences [127]. This is consistent with the observation that the glycosylases have broad substrate range. These enzymes also have very low turnover, presumably as the price to pay for their promiscuity. Recently, the *NEIL1* $^{-/-}$ mice have been reported to manifest a combination of diseases associated with the metabolic syndrome [188], the pathogenesis of this syndrome being attributed to ROS exposure [13]. As pointed out, neither OGG1 nor NTH1 null mice shows any strong phenotype [95, 183], although there was some accumulation of oxidative lesions in OGG1$^{-/-}$ mice and slower repair of thymine glycol in NTH1$^{-/-}$ mice. However, double-knockout mouse (OGG1$^{-/-}$ and MYH$^{-/-}$) showed susceptibility to tumor formation and have increased level of 8-oxoG [77]. MYH$^{-/-}$mice, did not have an increase in tumor formation compared to the wild type mice; however, these single knock-out mice had increased level of spontaneous mutations [199]. Like MMR-associated hereditary non-polyposis colorectal cancer (that accounts for 1–6% of all colorectal cancer), MYH-associated polyposis cororectal cancer has been described in the humans. This polyposis seems to be an autosomal recessive disorder (requiring mutations in both alleles of MYH) and accounts for ~1% of all colorectal cancer. Recent evidence suggest linkage of polymorphism in various DNA glycosylases to cancer susceptibility and other diseases. For example, polymorphism in NEIL1 have been linked to gastric cancer [171], while those in OGG1 associated with reduced activity have been related to higher risk for prostate cancer and smoking-related lung cancer [192, 195, 24, 57]. Also reduced activities of MPG and TDG towards εA and εC were linked to adenocarcinoma type of lung cancer [176]. In other studies, the activity and expression of NTH1 and OGG1 were found to be significantly altered during the acute (16–18 weeks) and early chronic (24 weeks) phases of hepatitis in the Long Evans Cinnamon (LEC) rat, an animal model for Wilson's disease which is said to cause a "genetically induced" oxidative condition with higher risk for liver cancer [26, 82]. Thus, in view of the presence of multiple DNA glycosylases further studies are required before we could evaluate their contribution to normal cellular functions or deficiency in human diseases.

9.13.2 APE1 is Essential in Mammalian Cells

The attempts of several groups to generate APE1-null mice failed because of very early (4–5 days post coitus) embryonic death. This strongly suggested that unlike in lower organisms, APE1 is essential at least for early embryonic development [198, 121]. Efforts to generate cell lines from APE1 null embryos were also uniformly unsuccessful which distinguishes it from Polβ or XRCC1 [184]. Null mutation for Polβ in mice is also embryonic lethal (although the embryos survived longer) [180]. However, Polβ null mouse embryonic fibroblast lines could be established which have normal viability [173]. We tested whether APE1 is essential in the postembryonic stage by introducing "floxed" hAPE1 transgene in the genome of APE1 heterozygous mice [85]. Our attempt to generate viable mouse embryos lacking endogenous APE1 but expressing ectopic hAPE1, by inbreeding the heterozygous, transgenic mice was unsuccessful. However, we were able to establish embryonic fibroblast lines from such crosses which lack the endogenous APE1 allele but have normal viability. Transgenic expression of Cre in these cells leading to deletion of the APE1 transgene triggered apoptosis. This observation and an independent study provide unequivocal evidence that APE1 is essential even for somatic cells [85, 56]. Kelley and his colleagues had earlier shown that APE1 downregulation triggers apoptosis in several human cell lines (reviewed in [49]). Further studies showed that both DNA repair and transcriptional regulation activity of APE1 are independently required for cell viability [85].

9.13.3 Accumulation of Single-Strand Breaks in the Genome of APE1-Null Cells

Essentiality of APE1 in mammalian cells but not in lower organisms suggests that some cytotoxic lesions normally repaired via APE1-dependent BER could be alternatively repaired by NER or other pathways in *E. coli*or yeast. We suggest that these include DNA strand breaks with phosphosugar fragment-containing 3′ termini which are efficiently repaired by the APE1-dependent pathway (Fig. 9.1). We had shown earlier that APE1 activity is limiting in repair of ROS-induced strand breaks but not of AP sites [83]. Using embryonic fibroblast line with conditional null mutation of APE1, we observed that APE1 inactivation led to significant increase in single-strand breaks as well as AP-sites [84, 130].

9.14 BER in Mitochondria

Mitochondrial DNA (mtDNA) contains higher steady state level of oxidative damage and its mutation rate is significantly greater than of the nuclear DNA, presumably because the mitochondria are the predominant site for production of endogenous ROS and increased production of toxic compounds [25]. In addition to their proximity to the sites of ROS generation, it is likely that the mitochondrial genomes are more prone to oxidative damage because histones and other chromatin-associated proteins, present in nuclear genomes and acting as scavengers of oxygen radicals, are absent in the mitochondria. BER is the only excision repair process active for mitochondrial genomes. All mtDNA repair proteins are nuclear encoded and imported into the mitochondrial matrix. Most mtDNA repair proteins so far discovered are isoforms of the nuclear BER proteins arising from differential splicing [179, 23] or truncation of the terminal sequences. The existence of repair of oxidative damage to mtDNA was established several years ago [47, 41, 102]. It was also reported that mitochondrial DNA polymerase γ (Polγ) and mtDNA ligase (Lig IIIα), which are involved in mtDNA replication are also functional in mitochondrial BER [152, 90]. Mitochondrial Lig IIIα is a splice variant of the nuclear Lig IIIα, whereas Polγ is unique for the mitochondria (reviewed in [186]). Similarly, splice variants of nuclear OGG1 and NTH1 have been characterized in mitochondria [93,

139]. Mitochondrial transport of proteins generally requires an N-terminal mitochondrial targeting sequence (MTS) which is cleaved off in the mitochondrial matrix by specific peptidases [150] although, some mitochondrial proteins lacking MTS, utilize an internal targeting sequence which is not removed during mitochondrial import [42]. Further, mitochondrial isozymes are generated primarily to eliminate the nuclear localization signal of such proteins, otherwise required for their nuclear targeting [150]. Although most mitochondrial BER enzymes were characterized and significant BER activity was reported in mitochondrial extracts several years ago [41, 102, 152], the identity of mtAPE, a key BER enzyme, was not established. Recently, our lab showed that the mtAPE lacks the N-terminal 33 amino acid residues including the NLS [23]. However, the full-length APE1 could also be detected in the mitochondria of some cells. This indicates that deletion of NLS is not absolutely required for mitochondrial import of APE1. On the other hand, we have shown that the extract of mitochondria (containing processing peptidases) but not of nuclei or cytosol, cleaves recombinant hAPE1 to generate an mtAPE-sized product, implying that the removal of NLS is necessary for mitochondrial import of APE. At the same time, deletion mapping of hAPE1 helped us identify a bipartite NLS as well as a nuclear export signal in the N-terminal segment of the APE1 polypeptide [86].

9.15 Regulation of BER Activity In Vivo in Response to Genotoxic Stress

BER activity is often modulated by genotoxins, such as oxidative stress, at various levels including transcriptional activation of repair genes, such as APE1 and NEIL1 [159, 30] and enhanced repair protein complex formation [31]. Regulation of BER may be further fine-tuned through posttranslational modifications of the BER proteins. Such modifications include phosphorylation, acetylation, mono- or poly-ADP-ribosylation, mono- and poly-ubiquitylation, sumoylation and methylation among others [97]. Covalent modifications could have multiple physiological effects on proteins, including modulation of stability, interaction, organelle targeting, and enzymatic activity etc. [170]. Such studies are still at an early stage. Some initial observations on modifications of mammalian DNA glycosylases are summarized as follows.

9.15.1 Sumoylation of TDG

Sumoylation and ubiquitylation alter activities of target proteins [135] SUMO-1 and -3 modify TDG, a glycosylase that removes uracil (or thymine) from its substrate U(T)•G pair [66]. TDG remains bound to the AP site after the excision of uracil and sumoylation reduces its affinity, thus releasing the product. The modified TDG is recycled by deconjugation, restoring its activity.

9.15.2 Acetylation of DNA Glycosylases

The presence of multiple species of OGG1 in 2-D gel analysis of HeLa extracts suggested its covalent modification, which we showed to be acetylation [16]. We also showed that about 20% of OGG1 is acetylated in HeLa cells, and identified two acetylation acceptor lysine residues (Lys 338 and Lys 341). We further showed that acetylation enhances OGG1 turnover by decreasing its affinity for the product. We observed earlier that APE1 enhances OGG1 turnover which is higher when OGG1 is acetylated [16, 76]. ROS induced an increase in the acetylated OGG1 (AcOGG1) level, indirectly through activation of p300. We have proposed that acetylation provides a mechanism for rapid cellular response to OGG1-mediated repair of the increased load of oxidized bases induced by ROS [16].

Studies from our lab showed that NEIL2 is also acetylated, predominantly at Lys 49 and 153 [15]. Lys 49 is conserved among Fpg family members and its mutation inactivates the enzyme. Lys 49 acetylation strongly inhibits NEIL2 activity. We have proposed that acetylation at Lys 49 could act as a regulatory switch for NEIL2's activity, and Lys 153 acetylation could regulate its interactions. It has been shown that TDG is acetylated both in vivo and in vitro by CBP/p300 which forms a functional complex with TDG [185]. Strong acetylation sites were identified at Lys residues 70, 94, 95 and 98. TDG acetylation does not affect binding and cleavage of G∉T, G∉U mispaired DNA but leads to release of CBP/p300 from the DNA bound complex and abrogates the interaction with APE suggesting a role in suppression of APE-dependent repair by blocking the next reaction that normally follows the removal of the mispaired base. Thus acetylation appears to deregulate TDG-coupled DNA repair and contribute to the genomic instability commonly associated with cancer.

9.16 Synopsis and Future Perspective

BER is distinct from the other excision repair pathways in that the first step of lesion recognition by a DNA glycosylase is unique. This review emphasizing early steps in base excision and strand break repair aims at highlighting the key roles of DNA glycosylases and APE1 (and other end-cleaning activities) in maintaining integrity of mammalian genomes. Furthermore, BER represents the predominant process for repairing endogenous DNA lesions mostly induced by ROS and also by RNS, alkylating agents and lipid peroxidation. While BER was earlier believed to be the simplest repair process, its complexities are now becoming increasingly evident. In this chapter, we have touched upon a few of these complexities including interaction of BER enzymes among themselves and with non-BER proteins, their covalent modifications, formation of pair-wise and multiprotein repair complexes and its regulation. Future research will undoubtedly be focused on these and other aspects of BER. For example, we have shown ROS-mediated activation of NEIL1 and APE1 [159, 30]. We have shown that ROS enhances NEIL2 complex formation [33]. The mechanistic bases of these phenomena are not clear. ROS also enhances modification of OGG1 [16] and possibly other BER proteins. How such modifications affect repair efficiency in vivo needs to be explored further. Preliminary studies showing preferential repair of oxidized bases in replicating genomes need to be confirmed. The linkage of BER to susceptibility to various diseases also warrants extensive investigation. Finally, the mechanism of recognition of BER substrates, which cause subtle deformations in DNA structure in chromatin, followed by their repair requiring access of the repair complex to the lesion needs to be explored. These challenges will keep the investigators busy for many years to come.

Acknowledgments The authors' research is supported by US Public Health Services Research grants R01 CA81063, R01 CA53791, (SM), R01 CA92306, RO1 CA113447 (RR), and P01 CA92584, P30 ES06676 (SM). The authors thank the Mitra and Roy laboratory members for scientific discussion and Ms Wanda Smith for expert secretarial assistance. Finally, we apologize for the omission of many relevant publications because of space limitation.

References

1. Adhikari S, Toretsky JA, Yuan L et al. (2006). Magnesium, essential for base excision repair enzymes, inhibits substrate binding of N-methylpurine-DNA glycosylase. J Biol Chem 281:29525–29532.
2. Adhikari S, Uren A, Roy R (2007). N-terminal extension of N-methylpurine DNA glycosylase is required for turnover in hypoxanthine excision reaction. J Biol Chem 282:30078–30084.
3. Adhikari S, Manthena PV, Uren A et al. (2008b). Expression, purification and characterization of codon-optimized human N-methylpurine-DNA glycosylase from Escherichia coli. Protein Expr Purif 58:257–262.
4. Adhikari S, Uren A, Roy R (2008a). Dipole-dipole interaction stabilizes the transition state of apurinic/apyrimidinic endonuclease – abasic site interaction. J Biol Chem 283:1334–1339.
5. Andersen JS, Lam YW, Leung AK et al. (2005). Nucleolar proteome dynamics. Nature 433:77–83.
6. Arai T, Kelly VP, Minowa O et al. (2002). High accumulation of oxidative DNA damage, 8-hydroxyguanine, in Mmh/Ogg1 deficient mice by chronic oxidative stress. Carcinogenesis 23:2005–2010.
7. Averbeck D (1989). Recent advances in psoralen phototoxicity mechanism. Photochem Photobiol 50:859–882.
8. Bailly V, Verly WG (1988). Possible roles of beta-elimination and delta-elimination reactions in the repair of DNA containing AP (apurinic/apyrimidinic) sites in mammalian cells. Biochem J 253:553–559.
9. Bailly V, Verly WG (1989). AP endonucleases and AP lyases. Nucleic Acids Res 17: 3617–3618.
10. Banerjee A, Yang W, Karplus M et al. (2005). Structure of a repair enzyme interrogating undamaged DNA elucidates recognition of damaged DNA. Nature 434:612–618.
11. Banerjee A, Verdine GL (2006). A nucleobase lesion remodels the interaction of its normal neighbor in a DNA glycosylase complex. Proc Natl Acad Sci USA 103: 15020–15025.
12. Banerjee A, Santos WL, Verdine GL (2006). Structure of a DNA glycosylase searching for lesions. Science 311:1153–1157.
13. Beckman M (2006). Mucking with metabolism. Sci Aging Knowl Environ 2006:nf6.
14. Berg OG, Winter RB, von Hippel PH (1981). Diffusion-driven mechanisms of protein translocation on nucleic acids: Models and theory. Biochemistry 20:6929–6948.

15. Bhakat KK, Hazra TK, Mitra S (2004). Acetylation of the human DNA glycosylase NEIL2 and inhibition of its activity. Nucleic Acids Res 32:3033–3039.
16. Bhakat KK, Mokkapati SK, Boldogh I et al. (2006). Acetylation of human 8-oxoguanine-DNA glycosylase by p300 and its role in 8-oxoguanine repair in vivo. Mol Cell Biol 26:1654–1665.
17. Biswas T, Clos LJ, SantaLucia J, Jr. et al. (2002). Binding of specific DNA base-pair mismatches by N-methylpurine-DNA glycosylase and its implication in initial damage recognition. J Mol Biol 320:503–513.
18. Blainey PC, van Oijen AM, Banerjee A et al. (2006). A base-excision DNA-repair protein finds intrahelical lesion bases by fast sliding in contact with DNA. Proc Natl Acad Sci U S A 103:5752–5757.
19. Boiteux S, Guillet M (2004). Abasic sites in DNA: repair and biological consequences in Saccharomyces cerevisiae. DNA Repair (Amst) 3:1–12.
20. Boldogh I, Milligan D, Lee MS et al. (2001). hMYH cell cycle-dependent expression, subcellular localization and association with replication foci: evidence suggesting replication-coupled repair of adenine:8-oxoguanine mispairs. Nucleic Acids Res 29:2802–2809.
21. Burkovics P, Szukacsov V, Unk I et al. (2006). Human Ape2 protein has a 3′-5′ exonuclease activity that acts preferentially on mismatched base pairs. Nucleic Acids Res 34:2508–2515.
22. Campalans A, Marsin S, Nakabeppu Y et al. (2005). XRCC1 interactions with multiple DNA glycosylases: a model for its recruitment to base excision repair. DNA Repair (Amst) 4:826–835.
23. Chattopadhyay R, Wiederhold L, Szczesny B et al. (2006). Identification and characterization of mitochondrial abasic (AP)-endonuclease in mammalian cells. Nucleic Acids Res 34:2067–2076.
24. Chen L, Elahi A, Pow-Sang J et al. (2003). Association between polymorphism of human oxoguanine glycosylase 1 and risk of prostate cancer. J Urol 170:2471–2474.
25. Chen R, Yang L, McIntyre T.M (2007). Cytotoxic phospholipid oxidation products. Cell death from mitochondrial damage and the intrinsic caspase cascade. J Biol Chem 282:24842–24850.
26. Choudhury S, Zhang R, Frenkel K et al. (2003). Evidence of alterations in base excision repair of oxidative DNA damage during spontaneous hepatocarcinogenesis in Long Evans Cinnamon rats. Cancer Res 63:7704–7707.
27. Choudhury S, Adhikari S, Cheema A et al. (2008). Evidence of complete cellular repair of 1,N^6-ethenoadenine, a mutagenic and potential damage for human cancer, revealed by a novel method. Mol Cell Biochem 313:19–28.
28. Cortazar D, Kunz C, Saito Y et al. (2007). The enigmatic thymine DNA glycosylase. DNA Repair (Amst) 6:489–504.
29. Crew KD, Gammon MD, Terry MB et al. (2007). Polymorphisms in nucleotide excision repair genes, polycyclic aromatic hydrocarbon-DNA adducts, and breast cancer risk. Cancer Epidemiol Biomarkers Prev 16:2033–2041.
30. Das A, Hazra TK, Boldogh I et al. (2005). Induction of the human oxidized base-specific DNA glycosylase NEIL1 by reactive oxygen species. J Biol Chem 280:35272–35280.
31. Das A, Wiederhold L, Leppard JB et al. (2006). NEIL2-initiated, APE-independent repair of oxidized bases in DNA: Evidence for a repair complex in human cells. DNA Repair (Amst) 5:1439–1448.
32. Das A, Boldogh I, Lee JW et al. (2007b). The human Werner syndrome protein stimulates repair of oxidative DNA base damage by the DNA glycosylase NEIL1. J Biol Chem 282:26591–26602.
33. Das S, Chattopadhyay R, Bhakat KK et al. (2007a). Stimulation of NEIL2-mediated oxidized base excision repair via YB-1 interaction during oxidative stress. J Biol Chem 282:28474–28484.
34. Dawson TL, Gores GJ, Nieminen AL et al. (1993). Mitochondria as a source of reactive oxygen species during reductive stress in rat hepatocytes. Am J Physiol 264:C961–C967.

35. Demple B, Linn S (1982). On the recognition and cleavage mechanism of Escherichia coli endodeoxyribonuclease V, a possible DNA repair enzyme. J Biol Chem 257:2848–2855.
36. Demple B, Johnson A, Fung D (1986). Exonuclease III and endonuclease IV remove 3′ blocks from DNA synthesis primers in H_2O_2-damaged Escherichia coli. Proc Natl Acad Sci USA 83:7731–7735.
37. Demple B, Herman T, Chen DS (1991). Cloning and expression of APE, the cDNA encoding the major human apurinic endonuclease: definition of a family of DNA repair enzymes. Proc Natl Acad Sci USA 88:11450–11454.
38. Demple B, Harrison L (1994). Repair of oxidative damage to DNA: enzymology and biology. Annu Rev Biochem 63:915–948.
39. Demple B, DeMott MS (2002). Dynamics and diversions in base excision DNA repair of oxidized abasic lesions. Oncogene 21:8926–8934.
40. Dianov G, Price A, Lindahl T (1992). Generation of single-nucleotide repair patches following excision of uracil residues from DNA. Mol Cell Biol 12:1605–1612.
41. Dianov GL, Souza-Pinto N, Nyaga SG et al. (2001). Base excision repair in nuclear and mitochondrial DNA. Prog Nucleic Acid Res Mol Biol 68:285–297.
42. Diekert K, Kispal G, Guiard B et al. (1999). An internal targeting signal directing proteins into the mitochondrial intermembrane space. Proc Natl Acad Sci USA 96:11752–11757.
43. Dizdaroglu M (2003). Substrate specificities and excision kinetics of DNA glycosylases involved in base-excision repair of oxidative DNA damage. Mutat Res 531:109–126.
44. Dodson ML, Lloyd RS (2002). Mechanistic comparisons among base excision repair glycosylases. Free Radic Biol Med 32:678–682.
45. Dou H, Mitra S, Hazra TK (2003). Repair of oxidized bases in DNA bubble structures by human DNA glycosylases NEIL1 and NEIL2. J Biol Chem 278:49679–49684.
46. Dou H, Theriot CA, Das A et al. (2008). Interaction of the human DNA glycosylase NEIL1 with proliferating cell nuclear antigen. The potential for replication-associated repair of oxidized bases in mammalian genomes. J Biol Chem 283:3130–3140.
47. Driggers WJ, LeDoux SP, Wilson GL (1993). Repair of oxidative damage within the mitochondrial DNA of RINr 38 cells. J Biol Chem 268:22042–22045.
48. Engelward BP, Weeda G, Wyatt MD et al. (1997). Base excision repair deficient mice lacking the Aag alkyladenine DNA glycosylase. Proc Natl Acad Sci U S A 94:13087–13092.
49. Evans AR, Limp-Foster M, Kelley MR (2000). Going APE over ref-1. Mutat Res 461: 83–108.
50. Ferguson LR, Pearson AE (1996). The clinical use of mutagenic anticancer drugs. Mutat Res 355:1–12.
51. Fishel ML, Seo YR, Smith ML et al. (2003). Imbalancing the DNA base excision repair pathway in the mitochondria; targeting and overexpressing N-methylpurine DNA glycosylase in mitochondria leads to enhanced cell killing. Cancer Res 63:608–615.
52. Foote RS, Mitra S, Pal BC (1980). Demethylation of O6-methylguanine in a synthetic DNA polymer by an inducible activity in Escherichia coli. Biochem Biophys Res Commun 97:654–659.
53. Friedberg EC, Aguilera A, Gellert M et al. (2006a). DNA repair: from molecular mechanism to human disease. DNA Repair (Amst) 5:986–996.
54. Friedberg EC, Walker GC, Siede W et al. (2006b). DNA Repair and Mutagenesis. 2nd Edition. ASM Press, Washington
55. Frosina G, Fortini P, Rossi O et al. (1996). Two pathways for base excision repair in mammalian cells. J Biol Chem 271:9573–9578.
56. Fung H, Demple B (2005). A vital role for Ape1/Ref1 protein in repairing spontaneous DNA damage in human cells. Mol Cell 17:463–470.
57. Gackowski D, Speina E, Zielinska M et al. (2003). Products of oxidative DNA damage and repair as possible biomarkers of susceptibility to lung cancer. Cancer Res 63:4899–4902.
58. Gedik CM, Collins A (2005). Establishing the background level of base oxidation in human lymphocyte DNA: results of an interlaboratory validation study. FASEB J 19:82–84.

59. Glassner BJ, Rasmussen LJ, Najarian MT et al. (1998). Generation of a strong mutator phenotype in yeast by imbalanced base excision repair. Proc Natl Acad Sci USA 95:9997–10002.
60. Griendling KK, Sorescu D, Ushio-Fukai M (2000). NAD(P)H oxidase: role in cardiovascular biology and disease. Circ Res 86:494–501.
61. Gu H, Marth JD, Orban PC et al. (1994). Deletion of a DNA polymerase beta gene segment in T cells using cell type-specific gene targeting. Science 265:103–106.
62. Guillet M, Boiteux S (2002). Endogenous DNA abasic sites cause cell death in the absence of Apn1, Apn2 and Rad1/Rad10 in Saccharomyces cerevisiae. EMBO J 21:2833–2841.
63. Hadi MZ, Wilson DM, III (2000). Second human protein with homology to the Escherichia coli abasic endonuclease exonuclease III. Environ Mol Mutagen 36:312–324.
64. Halford SE, Marko JF (2004). How do site-specific DNA-binding proteins find their targets? Nucleic Acids Res 32:3040–3052.
65. Hanna M, Chow BL, Morey NJ et al. (2004). Involvement of two endonuclease III homologs in the base excision repair pathway for the processing of DNA alkylation damage in Saccharomyces cerevisiae. DNA Repair (Amst) 3:51–59.
66. Hardeland U, Steinacher R, Jiricny J et al. (2002). Modification of the human thymine-DNA glycosylase by ubiquitin-like proteins facilitates enzymatic turnover. EMBO J 21: 1456–1464.
67. Haring M, Rudiger H, Demple B et al. (1994). Recognition of oxidized abasic sites by repair endonucleases. Nucleic Acids Res 22:2010–2015.
68. Hazra TK, Izumi T, Boldogh I et al. (2002a). Identification and characterization of a human DNA glycosylase for repair of modified bases in oxidatively damaged DNA. Proc Natl Acad Sci USA 99:3523–3528.
69. Hazra TK, Kow YW, Hatahet Z et al. (2002b) Identification and characterization of a novel human DNA glycosylase for repair of cytosine-derived lesions. J Biol Chem 277: 30417–30420.
70. Hazra TK, Das A, Das S et al. (2007). Oxidative DNA damage repair in mammalian cells: a new perspective. DNA Repair (Amst) 6:470–480.
71. Hegde ML, Theriot CA, Das A et al. (2008b). Physical and functional interaction between human oxidized base-specific DNA glycosylase NEIL1 and flap endonuclease 1. J Biol Chem 283:27028–27037.
72. Hegde ML, Hazra TK, Mitra S (2008a). Early steps in the DNA base excision/single-strand interruption repair pathway in mammalian cells. Cell Res 18:27–47.
73. Henle ES, Linn S (1997). Formation, prevention, and repair of DNA damage by iron/hydrogen peroxide. J Biol Chem 272:19095–19098.
74. Hickman MJ, Samson LD (1999). Role of DNA mismatch repair and p53 in signaling induction of apoptosis by alkylating agents. Proc Natl Acad Sci USA 96: 10764–10769.
75. Hickman MJ, Samson LD (2004). Apoptotic signaling in response to a single type of DNA lesion, O^6-methylguanine. Mol Cell 14:105–116.
76. Hill JW, Hazra TK, Izumi T et al. (2001). Stimulation of human 8-oxoguanine-DNA glycosylase by AP-endonuclease: potential coordination of the initial steps in base excision repair. Nucleic Acids Res 29:430–438.
77. Hirano S, Tominaga Y, Ichinoe A et al. (2003). Mutator phenotype of MUTYH-null mouse embryonic stem cells. J Biol Chem 278:38121–38124.
78. Hollis T, Lau A, Ellenberger T (2001). Crystallizing thoughts about DNA base excision repair. Prog Nucleic Acid Res Mol Biol 68:305–314.
79. Horton JK, Prasad R, Hou E et al. (2000). Protection against methylation-induced cytotoxicity by DNA polymerase beta-dependent long patch base excision repair. J Biol Chem 275:2211–2218.
80. Hu T, Grosberg AY, Shklovskii BI (2006). How proteins search for their specific sites on DNA: the role of DNA conformation. Biophys J 90:2731–2744.

81. Huffman JL, Sundheim O, Tainer JA (2005). DNA base damage recognition and removal: new twists and grooves. Mutat Res 577:55–76.
82. Hussain SP, Raja K, Amstad PA et al. (2000). Increased p53 mutation load in nontumorous human liver of wilson disease and hemochromatosis: oxyradical overload diseases. Proc Natl Acad Sci USA 97:12770–12775.
83. Izumi T, Hazra TK, Boldogh I et al. (2000). Requirement for human AP endonuclease 1 for repair of 3′-blocking damage at DNA single-strand breaks induced by reactive oxygen species. Carcinogenesis 21:1329–1334.
84. Izumi T, Wiederhold LR, Roy G et al. (2003). Mammalian DNA base excision repair proteins: their interactions and role in repair of oxidative DNA damage. Toxicology 193:43–65.
85. Izumi T, Brown DB, Naidu CV et al. (2005). Two essential but distinct functions of the mammalian abasic endonuclease. Proc Natl Acad Sci USA 102:5739–5743.
86. Jackson EB, Theriot CA, Chattopadhyay R et al. (2005). Analysis of nuclear transport signals in the human apurinic/apyrimidinic endonuclease (APE1/Ref1). Nucleic Acids Res 33:3303–3312.
87. Jaiswal AS, Bloom LB, Narayan S (2002). Long-patch base excision repair of apurinic/apyrimidinic site DNA is decreased in mouse embryonic fibroblast cell lines treated with plumbagin: involvement of cyclin-dependent kinase inhibitor p21Waf-1/Cip-1. Oncogene 21:5912–5922.
88. Jaiswal M, Lipinski LJ, Bohr VA et al. (1998). Efficient in vitro repair of 7-hydro-8-oxodeoxyguanosine by human cell extracts: involvement of multiple pathways. Nucleic Acids Res 26:2184–2191.
89. Johnson RE, Torres-Ramos CA, Izumi T et al. (1998). Identification of APN2, the Saccharomyces cerevisiae homolog of the major human AP endonuclease HAP1, and its role in the repair of abasic sites. Genes Dev 12:3137–3143.
90. Kaguni LS (2004). DNA polymerase gamma, the mitochondrial replicase. Annu Rev Biochem 73:293–320.
91. Kaina B, Christmann M, Naumann S et al. (2007). MGMT: key node in the battle against genotoxicity, carcinogenicity and apoptosis induced by alkylating agents. DNA Repair (Amst) 6:1079–1099.
92. Kanno S, Kuzuoka H, Sasao S et al. (2007). A novel human AP endonuclease with conserved zinc-finger-like motifs involved in DNA strand break responses. EMBO J 26:2094–2103.
93. Karahalil B, Souza-Pinto NC, Parsons JL et al. (2003). Compromised incision of oxidized pyrimidines in liver mitochondria of mice deficient in NTH1 and OGG1 glycosylases. J Biol Chem 278:33701–33707.
94. Klungland A, Lindahl T (1997). Second pathway for completion of human DNA base excision-repair: reconstitution with purified proteins and requirement for DNase IV (FEN1). EMBO J 16:3341–3348.
95. Klungland A, Rosewell I, Hollenbach S et al. (1999). Accumulation of premutagenic DNA lesions in mice defective in removal of oxidative base damage. Proc Natl Acad Sci USA 96:13300–13305.
96. Krokan HE, Drablos F, Slupphaug G (2002). Uracil in DNA – occurrence, consequences and repair. Oncogene 21:8935–8948.
97. Krueger KE, Srivastava S (2006). Posttranslational protein modifications: current implications for cancer detection, prevention, and therapeutics. Mol Cell Proteomics 5:1799–1810.
98. Kubota Y, Nash RA, Klungland A et al. (1996). Reconstitution of DNA base excision-repair with purified human proteins: interaction between DNA polymerase beta and the XRCC1 protein. EMBO J 15:6662–6670.
99. Lakshmipathy U, Campbell C (1999). The human DNA ligase III gene encodes nuclear and mitochondrial proteins 1. Mol Cell Biol 19:3869–3876.
100. Lawley PD (1989). Mutagens as carcinogens: development of current concepts. Mutat Res 213:3–25.

101. Lawley PD, Phillips DH (1996). DNA adducts from chemotherapeutic agents. Mutat Res 355:13–40.
102. LeDoux SP, Wilson GL (2001). Base excision repair of mitochondrial DNA damage in mammalian cells. Prog Nucleic Acid Res Mol Biol 68:273–284.
103. Levin DS, McKenna AE, Motycka TA et al. (2000). Interaction between PCNA and DNA ligase I is critical for joining of Okazaki fragments and long-patch base-excision repair. Curr Biol 10:919–922.
104. Levin DS, Vijayakumar S, Liu X et al. (2004). A conserved interaction between the replicative clamp loader and DNA ligase in eukaryotes: implications for Okazaki fragment joining. J Biol Chem 279:55196–55201.
105. Lindahl T (1974). An N-glycosidase from Escherichia coli that releases free uracil from DNA containing deaminated cytosine residues. Proc Natl Acad Sci USA 71:3649–3653.
106. Lindahl T (1979). DNA glycosylases, endonucleases for apurinic/apyrimidinic sites, and base excision-repair. Prog Nucleic Acid Res Mol Biol 22:135–192.
107. Liu X, Roy R (2002). Truncation of amino-terminal tail stimulates activity of human endonuclease III (hNTH1). J Mol Biol 321:265–276.
108. Liu Y, Kao HI, Bambara RA (2004). Flap endonuclease 1: a central component of DNA metabolism. Annu Rev Biochem 73:589–615.
109. Ljungquist S (1977). A new endonuclease from Escherichia coli acting at apurinic sites in DNA. J Biol Chem 252:2808–2814.
110. Longley MJ, Prasad R, Srivastava DK et al. (1998). Identification of 5′-deoxyribose phosphate lyase activity in human DNA polymerase gamma and its role in mitochondrial base excision repair in vitro. Proc Natl Acad Sci USA 95:12244–12248.
111. Lutz WK (1990). Endogenous genotoxic agents and processes as a basis of spontaneous carcinogenesis. Mutat Res 238:287–295.
112. Marenstein DR, Ocampo MT, Chan MK et al. (2001). Stimulation of human endonuclease III by Y box-binding protein 1 (DNA-binding protein B). Interaction between a base excision repair enzyme and a transcription factor. J Biol Chem 276:21242–21249.
113. Marenstein DR, Chan MK, Altamirano A et al. (2003). Substrate specificity of human endonuclease III (hNTH1). Effect of human APE1 on hNTH1 activity. J Biol Chem 278:9005–9012.
114. Marnett LJ (2000). Oxyradicals and DNA damage. Carcinogenesis 21:361–370.
115. Marsin S, Vidal AE, Sossou M et al. (2003). Role of XRCC1 in the coordination and stimulation of oxidative DNA damage repair initiated by the DNA glycosylase hOGG1. J Biol Chem 278:44068–44074.
116. Matsumoto Y, Kim K (1995). Excision of deoxyribose phosphate residues by DNA polymerase beta during DNA repair. Science 269:699–702.
117. Matsumoto Y, Kim K, Katz DS et al. (1998). Catalytic center of DNA polymerase beta for excision of deoxyribose phosphate groups. Biochemistry 37:6456–6464.
118. Matsumoto Y (2001). Molecular mechanism of PCNA-dependent base excision repair. Prog Nucleic Acid Res Mol Biol 68:129–138.
119. McCullough AK, Dodson ML, Lloyd RS (1999). Initiation of base excision repair: glycosylase mechanisms and structures. Annu Rev Biochem 68:255–285.
120. Meijer M, Karimi-Busheri F, Huang TY et al. (2002). Pnk1, a DNA kinase/phosphatase required for normal response to DNA damage by gamma-radiation or camptothecin in Schizosaccharomyces pombe. J Biol Chem 277:4050–4055.
121. Meira LB, Devaraj S, Kisby GE et al. (2001). Heterozygosity for the mouse Apex gene results in phenotypes associated with oxidative stress. Cancer Res 61:5552–5557.
122. Meira LB, Bugni JM, Green SL et al. (2008). DNA damage induced by chronic inflammation contributes to colon carcinogenesis in mice. J Clin Invest 118:2516–2525.
123. Minowa O, Arai T, Hirano M et al. (2000). Mmh/Ogg1 gene inactivation results in accumulation of 8-hydroxyguanine in mice. Proc Natl Acad Sci USA 97: 4156–4161.

124. Mitra S, Kaina B (1993). Regulation of repair of alkylation damage in mammalian genomes. Prog Nucleic Acid Res Mol Biol 44:109–142.
125. Mitra S, Hazra TK, Roy R et al. (1997). Complexities of DNA base excision repair in mammalian cells. Mol Cells 7:305–312.
126. Mitra S, Boldogh I, Izumi T et al. (2001). Complexities of the DNA base excision repair pathway for repair of oxidative DNA damage. Environ Mol Mutagen 38:180–190.
127. Mitra S, Izumi T, Boldogh I et al. (2002). Choreography of oxidative damage repair in mammalian genomes. Free Radic Biol Med 33:15–28.
128. Mitra S (2007a) MGMT: Reversal of fortune – from basic biochemistry to clinical exploitation of the DNA damage reversal protein, O^6-methylguanine-DNA methyltransferase. In: Kaina B, Margison GP (eds) DNA Repair (Amst) 6:1063–1228.
129. Mitra S (2007b). MGMT: a personal perspective. DNA Repair (Amst) 6:1064–1070.
130. Mitra S, Izumi T, Boldogh I et al. (2007). Intracellular trafficking and regulation of mammalian AP-endonuclease 1 (APE1), an essential DNA repair protein. DNA Repair (Amst) 6:461–469.
131. Modrich P, Lahue R (1996). Mismatch repair in replication fidelity, genetic recombination, and cancer biology. Annu Rev Biochem 65:101–133.
132. Mokkapati SK, Wiederhold L, Hazra TK et al. (2004). Stimulation of DNA glycosylase activity of OGG1 by NEIL1: functional collaboration between two human DNA glycosylases. Biochemistry 43:11596–11604.
133. Mol CD, Izumi T, Mitra S et al. (2000). DNA-bound structures and mutants reveal abasic DNA binding by APE1 and DNA repair coordination [corrected]. Nature 403: 451–456.
134. Muftuoglu M, Wong HK, Imam SZ et al. (2006). Telomere repeat binding factor 2 interacts with base excision repair proteins and stimulates DNA synthesis by DNA polymerase beta. Cancer Res 66:113–124.
135. Muller S, Hoege C, Pyrowolakis G et al. (2001). SUMO, ubiquitin's mysterious cousin. Nat Rev Mol Cell Biol 2:202–210.
136. Nagelhus TA, Haug T, Singh KK et al. (1997). A sequence in the N-terminal region of human uracil-DNA glycosylase with homology to XPA interacts with the C-terminal part of the 34-kDa subunit of replication protein A. J Biol Chem 272:6561–6566.
137. Narayan S, Jaiswal AS, Balusu R (2005). Tumor suppressor APC blocks DNA polymerase beta-dependent strand displacement synthesis during long patch but not short patch base excision repair and increases sensitivity to methylmethane sulfonate. J Biol Chem 280: 6942–6949.
138. Nilsen H, Rosewell I, Robins P et al. (2000). Uracil-DNA glycosylase (UNG)-deficient mice reveal a primary role of the enzyme during DNA replication. Mol Cell 5:1059–1065.
139. Nishioka K, Ohtsubo T, Oda H et al. (1999). Expression and differential intracellular localization of two major forms of human 8-oxoguanine DNA glycosylase encoded by alternatively spliced OGG1 mRNAs. Mol Biol Cell 10:1637–1652.
140. Ochs K, Sobol RW, Wilson SH et al. (1999). Cells deficient in DNA polymerase beta are hypersensitive to alkylating agent-induced apoptosis and chromosomal breakage. Cancer Res 59:1544–1551.
141. Oei SL, Ziegler M (2000). ATP for the DNA ligation step in base excision repair is generated from poly(ADP-ribose). J Biol Chem 275:23234–23239.
142. Olsson M, Lindahl T (1980). Repair of alkylated DNA in Escherichia coli. Methyl group transfer from O^6-methylguanine to a protein cysteine residue. J Biol Chem 255: 10569–10571.
143. Osterod M, Hollenbach S, Hengstler JG et al. (2001). Age-related and tissue-specific accumulation of oxidative DNA base damage in 7,8-dihydro-8-oxoguanine-DNA glycosylase (Ogg1) deficient mice. Carcinogenesis 22:1459–1463.
144. Otterlei M, Warbrick E, Nagelhus TA et al. (1999). Post-replicative base excision repair in replication foci. EMBO J 18:3834–3844.

145. Parker A, Gu Y, Mahoney W et al. (2001). Human homolog of the MutY repair protein (hMYH) physically interacts with proteins involved in long patch DNA base excision repair. J Biol Chem 276:5547–5555.
146. Parker JB, Bianchet MA, Krosky DJ et al. (2007). Enzymatic capture of an extrahelical thymine in the search for uracil in DNA. Nature 449:433–437.
147. Parlanti E, Locatelli G, Maga G et al. (2007). Human base excision repair complex is physically associated to DNA replication and cell cycle regulatory proteins. Nucleic Acids Res 35:1569–1577.
148. Petermann E, Ziegler M, Oei SL (2003). ATP-dependent selection between single nucleotide and long patch base excision repair. DNA Repair (Amst) 2:1101–1114.
149. Petermann E, Keil C, Oei SL (2006). Roles of DNA ligase III and XRCC1 in regulating the switch between short patch and long patch BER. DNA Repair (Amst) 5:544–555.
150. Pfanner N (2000). Protein sorting: recognizing mitochondrial presequences. Curr Biol 10:R412–R415.
151. Piersen CE, Prasad R, Wilson SH et al. (1996). Evidence for an imino intermediate in the DNA polymerase beta deoxyribose phosphate excision reaction. J Biol Chem 271: 17811–17815.
152. Pinz KG, Bogenhagen DF (1998). Efficient repair of abasic sites in DNA by mitochondrial enzymes. Mol Cell Biol 18:1257–1265.
153. Popoff SC, Spira AI, Johnson AW et al. (1990). Yeast structural gene (APN1) for the major apurinic endonuclease: homology to Escherichia coli endonuclease IV. Proc Natl Acad Sci USA 87:4193–4197.
154. Porecha RH, Stivers JT (2008). Uracil DNA glycosylase uses DNA hopping and short-range sliding to trap extrahelical uracils. Proc Natl Acad Sci USA 105:10791–10796.
155. Posnick LM, Samson LD (1999). Imbalanced base excision repair increases spontaneous mutation and alkylation sensitivity in Escherichia coli. J Bacteriol 181:6763–6771.
156. Prasad R, Beard WA, Chyan JY et al. (1998). Functional analysis of the amino-terminal 8-kDa domain of DNA polymerase beta as revealed by site-directed mutagenesis. DNA binding and 5′-deoxyribose phosphate lyase activities. J Biol Chem 273:11121–11126.
157. Prasad R, Dianov GL, Bohr VA et al. (2000). FEN1 stimulation of DNA polymerase beta mediates an excision step in mammalian long patch base excision repair. J Biol Chem 275:4460–4466.
158. Radicella JP, Dherin C, Desmaze C et al. (1997). Cloning and characterization of hOGG1, a human homolog of the OGG1 gene of Saccharomyces cerevisiae. Proc Natl Acad Sci USA 94:8010–8015.
159. Ramana CV, Boldogh I, Izumi T et al. (1998). Activation of apurinic/apyrimidinic endonuclease in human cells by reactive oxygen species and its correlation with their adaptive response to genotoxicity of free radicals. Proc Natl Acad Sci USA 95:5061–5066.
160. Ribar B, Izumi T, Mitra S (2004). The major role of human AP-endonuclease homolog Apn2 in repair of abasic sites in Schizosaccharomyces pombe. Nucleic Acids Res 32: 115–126.
161. Rinne M, Caldwell D, Kelley MR (2004). Transient adenoviral N-methylpurine DNA glycosylase overexpression imparts chemotherapeutic sensitivity to human breast cancer cells. Mol Cancer Ther 3:955–967.
162. Rinne ML, He Y, Pachkowski BF et al. (2005). N-methylpurine DNA glycosylase overexpression increases alkylation sensitivity by rapidly removing non-toxic 7-methylguanine adducts. Nucleic Acids Res 33:2859–2867.
163. Roos WP, Kaina B (2006). DNA damage-induced cell death by apoptosis. Trends Mol Med 12:440–450.
164. Roth RB, Samson LD (2002). 3-Methyladenine DNA glycosylase-deficient Aag null mice display unexpected bone marrow alkylation resistance. Cancer Res 62:656–660.
165. Roy R, Kennel SJ, Mitra S (1996). Distinct substrate preference of human and mouse N-methylpurine-DNA glycosylases. Carcinogenesis 17:2177–2182.

166. Roy R, Biswas T, Hazra TK et al. (1998). Specific interaction of wild-type and truncated mouse N-methylpurine-DNA glycosylase with ethenoadenine-containing DNA. Biochemistry 37:580–589.
167. Samson L, Cairns J (1977). A new pathway for DNA repair in Escherichia coli. Nature 267:281–283.
168. Sattler U, Frit P, Salles B et al. (2003). Long-patch DNA repair synthesis during base excision repair in mammalian cells. EMBO Rep 4:363–367.
169. Sedgwick B, Bates PA, Paik J et al. (2007). Repair of alkylated DNA: recent advances. DNA Repair (Amst) 6:429–442.
170. Seet BT, Dikic I, Zhou MM et al. (2006). Reading protein modifications with interaction domains. Nat Rev Mol Cell Biol 7:473–483.
171. Shinmura K, Tao H, Goto M et al. (2004). Inactivating mutations of the human base excision repair gene NEIL1 in gastric cancer. Carcinogenesis 25:2311–2317.
172. Singh P, Zheng L, Chavez V et al. (2007). Concerted action of exonuclease and Gap-dependent endonuclease activities of FEN-1 contributes to the resolution of triplet repeat sequences (CTG)n- and (GAA)n-derived secondary structures formed during maturation of Okazaki fragments. J Biol Chem 282:3465–3477.
173. Sobol RW, Horton JK, Kuhn R et al. (1996). Requirement of mammalian DNA polymerase-beta in base-excision repair. Nature 379:183–186.
174. Sobol RW, Kartalou M, Almeida KH et al. (2003). Base excision repair intermediates induce p53-independent cytotoxic and genotoxic responses. J Biol Chem 278:39951–39959.
175. Sokhansanj BA, Rodrigue GR, Fitch JP et al. (2002). A quantitative model of human DNA base excision repair. I. Mechanistic insights Nucleic Acids Res 30:1817–1825.
176. Speina E, Zielinska M, Barbin A et al. (2003). Decreased repair activities of 1,N^6-ethenoadenine and 3,N(4)-ethenocytosine in lung adenocarcinoma patients. Cancer Res 63:4351–4357.
177. Srivastava DK, Berg BJ, Prasad R et al. (1998). Mammalian abasic site base excision repair. Identification of the reaction sequence and rate-determining steps. J Biol Chem 273: 21203–21209.
178. Stivers JT (2008). Extrahelical damaged base recognition by DNA glycosylase enzymes. Chemistry 14:786–793.
179. Stuart JA, Hashiguchi K, Wilson DM III, et al. (2004). DNA base excision repair activities and pathway function in mitochondrial and cellular lysates from cells lacking mitochondrial DNA. Nucleic Acids Res 32:2181–2192.
180. Sugo N, Aratani Y, Nagashima Y et al. (2000). Neonatal lethality with abnormal neurogenesis in mice deficient in DNA polymerase beta EMBO J 19:1397–1404.
181. Sung JS, Mosbaugh DW (2003). Escherichia coli uracil- and ethenocytosine-initiated base excision DNA repair: rate-limiting step and patch size distribution Biochemistry 42: 4613–4625.
182. Szczesny B, Tann AW, Longley MJ et al. (2008). Long patch base excision repair in mammalian mitochondrial genomes. J Biol Chem 283:26349–26356.
183. Takao M, Kanno S, Shiromoto T et al. (2002). Novel nuclear and mitochondrial glycosylases revealed by disruption of the mouse Nth1 gene encoding an endonuclease III homolog for repair of thymine glycols. EMBO J 21:3486–3493.
184. Tebbs RS, Flannery ML, Meneses JJ et al. (1999). Requirement for the Xrcc1 DNA base excision repair gene during early mouse development. Dev Biol 208:513–529.
185. Tini M, Benecke A, Um SJ et al. (2002). Association of CBP/p300 acetylase and thymine DNA glycosylase links DNA repair and transcription. Mol Cell 9:265–277.
186. Tomkinson AE, Mackey ZB (1998). Structure and function of mammalian DNA ligases. Mutat Res 407:1–9.
187. Tsuchimoto D, Sakai Y, Sakumi K et al. (2001). Human APE2 protein is mostly localized in the nuclei and to some extent in the mitochondria, while nuclear APE2 is partly associated with proliferating cell nuclear antigen. Nucleic Acids Res 29:2349–2360.

188. Vartanian V, Lowell B, Minko IG et al. (2006). The metabolic syndrome resulting from a knockout of the NEIL1 DNA glycosylase. Proc Natl Acad Sci USA 103:1864–1869.
189. Vidal AE, Hickson ID, Boiteux S, et al. (2001). Mechanism of stimulation of the DNA glycosylase activity of hOGG1 by the major human AP endonuclease: bypass of the AP lyase activity step. Nucleic Acids Res 29:1285–1292.
190. von Hippel PH, Revzin A, Gross CA et al. (1974). Non-specific DNA binding of genome regulating proteins as a biological control mechanism: I. The lac operon: equilibrium aspects. Proc Natl Acad Sci USA 71:4808–4812.
191. Ward JF (1985). Biochemistry of DNA lesions. Radiat Res Suppl 8:S103–S111.
192. Weiss JM, Goode EL, Ladiges WC et al. (2005). Polymorphic variation in hOGG1 and risk of cancer: a review of the functional and epidemiologic literature. Mol Carcinog 42:127–141.
193. Whitehouse CJ, Taylor RM, Thistlethwaite A et al. (2001). XRCC1 stimulates human polynucleotide kinase activity at damaged DNA termini and accelerates DNA single-strand break repair. Cell 104:107–117.
194. Wiederhold L, Leppard JB, Kedar P et al. (2004). AP endonuclease-independent DNA base excision repair in human cells. Mol Cell 15:209–220.
195. Wikman H, Risch A, Klimek F et al. (2000). hOGG1 polymorphism and loss of heterozygosity (LOH): significance for lung cancer susceptibility in a caucasian population. Int J Cancer 88:932–937.
196. Wilson DM, III, Barsky D (2001). The major human abasic endonuclease: formation, consequences and repair of abasic lesions in DNA. Mutat Res 485:283–307.
197. Wilson SH, Kunkel TA (2000). Passing the baton in base excision repair. Nat Struct Biol 7:176–178.
198. Xanthoudakis S, Smeyne RJ, Wallace JD et al. (1996). The redox/DNA repair protein, Ref-1, is essential for early embryonic development in mice. Proc Natl Acad Sci USA 93: 8919–8923.
199. Xie Y, Yang H, Cunanan C et al. (2004). Deficiencies in mouse Myh and Ogg1 result in tumor predisposition and G to T mutations in codon 12 of the K-ras oncogene in lung tumors. Cancer Res 64:3096–3102.
200. Zharkov DO, Shoham G, Grollman AP (2003). Structural characterization of the Fpg family of DNA glycosylases. DNA Repair (Amst) 2:839–862.

Chapter 10
DNA Damage Tolerance and Translesion Synthesis

Alan R. Lehmann

Abstract When the replication machinery encounters a DNA lesion, it is able to continue to replicate the DNA either by damage avoidance processes involving switching templates, or by translesion synthesis past the lesion using specialized DNA polymerases. Most of these polymerases are in the Y-family and have open structures that enable them to accommodate particular damaged bases in their active sites. Translesion synthesis can be error-free or error-prone and defective DNA polymerase η results in the variant form of the highly skin-cancer prone disorder xeroderma pigmentosum. Single-stranded regions of DNA exposed at sites of stalled replication forks trigger the ubiquitination of the sliding clamp protein proliferating cell nuclear antigen (PCNA). This increases the affinity of the Y-family polymerases for PCNA, as they all contain ubiquitin-binding domains, and provides a mechanism for the recruitment of these enzymes to stalled replication forks.

Keywords DNA polymerase · Y-family · Ultraviolet light · Xeroderma pigmentosum

Abbreviations

BP	benzo[a]pyrene
CPD	cyclobutane pyrimidine dimers
MMS	methyl methanesulfonate
SHM	somatic hypermutation
NER	nucleotide excision repair
TLS	Translesion synthesis
UV	ultraviolet
XP	xeroderma pigmentosum
XP-V	xeroderma pigmentosum variant

A.R. Lehmann (✉)
Genome Damage and Stability Centre, University of Sussex, Falmer, Brighton BN1 9RQ, UK
e-mail: a.r.lehmann@sussex.ac.uk

K.K. Khanna, Y. Shiloh (eds.), *The DNA Damage Response: Implications on Cancer Formation and Treatment*, DOI 10.1007/978-90-481-2561-6_10,

10.1 Introduction

All cells are able to remove a plethora of different types of damage from cellular DNA using the variety of repair processes that have evolved to deal with damage from many different sources [30]. In proliferating cells, however, not all damage is removed before the DNA is replicated. Most types of DNA damage block the progress of the replication fork because the active sites of the replicative polymerases are highly stringent and cannot accommodate most damaged bases. In order to deal with this situation, cells have evolved various tolerance processes with which they are able to bypass or avoid the damage during or after replication. Together these processes are sometimes termed postreplication repair (PRR). Prior to 1999 our understanding of these tolerance processes was based largely on genetics and cell biology, so we knew a lot about the genes involved, but understood very little of what the gene products actually did. A major advance took place in 1999 with the discovery of the Y-family of DNA polymerases, which are key players in replication past DNA damage. These specialised polymerases are designed to replicate past damaged bases (a process termed translesion synthesis – TLS). They have active sites that are less stringent than those of replicative polymerases, and can accommodate damaged bases at the expense of decreased polymerase efficiency and fidelity [154]. One of these polymerases, DNA polymerase η, is defective in the variant form of the highly skin-cancer prone disorder xeroderma pigmentosum (XP), demonstrating that TLS using this polymerase provides an important protective mechanism against ultraviolet (UV)-induced carcinogenesis. In this chapter I will review our current understanding of DNA damage tolerance mechanisms in eukaryotes, drawing on models from *Escherichia coli* where appropriate. Other recent reviews include [2, 28, 50, 65, 112, 154].

10.2 In the Wilderness – Pre 1999

The earliest evidence for DNA damage tolerance was the observation that *uvrA* strains of *E. coli* that were unable to remove UV photoproducts from their DNA, were still able to survive with about 50 photoproducts in their DNA. During replication of these strains after UV-irradiation, the daughter strands were smaller than in unirradiated cells and this small DNA subsequently increased in size. These results were interpreted as evidence for gaps being left opposite the damage, and these gaps were subsequently sealed [117]. This mechanism necessitates restarting replication beyond the site of damage to leave the gap, a topic I shall return to later. In *E. coli*, the major mechanism for repairing these gaps involved exchanges of material between daughter and parental strands [118] in a process that we now describe as damage-avoidance and is expected to be error-free. This process is absolutely dependent on the *recA* gene and indeed a *uvrArecA* double mutant lacking both NER and this daughter-strand gap repair is exquisitely sensitive and is killed by a single UV photoproduct [117].

Despite the expected high fidelity of the damage avoidance process, other genetic experiments showed that UV-induced mutations were generated during replication of damaged DNA. Key genes required for mutations induced by UV and many other mutagens were *umuC* and *D*. However, mutants lacking these genes were only marginally sensitive to killing by UV despite being absolutely required for mutagenesis. This gave rise, in the 1970's, to the widely accepted belief that UmuDC controlled a minor pathway, which was for some reason refractory to the recombination repair mechanism, was error-prone and probably involved some kind of TLS pathway. Despite the demonstration that UmuC and D formed a $UmuD_2C$ heterotrimer and that the first 24 aa of UmuD were cleaved off during the SOS response to DNA damage to activate the trimeric protein [126], it was another 25 years before the true the nature of the UmuDC gene product was revealed as a DNA polymerase [115, 136].

In eukaryotic systems, replication of damage has been extensively studied in mammalian cells and in the yeast *Saccharomyces cerevisiae* and more recently in the chicken DT40 system. In both mammals and yeast, evidence was obtained for daughter strand gap repair but recombinational exchanges did not seem to play a major role [63, 111].

The genetic disorder XP is characterised by abnormal pigmentation including freckling, hypo- and hyperpigmentation and multiple skin cancers on unprotected areas of the skin. In 80% of XP patients, the defect is in the ability of the cells to remove photoproducts from exposed DNA by NER [60]. In 20% of cases, so-called XP variants (XP-V), NER is normal but the cells have problems replicating UV-irradiated DNA. The newly synthesised strands are smaller than in normal cells and it takes much longer for this DNA to attain the size of intact DNA strands [64]. Again however 25 years were to elapse before the gene product defective in XP-V cells was identified [53, 79].

Genetic studies in *S. cerevisiae* identified a series of genes involved in replicating DNA damage – these genes were designated the *RAD6* epistasis group [30]. The genes *RAD6* and *RAD18* were required for all processes and mutants in these genes completely prevented the conversion of small pieces of newly synthesised DNA in UV-irradiated cells into intact molecules [111]. Other genes were assigned to either an error-free (*MMS2, UBC13, RAD5*) or an error-prone (*REV1,3,7*) sub-pathway [151]. Subsequent work showed that the gene products of several of these genes were involved in ubiquitin conjugation pathways [142], whereas Rev3 together with its herodimeric partner Rev7 was found to be a DNA polymerase of the B-family, designated polζ [90].

10.3 1999 – Light at the End of the Tunnel – Y Family Polymerases Discovered

The following major discoveries were made independently in 1999

1. $UmuD_2C$ was shown to be a DNA polymerase, designated polV [115, 136].
2. The *E. coli* DinB gene product, required for certain types of mutagenesis was found to be a related polymerase and designated PolIV [147]

3. The product of the *S. cerevisiae RAD30* gene, known to be related to UmuC, was also found to be a DNA polymerase [54]
4. The gene defective in XP-V cells was finally identified as a human ortholog of *RAD30* and its product was a DNA polymerase, designated polη [53, 79]
5. A paralog of polη was discovered and is designated polι [82]
6. The human ortholog of DinB was identified and designated polκ [33, 96].

These new polymerases are all related to each other but have no sequence similarity to classical DNA polymerases, although subsequent crystal structures have revealed domains common to all polymerases (see below). This polymerase family is designated the Y-family [101]. The Y-family has two members in *E. coli* (polIV and V), two members in *S. cerevisiae* (polη and Rev1) and four members in mammals (polη, ι and κ and Rev1). Many biochemical studies have been carried out on the ability of these polymerases to carry out TLS past substrates containing different lesions and these have been extensively reviewed elsewhere [154, 112]. Crystal structures of several Y-polymerases have been solved. These have been reviewed in detail recently [154] and their major features will be summarised here.

10.4 Structures of Y-Family Polymerases

All family members have a conserved catalytic domain of 250–450 aa, in most cases close to the N-terminus of the protein (Fig. 10.1.). The catalytic domain consists of the characteristic palm, thumb and finger domains found in all DNA polymerases, together with a fourth domain variously described as little finger [72] or PAD [140], a domain not found in other polymerase families. The major difference between the Y and other polymerases is the more open structure of the former, caused by the finger and thumb domains being unusually short [154]. This open structure confers their unique property of being able to accommodate damaged bases in their active sites. It enables the polymerases to replicate past different lesions, but their concomitant lower stringency results in lower fidelity when copying undamaged DNA and a much reduced processivity compared to replicative polymerases. The Y-family polymerase Dpo4 from the archaea *Sulfolobus solfataricus*was the first to be crystallised [72] and can accommodate many different lesions in its active site [71, 153]. The different geometries of the active sites confer different specificities to the polymerases for bypassing various lesions.

10.5 Functions of Polymerases in TLS

10.5.1 Polη

The discovery that polη is the product of the gene defective in XP variants [53, 79] and is able in vitro to bypass the major UV photoproduct, the cyclobutane pyrimidine dimer (CPD) efficiently [78], immediately suggested an important role in

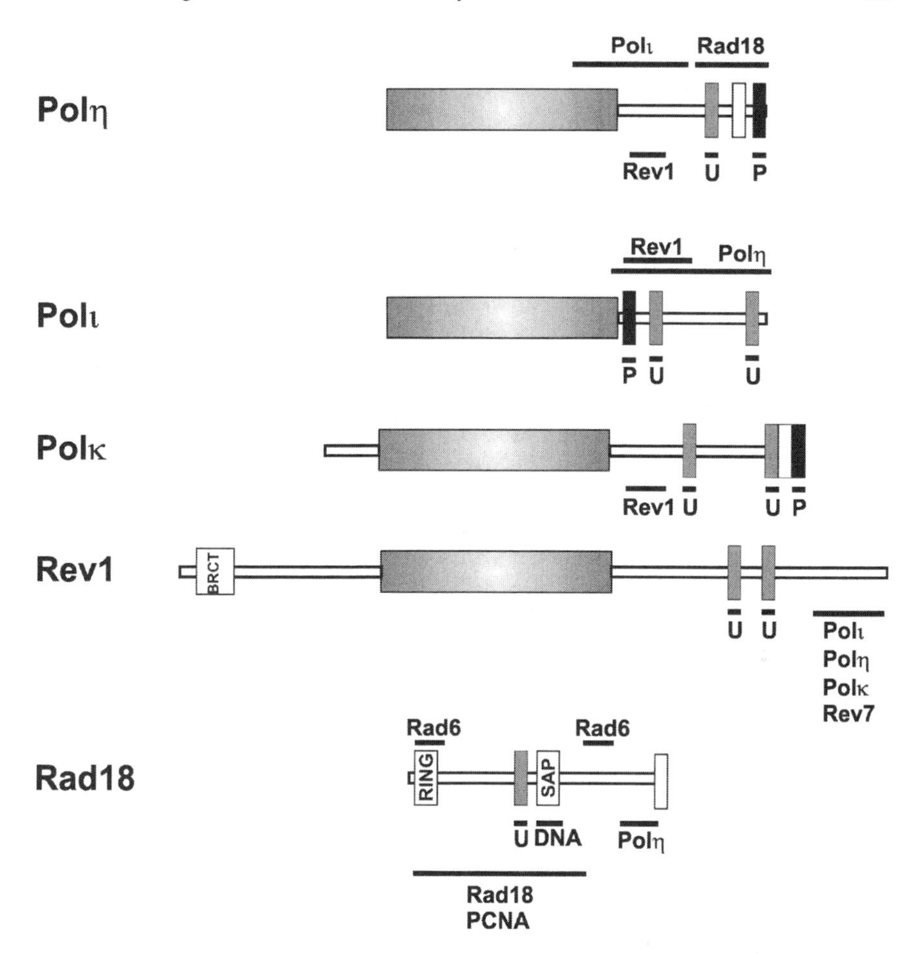

Fig. 10.1 Y-family polymerase and Rad18 domain structure and interactions. All structures represent the human proteins. Large boxes, polymerase domain; small unmarked boxes: white, NLS; grey, ubiquitin-binding domain; black, PIP PCNA binding motif. Bars represent regions of interactions with indicated proteins; U, ubiquitin; P, PCNA

cancer protection. XP-V patients have a high incidence of UV-induced skin cancers, and fibroblasts from these patients, though barely sensitive to killing by UV-irradiation, are extremely sensitive to its mutagenic effects [75, 133]. This gave rise to the idea that polη can replicate past a CPD, in general inserting the correct nucleotides opposite the damaged bases. In its absence in XP-V cells, a polymerase of lower fidelity and lower efficiency can substitute, resulting in elevated UV-induced mutations, which are likely to be responsible for the very high incidence of skin cancers. Experimental data support this model [46, 78, 80]. The 6–4 photoproduct, generated by UV at about 25% the frequency of CPDs, generates a much greater distortion in DNA than the CPD and no polymerase appears able to bypass 6–4 photoproducts on its own. Using a single lesion plasmid replicating in yeast cells, Bresson and Fuchs showed that polη was not involved in error-free TLS past these lesions in *S. cerevisiae*, but was required for error-prone TLS. They suggested, based on the properties of polη in vitro, that polη may insert a GMP residue opposite the 3' thymine of the photoproduct [11]. A similar conclusion was reached by Zhang and Siede using a reversion assay in the yeast genome [159]. Another polymerase, probably polζ (see below) is required to extend from this inserted GMP. In contrast, in mammalian cells a plasmid containing a single 6–4 photoproduct generated similar high frequencies of mutations irrespective of whether the cells were proficient or deficient for polη [46]. Lack of polη also leads to an increased level of UV-induced recombination, manifested as sister chromatid exchanges and accumulation of Mre11 foci in S phase following UV-irradiation [20, 69]. These events are more marked in cells with inactive p53 [20] and it is not clear if they represent attempts to repair secondary damage generated during replication in XP-V cells or whether they are a pathological consequence of such damage.

Though XP-V cells are only marginally sensitive to UV-induced cell killing, this sensitivity can be increased by incubating the cells in caffeine after irradiation [4]. Despite this phenomenon having been first described in 1975, the mechanism of the caffeine sensitization remains something of a mystery. As caffeine is an inhibitor of the checkpoint kinases ATM and ATR, it is often thought that the caffeine sensitivity of XP-V cells is caused by checkpoint inhibition. However this is unlikely to be the case as the dose of caffeine needed to sensitise XP-V cells to UV is much lower than that needed for checkpoint inhibition. It is more likely to be a direct effect of inhibiting a TLS or damage avoidance process. With a post-treatment with caffeine, XP-V cells are also sensitive to treatment with cisplatin, which generates mainly intrastrand crosslinks between two adjacent guanines [1, 16]. Using a reporter plasmid containing a single mitomycin interstrand crosslink, reactivation of the plasmid was reduced more than tenfold in NER-deficient cells and about 4-fold in XP-V cells, consistent with a model in which one arm of the crosslink is unhooked by NER, and subsequently polη, possibly in association with polζ (see below), participates in TLS past the unhooked crosslink to repair the gap opposite the lesion [160].

Despite its low fidelity on undamaged DNA, overexpression of polη in human or yeast cells has only a slight or no effect on the spontaneous mutation frequency suggesting that access by polη to the replication fork during S phase is tightly controlled [38, 59, 108].

Several groups have generated mice in which the polη gene has been disrupted. These mice mimic the features of XP-V patients, in that they have an increased susceptibility to UV-induced tumours [70, 100]. The incidence of UV-induced mutations in these mice is increased as in XP-V cells, but the spontaneous mutant frequency in several organs is similar to that in wild-type mice [13].

Mice deficient in different polymerases have been used to investigate their roles in somatic hypermutation (SHM) during development of the immune response. This is a specialised form of mutation that is generated in only a very small section of the variable regions of immunoglobulin genes. XP-V patients have normal levels of SHM, but the mutation spectrum is different from normal individuals with a decrease in SHM at A:T bases and an increase at G:C sites [158]. Very similar results were obtained with polη deficient mice [23, 77]. This suggests that polη has an important and possibly essential role in the generation of mutations at A:T base-pairs during the SHM process.

10.5.2 Polι

Polι, like polη, is an ortholog of *S. cerevisiae* Rad30, but as yet we understand little of its function. Its most unusual property is its insertion of G in preference to A opposite a template T [139]. It has been suggested from the crystal structure that this is likely to be the result of the structure forcing the template base into the syn conformation, and base selection is then effected using Hoogsteen base-pairing [87]. In vitro, polι can insert nucleotides opposite several different lesions, but is unable to extend further from these inserted lesions. No patient has been identified with defective polι, but the commonly used 129 mouse strain has an amber mutation at aa27, close to the N-terminus of polι, so that this strain of mice is essentially a polι null [81]. It has no manifest phenotype and SHM levels and spectra are normal, even after crossing this mutation into the C57BL/6 background [76]. In contrast to these findings, Weill and coworkers have used the BL2 Burkitt's lymphoma cell line in which somatic hypermutation can be induced by cytokines. They made a polι knock-out derivative of BL2 and found a substantial decrease in SHM compared to the parental line, suggesting, in contrast to the mouse data, that polι is required for SHM. This discrepancy needs to be resolved.

Two important issues that have been addressed are the nature of lesions that polι bypasses in vivo, and whether it can partially substitute for polη in cells in which polη is either absent or depleted. One group used fibroblasts from polι$^{-/-}$mice and found that the cells were slightly UV-sensitive, and that most of the UV-induced mutations at the *Hprt* locus, either in the presence or absence of polη were dependent on polι [25]. In the BL2 cell system, polι-defective cells were not themselves UV-sensitive but conferred extra sensitivity to polη-defective cells. However the lack of polι had no effect on the mutation frequency in the lacZ gene of a UV-irradiated shuttle vector [38]. Finally no effect of depleting cells of polι using siRNA was found on either UV survival or UV-induced mutations in the *supF* gene in a transfected plasmid [18]. At the time of writing therefore there appears to be some evidence that in the absence of polη, polι can participate in TLS past UV lesions thereby generating mutations, but this remains controversial.

10.5.3 Polκ

Polκ differs from the other Y-family pols by having a short N-terminal extension of about 75 aa that precedes the catalytic core and is required for polymerase activity. This extension forms an alpha-helical "clasp" that helps to lock the polymerase around the DNA [73].

Embryonic fibroblasts from polκ deficient mice are sensitive to both killing and mutagenesis by benzo[a]pyrene-diolepoxide [98] and have a threefold reduced ability to bypass benzo[a]pyrene (BP) -G adducts in a gapped shuttle plasmid [5]. The residual bypass activity was more mutagenic than found in polκ $^{+/+}$ cells, (50% as opposed to 29%), suggesting that polκ carries out relatively accurate TLS past BP-G lesions. Evidence has also been provided from both chicken and mouse cells for a role for polκ in bypassing methylation damage [135]. Polκ is highly expressed in the adrenal cortex and it has been proposed that its natural TLS substrate is a site where a steroid molecules is bound to DNA [145].

Overexpression of polκ in mouse or hamster cells resulted in increases in point mutations [96], homologous recombination, loss of heterozygosity, aneuploidy and tumourigenesis [6]. However polκ $^{-/-}$ mice have no clear phenotype, but the spontaneous mutation rate at tandem repeat loci is about twofold elevated [12]. However SHM is unaffected [123, 125]. Apart from its role in TLS, polκ is, surprisingly, also involved in nucleotide excision repair [97].

10.5.4 Rev1 and polζ

The *REV1, 3* and *7* genes of *S. cerevisiae* were identified in a screen for mutants that could not be mutated by UV [66]. Rev3 and 7 are subunits of a heterodimeric DNA polymerase of the B-family designated polζ [89]. Rev1 is a member of the Y-family of polymerases but does not possess polymerase activity. It is a template-dependent dCMP transferase [89], which can add dCMP to the end of primers opposite either G residues or abasic sites. This transferase activity has been attributed to the ability of an arginine residue in the active site to pair specifically with an incoming dCTP [86]. The transferase activity is dispensable for the role of Rev1 in mutagenesis. Rev1 has an N-terminal extension containing a BRCT domain (Fig. 10.1). Mutations in any of the three *REV* genes generally confer very similar properties, implying that the activities of Rev1 and polζ are closely co-ordinated. Using a yeast plasmid system, Yeast Rev1 and Rev3 are required for replication past abasic sites [34, 104] and 6–4 photoproducts [34, 103] and a similar conclusion was reached for human cells [88]. This is partially dependent on both the BRCT domain and the catalytic activity of Rev1 [103]. Chicken DT40 cells in which the Rev1 or Rev3 gene is deleted are sensitive to a wide range of DNA damaging agents [102, 128, 131]. This sensitivity of Rev1-deficient DT40 cells was rescued by human Rev1 and seemed to require neither the BRCT domain nor the catalytic activity [116]. Rev1 was required for non-templated mutations in a process akin to mammalian SHM [128].

As with yeast mutants in these genes, depletion of any of their human orthologs with either antisense or ribozyme technology had little effect on sensitivity to UV, but markedly reduced their mutability by UV [19, 24, 35, 36], BP-G adducts [68] or methyl methanesulfonate (MMS) [110]. In BL2 cells, however, in which exon 5 of the *REV3* gene was deleted, the cells were refractory to UV mutagenesis, as in the other studies, but they were also, as in DT40 cells, highly sensitive to killing by UV [38].

A *Rev1* mouse has been generated in which the BRCT domain was deleted. SHM was normal in these mice. Cells derived from them showed slight sensitivity to UV and 4-nitroquinoline-1-oxide and a substantial reduction in UV mutagenesis [52]. A second mouse was generated in which exon 10 containing the catalytic site was deleted. These mice were small at birth and remained small during their first year of life. The level of SHM was not affected in these mice but there was an absence of C to G transversions. These data suggested that Rev1 was responsible for the insertion of dCMP opposite abasic sites during SHM [51].

Polζ from yeast is a DNA polymerase which in vitro is able to carry out TLS past CPDs [90], though it is likely that in general this is carried out in vivo by polη. Mammalian Rev3 is almost twice the size of its yeast ortholog [35] and has remained refractory to isolation for biochemical analysis, and attempts to express it at detectable levels in any organism have been uniformly unsuccessful. Although *rev3* mutants in yeast are viable, the *rev3-/-* mice are embryonic lethal. However a few groups have managed to establish $rev3^{-/-}$ cell lines. They grow slowly and are sensitive to UV, cisplatin [157] and other DNA damaging agents [38, 149]. Rev1 and Rev3 mutants of both mouse and chicken cells are highly sensitive to cross-linking agents [131] and data from yeast and from these organisms suggest that Rev1 and polζ are recruited to sites of unhooked crosslinks where they participate in TLS past the unhooked crosslink [122, 124]. As mentioned above, polη also appears to be involved in this process in human cells [160], but in yeast this appears not to be the case [122].

10.6 Localisation and Protein-Protein Interactions of TLS Polymerases

The catalytic domains of the polymerases are conserved between the Y-family members and play important roles in substrate selection and in conferring specificity for the types of damaged bases that they are able to bypass. The C-terminal extensions of the polymerases are important for localisation and protein-protein interactions that are vital for their regulation. These extensions are not conserved at all and much of them appears to be unstructured. There are however three interaction motifs that are found in most of the TLS polymerases, namely nuclear localisation signals, PCNA binding motifs (PIP boxes) and ubiquitin-binding motifs (Table 10.1) (Fig. 10.1). These motifs play two important roles, firstly in the localisation of the molecules in the nuclei and, within the nucleus, in replication factories that can be visualised as bright foci that contain many replication proteins; and secondly in the interaction with PCNA at replication forks stalled at sites of DNA damage. PCNA is a sliding clamp accessory protein formed of a homotrimeric ring that tethers polymerases to the DNA and can increase their processivity. This ring is structurally very similar to the homodimeric beta clamp associated with *E. coli* DNA polymerase III [84].

Polη, ι and Rev1 are constitutively localised in replication factories during S phase and, following UV-irradiation, the number of nuclei with foci in which these

Table 10. 1 Motifs found in the C-termini of human TLS polymerases

Polymerase	Size (aa)	NLS	PIP	Ubiquitin binding[b]
Polη	713	Bipartite 682–698	702–709	1 UBZ (C_2H_2) 635–658
Polι	715[a]	Non-classical	420–428	2 UBM 498–522, 679–703
Polκ	880	Bipartite 842–868	871–879	2 UBZ (C_2HC) 624–646, 779–802
Rev1	1251		–	2 UBM 934–958, 1012–1036
Rev3	3130			

[a]Major form. A minor form exists with a 24 aa N-terminal extension; [b]From [9]

polymerases accumulate increases substantially [56, 57, 138]. When DNA damage blocks or slows down the progress of the replication fork, the passage of cells through S phase is prolonged. This will result in an increase in the proportion of S-phase cells in the culture, which may account for much if not all in the increase in the proportion of cells with replication factories containing the polymerases. In contrast, polκ is only localised in foci in a minor proportion of S phase nuclei [95]. The accumulation into foci of polη, ι and κ is dependent on all three conserved motifs, namely the NLS, PIP box and ubiquitin binding motifs [9, 109]. Mutating or deleting any of these elements prevents accumulation in foci.

Recent studies on the dynamics of polη and ι have shown that they are highly mobile within the nucleus. Even when localised in replication foci, they have a residence time of less than one second, the data suggesting that the polymerases enter the foci, probe the chromatin transiently and then exit [119a].

All five DNA polymerases of *E. coli* can interact with the beta clamp, and Fuchs and coworkers have shown that this interaction is essential for TLS by polIV and polV [7]. The PIP boxes in the eukaryotic Y-family polymerases suggest that they physically interact with PCNA and indeed this is the case [40, 43, 44, 45, 146]. The interaction inside cells is weak but is enhanced when PCNA is ubiquitinated following UV-irradiation (see below) [58, 148]. This increased affinity is mediated by specific ubiquitin-binding domains in all the polymerases [9, 41, 42, 109]. The significance of these interactions will be discussed in the following section.

As well as interacting with PCNA, several of the polymerases can interact physically with each other (Fig. 10.1). Rev1 interacts with the other three Y-polymerases and the Rev7 subunit of polζ via its C-terminal 150 aa [39, 99, 138]. This has led to the suggestion that Rev1 can act as a platform on which different polymerases can exchange at the replication fork [29]. Polη and ι can also physically interact [57], although in the cell their dynamics are different, suggesting that any interactions must be transient [119a].

Yuasa and coworkers investigated interactions of polη using immunoprecipitation and mass spectroscopy [156]. Many interacting proteins were identified, among which were Rev1 and Rad18 (see below), the latter confirming an earlier finding in which it was shown that this interaction was essential for the accumulation of polη into replication factories [148].

Stalling of the replication fork triggers the recruitment of polymerases to carry out TLS and the activation of cell cycle checkpoints to arrest cell cycle progression. It is not clear if there is any cross-talk between these two important processes. Using yeast systems, a physical interaction between polζ and the 9-1-1 checkpoint complex was reported in *S. cerevisiae* [120] and Rev1 is phosphorylated by the checkpoint damage sensor Mec1 [119]. In *S. pombe*, following treatment with MMS, the polκ ortholog DinB was found to associate with the 9-1-1 clamp components Hus1 and Rad1, this interaction requiring the Rad17 checkpoint clamp loader to load the 9-1-1 clamp onto the chromatin [55]. In both cases, failure to activate the checkpoint reduced the mutagenic processes mediated by the polymerase under study. This is consistent with an earlier study suggesting that checkpoint mutants were defective in UV-induced mutagenesis, suggesting a possible role in TLS-related processes [107].

10.7 Polymerase Switching

10.7.1 Ubiquitination of PCNA

Genetic studies in *S. cerevisiae* had identified many genes involved in the replication of damaged DNA, these genes comprising the *RAD6* epistasis group (reviewed in [2]). Rad6 itself together with Rad18 regulated the whole process, deletions of these genes resulting in profound UV sensitivity. Two sub-pathways were subsequently delineated into an error-free process requiring Mms2, Ubc13 and Rad5, and an error-prone pathway requiring Rev1, 3 and 7 [151]. Rad6, Rad18, Mms2, Ubc13 and Rad5 were identified as members of the ubiquitin conjugating system that transfers ubiquitin onto target proteins via a three step process of E1-mediated ubiquitin activation, E2 conjugation and E3 ubiquitin ligase. The target of these proteins was identified as PCNA in 2002 [48]. The first ubiquitin is attached to lysine-164 of PCNA via the E2 activity of Rad6 and the E3 activity of Rad18, and this mono-ubiquitination of PCNA is involved in recruiting TLS polymerases [48, 134]. Subsequent ubiquitins are added to form a lysine-63-linked polyubiquitin chain, via the E2 activity of the Ubc13-Mms2 heterodimer and the E3 activity of Rad5. Since the latter group have been assigned to the error-free pathway, polyubiquitination is thought to channel the DNA into an error-free damage avoidance pathway of unknown mechanism. In human cells the same situation pertains, though mono-ubiquitination is the predominant modification [58]. We now know that mono-ubiquitination of PCNA increases its affinity for the TLS polymerases [58, 148], and that this is mediated via their ubiquitin binding domains (see above).

PCNA ubiquitination is triggered by DNA damaging agents or replication inhibitors that slow the passage of the replication fork and result in exposure of single-stranded DNA [8, 16, 58, 91a, 132] (see below). It appears to be independent of checkpoint activation in both budding and fission yeasts, *Xenopus laevis*oocytes and human cells [14, 22, 27, 91a, 155]. Interestingly however PCNA ubiquitination is dependent on the checkpoint kinase Chk1, its cofactor claspin and the claspin-interacting protein timeless [155] as well as PTIP, a six BRCT-domain-containing protein [37]. The role that these proteins play is not yet clear.

The ubiquitin modification can be removed from PCNA by the de-ubiquitinating enzyme USP1, and this plays a further regulatory role in PCNA ubiquitination. USP1 is continuously turned over by an autocleavage mechanism, but following UV-irradiation it disappears from the cell because its transcription is inhibited [21, 49, 94]. This contributes to the increased ubiquitination of PCNA in response to UV and to its persistence for many hours, even after the damage has been removed [91a]. Curiously however, USP1 does not disappear following other DNA damaging treatments, even though PCNA ubiquitination is triggered and persists for a long time [91a].

Polyubiquitination of PCNA is easy to detect in *S. cerevisiae* and *S. pombe* [27, 48] but is much more difficult to detect in mammalian cells, in which it forms a quantitatively much smaller proportion of the modified PCNA population [17]. Nevertheless the genes involved in polyubiquitination in the yeasts have orthologs in human cells and genetic studies suggest that, as in *S. cerevisiae,* they control an error-free pathway. Thus transfecting cells with either an antisense construct to *MMS2* or over expressing ubiquitin mutated at lysine 63 to prevent the formation of lysine-63 linked polyubiquitin chains resulted in an increase in mutations induced by UV or BP-G lesions [17, 62, 67].

PCNA ubiquitination is required for SHM in DT40 cells [3], but in mouse cells, in which lysine 164 of PCNA has been mutated to arginine to prevent ubiquitination, the SHM frequency is unaffected, but the mutation spectrum is altered [61]. The altered pattern is very similar to what is observed in the polη mutant (see above) and in Msh2 mismatch repair-deficient mutants [114], i.e. very few A:T mutations. These observations have given rise to the following model for the generation of A:T mutations during SHM. SHM is triggered by the deamination of cytosine to uracil in immunoglobulin genes by the enzyme AID (Activation-induced cytidine deaminase). The resulting U:G mismatch is recognised by Msh2-6. One of the strands is digested by a nuclease, possibly exoI. The resulting single stranded gap triggers PCNA ubiquitination and recruitment of polη to fill in the gap [121]. This model nicely accounts for the genetic requirements for A:T mutagenesis, but leaves open the question as to why the cell processes the U:G lesion by this pathway.

10.7.2 Rad18 and Rad5

Rad18 plays a central and complex regulatory role at stalled forks. This protein has a RING finger motif close to the N-terminus, UBZ zinc finger motif and SAP domains in the middle. (Fig. 10.1) [93, 141]. The RING finger motif confers E3 ligase activity that is responsible for ubiquitination of PCNA. Rad18 binds to single-stranded DNA and especially to forked DNA structures via its SAP domain [141] and to the single-stranded binding protein RPA [22]. It associates with its cognate E2, Rad6 and orthologs, via domains near the C-terminus and the N-terminus, and it dimerises via the N-terminal half to form a heterotetramer. It also has a binding site for polη close to the C-terminus. The SAP domain is crucial for many of the activities of Rad18 including ubiquitination of PCNA, localisation in replication factories and transporting polη to replication factories [141]. The UBZ domain enables Rad18 to bind to ubiquitinated proteins [93] and, like the Y-polymerases [9], it becomes ubiquitinated itself [83]. Rad18-deficient mice and chicken cells have been generated. These cells are sensitive to UV and other DNA damaging agents [137, 152], and PCNA ubiquitination is abolished in the mouse cells and in human cells in which Rad18 is depleted by siRNA [58]. In the DT40 system however a fraction of the PCNA ubiquitination is independent of Rad18 [127].

Apart from PCNA, in *S. cerevisiae*, Rad6-Rad18 can ubiquitinate the 9-1-1 PCNA-like checkpoint clamp component Rad17 on lysine-197 [31]. This modification is involved in activation of the transcriptional response to DNA damage. However, it remains to be established if this interesting finding can be extrapolated to other organisms, since lysine-197 is not well conserved and the transcriptional response is more marked in *S. cerevisiae* than in other organisms.

Rad5 is a protein that belongs to the Swi2/Snf2 family of DNA-dependent ATPases and genetically it has been assigned to the error-free pathway of damage avoidance [151]. It has a RING finger motif characteristic of E3 ubiquitin ligases and seven helicase domains, although the Swi2/Snf2 family does not have helicase activity. Together with Ubc13-Mms2, it polyubiquitinates PCNA [48]. It appears however to have other roles in the replication of UV-irradiated DNA, as a *RAD5* deletion is more sensitive than a *UBC13* deletion, and mutation of the ATPase motif of Rad5 confers increased sensitivity to a *UBC13* deletion strain [15]. Two human orthologs of Rad5 have been identified, SHPRH and HLTF. SHPRH associates with PCNA, Rad18 and Ubc13 and stimulates PCNA polyubiquitination following MMS treatment. Depletion by siRNA confers sensitivity to MMS [85, 144]. Similarly, depletion of HLTF by siRNA results in UV sensitisation, and overexpression of HLTF in human cells results in increased polyubiquitination of PCNA [143].

The error-free pathway has been proposed to involve a template-switch mechanism, first proposed in 1976 [47]. Several models have been suggested for this template switching, whereby instead of copying the damaged template strand, the nascent daughter strand switches to copying the daughter-strand of the sister duplex. One way that this might happen is by reversal of the growing fork to generate a so-called chicken-foot structure (Fig. 10.2, Mode 5). Rad5 is able to reverse model fork structures making it a suitable candidate for involvement in this mechanism [10]. However a caveat to this work is the suggestion by Lopes et al. that such chicken-foot structures are rarely found in vivo and may be pathological structures rather than repair intermediates [129].

The model that has developed from these studies is that depending on the strand being replicated, polδ (lagging strand) [91] or polε (leading strand) [113] is blocked by damaged DNA. In the case of a lesion on the lagging strand, this should not lead

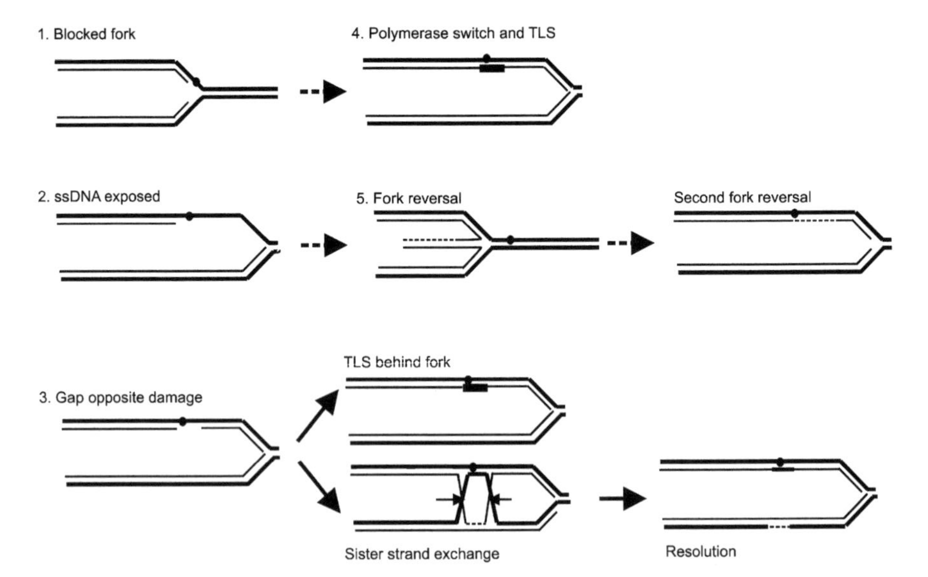

Fig. 10.2 Possible scenarios when a replication fork encounters damage. See text for discussion. Filled circle, damaged site. Thin lines, daughter strands; thicker lines, parental strands; very thick lines, DNA synthesised during TLS; dashed lines DNA synthesised by copy choice using other daughter strand as template

to inhibition of fork progression, but the Okazaki fragment opposite the lesion will not be completed, leaving a gap (Fig. 10.2, mode 3). On the leading strand, single-stranded DNA can be exposed in two ways. If re-initiation occurs beyond the lesion, a gap would again be left. Alternatively progression on the leading strand could be blocked while synthesis on the lagging strand continued, resulting in an exposed single-stranded region on the leading strand (Fig. 10.2, mode 2). Recent data using *S. cerevisiae* suggest that both these situations can pertain [74].

The exposed single-stranded DNA at the site of the blocking lesion will bind RPA and Rad18, which together with Rad6 and an E1 bring about the mono-ubiquitination of PCNA (Fig. 10.3). This in turn will increase the affinity for the Y-family polymerases by virtue of their ubiquitin-binding domains and recruit them to the site of the stalled replication machinery [9, 42, 58, 106, 148]. Recent in vitro data using a reconstituted system are entirely consistent with this model. Using primed M13 DNA, ubiquitination of PCNA had no effect on the processivity of synthesis with polδ, but facilitated bypass of abasic sites by polη and Rev1 [32] For Rev1 this was much reduced by mutations in the UBM motif [150]. When polδ was artificially stalled by only providing two dNTPs, it was displaced by polη. This exchange was largely dependent on PCNA being ubiquitinated and both PIP box and UBZ motifs of polη contributed to the exchange. Intriguingly ubiquitinated PCNA did not permit the reverse exchange of polη to polδ [161]. If this were also true in vivo, since PCNA appears to remain ubiquitinated for a long time, these results are not compatible with TLS followed by reversion to replicative polymerase (Fig. 10.2, mode 4) but rather favour the model in which replication restarts beyond the lesion

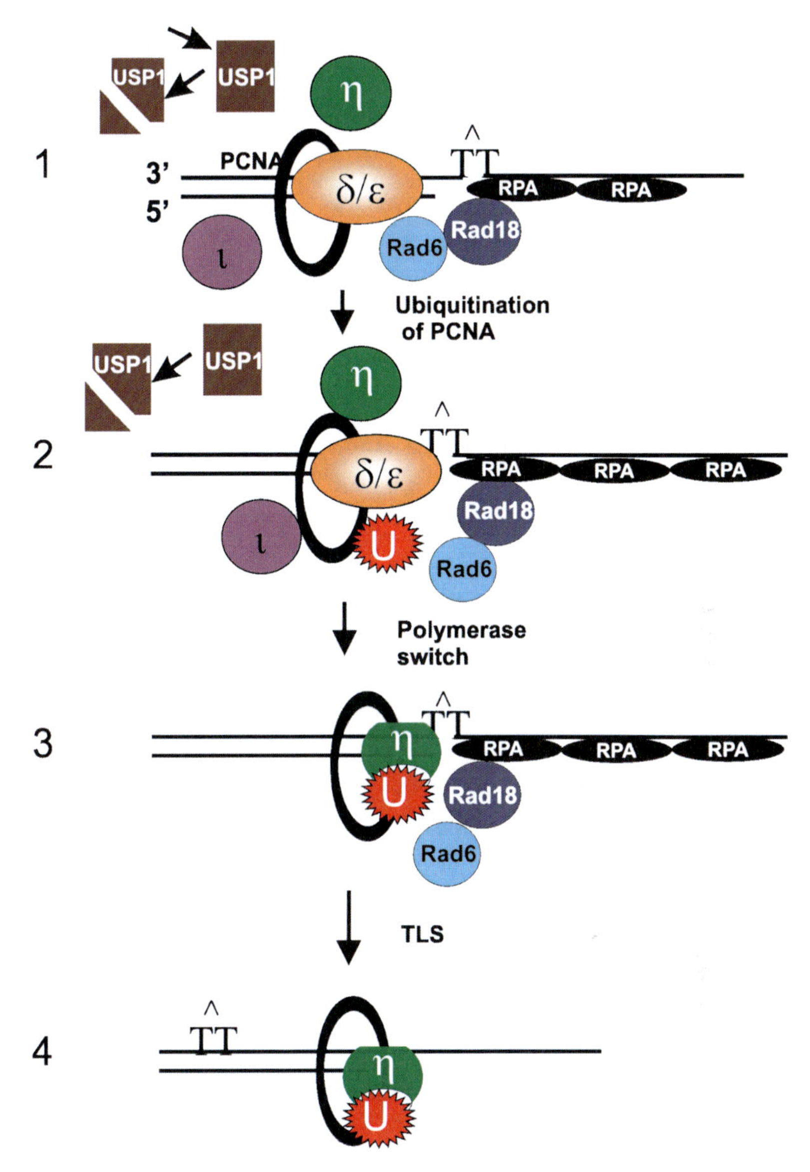

Fig. 10.3 Model for translesion synthesis. (1) The replication machinery including PCNA and polδ or ε is stalled at a CPD. Exposure of single-stranded DNA coated with RPA recruits Rad18-Rad6, USP1 is cleaved and PCNA is mono-ubiquitinated (2). For clarity only one PCNA monomer is shown to be ubiquitinated, although in reality, all three monomers of one trimer are probably ubiquitinated [58]. (3) This increases the affinity for Y-family polymerases, in this case polη, which carries out TLS past the CPD (4)

and the gap is subsequently filled in by the TLS polymerase (mode 3). This latter model does not require a second switching process.

In contrast to these studies, it should be mentioned that a couple of studies do not support the importance of PCNA ubiquitination in TLS. Using an a partially reconstituted in vitro system in which SV40 DNA contained a single CPD, TLS was equally effective with wild-type and K164R mutated PCNA. [92].

10.8 Events at Stalled Forks

Despite the many advances of the last decade, many uncertainties remain about replication of DNA damage. There are many possible events (which are not mutually exclusive) that can happen at the blocked fork (Fig. 10.2):

1. When the fork is blocked by the damage, there is a polymerase switch and TLS is carried out, followed by a second switch to reinstate the replicative polymerase, thereby alleviating the fork block.
2. DNA synthesis may be blocked on the damaged strand, but fork progression and synthesis continues on the undamaged strand, leaving long single-stranded regions on the damaged strand.
3. As 2. but synthesis re-initiates beyond the lesion leaving a short single-strand gap opposite the damaged bases. This gap can subsequently be filled by TLS or sister-strand exchanges (damage avoidance) behind the fork.
4. After scenario 2, fork reversal occurs to generate a chicken foot structure. This allows synthesis to continue on the stalled strand by copying the sister. A second fork reversal re-instates the normal replication mechanism and the damage is avoided.
5. The fork may be blocked on both strands e.g. by an interstrand crosslink or clustered damage.

There is accumulating evidence for scenarios 2 and 3. Pages and Fuchs first showed that introduction of damage in the leading strand led to the uncoupling of synthesis of leading and lagging strands in *E. coli,* with single-stranded regions of up to 3 kb generated on the damaged leading strand while synthesis continued on the undamaged lagging strand [105]. Similarly in *S. cerevisiae*, single stranded regions immediately behind the replication fork were revealed by electron microscopy [74]. This same study also was able to visualise short single-stranded gaps several kilobases behind the replication fork, consistent with early studies that indirectly implicated gaps opposite lesions in *E. coli* [117], *S. cerevisiae* [111] and mammalian cells [63]. A recent study using DT40 cells suggested that gap-filling required ubiquitination of PCNA and polη but was less dependent on Rev1 [26].

Irrespective of whether the damage is bypassed at or behind the fork, an important question concerns the selection of the polymerase to carry out TLS past a blocking lesion. Does each polymerase have a cognate lesion, or do the polymerases compete by mass action and the enzyme best able to do the job wins the competition? Several findings tend to favour the latter hypothesis. (1) There are weak interactions between the different polymerases themselves; (2) there are weak interactions between the polymerases and PCNA, which are increased when PCNA is ubiquitinated; (3) the homotrimeric ring structure of PCNA could allow up to three polymerases to interact with PCNA simultaneously; (4) the residence time of polη and polι in replication foci is very short [119a, 130]. Taken together these results suggest a dynamic scenario with the different polymerases binding briefly to PCNA, somewhat longer if it is ubiquitinated. If PCNA is sited at a blocking lesion, an associated polymerase will attempt to replicate past the lesion, and if the lesion can be accommodated in its active site it will carry out TLS successfully. Alternatively it may be able to insert nucleotides opposite the lesion but not to extend from the lesion, as found for polι with certain lesions in vitro. In this case, a second switch would be necessary to a polymerase that can extend from a nucleotide inserted opposite a lesion, a possible role for polζ. If the polymerase cannot accommodate the lesion in its active site, it will dissociate from the PCNA.

The ability of Rev1 to bind all other TLS polymerases as well as to ubiquitinated PCNA (Fig. 10.1), together with its role in TLS being independent of its catalytic activity has led to the suggestion that it too may play a role in polymerase switching by acting as a platform for different polymerases [29]. Its biological properties suggest that it may have a role in switching events that involve polζ, for example in a two-stage event with one polymerase carrying out the insertion opposite the lesion, and a second switching event facilitating extension from the inserted base by polζ.

10.9 Concluding Remarks

As well as the issues discussed above many further gaps remain in our understanding of DNA damage tolerance. For example: (1) What are the functions of polι, polκ, Rev1 and polζ in vivo? (2) What further factors are involved in the control of PCNA ubiquitination and de-ubiquitination? (3) How is PCNA polyubiquitination induced and/or prevented, and how does it channel lesions into the error-free pathway that it is thought to mediate? (4) What is the nature of this error-free pathway? (5) Does chromatin structure play any part in all these processes?

Acknowledgements I am grateful to Julian Sale for helpful comments on the manuscript. Work in my laboratory is supported by the Medical Research Council, ESF Eurodyna programme and the European Community integrated project on DNA repair.

References

1. Albertella, M. R., Green, C. M., Lehmann, A. R., and O'Connor, M. J. (2005). A role for polymerase eta in the cellular tolerance to cisplatin-induced damage. Cancer Res *65*, 9799–9806.
2. Andersen, P. L., Xu, F., and Xiao, W. (2008). Eukaryotic DNA damage tolerance and translesion synthesis through covalent modifications of PCNA. Cell Res *18*, 162–173.
3. Arakawa, H., Moldovan, G. L., Saribasak, H., Saribasak, N. N., Jentsch, S., and Buerstedde, J. M. (2006). A role for PCNA ubiquitination in immunoglobulin hypermutation. PLoS Biol *4*, e366.

4. Arlett, C. F., Harcourt, S. A., and Broughton, B. C. (1975). The influence of caffeine on cell survival in excision- proficient and excision-deficient xeroderma pigmentosum and normal human cell strains following ultraviolet light irradiation. Mutat. Res. *33*, 341–346.
5. Avkin, S., Goldsmith, M., Velasco-Miguel, S., Geacintov, N., Friedberg, E. C., and Livneh, Z. (2004). Quantitative analysis of translesion DNA synthesis across a benzo[a]pyrene-guanine adduct in mammalian cells: the role of DNA polymeraseκ. J Biol Chem *279*, 53298–58305.
6. Bavoux, C., Leopoldino, A. M., Bergoglio, V., J, O. W., Ogi, T., Bieth, A., Judde, J. G., Pena, S. D., Poupon, M. F., Helleday, T., et al. (2005). Up-regulation of the error-prone DNA polymerase κ promotes pleiotropic genetic alterations and tumorigenesis. Cancer Res *65*, 325–330.
7. Becherel, O. J., Fuchs, R. P. P., and Wagner, J. (2002). Pivotal role of the β-clamp in translesion DNA synthesis and mutagenesis in E. coli cells. DNA Repair *1*, 703–708.
8. Bi, X., Barkley, L. R., Slater, D. M., Tateishi, S., Yamaizumi, M., Ohmori, H., and Vaziri, C. (2006). Rad18 regulates DNA polymerase κ and is required for recovery from S-phase checkpoint-mediated arrest. Mol Cell Biol *26*, 3527–3540.
9. Bienko, M., Green, C. M., Crosetto, N., Rudolf, F., Zapart, G., Coull, B., Kannouche, P., Wider, G., Peter, M., Lehmann, A. R., et al. (2005). Ubiquitin-binding domains in translesion synthesis polymerases. Science *310*, 1821–1824.
10. Blastyak, A., Pinter, L., Unk, I., Prakash, L., Prakash, S., and Haracska, L. (2007). Yeast Rad5 protein required for postreplication repair has a DNA helicase activity specific for replication fork regression. Mol Cell *28*, 167–175.
11. Bresson, A., and Fuchs, R. P. (2002). Lesion bypass in yeast cells: Polη participates in a multi-DNA polymerase process. Embo J *21*, 3881–3887.
12. Burr, K. L., Velasco-Miguel, S., Duvvuri, V. S., McDaniel, L. D., Friedberg, E. C., and Dubrova, Y. E. (2006). Elevated mutation rates in the germline of Polkappa mutant male mice. DNA Repair (Amst) *5*, 860–862.
13. Busuttil, R. A., Lin, Q., Stambrook, P. J., Kucherlapati, R., and Vijg, J. (2008). Mutation frequencies and spectra in DNA polymerase η-deficient mice. Cancer Res *68*, 2081–2084.
14. Chang, D. J., Lupardus, P. J., and Cimprich, K. A. (2006). Monoubiquitination of proliferating cell nuclear antigen induced by stalled replication requires uncoupling of DNA polymerase and mini-chromosome maintenance helicase activities. J Biol Chem *281*, 32081–32088.
15. Chen, S., Davies, A. A., Sagan, D., and Ulrich, H. D. (2005). The RING finger ATPase Rad5p of Saccharomyces cerevisiae contributes to DNA double-strand break repair in a ubiquitin-independent manner. Nucleic Acids Res *33*, 5878–5886.
16. Chen, Y. W., Cleaver, J. E., Hanaoka, F., Chang, C. F., and Chou, K. M. (2006). A novel role of DNA polymerase eta in modulating cellular sensitivity to chemotherapeutic agents. Mol Cancer Res *4*, 257–265.
17. Chiu, R. K., Brun, J., Ramaekers, C., Theys, J., Werng, L., Lambin, P., Gray, D. A., and Wouters, B. G. (2006). Lysine 63-polyubiquitination guards against translesion synthesis-induced mutations. PLOS Genet *2*, e116.
18. Choi, J. H., Besaratinia, A., Lee, D. H., Lee, C. S., and Pfeifer, G. P. (2006). The role of DNA polymerase iota in UV mutational spectra. Mutat Res *599*, 58–65.
19. Clark, D. R., Zacharias, W., Panaitescu, L., and McGregor, W. G. (2003). Ribozyme-mediated REV1 inhibition reduces the frequency of UV-induced mutations in the human HPRT gene. Nucleic Acids Res *31*, 4981–4988.
20. Cleaver, J. E., Afzal, V., Feeney, L., McDowell, M., Sadinski, W., Volpe, J. P., Busch, D. B., Coleman, D. M., Ziffer, D. W., Yu, Y., et al. (1999). Increased ultraviolet sensitivity and chromosomal instability related to p53 function in the xeroderma pigmentosum variant. Cancer Res *59*, 1102–1108.

21. Cohn, M. A., Kowal, P., Yang, K., Haas, W., Huang, T. T., Gygi, S. P., and D'Andrea, A. D. (2007). A UAF1-containing multisubunit protein complex regulates the Fanconi anemia pathway. Mol Cell *28*, 786–797.
22. Davies, A. A., Huttner, D., Daigaku, Y., Chen, S., and Ulrich, H. D. (2008). Activation of ubiquitin-dependent DNA damage bypass is mediated by replication protein A. Molecular Cell *29*, 625–636.
23. Delbos, F., Aoufouchi, S., Faili, A., Weill, J. C., and Reynaud, C. A. (2007). DNA polymerase η is the sole contributor of A/T modifications during immunoglobulin gene hypermutation in the mouse. J Exp Med *204*, 17–23.
24. Diaz, M., Watson, N. B., Turkington, G., Verkoczy, L. K., Klinman, N. R., and McGregor, W. G. (2003). Decreased frequency and highly aberrant spectrum of ultraviolet-induced mutations in the hprt gene of mouse fibroblasts expressing antisense RNA to DNA polymerase ζ. Mol Cancer Res *1*, 836–847.
25. Dumstorf, C., Clark, A. B., Q., L., Kissling, G. E., Yuan, T., Kucherlapati, R., McGregor, W. G., and Kunkel, T. A. (2006). Participation of mouse DNA polymerase ι in strand-biased mutagenic bypass of UV photoproducts and suppression of skin cancer. Proc Natl Acad Sci, USA *103*, 18083–18088.
26. Edmunds, C. E., Simpson, L. J., and Sale, J. E. (2008). PCNA ubiquitination and REV1 define temporally distinct mechanisms for controlling translesion synthesis in the avian cell line DT40. Mol Cell *30*, 519–529.
27. Frampton, J., Irmisch, A., Green, C. M., Neiss, A., Trickey, M., Ulrich, H. D., Furuya, K., Watts, F. Z., Carr, A. M., and Lehmann, A. R. (2006). Postreplication repair and PCNA modification in Schizosaccharomyces pombe. Mol Biol Cell *17*, 2976–2985.
28. Friedberg, E. C. (2005). Suffering in silence: the tolerance of DNA damage. Nat Rev Mol Cell Biol *6*, 943–953.
29. Friedberg, E. C., Lehmann, A. R., and Fuchs, R. P. (2005a). Trading places: how do DNA polymerases switch during translesion DNA synthesis? Mol Cell *18*, 499–505.
30. Friedberg, E. C., Walker, G. C., Siede, W., Wood, R. D., Schultz, R. A., and Ellenberger, T. (2005b). DNA Repair and Mutagenesis, second edition (Washington DC: ASM Press).
31. Fu, Y., Zhu, Y., Zhang, K., Yeung, M., Durocher, D., and Xiao, W. (2008). Rad6-Rad18 mediates a eukaryotic SOS response by ubiquitinating the 9-1-1 checkpoint clamp. Cell *133*, 601–611.
32. Garg, P., and Burgers, P. M. (2005). Ubiquitinated proliferating cell nuclear antigen activates translesion DNA polymerases η and REV1. Proc Natl Acad Sci USA *102*, 18361–18366.
33. Gerlach, V. L., Aravind, L., Gotway, G., Schultz, R. A., Koonin, E. V., and Friedberg, E. C. (1999). Human and mouse homologs of *Escherichia coli* DinB (DNA polymerase IV), members of the UmuC/DinB superfamily. Proc Natl Acad Sci USA *96*, 11922–11927.
34. Gibbs, P. E., McDonald, J., Woodgate, R., and Lawrence, C. W. (2005). The relative roles in vivo of Saccharomyces cerevisiae Pol η, Pol ζ, Rev1 protein and Pol32 in the bypass and mutation induction of an abasic site, T-T (6-4) photoadduct and T-T cis-syn cyclobutane dimer. Genetics *169*, 575–582.
35. Gibbs, P. E. M., McGregor, W. G., Maher, V. M., Nisson, P., and Lawrence, C. W. (1998). A human homolog of the *Saccharomyces cerevisiae REV3* gene, which encodes the catalytic subunit of DNA polymerase ζ. Proc Natl Acad Sci USA *95*, 6876–6880.
36. Gibbs, P. E. M., Wang, X.-D., Li, Z., McManus, T. P., McGregor, G., Lawrence, C. W., and Maher, V. M. (2000). The function of the human homolog of *Saccharomyces cerevisiae REV1* is required for mutagenesis induced by ultraviolet light. Proc Natl Acad Sci USA *97*, 4186–4191.
37. Goehler, T., Munoz, I. M., Rouse, J., and Blow, J. J. (2008). PTIP/Swift is required for efficient PCNA ubiquitination in response to DNA damage. DNA Repair (Amst) *7*, 775–787.
38. Gueranger, Q., Stary, A., Aoufouchi, S., Faili, A., Sarasin, A., Reynaud, C. A., and Weill, J. C. (2008). Role of DNA polymerases η, ι and ζ in UV resistance and UV-induced mutagenesis in a human cell line. DNA Repair (Amst) *7*, 1551–1562.

39. Guo, C., Fischhaber, P. L., Luk-Paszyc, M. J., Masuda, Y., Zhou, J., Kamiya, K., Kisker, C., and Friedberg, E. C. (2003). Mouse Rev1 protein interacts with multiple DNA polymerases involved in translesion DNA synthesis. Embo J *22*, 6621–6630.
40. Guo, C., Sonoda, E., Tang, T. S., Parker, J. L., Bielen, A. B., Takeda, S., Ulrich, H. D., and Friedberg, E. C. (2006a). REV1 protein interacts with PCNA: significance of the REV1 BRCT domain in vitro and in vivo. Mol Cell *23*, 265–271.
41. Guo, C., Tang, T. S., Bienko, M., Dikic, I., and Friedberg, E. C. (2008). Requirements for the interaction of mouse Polκ with ubiquitin, and its biological significance. J Biol Chem *283*, 4658–4664.
42. Guo, C., Tang, T. S., Bienko, M., Parker, J. L., Bielen, A. B., Sonoda, E., Takeda, S., Ulrich, H. D., Dikic, I., and Friedberg, E. C. (2006b). Ubiquitin-Binding Motifs in REV1 protein are required for its role in the tolerance of DNA damage. Mol Cell Biol *26*, 8892–8900.
43. Haracska, L., Johnson, R. E., Unk, I., Phillips, B., Hurwitz, J., Prakash, L., and Prakash, S. (2001a). Physical and functional interactions of human DNA polymerase η with PCNA. Mol Cell Biol *21*, 7199–7206.
44. Haracska, L., Johnson, R. E., Unk, I., Phillips, B. B., Hurwitz, J., Prakash, L., and Prakash, S. (2001b). Targeting of human DNA polymerase ι to the replication machinery via interaction with PCNA. Proc Natl Acad Sci USA *98*, 14256–14261.
45. Haracska, L., Unk, I., Johnson, R. E., Phillips, B. B., Hurwitz, J., Prakash, L., and Prakash, S. (2002). Stimulation of DNA synthesis activity of human DNA polymerase κ by PCNA. Mol Cell Biol *22*, 784–791.
46. Hendel, A., Ziv, O., Gueranger, Q., Geacintov, N., and Livneh, Z. (2008). Reduced efficiency and increased mutagenicity of translesion DNA synthesis across a TT cyclobutane pyrimidine dimer, but not a TT 6-4 photoproduct, in human cells lacking DNA polymerase η. DNA Repair (Amst) *7*, 1636–1646.
47. Higgins, N. P., Kato, K., and Strauss, B. (1976). A model for replication repair in mammalian cells. J Mol Biol *101*, 417–425.
48. Hoege, C., Pfander, B., Moldovan, G.-L., Pyrolowakis, G., and Jentsch, S. (2002). *RAD6*-dependent DNA repair is linked to modification of PCNA by ubiquitin and SUMO. Nature *419*, 135–141.
49. Huang, T. T., Nijman, S. M., Mirchandani, K. D., Galardy, P. J., Cohn, M. A., Haas, W., Gygi, S. P., Ploegh, H. L., Bernards, R., and D'Andrea, A. D. (2006). Regulation of monoubiquitinated PCNA by DUB autocleavage. Nat Cell Biol *8*, 341–347.
50. Jansen, J. G., Fousteri, M. I., and de Wind, N. (2007). Send in the clamps: control of DNA translesion synthesis in eukaryotes. Mol Cell *28*, 522–529.
51. Jansen, J. G., Langerak, P., Tsaalbi-Shtylik, A., van den Berk, P., Jacobs, H., and de Wind, N. (2006). Strand-biased defect in C/G transversions in hypermutating immunoglobulin genes in Rev1-deficient mice. J Exp Med *203*, 319–323.
52. Jansen, J. G., Tsaalbi-Shtylik, A., Langerak, P., Calleja, F., Meijers, C. M., Jacobs, H., and de Wind, N. (2005). The BRCT domain of mammalian Rev1 is involved in regulating DNA translesion synthesis. Nucleic Acids Res *33*, 356–365.
53. Johnson, R. E., Kondratick, C. M., Prakash, S., and Prakash, L. (1999a). *hRAD30* mutations in the variant form of xeroderma pigmentosum. Science *285*, 263–265.
54. Johnson, R. E., Prakash, S., and Prakash, L. (1999b). Efficient bypass of a thymine-thymine dimer by yeast DNA polymerase, Polη. Science *283*, 1001–1004.
55. Kai, M., and Wang, T. (2003). Checkpoint activation regulates mutagenic translesion synthesis. Genes Dev *17*, 64–76.
56. Kannouche, P., Broughton, B. C., Volker, M., Hanaoka, F., Mullenders, L. H. F., and Lehmann, A. R. (2001). Domain structure, localization and function of DNA polymerase η, defective in xeroderma pigmentosum variant cells. Genes Dev *15*, 158–172.
57. Kannouche, P., Fernandez de Henestrosa, A. R., Coull, B., Vidal, A. E., Gray, C., Zicha, D., Woodgate, R., and Lehmann, A. R. (2003). Localization of DNA polymerases η and ι to the replication machinery is tightly co-ordinated in human cells. Embo J *22*, 1223–1233.

58. Kannouche, P. L., Wing, J., and Lehmann, A. R. (2004). Interaction of Human DNA Polymerase η with Monoubiquitinated PCNA; A Possible Mechanism for the Polymerase Switch in Response to DNA Damage. Mol Cell *14*, 491–500.
59. King, N. M., Nikolaishvili-Feinberg, N., Bryant, M. F., Luche, D. D., Heffernan, T. P., Simpson, D. A., Hanaoka, F., Kaufmann, W. K., and Cordeiro-Stone, M. (2005). Overproduction of DNA polymerase η does not raise the spontaneous mutation rate in diploid human fibroblasts. DNA Repair (Amst) *4*, 714–724.
60. Kraemer, K. H., Lee, M. M., and Scotto, J. (1987). Xeroderma Pigmentosum. Cutaneous, ocular and neurologic abnormalities in 830 published cases. Arch Dermatol *123*, 241–250.
61. Langerak, P., Nygren, A. O., Krijger, P. H., van den Berk, P. C., and Jacobs, H. (2007). A/T mutagenesis in hypermutated immunoglobulin genes strongly depends on PCNAK164 modification. J Exp Med *204*, 1989–1998.
62. Langie, S. A., Knaapen, A. M., Ramaekers, C. H., Theys, J., Brun, J., Godschalk, R. W., van Schooten, F. J., Lambin, P., Gray, D. A., Wouters, B. G., and Chiu, R. K. (2007). Formation of lysine 63-linked poly-ubiquitin chains protects human lung cells against benzo[a]pyrene-diol-epoxide-induced mutagenicity. DNA Repair (Amst) *6*, 852–862.
63. Lehmann, A. R. (1972). Postreplication repair of DNA in ultraviolet-irradiated mammalian cells. J Mol Biol *66*, 319–337.
64. Lehmann, A. R., Kirk-Bell, S., Arlett, C. F., Paterson, M. C., Lohman, P. H. M., de Weerd-Kastelein, E. A., and Bootsma, D. (1975). Xeroderma pigmentosum cells with normal levels of excision repair have a defect in DNA synthesis after UV-irradiation. Proc Natl Acad Sci USA *72*, 219–223.
65. Lehmann, A. R., Niimi, A., Ogi, T., Brown, S., Sabbioneda, S., Wing, J. F., Kannouche, P. L., and Green, C. M. (2007). Translesion synthesis: Y-family polymerases and the polymerase switch. DNA Repair (Amst) *6*, 891–899.
66. Lemontt, J. F. (1971). Mutants of yeast defective in mutation induced by ultraviolet light. Adv Genet *68*, 21–33.
67. Li, Z., Xiao, W., McCormick, J. J., and Maher, V. M. (2002a). Identification of a protein essential for a major pathway used by human cells to avoid UV- induced DNA damage. Proc Natl Acad Sci USA *99*, 4459–4464.
68. Li, Z., Zhang, H., McManus, T. P., McCormick, J. J., Lawrence, C. W., and Maher, V. M. (2002b). hREV3 is essential for error-prone translesion synthesis past UV or benzo[a]pyrene diol epoxide-induced DNA lesions in human fibroblasts. Mutat Res *510*, 71–80.
69. Limoli, C. L., Giedzinski, E., Bonner, W. M., and Cleaver, J. E. (2002). UV-induced replication arrest in the xeroderma pigmentosum variant leads to DNA double-strand breaks, γ -H2AX formation, and Mre11 relocalization. Proc Natl Acad Sci USA *99*, 233–238.
70. Lin, Q., Clark, A. B., McCulloch, S. D., Yuan, T., Bronson, R. T., Kunkel, T. A., and Kucherlapati, R. (2006). Increased susceptibility to UV-induced skin carcinogenesis in polymerase eta-deficient mice. Cancer Res *66*, 87–94.
71. Ling, H., Boudsocq, F., Plosky, B. S., Woodgate, R., and Yang, W. (2003). Replication of a cis-syn thymine dimer at atomic resolution. Nature *424*, 1083–1087.
72. Ling, H., Boudsocq, F., Woodgate, R., and Yang, W. (2001). Crystal structure of a Y-family DNA polymerase in action: a mechanism for error-prone and lesion-bypass replication. Cell *107*, 91–102.
73. Lone, S., Townson, S. A., Uljon, S. N., Johnson, R. E., Brahma, A., Nair, D. T., Prakash, S., Prakash, L., and Aggarwal, A. K. (2007). Human DNA polymerase κ encircles DNA: implications for mismatch extension and lesion bypass. Mol Cell *25*, 601–614.
74. Lopes, M., Foiani, M., and Sogo, J. M. (2006). Multiple mechanisms control chromosome integrity after replication fork uncoupling and restart at irreparable UV lesions. Mol Cell *21*, 15–27.
75. Maher, V. M., Ouellette, L. M., Curren, R. D., and McCormick, J. J. (1976). Frequency of ultraviolet light-induced mutations is higher in xeroderma pigmentosum variant cells than in normal human cells. Nature *261*, 593–595.

76. Martomo, S. A., Yang, W. W., Vaisman, A., Maas, A., Yokoi, M., Hoeijmakers, J. H., Hanaoka, F., Woodgate, R., and Gearhart, P. J. (2006). Normal hypermutation in antibody genes from congenic mice defective for DNA polymerase ι. DNA Repair (Amst) *5*, 392–398.
77. Martomo, S. A., Yang, W. W., Wersto, R. P., Ohkumo, T., Kondo, Y., Yokoi, M., Masutani, C., Hanaoka, F., and Gearhart, P. J. (2005). Different mutation signatures in DNA polymerase η- and MSH6-deficient mice suggest separate roles in antibody diversification. Proc Natl Acad Sci USA *102*, 8656–8661.
78. Masutani, C., Kusumoto, R., Iwai, S., and Hanaoka, F. (2000). Accurate translesion synthesis by human DNA polymerase η. EMBO J *19*, 3100–3109.
79. Masutani, C., Kusumoto, R., Yamada, A., Dohmae, N., Yokoi, M., Yuasa, M., Araki, M., Iwai, S., Takio, K., and Hanaoka, F. (1999). The XPV (xeroderma pigmentosum variant) gene encodes human DNA polymerase η. Nature *399*, 700–704.
80. McCulloch, S. D., Kokoska, R. J., Masutani, C., Iwai, S., Hanaoka, F., and Kunkel, T. A. (2004). Preferential cis-syn thymine dimer bypass by DNA polymerase η occurs with biased fidelity. Nature *428*, 97–100.
81. McDonald, J. P., Frank, E. G., Plosky, B. S., Rogozin, I. B., Masutani, C., Hanaoka, F., Woodgate, R., and Gearhart, P. J. (2003). 129-derived strains of mice are deficient in DNA polymerase ι and have normal immunoglobulin hypermutation. J Exp Med *198*, 635–643.
82. McDonald, J. P., Rapic-Otrin, V., Epstein, J. A., Broughton, B. C., Wang, X., Lehmann, A. R., Wolgemuth, D. J., and Woodgate, R. (1999). Novel human and mouse homologs of *Saccharomyces cerevisiae* DNA polymerase η. Genomics *60*, 20–30.
83. Miyase, S., Tateishi, S., Watanabe, K., Tomita, K., Suzuki, K., Inoue, H., and Yamaizumi, M. (2005). Differential regulation of Rad18 through Rad6-dependent mono- and polyubiquitination. J Biol Chem *280*, 515–524.
84. Moldovan, G. L., Pfander, B., and Jentsch, S. (2007). PCNA, the maestro of the replication fork. Cell *129*, 665–679.
85. Motegi, A., Sood, R., Moinova, H., Markowitz, S. D., Liu, P. P., and Myung, K. (2006). Human SHPRH suppresses genomic instability through PCNA polyubiquitination. J Cell Biol *175*, 703–708.
86. Nair, D. T., Johnson, R. E., Prakash, L., Prakash, S., and Aggarwal, A. K. (2005). Rev1 employs a novel mechanism of DNA synthesis using a protein template. Science *309*, 2219–2222.
87. Nair, D. T., Johnson, R. E., Prakash, S., Prakash, L., and Aggarwal, A. K. (2004). Replication by human DNA polymerase-ι occurs by Hoogsteen base-pairing. Nature *430*, 377–380.
88. Nakajima, S., Lan, L., Kanno, S., Takao, M., Yamamoto, K., Eker, A. P., and Yasui, A. (2004). UV light-induced DNA damage and tolerance for the survival of nucleotide excision repair-deficient human cells. J Biol Chem *279*, 46674–46677.
89. Nelson, J. R., Lawrence, C. W., and Hinkle, D. C. (1996a). Deoxycytidyl transferase activity of yeast *REV1* protein. Nature *382*, 729–731.
90. Nelson, J. R., Lawrence, C. W., and Hinkle, D. C. (1996b). Thymine-thymine dimer bypass by yeast DNA polymerase ζ. Science *272*, 1646–1649.
91. Nick McElhinny, S. A., Gordenin, D. A., Stith, C. M., Burgers, P. M., and Kunkel, T. A. (2008). Division of labor at the eukaryotic replication fork. Mol Cell *30*, 137–144.
91a. Niimi, A., Brown, S., Sabbioneda, S., Kannouche, P. L., Scott, A., Yasui, A., Green, C. M., and Lehmann, A. R. (2008). Regulation of proliferating cell nuclear antigen ubiquitination in mammalian cells. Proc Natl Acad Sci USA *105*, 16125–16130.
92. Nikolaishvili-Feinberg, N., Jenkins, G. S., Nevis, K. R., Staus, D. P., Scarlett, C. O., Unsal-Kacmaz, K., Kaufmann, W. K., and Cordeiro-Stone, M. (2008). Ubiquitylation of proliferating cell nuclear antigen and recruitment of human DNA polymerase eta. Biochemistry *47*, 4141–4150.
93. Notenboom, V., Hibbert, R. G., van Rossum-Fikkert, S. E., Olsen, J. V., Mann, M., and Sixma, T. K. (2007). Functional characterization of Rad18 domains for Rad6, ubiquitin, DNA binding and PCNA modification. Nucleic Acids Res *35*, 5819–5830.

94. Oestergaard, V. H., Langevin, F., Kuiken, H. J., Pace, P., Niedzwiedz, W., Simpson, L. J., Ohzeki, M., Takata, M., Sale, J. E., and Patel, K. J. (2007). Deubiquitination of FANCD2 is required for DNA crosslink repair. Mol Cell *28*, 798–809.
95. Ogi, T., Kannouche, P., and Lehmann, A. R. (2005). Localization of human DNA polymerase κ (polκ), a Y-family DNA polymerase: relationship to PCNA foci. J Cell Sci *118*, 129–136.
96. Ogi, T., Kato, T., Kato, T., and Ohmori, H. (1999). Mutation enhancement by DINB1, a mammalian homologue of the *Escherichia coli* mutagenesis protein dinB. Genes Cells *4*, 607–618.
97. Ogi, T., and Lehmann, A. R. (2006). The Y-family DNA polymerase κ (pol κ) functions in mammalian nucleotide-excision repair. Nat Cell Biol *8*, 640–642.
98. Ogi, T., Shinkai, Y., Tanaka, K., and Ohmori, H. (2002). Pol κ protects mammalian cells against the lethal and mutagenic effects of benzo[a]pyrene. Proc Natl Acad Sci USA *99*, 15548–15553.
99. Ohashi, E., Murakumo, Y., Kanjo, N., Akagi, J., Masutani, C., Hanaoka, F., and Ohmori, H. (2004). Interaction of hREV1 with three human Y-family DNA polymerases. Genes Cells *9*, 523–531.
100. Ohkumo, T., Kondo, Y., Yokoi, M., Tsukamoto, T., Yamada, A., Sugimoto, T., Kanao, R., Higashi, Y., Kondoh, H., Tatematsu, M., et al. (2006). UV-B radiation induces epithelial tumors in mice lacking DNA polymerase η and mesenchymal tumors in mice deficient for DNA polymerase iota. Mol Cell Biol *26*, 7696–7706.
101. Ohmori, H., Friedberg, E. C., Fuchs, R. P. P., Goodman, M. F., Hanaoka, F., Hinkle, D., Kunkel, T. A., Lawrence, C. W., Livneh, Z., Nohmi, T., et al. (2001). The Y-family of DNA polymerases. Molecular Cell *8*, 7–8.
102. Okada, T., Sonoda, E., Yoshimura, M., Kawano, Y., Saya, H., Kohzaki, M., and Takeda, S. (2005). Multiple roles of vertebrate REV genes in DNA repair and recombination. Mol Cell Biol *25*, 6103–6111.
103. Otsuka, C., Kunitomi, N., Iwai, S., Loakes, D., and Negishi, K. (2005). Roles of the polymerase and BRCT domains of Rev1 protein in translesion DNA synthesis in yeast in vivo. Mutat Res *578*, 79–87.
104. Otsuka, C., Sanadai, S., Hata, Y., Okuto, H., Noskov, V. N., Loakes, D., and Negishi, K. (2002). Difference between deoxyribose- and tetrahydrofuran-type abasic sites in the in vivo mutagenic responses in yeast. Nucleic Acids Res *30*, 5129–5135.
105. Pages, V., and Fuchs, R. P. (2003). Uncoupling of leading- and lagging-strand DNA replication during lesion bypass in vivo. Science *300*, 1300–1303.
106. Parker, J. L., Bielen, A. B., Dikic, I., and Ulrich, H. D. (2007). Contributions of ubiquitin- and PCNA-binding domains to the activity of Polymerase eta in Saccharomyces cerevisiae. Nucleic Acids Res *35*, 881–889.
107. Paulovich, A. G., Armour, C. D., and Hartwell, L. H. (1998). The *Saccharomyces cerevisiae* RAD9, RAD17, RAD24 and MEC3 genes are required for tolerating irreparable, ultraviolet-induced DNA damage. Genetics *150*, 75–93.
108. Pavlov, Y. I., Nguyen, D., and Kunkel, T. A. (2001). Mutator effects of overproducing DNA polymerase η (Rad30) and its catalytically inactive variant in yeast. Mutat Res *478*, 129–139.
109. Plosky, B. S., Vidal, A. E., Fernandez de Henestrosa, A. R., McLenigan, M. P., McDonald, J. P., Mead, S., and Woodgate, R. (2006). Controlling the subcellular localization of DNA polymerases ι and η via interactions with ubiquitin. Embo J *25*, 2847–2855.
110. Poltoratsky, V., Horton, J. K., Prasad, R., and Wilson, S. H. (2005). REV1 mediated mutagenesis in base excision repair deficient mouse fibroblast. DNA Repair (Amst) *4*, 1182–1188.
111. Prakash, L. (1981). Characterization of postreplication repair in *Saccharomyces cerevisiae* and effects of *rad6, rad18, rev3* and *rad52* mutations. Mol Gen Genet *184*, 471–478.
112. Prakash, S., Johnson, R. E., and Prakash, L. (2005). Eukaryotic translesion synthesis DNA polymerases: specificity of structure and function. Annu Rev Biochem *74*, 317–353.

113. Pursell, Z. F., Isoz, I., Lundstrom, E. B., Johansson, E., and Kunkel, T. A. (2007). Yeast DNA polymerase ε participates in leading-strand DNA replication. Science *317*, 127–130.
114. Rada, C., Di Noia, J. M., and Neuberger, M. S. (2004). Mismatch recognition and uracil excision provide complementary paths to both Ig switching and the A/T-focused phase of somatic mutation. Mol Cell *16*, 163–171.
115. Reuven, N. B., Arad, G., Maor-Shoshani, A., and Livneh, Z. (1999). The mutagenesis protein UmuC is a DNA polymerase activated by UmuD', RecA, and SSB and Is specialized for translesion replication. J Biol Chem *274*, 31763–31766.
116. Ross, A. L., Simpson, L. J., and Sale, J. E. (2005). Vertebrate DNA damage tolerance requires the C-terminus but not BRCT or transferase domains of REV1. Nucleic Acids Res *33*, 1280–1289.
117. Rupp, W. D., and Howard-Flanders, P. (1968). Discontinuities in the DNA synthesized in an excision-defective strain of *Escherichia coli* following ultraviolet irradiation. J Mol Biol *31*, 291–304.
118. Rupp, W. D., Wilde, C. E., Reno, D. L., and Howard-Flanders, P. (1971). Exchanges between DNA strands in ultraviolet irradiated *Escherichia coli*. J Mol Biol *61*, 25–44.
119. Sabbioneda, S., Bortolomai, I., Giannattasio, M., Plevani, P., and Muzi-Falconi, M. (2007). Yeast Rev1 is cell cycle regulated, phosphorylated in response to DNA damage and its binding to chromosomes is dependent upon MEC1. DNA Repair (Amst) *6*, 121–127.
119a. Sabbioneda, S., Gourdin, A. M., Green, C. M., Zotter, A., Giglia-Mari, G., Houtsmuller, A., Vermeulen, W., and Lehmann, A. R. (2008). Effect of Proliferating Cell Nuclear Antigen ubiquitination and chromatin structure on the dynamic properties of the Y-family DNA polymerases. Mol Biol Cell *19*, 5193–5202.
120. Sabbioneda, S., Minesinger, B. K., Giannattasio, M., Plevani, P., Muzi-Falconi, M., and Jinks-Robertson, S. (2005). The 9-1-1 checkpoint clamp physically interacts with polzeta and is partially required for spontaneous polζ-dependent mutagenesis in Saccharomyces cerevisiae. J Biol Chem *280*, 38657–38665.
121. Sale, J. E., Batters, C., Edmunds, C. E., Phillips, L. G., Simpson, L. J., and Szuts, D. (2008). Timing matters: error-prone gap filling and translesion synthesis in immunoglobulin gene hypermutation. Philos Trans R Soc Lond B *364*, 595–603.
122. Sarkar, S., Davies, A. A., Ulrich, H. D., and McHugh, P. J. (2006). DNA interstrand crosslink repair during G1 involves nucleotide excision repair and DNA polymerase ζ. Embo J *25*, 1285–1294.
123. Schenten, D., Gerlach, V. L., Guo, C., Velasco-Miguel, S., Hladik, C. L., White, C. L., Friedberg, E. C., Rajewsky, K., and Esposito, G. (2002). DNA polymerase κ deficiency does not affect somatic hypermutation in mice. Eur J Immunol *32*, 3152–3160.
124. Shen, X., Jun, S., O'Neal, L. E., Sonoda, E., Bemark, M., Sale, J. E., and Li, L. (2006). REV3 and REV1 play major roles in recombination-independent repair of DNA interstrand cross-links mediated by monoubiquitinated proliferating cell nuclear antigen (PCNA). J Biol Chem *281*, 13869–13872.
125. Shimizu, T., Shinkai, Y., Ogi, T., Ohmori, H., and Azuma, T. (2003). The absence of DNA polymerase κ does not affect somatic hypermutation of the mouse immunoglobulin heavy chain gene. Immunol Lett *86*, 265–270.
126. Shinagawa, H., Iwasaki, H., Kato, T., and Nakata, A. (1988). RecA protein-dependent cleavage of UmuD protein and SOS mutagenesis. Proc Natl Acad Sci USA *85*, 1806–1810.
127. Simpson, L. J., Ross, A. L., Szuts, D., Alviani, C. A., Oestergaard, V. H., Patel, K. J., and Sale, J. E. (2006). RAD18-independent ubiquitination of proliferating-cell nuclear antigen in the avian cell line DT40. EMBO Rep *7*, 927–932.
128. Simpson, L. J., and Sale, J. E. (2003). Rev1 is essential for DNA damage tolerance and non-templated immunoglobulin gene mutation in a vertebrate cell line. Embo J *22*, 1654–1664.
129. Sogo, J. M., Lopes, M., and Foiani, M. (2002). Fork reversal and ssDNA accumulation at stalled replication forks owing to checkpoint defects. Science *279*, 599–602.

130. Solovjeva, L., Svetlova, M., Sasina, L., Tanaka, K., Saijo, M., Nazarov, I., Bradbury, M., and Tomilin, N. (2005). High mobility of flap endonuclease 1 and DNA polymerase η associated with replication foci in mammalian S-phase nucleus. Mol Biol Cell *16*, 2518–2528.
131. Sonoda, E., Okada, T., Zhao, G. Y., Tateishi, S., Araki, K., Yamaizumi, M., Yagi, T., Verkaik, N. S., van Gent, D. C., Takata, M., and Takeda, S. (2003). Multiple roles of Rev3, the catalytic subunit of polzeta in maintaining genome stability in vertebrates. Embo J *22*, 3188–3197.
132. Soria, G., Podhajcer, O., Prives, C., and Gottifredi, V. (2006). P21(Cip1/WAF1) downregulation is required for efficient PCNA ubiquitination after UV irradiation. Oncogene *25*, 2829–2838.
133. Stary, A., Kannouche, P., Lehmann, A. R., and Sarasin, A. (2003). Role of DNA polymerase η in the UV-mutation spectrum in human cells. J Biol Chem *278*, 18767–18775.
134. Stelter, P., and Ulrich, H. D. (2003). Control of spontaneous and damage-induced mutagenesis by SUMO and ubiquitin conjugation. Nature *425*, 188–191.
135. Takenaka, K., Ogi, T., Okada, T., Sonoda, E., Guo, C., Friedberg, E. C., and Takeda, S. (2006). Involvement of vertebrate Polκ in translesion DNA synthesis across DNA monoalkylation damage. J Biol Chem *281*, 2000–2004.
136. Tang, M., Shen, X., Frank, E. G., O'Donnell, M., Woodgate, R., and Goodman, M. F. (1999). UmuD'(2)C is an error-prone DNA polymerase, *Escherichia coli* pol V. Proc Natl Acad Sci USA *96*, 8919–8924.
137. Tateishi, S., Niwa, H., Miyazaki, J., Fujimoto, S., Inoue, H., and Yamaizumi, M. (2003). Enhanced genomic instability and defective postreplication repair in RAD18 knockout mouse embryonic stem cells. Mol Cell Biol *23*, 474–481.
138. Tissier, A., Kannouche, P., Reck, M.-P., Lehmann, A. R., Fuchs, R. P. P., and Cordonnier, A. (2004). Co-localization in replication foci and interaction of human Y-family members, DNA polymerase polη and Rev1 protein. DNA repair *3*, 1503–1514.
139. Tissier, A., McDonald, J. P., Frank, E. G., and Woodgate, R. (2000). Polι, a remarkably error-prone human DNA polymerase. Genes Dev *14*, 1642–1650.
140. Trincao, J., Johnson, R. E., Escalante, C. R., Prakash, S., Prakash, L., and Aggarwal, A. K. (2001). Structure of the catalytic core of *S. cerevisiae* DNA polymerase η: implications for translesion DNA synthesis. Mol Cell *8*, 417–426.
141. Tsuji, Y., Watanabe, K., Araki, K., Shinohara, M., Yamagata, Y., Tsurimoto, T., Hanaoka, F., Yamamura, K., Yamaizumi, M., and Tateishi, S. (2008). Recognition of forked and single-stranded DNA structures by human RAD18 complexed with RAD6B protein triggers its recruitment to stalled replication forks. Genes Cells *13*, 343–354.
142. Ulrich, H. D., and Jentsch, S. (2000). Two RING finger proteins mediate cooperation between ubiquitin-conjugating enzymes in DNA repair. Embo J *19*, 3388–3397.
143. Unk, I., Hajdu, I., Fatyol, K., Hurwitz, J., Yoon, J. H., Prakash, L., Prakash, S., and Haracska, L. (2008). Human HLTF functions as a ubiquitin ligase for proliferating cell nuclear antigen polyubiquitination. Proc Natl Acad Sci USA *105*, 3768–3773.
144. Unk, I., Hajdu, I., Fatyol, K., Szakal, B., Blastyak, A., Bermudez, V., Hurwitz, J., Prakash, L., Prakash, S., and Haracska, L. (2006). Human SHPRH is a ubiquitin ligase for Mms2-Ubc13-dependent polyubiquitylation of proliferating cell nuclear antigen. Proc Natl Acad Sci USA *103*, 18107–18112.
145. Velasco-Miguel, S., Richardson, J. A., Gerlach, V. L., Lai, W. C., Gao, T., Russell, L. D., Hladik, C. L., White, C. L., and Friedberg, E. C. (2003). Constitutive and regulated expression of the mouse Dinb (Polkappa) gene encoding DNA polymerase kappa. DNA Repair (Amst) *2*, 91–106.
146. Vidal, A. E., Kannouche, P. P., Podust, V. N., Yang, W., Lehmann, A. R., and Woodgate, R. (2004). PCNA-dependent coordination of the biological functions of human DNA polymerase ι. J Biol Chem *279*, 48360–48368.
147. Wagner, J., Gruz, P., Kim, S. R., Yamada, M., Matsui, K., Fuchs, R. P., and Nohmi, T. (1999). The dinB gene encodes a novel E. coli DNA polymerase, DNA pol IV, involved in mutagenesis. Mol Cell *4*, 281–286.

148. Watanabe, K., Tateishi, S., Kawasuji, M., Tsurimoto, T., Inoue, H., and Yamaizumi, M. (2004). Rad18 guides polη to replication stalling sites through physical interaction and PCNA monoubiquitination. Embo J *28*, 3886–3896.
149. Wittschieben, J. P., Reshmi, S. C., Gollin, S. M., and Wood, R. D. (2006). Loss of DNA polymerase zeta causes chromosomal instability in mammalian cells. Cancer Res *66*, 134–142.
150. Wood, A., Garg, P., and Burgers, P. M. (2007). A ubiquitin-binding motif in the translesion DNA polymerase Rev1 mediates its essential functional interaction with ubiquitinated proliferating cell nuclear antigen in response to DNA damage. J Biol Chem *282*, 20256–20263.
151. Xiao, W., Chow, B. L., Broomfield, S., and Hanna, M. (2000). The Saccharomyces cerevisiae RAD6 group is composed of an error-prone and two error-free postreplication repair pathways. Genetics *155*, 1633–1641.
152. Yamashita, Y. M., Okada, T., Matsusaka, T., Sonoda, E., Zhao, G. Y., Araki, K., Tateishi, S., Yamaizumi, M., and Takeda, S. (2002). RAD18 and RAD54 cooperatively contribute to maintenance of genomic stability in vertebrate cells. Embo J *21*, 5558–5566.
153. Yang, W. (2005). Portraits of a Y-family DNA polymerase. FEBS Lett *579*, 868–872.
154. Yang, W., and Woodgate, R. (2007). What a difference a decade makes: insights into translesion DNA synthesis. Proc Natl Acad Sci USA *104*, 15591–15598.
155. Yang, X. H., Shiotani, B., Classon, M., and Zou, L. (2008). Chk1 and Claspin potentiate PCNA ubiquitination. Genes Dev *22*, 1147–1152.
156. Yuasa, M. S., Masutani, C., Hirano, A., Cohn, M. A., Yamaizumi, M., Nakatani, Y., and Hanaoka, F. (2006). A human DNA polymerase eta complex containing Rad18, Rad6 and Rev1; proteomic analysis and targeting of the complex to the chromatin-bound fraction of cells undergoing replication fork arrest. Genes Cells *11*, 731–744.
157. Zander, L., and Bemark, M. (2004). Immortalized mouse cell lines that lack a functional Rev3 gene are hypersensitive to UV irradiation and cisplatin treatment. DNA Repair (Amst) *3*, 743–752.
158. Zeng, X., Winter, D. B., Kasmer, C., Kraemer, K. H., Lehmann, A. R., and Gearhart, P. J. (2001). DNA polymerase η is an A-T mutator in somatic hypermutation of immunoglobulin variable genes. Nat Immunol *2*, 537–541.
159. Zhang, H., and Siede, W. (2002). UV-induced T–>C transition at a TT photoproduct site is dependent on Saccharomyces cerevisiae polymerase η in vivo. Nucleic Acids Res *30*, 1262–1267.
160. Zheng, H., Wang, X., Warren, A. J., Legerski, R. J., Nairn, R. S., Hamilton, J. W., and Li, L. (2003). Nucleotide excision repair- and polymerase η-mediated error-prone removal of mitomycin C interstrand cross-links. Mol Cell Biol *23*, 754–761.
161. Zhuang, Z., Johnson, R. E., Haracska, L., Prakash, L., Prakash, S., and Benkovic, S. J. (2008). Regulation of polymerase exchange between Polη and Polδ by monoubiquitination of PCNA and the movement of DNA polymerase holoenzyme. Proc Natl Acad Sci USA *105*, 5361–5366.

Chapter 11
Nucleotide Excision Repair: from DNA Damage Processing to Human Disease

Mischa G. Vrouwe and Leon H.F. Mullenders

Abstract A network of DNA damage surveillance systems warrants genomic stability under conditions where cells and organisms are continuously exposed to DNA damaging agents. This network includes DNA repair pathways, but also signaling pathways that activate cell cycle checkpoints, apoptosis, transcription, and chromatin remodeling. Among the various repair pathways nucleotide excision repair (NER) is a highly versatile and evolutionary conserved pathway with an intriguing wide substrate specificity; this pathway removes structurally unrelated bulky DNA lesions from the genome such as sunlight induced photolesions, bulky adducts formed by polycyclic aromatic hydrocarbons, cisplatin intrastrand crosslinks, and alkylation products. The common features of these lesions are the variable degree of DNA helix distortion inflicted and their potency to block replication and transcription. The importance of functional NER for human health is highlighted by the existence of rare autosomal recessive human disorders such as xeroderma pigmentosum (XP). Affected individuals, characterized by a defect in NER, suffer from hypersensitivity to sunlight and display strongly enhanced cancer susceptibility in sunlight exposed parts of the skin. Mammalian NER involves multiple proteins (in excess of 30) and carries out the repair reaction in a highly orchestrated fashion.

In this book chapter, we discuss the current knowledge of the molecular mechanisms underlying NER, its relation with human disease and the translation of knowledge to clinical use.

Keywords Ultraviolet light (UV) · Nucleotide excision repair (NER) · Global genome repair · Transcription coupled repair · UV-photolesions · Xeroderma pigmentosum · Cockayne Syndrome · Trichothiodystrophy · DNA damage recognition · Dual incision · Repair synthesis · DNA damage signalling · UV induced skin cancer · Chromatin · Mouse models · UV-induced mutations · Ubiquitylation · ATM/ATR kinases

---

L.H.F. Mullenders (✉)
Department of Toxicogenetics, Leiden University Medical Center, Leiden, The Netherlands
e-mail: l.mullenders@lumc.nl

K.K. Khanna, Y. Shiloh (eds.), *The DNA Damage Response: Implications on Cancer Formation and Treatment*, DOI 10.1007/978-90-481-2561-6_11,

11.1 Introduction

Through evolution a network of DNA damage surveillance systems has evolved to warrant genomic stability under conditions where cells and organisms are continuously exposed to genotoxic agents present within the environment or exerted by endogenous processes. This network not only includes DNA repair pathways, but also signaling pathways that activate cell cycle checkpoints, apoptosis, transcription, and chromatin remodelling. The mechanisms by which eukaryotic cells sense DNA damage and activate signaling pathways are still poorly understood. One of the challenges is to understand how cells are capable to sense, recognize and repair low levels of different DNA lesions in their genomes at various stages of the cell cycle and in different chromatin environments.

A limited set of DNA repair pathways is capable to repair the large variety of structurally different DNA lesions that are formed in the genome. Nucleotide excision repair (NER) is a highly versatile and evolutionary conserved repair pathway that removes structurally unrelated bulky DNA lesions from the genome such as bulky adducts formed by polycyclic aromatic hydrocarbons, cisplatin intrastrand crosslinks and alkylation products. The common features of these lesions are the variable degree of DNA helix distortion inflicted and their potency to block replication and transcription elongation. In fact, NER is the only repair pathway in humans to remove the toxic and mutagenic photodimers from sunlight exposed parts of the skin. Mammalian NER involves multiple proteins (in excess of 30) and carries out the repair reaction in a highly orchestrated fashion involving a number of defined steps: (I) lesion recognition, (II) DNA unwinding and lesion demarcation, (III) dual incision and release of the incised fragment and (IV) gap filling by repair synthesis and ligation.

Two mechanistically distinct NER subpathways have been identified: global genome NER (GG-NER) is capable of repairing DNA lesions in chromatin of different compaction levels and different functional states throughout the cell cycle, while a subpathway of NER designated transcription-coupled repair (TC-NER) enables efficient repair of RNA polymerase II (RNAPII) blocking DNA lesions and allows quick resumption of transcription. The existence of three rare autosomal recessive human disorders i.e. xeroderma pigmentosum (XP), Cockayne syndrome (CS) and trichothiodystrophy (TTD) all associated with sensitivity to sunlight and NER deficiency, highlights the importance of functional NER for human health. Cells from XP patients are sensitive to UV (ultraviolet)-light and chemicals inducing bulky DNA lesions, and complementation studies revealed eight genes involved in the disease (XPA–XPG and XP Variant; see also Chapter 10). Complementation studies have revealed two CS complementation groups, CS-A and CS-B. In addition to XP and CS patients, a third group of UV sensitive and cancer prone patients has been identified that encompasses individuals exhibiting both XP and CS symptoms.

Although NER removes a variety of structurally unrelated lesions from the genome, we will concentrate on NER in UV-irradiated mammalian cells since UV-induced pyrimidine-pyrimidone (6-4) photoproducts (6-4PP) and cyclobutane pyrimidine dimers (CPD) (Fig. 1) are the lesions most intensively studied and as such paradigmatic for NER.

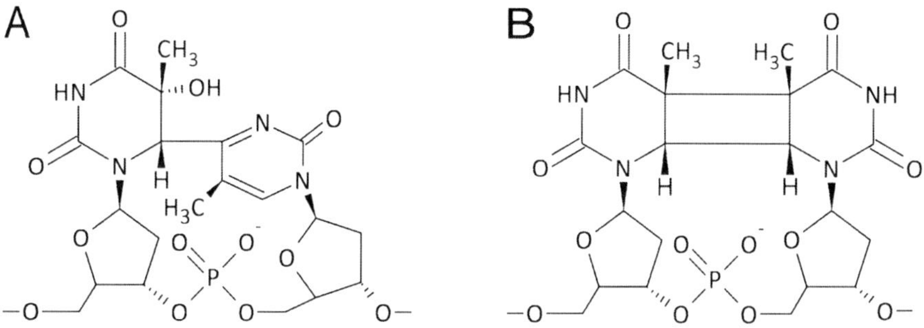

Fig. 1 Structure of UV induced DNA photolesions. (**A**) pyrimidine-pyrimidone (6-4) photoproduct (6-4PP). (**B**) *cis-syn* cyclobutane thymine (pyrimidine) dimer (CPD)

11.2 Global Genome Repair

Cells of all XP patients (except XP variant) were found to be defective in global genome repair of UV-induced photolesions. The identification of the different XP complementation groups led to the isolation of the XP genes and encoded proteins and allowed reconstitution of the process in vitro using purified proteins and naked DNA harboring a specific lesion. The first steps in NER, i.e. DNA lesion recognition and dual incision, require all XP factors (XPC, XPA, XPG, ERCC1-XPF) together with other factors (RPA, TFIIH). In addition, the in vitro reaction requires RF-C, PCNA, DNA polymerase ε and DNA ligase I for repair synthesis [1]. This scheme represents the proteins to perform NER in vitro; additional factors are required for in vivo NER on chromatinized DNA templates.

11.2.1 DNA Lesion Recognition in GG-NER

Although XPA was originally proposed to be the principal damage recognition protein, it is now well established that GG-NER is initiated by the XPC protein which forms a trimeric complex with hHR23B (the human homologue of Rad23) and CEN2 [50, 90, 97, 3]. In vivo the recruitment of NER proteins to UV damage is abolished in XPC deficient cells, indicating that assembly of the NER complex is strictly XPC dependent. Mobility studies on GFP-tagged XPC suggest that the majority of XPC-hHR23B molecules (>90%) transiently interact non-specifically with genomic DNA [33, 70]. In fact, the general affinity of the complex for DNA and its specific affinity for photolesions such as CPD and 6-4PP are relatively low in vivo [58]. The capacity of XPC-hHR23B to recognize a broad spectrum of structurally unrelated lesions might be understood from the observation that XPC binds to the accessible non-damaged DNA strand opposite to a DNA injury [89, 54].

Although the XPC-hHR23B complex acts as the principle initiator of NER, its action is preceded by the heterodimeric UV-DDB complex consisting of the p48 and p127 proteins, products of the DDB2 and DDB1 genes respectively [40]. In fact, repair of CPDs requires functional UV-DDB [92] and, in addition, UV-DDB significantly stimulates the repair of 6-4PPs, particularly at low UV doses [34, 58]. Microinjection of purified UV-DDB was originally found to restore the repair defect of XP-E cells as measured by unscheduled DNA synthesis. Subsequently it was found that the XP group E phenotype is caused by mutations in the DDB2 gene, encoding the p48 protein (XPE). The general affinity of UV-DDB for DNA is much higher (100–1000-fold) than that of XPC-hHR23B, while the specific affinity for 6-4PP is comparable [5]. DDB2 is part of a functional CUL4A-based E3 ubiquitin ligase through its interaction with DDB1 [27] and also binds to UV lesions as an active E3 ubiquitin ligase independent of XPC. UV irradiation activates the E3 ubiquitin ligase activity of the DDB2 complex by binding of the ubiquitin-like protein Nedd 8 to the Cullin 4A. Several substrates for ubiquitylation were identified including DDB2 itself, XPC and histones H2A, H3 and H4 [91, 39, 99]. The current view is that ubiquitylation (at least partly) facilitates NER: ubiquitylation of XPC enhances its affinity for DNA (both damaged and non-damaged DNA) whereas ubiquitylation of histones facilitates the access of repair proteins to DNA damage in chromatin by weakening the histone-DNA association. Interestingly and less well understood, ubiquitylated DDB2 is quickly targeted for degradation after UV [74, 91] even in the presence of large numbers of unrepaired photolesions.

As a general mechanism it was proposed [58] that UV-DDB forms a stable complex when bound to DNA damage, such as UV-induced 6-4PP, allowing subsequent repair proteins, starting with XPC-hHR23B, to accumulate and to verify the lesion, ultimately resulting in efficient repair. The fraction of 6-4PP that can be bound by UV-DDB is limited due to the low cellular quantity and fast UV dependent degradation of DDB2. In cells lacking UV-DDB a slow XPC-hHR23B dependent pathway is capable of repairing 6-4PP whereas repair of CPD is virtually absent.

11.2.2 Assembly of the Preincision Complex

Upon the recognition of lesions by a concerted action of UV-DDB and XPC-hHR23B, the latter recruits the multiprotein transcription factor TFIIH via direct protein-protein interactions (Fig. 2). The TFIIH complex is composed of a seven-subunit core containing two XP factors (XPB, XPD, TTD, p34, p44, p52, p62) and a three-subunit kinase complex (Cdk7, cyclin H and MAT1) termed the CAK unit. The complex exhibits dual functions i.e. it plays a role in RNAPI and RNAPII driven transcription as well as in NER. The XPB subunit of TFIIH is an ATP-dependent helicase that mediates unwinding of promoter DNA in a 3′-5′ orientation during transcription initiation, whereas the XPD subunit of TFIIH is a 5′-3′ATP dependent helicase. Interestingly, the unwinding step of the damaged DNA during NER requires only the XPD helicase activity, whereas the ATPase activity of XPB is dispensable for NER [12]. A two-step mechanism underlies the opening of the damaged DNA to allow assembly of the NER preincision complex: TFIIH mediates the initial opening after which RPA, XPA and XPG bind to obtain full opening of approximately 30 nucleotides around the lesion [18]. XPA stimulates the ATPase activity of TFIIH whereas RPA and XPG stabilize the repair intermediate and contribute to full opening around the lesion. Since TFIIH functions in both transcription

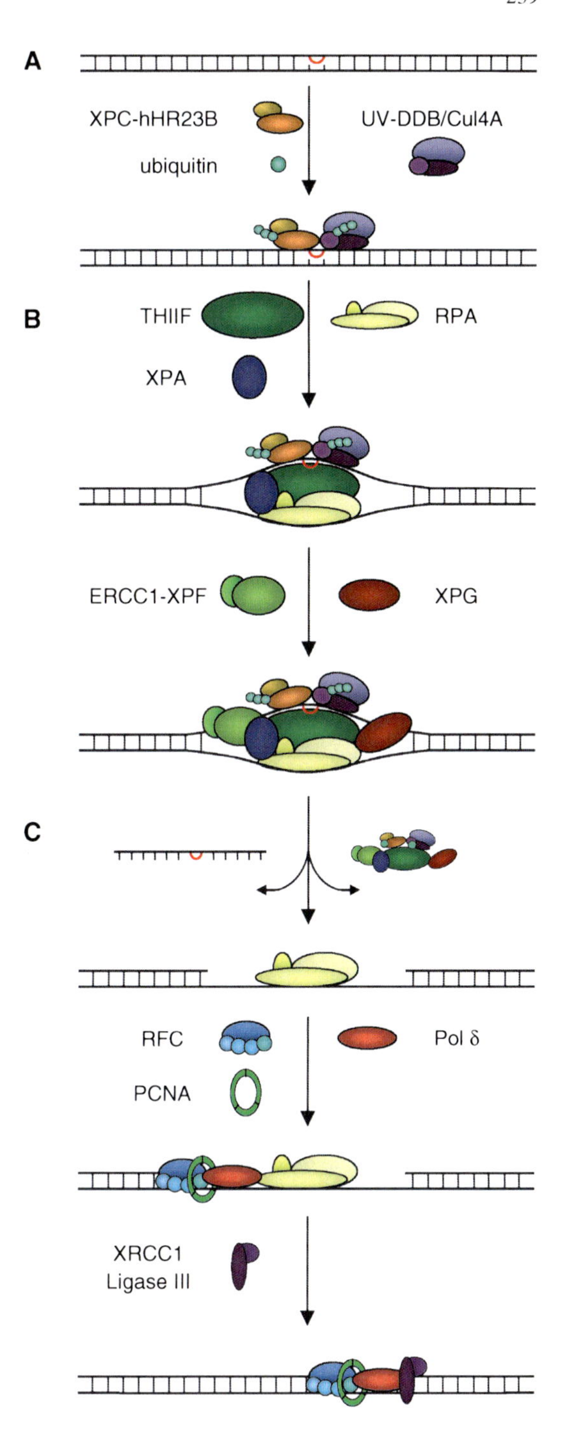

Fig. 2 Model for global genome NER. (**A**) DNA damage recognition by UV-DDB and XPC-hHR23B. Ubiquitylation of damaged bound XPC and DDB2 is mediated via DDB1/Cul4A. (**B**) Assembly of the preincision complex. TFIIH is recruited to the lesion by the XPC-hHR23B complex, opening up the DNA though its helicase activity. The association of RPA further stimulates the unwinding. XPA binding contributes to damage verification and recruits the ERCC1-XPF complex. The ERCC1-XPF and XPG endonucleases incise the damaged DNA strand both 5′ and 3′ of the lesion respectively. In addition to its incision activity XPG also serves to stabilize the open DNA bubble structure. (**C**) Gap filling and ligation. Following dual incision the single stranded DNA patch is filled by the concerted action of RFC, PCNA and DNA polymerase δ, whereas XRCC1/Ligase III performs the final DNA ligation step. Note that in cycling cells DNA polymerase ε and DNA Ligase I also contribute to repair in addition to DNA polymerase δ and XRCC1/Ligase III

initiation and NER it has been proposed that the recruitment of TFIIH to UV damage abolishes transcription initiation in UV-irradiated cells by a trans mechanism [60, 56]. However, inhibition of transcription by UV irradiation only occurs by direct interference of damage with the transcription machinery [56] and does not otherwise affect the engagement of TFIIH in transcription [32].

Replication protein A (RPA) consists of three subunits and is an abundant single-stranded DNA-binding protein that binds optimally to approximately 30 nucleotides [14]. During the formation of the preincision complex RPA associates with the undamaged DNA strand partially unwound by TFIIH (~8–10 nucleotides) and subsequently extends it association to a 30-nucleotide region. This action leads to a separation of the DNA strands around the lesion. Importantly, RPA has been shown to interact with several core NER proteins including XPA, XPG and ERCC1-XPF. In living cells RPA can assemble into the pre-incision complex (consisting of XPC, TFIIH and XPG) in the absence of XPA. This complex, however, is insufficient to stimulate the 3′ incision by XPG and incapable to recruit the 5′ XRCC1-XPF endonuclease; the latter event requires recruitment of XPA in order to assemble the complete pre-incision complex [72]. XPA is essential for GG-NER and it is likely that RPA is required to recruit the XPA protein, although direct in vivo evidence is lacking. The multiple protein interactions of XPA, i.e. the association with RPA and XRCC1-XPF, were demonstrated by the finding that the N-terminus of XPA binds to RPA and ERCC1 whereas the C-terminus interacts with TFIIH [68], consistent with a central role for XPA in the formation of the NER pre-incision complex. Most notably, XPA deficient cells completely lack incision activity indicating that XPA plays an important role in the coordination of dual incision. The observation that XPA binds preferentially to bent or kinked DNA duplexes [9, 10] sheds light on its function in NER and links initial damage recognition by UV-DDB and XPC-hR23B to XPA recruitment. Both XPC and DDB2 introduce kinks in the DNA upon binding and hence might stimulate binding of XPA [36]. The binding of XPA most likely contributes to DNA damage verification in the pre-incision complex: the interaction between RPA bound to the undamaged strand and XPA with the kinked DNA duplex provide the molecular tools that allow identification of the DNA lesion in the pre-incision complex.

RPA plays a key role at the interface of the pre- and post-incision step of NER as the protein precipitates in chromatin immunoprecipitation (ChIP) reactions with antibodies raised against pre-incision and post-incision proteins [57]. Analysis of the assembly and disassembly of repair proteins on immobilized damaged-DNA templates in vitro revealed that RPA remains bound after dual incision and initiates the assembly of DNA synthesis factors such as PCNA [76].

11.2.3 Dual Incision Step

The two structure specific endonucleases XPG and ERCC1-XPF are involved in dual incision 3′ and 5′ of the lesion respectively. In the presence of both proteins, both 5′ and 3′ uncoupled incisions have been observed indicating that both incisions are made simultaneously [55]. The XPG protein specifically incises DNA at the side of the junction between single-stranded DNA and double stranded DNA [65] approximately 2–8 nucleotides from the 3′ side of the lesion. The protein interacts with RPA and TFIIH and the recruitment of XPG to the preincision complex was shown to depend on functional TFIIH. However, the presence of XPG in the pre-incision complex was shown to be required for stabilizing the open DNA bubble structure containing the DNA lesion, allowing binding and 5′ incision by XRCC1-XPF [98]. Hence, XPG also has a structural role in NER and this goes along with the recent finding that XPG may act as a major stabilizing factor by associating with TFIIH [35], although dynamic measurements support separate moieties rather than a joined complex in vivo [104]. In cells of XPG patients with a combined XP and CS phenotype, XPG fails to associate with TFIIH and as a consequence the CAK subunit dissociates from core TFIIH. Deletion mutant analysis of XPG revealed that the so-called spacer region within the protein (which is not required for endonuclease activity) contributes to the substrate specificity of XPG and is required for the interaction with TFIIH and for NER activity in vitro and in vivo [17].

The 5′ junction between single-stranded and double-stranded DNA is cleaved by the heterodimeric endonuclease ERCC1-XPF approximately 15–24 nucleotides away from the 5′ side of the lesion [52]. The two proteins cannot be isolated as separate entities indicating that complex formation underlies the stability of the dimeric endonuclease [84]. Both in vivo and in vitro it has been shown that the interaction of ERCC1-XPF with XPA is essential for NER and that XPA recruits ERCC1-XPF to the pre-incision complex [97]. The incision activity of ERCC1-XPF is stimulated by direct interactions with RPA in model substrates [14]. Mutations in XPF are associated with mild XP; however, a unique XP-F patient with a severe phenotype was recently described displaying signs of accelerated aging [64]. Moreover, mutations in ERCC1 have so far only been reported for one patient with severe clinical features but only a mild repair defect at the cellular level [38]. These latter findings suggest additional functions for ERCC1 and XPF. Indeed, ERCC1-XPF is involved in several other processes such as homologous recombination, repair of interstrand cross-links and telomere maintenance.

11.2.4 The Post-Incision Step in NER

Dual incision and removal of the lesion containing single stranded DNA fragment is followed by gap filling and ligation, generally termed repair synthesis. The transition between dual incision and repair synthesis needs to be coordinated to omit activation of DNA damage signaling and to prevent recombination, the formation of deletions etc. Conceivably the incision reactions might not occur simultaneously and might be initiated by ERCC1-XPF to start DNA synthesis before XPG cutting takes place [24]. An alternative mechanism to prevent undesired processing is that one key factor is partner in the pre- and post-incision stages of NER and remains bound to the DNA. The two stages of NER can be separated in vitro [76] and pre- and post-incision complexes have been isolated from living cells [57]. These analyses revealed RPA as common factor in the reaction and showed that RPA remains associated with the DNA upon dual incision. In addition to RPA, repair synthesis requires RF-C, PCNA, DNA polymerases ε and δ as well as Ligase I in vitro [82, 1, 83]; the recruitment of the post-incision factors is entirely depending on dual incision. PCNA is a homotrimeric sliding clamp that encircles the DNA and acts as a template to allocate DNA polymerases ε and δ to the DNA [47]. Loading of PCNA on the DNA requires the clamp loader RF-C and ATP. Recent in vivo experiments showed that predominantly DNA polymerase δ is recruited to repair patches upon UV irradiation in replicating and quiescent cells and that the role of DNA polymerase ε is restricted to S-phase cells [57]. Moreover, the surprising finding was recently made that under certain conditions (G_0 cells, DNA synthesis inhibitors) the translesion synthesis DNA polymerase κ may also play a role in repair synthesis during NER [67], emphasizing the need to confirm the roles of these late factors in NER in vivo.

The NER reaction is completed by ligation of the 5′ end of the newly synthesized DNA to the original sequence. Although Ligase I is sufficient for sealing nicks during in vitro repair synthesis, XRCC1-LigIIIα appears to be indispensable for ligation of NER-induced breaks. Two distinct complexes were identified that differentially carry out gap filling in NER [57]. XRCC1-LigIIIα and DNA polymerase δ co-localize and interact with NER components in a UV- and incision-dependent manner throughout the cell cycle. In contrast, DNA Ligase I and DNA polymerase ε are recruited to UV-damage sites only in proliferating cells. These findings indicate that cells have differential requirements for ligases and polymerases in repair synthesis depending on the cell cycle.

Finally, the progression of NER seems to be controlled and requires the completion of the post-incision step in NER. Inhibition of DNA polymerases δ and ε in nondividing normal human cells by the DNA polymerase inhibitors HU/AraC leads to accumulation of DNA strand breaks, DNA damage signaling (H2AX signaling) but also to strong retardation of repair of UV induced photolesions such as 6-4PP [57]. Obviously, efficient gap filling by DNA synthesis and ligation of the repair patch is required to drive NER to completeness and implicates either the existence of efficient cellular control mechanisms or factors that limit the number of (pre-) incision events.

11.2.5 Damage Signaling in NER

It has been long acknowledged that exposure of cells to UV light not only activates NER, but also modulates other DNA damage responses impacting cell cycle progression and apoptosis. The exact mechanisms underlying the decision in cell fate have long remained obscure. However, recent works have begun to uncover the molecular mechanisms determining cell fate following UV exposure and demonstrate links between NER and other pathways in the DNA damage response. One of the most prominent players in the UV-induced DNA damage response would be the ataxia telangiectasia mutated and Rad3-related (ATR) protein. ATR is a member of the phosphoinositol-3-kinase like kinase family that also includes the ataxia telangiectasia mutated (ATM) protein. It has now become clear that both proteins act as one of the earliest components in the damage response, their main function being the phosphorylation of various proteins to effectively propagate the damage signaling. While ATM and ATR share many substrates for phosphorylation, the activating structures for these kinases themselves differ. ATM is activated by double stranded DNA breaks [79], whereas it is RPA bound to single stranded DNA what activates the ATR kinase at stalled replication forks [105]. Although it was initially believed that the capacity to activate ATR was restricted to cells in S phase, it was later demonstrated that H2AX is phosphorylated in an ATR dependent manner following UV exposure of non-replicating cells [66]. The origin of this signaling lies in the formation of single stranded DNA patches following the excision of the damage containing oligo by GG-NER [66, 48, 49, 51]. Such RPA containing ssDNA patches would resemble the structures formed after replication fork stalling and hence allow ATR signaling via a common mechanism. This allows not only the phosphorylation of the many substrates of this kinase, but also serves as a prerequisite for ubiquitylation of histone H2A [7].

Normally the activation of ATR is associated with cell cycle checkpoint arrest and NER dependent activation of ATR indeed is able to induce cell cycle arrest outside the S phase [87]. Whether other processes are affected by this signaling is currently unknown. However, one possibility would be that ATR activation could regulate the levels of checkpoint protein p53. As a transcriptional regulator p53 mediates the expression of genes involved in DNA repair, cell cycle arrest and apoptosis. The fact that both DDB2 and XPC expression are regulated in a p53 dependent manner allows for the possibility that ATR activation could enhance the NER capacity following damage induction. Although the existence of such a regulatory mechanism remains to be demonstrated, support comes from the observation that cells lacking functional p53 have a deficiency in GG-NER [19]. Nevertheless, recent experiments indicate that ATR is indeed important for efficient repair of photolesions as ATR deficient cells are profoundly defective in GG-NER but, surprisingly, only during S-phase [4]. The mechanism underlying this cell cycle specific regulation of repair remains to be clarified. Although GG-NER mediated signaling is now well established, the first demonstration that UV exposure induced a checkpoint response [101] was in TC-NER deficient cells. Here checkpoint activation does not depend on processing of UV lesions by NER, but rather it is the absence of repair that results in enhanced checkpoint activation. DNA lesions that block RNA polymerase II, such as UV lesions, cause a dramatic increase in the levels of both normal and phosphorylated p53 when cells are deficient in TC-NER. Despite the fact that p53 induction has long been associated with stalled transcription, the molecular mechanisms that underlie it remain enigmatic. The identification of ATR as the kinase that, in conjunction with RPA, phosphorylates p53 has begun to shed some light on this matter [16]. However, this discovery itself raises a question about the mechanism of ATR activation at stalled RNA polymerases. The archetypical activating structure for ATR is believed to be a single stranded DNA gap and activation of the kinase depends on additional proteins (e.g. TopBP1, Rad9) that are independently recruited to such substrates [41, 15]. It is questionable whether a stalled RNA polymerase confers a structure that resembles gapped DNA, and as such it would be of interest to investigate the participation of other factors normally associated with ATR signaling in the context of RNA polymerase II mediated signaling.

11.2.6 Chromatin Structure and NER

In general the condensed structure of chromatin poses problems to DNA metabolizing processes; notably, NER in a chromatin context is severely inhibited compared to naked DNA [28]. To overcome this barrier, different mechanisms have evolved to remodel chromatin enhancing the accessibility of damaged DNA for repair proteins. In addition, following removal of the lesion and completing of the post incision stage of NER, cells need to restore the original chromatin structure to maintain the epigenetic information [26]. Finally, there is clear evidence that repair efficiencies differ greatly in various chromatin environments but the underlying mechanism is not well understood [62]. Two major mechanisms may alter chromatin structure: posttranslational modification of histone tails and ATP dependent chromatin remodelling.

As mentioned in Section 11.2.1, the role of UV-DDB in NER has revealed unexpected complexities as this complex associates with proteins that are involved in chromatin remodeling (acetylation) and ubiquitylation [13, 27]; the latter activity is related to the participation of DDB1 and DDB2 in a large complex making up a ubiquitin ligase together with Cul4A and Roc1. The ligase activity of this complex is regulated by the COP9 signalosome (CSN). The ubiquitin ligase activity is stimulated by UV (at least with respect to GG-NER) leading to poly-ubiquitylation and subsequent degradation of DDB2 itself; importantly, ubiquitylation of XPC does not serve as a signal for degradation, but merely stimulates the activity of XPC–HR23B by an unknown mechanism. It is conceivable that the ubiquitylation of histones and DDB2 may lead to increased accessibility of the site of damage by removal and/or loosening of DNA–histone contacts and the displacement of UV-DDB from the lesion.

One of the important changes after UV irradiation is the appearance of hyperacetylated histones, most notably shown by the inhibition of histone deacetylases (HDAC) that trigger genome-wide histone hyperacetylation at both histone H3 and H4 upon UV irradiation. Recently, histone acetyl transferases (HAT) such as the HAT p300 [20] and Gcn5 [103] have been suggested to play a role in increasing the accessibility of chromatin to NER proteins. A role for p300 in NER is suggested by interactions of p300 with the repair factors UV-DDB [74] and PCNA [30], stimulation of repair by p300 in vitro [21] and enhancement of NER by the histone deacetylase inhibitor sodium butyrate [73]. Genetic approaches [85, 61] revealed that histone acetylation is not only important for GG-NER but also for TC-NER, although transcriptionally active genes themselves are enriched for acetylated histones.

In vitro NER assays using chromatin substrates with defined lesions, generally reveal that repair is slow in the nucleosomal DNA with no movement or disruption of nucleosomes [22]. Repair measurement of a defined DNA lesion (i.e. 6-4PP) located in a dinucleosome chromatin template demonstrated that ATP-dependent remodeling might enhance the pre- and post-incision steps of NER as dual incision is facilitated by ACF, an ATP-dependent chromatin remodeling factor [93]. The ACF protein moves nucleosomes rather than displacing them. Also incubation with the nucleosome remodeling complex SWI/SNF and ATP altered the conformation of nucleosomal DNA and promoted more homogeneous repair by nucleosome sliding, thereby increasing accessibility to DNA [22, 29]. In vivo data in yeast suggest that SWI/SNF has a significant role in modulating the accessibility of UV induced photolesions for the NER repair machinery thereby enhancing repair [103]. Interestingly, the SWI/SNF complex and the abovementioned Gcn5 histone acetyl transferase facilitate chromatin modifications independent of functional NER, indicating that chromatin remodeling precedes NER. In spite of this, and unexpectedly, the homologues of XPC-HR23B in yeast (Rad4-Rad23) directly interact with the SWI/SNF remodeling complex via two subunits and this interaction was shown to be enhanced following UV irradiation [25]. Taken together the limited data available to date, suggest important roles for histone modifying enzymes such as HATs and ATP-dependent chromatin remodelers, yet mechanistic understanding of their impact on NER awaits further experimentation particularly to clarify their roles in mammalian NER.

Several studies have provided evidence for the involvement of the acidic HMG proteins that destabilize higher order chromatin structures, in the response to bulky DNA lesions. High mobility group protein B1 (HMGB1) binds to and bends damaged DNA and recent evidence demonstrates that mouse cells lacking HMGB1 are hypersensitive to the toxic effects of UVC radiation and may display reduced NER [43]. HMGN1 was demonstrated to be recruited to TCR complexes [20] and is exclusively involved in TC-NER as HMGN1-deficient mouse cells showed decreased rates of CPDs removal in actively transcribed genes [8]. HMGN1 proteins directly compete for DNA binding sites with histone H1, elevate the level of histone H3 acetylation [45] and modulate the level of histone H3 phosphorylation [44]. It is feasible that the loss of H1 in concert with histone modifications might enhance the DNA damage response following UV irradiation, but surprisingly this only affects TC-NER.

The current models of NER propose that chromatin structure is transiently disrupted during the various stages of repair to facilitate access of the repair machinery to DNA lesions and to carry out the subsequent steps. As a final step it is then necessary to restore the preexisting chromatin structure. A central question is whether chromatin restoration involves recycling of parental histones or new histone incorporation. The chromatin assembly factor (CAF-1), a key factor involved in histone deposition, plays a role in the restoration of chromatin following gap filling and ligation. In living cells this protein is recruited to sites of UV-induced DNA damage in a NER-dependent manner [26]; a process that is possibly mediated by PCNA [23]. The role of CAF-1 as chromatin assembly factor was further highlighted by the observation that histone variant H3.1 was assembled de novo at repair sites, reflecting a chromatin restoration step following NER [71]. Hence, chromatin restoration after DNA damage is more than recycling of histones and may represent an imprint for newly repaired chromatin.

Taken together, it appears that repair proteins, DNA and histone binding proteins and chromatin modifiers play a key role in modulating the accessibility of nucleosomal DNA to the repair machinery as well as in the restoration of the chromatin state following repair. However, it is also evident that we are only beginning to understand the modifications that are required to allow NER in different chromatin environments.

11.3 Transcription Coupled Repair

As pointed out, stalled transcription elongation by a DNA lesion is counteracted by the activation of a specialized NER subpathway named transcription coupled repair (TC-NER). A hallmark of TC-NER is the accelerated repair of DNA lesions (most notably demonstrated for UV-induced CPD) in the transcribed strand of active genes and the inability of TC-NER deficient cells to resume DNA damage-inhibited DNA and RNA synthesis [53, 96]. Obviously, the elongating RNA polymerase II (RNAPIIo) when stalled at a lesion efficiently triggers the recruitment of TC-NER specific factors and NER proteins. Once the lesion has been recognized, all subsequent steps leading to assembly of a functional NER complex require the same NER core factors as described for GG-NER. TC-NER is a strongly conserved repair pathway identified in a variety of organisms including bacteria, yeast and mammals. In UV-irradiated *E. coli* cells, a 130 kDa protein encoded by the mfd gene (termed TRCF: transcription-repair coupling factor) was found to be essential for TC-NER [81]. This protein releases the RNA polymerase and transcript from the DNA in an ATP dependent manner and also facilitates repair of DNA damage by attracting NER factors, in particular UvrA. Also in mammalian cells specific factors for TC-NER have been identified. Measurements of UV–photolesions in transcriptionally active genes of cells derived from various UV sensitive patients identified impaired TC-NER in cells from individuals suffering from Cockayne syndrome (CS). CS is a rare disorder that is associated with a wide variety of clinical symptoms including dwarfism, mental retardation, cataract and eye abnormalities as well as photosensitivity, but no enhanced susceptibility to cancer. As a consequence, these patients die at an early age and CS has been classified as a premature aging syndrome. Complementation studies have identified two CS complementation groups, CS-A and CS-B. A third group encompasses patients with mutations in XPB, XPD or XPG genes exhibiting both XP and CS symptoms. The CSB gene encodes a 168 kDa protein that contains helicase domains (strong homology to similar domains in SNF2-like proteins) and that displays DNA-dependent ATPase and DNA binding activity, but no helicase activity. Also, the bacterial and yeast counterparts of CSB, i.e. Mfd and Rad26 respectively, are DNA dependent ATPases. In addition, CSB has nucleosome remodelling activity and binds to core histone proteins in vitro [11] and transcriptome analysis of CS-B cells revealed deregulation of gene expression similar to that caused by agents that disrupt chromatin structure [63].

The CSA protein contains WD-40 repeats (a motif involved in protein-protein interactions) and is part of an E3-ubiquitin ligase (E3-ub ligase) complex consisting of DDB1, Cullin 4A and ROC1/Rbx1 proteins [27]. In response to UV the COP9 signalosome (CSN) was found to associate with the CSA complex resulting in the inactivation of the ubiquitin ligase activity of the CSA complex in TC-NER.

XAB2 is an XPA binding protein and an essential factor in TCR, but so far mutations in XAB2 have not been associated with UV sensitive patients. The protein is involved in pre-mRNA splicing and transcription, interacts with chromatin bound stalled RNAPIIo complex in a UV- and CS-dependent manner and might function as a scaffold for protein complex formation in TC-NER [42]. Finally, deficiency in HMGN1 (a nucleosome binding protein) leads to UV-B sensitivity in HMGN1 knock out mice and impairs TC-NER in UV-C irradiated mouse embryonic fibroblasts. Interestingly, HMGN1 interacts with UV-stalled RNAPIIo and this interaction depends on CS proteins [20].

11.3.1 Molecular Models for TC-NER

The additional involvement of RNAPII in TC-NER replaces the requirement for XPC-HR23B and UV-DDB to identify DNA lesions, as is the case in GG-NER. Instead the system utilises a factor that is capable to couple blockage of transcription by DNA damage to efficient DNA damage recognition and repair. The transcription-repair coupling factor in mammalian cells appears to be CSB: a DNA dependent ATPase that interacts with RNAPII even in undamaged cells [94]. Mu and Sancar [59] showed that purified human excision repair factors and a DNA substrate analogous to a transcription bubble terminating at a CPD, are capable to excise the lesion independent of XPC. Hence the transcription bubble may substitute for XPC function, which in GG-NER causes the two damaged base pairs to flip out of the double helix.

Persistent blockage of transcription activates a stress response leading to stabilization of p53 and specific modifications of p53 at Ser 15 providing a strong signal for apoptosis in cultured cells and in the epidermis of mice [46, 95]. To relieve the strong apoptotic signal the cell has to remove the transcription blockage; however, the stalled RNAPII is likely to shield the DNA lesion and prevents access to the NER machinery. Two scenarios exist to cope with this problem. One potential mechanism would be that the RNAPII is displaced from the DNA or removed by ubiquitylation and subsequent degradation by the proteosome thereby making the lesion available for the NER machinery. This mechanism has been described for bacterial TC-NER as Mfd releases RNAP and recruits repair proteins [80]. In mammals such a scenario would require the recruitment of NER proteins by the action of CS proteins.

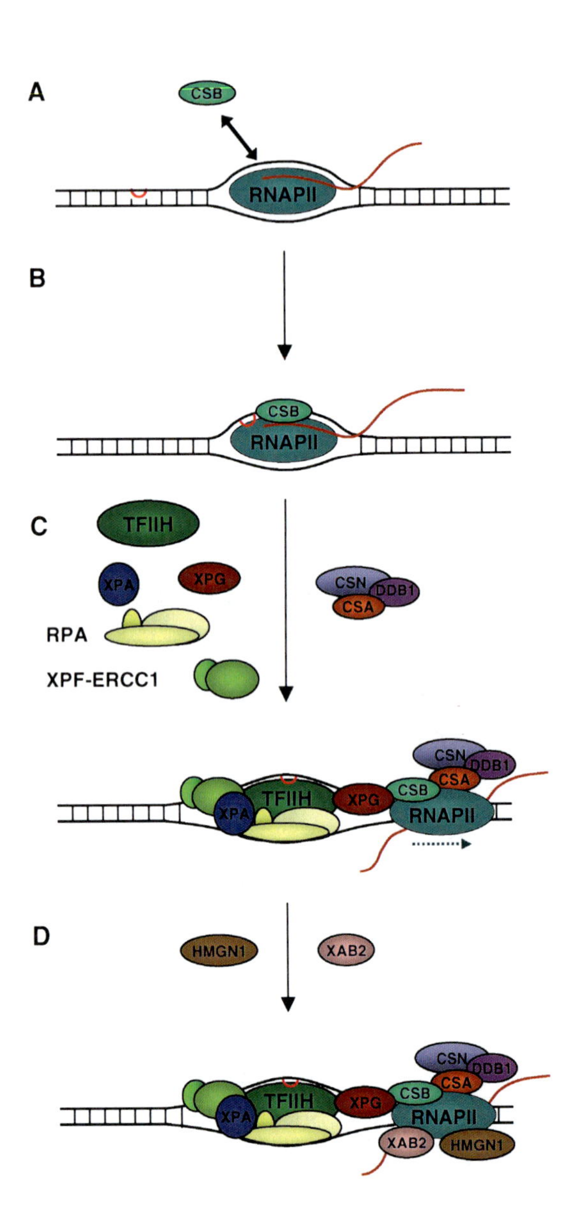

Fig. 3 Model for transcription coupled NER. (**A**) During transcription CSB dynamically interacts with elongating RNAPII. (**B**) Stalling of RNAPII at DNA lesions stabilizes the interaction of CSB with the RNA polymerase. (**C**) The stalled RNAPII/CSB complex allows for the recruitment of the core NER factors as well as the CSA/DDB1/CSN ubiquitin ligase complex. Conformational changes to the stalled RNAPII complex imposed by TFIIH and XPG would allow access to the damaged DNA strand. (**D**) The association of CSA with the stalled polymerase enables the recruitment of additional repair factors like HMGN1 and XAB2

Another possible mechanism is that TC-NER occurs without displacement/removal of RNAPIIo but requires conformational changes of RNAPII to allow access to the DNA lesion and resume transcription. Particularly the XPG endonuclease in concert with the basal transcription factor TFIIH have been implicated in an ATP dependent remodelling of the arrested RNAPII, allowing incision 3′ of the lesion without the need for CSB [78]. Although in vitro experiments indicate a prominent role of XPG in the early stages of TC-NER, recruitment of XPG to stalled RNAPII in intact cells requires functional CSB [20]. Upon binding to stalled RNAPIIo, CSB functions as a coupling factor that mediates the recruitment of subsequent NER repair factors TFIIH, XPG, RPA and ERCC1-XPF (Fig. 3). Indeed, live cell imaging revealed that GFP-tagged CSB interacts with the transcription machinery in the presence of DNA damage. Recruitment of CSA is CSB dependent and required for binding of both HMGN1 and XAB2 but is dispensable for the recruitment of pre-incision NER proteins. The emerging picture of TC-NER is rather complex and not well understood at the molecular level. Most strikingly, repair of transcription blocking lesions in mammalian cells occurs without displacement of the stalled RNAPIIo and requires at least two essential assembly factors with differential modes of action: CSB as a repair–transcription coupling factor to attract the core NER pre-incision factors and CSA to recruit chromatin remodelers. However, the precise role of the CSB ATP-ase activity and the CSA associated E3-ubiquitin ligase complex in TC-NER are not known.

11.4 NER Deficiencies and Cancer

As pointed out, inherited defects in the NER pathway are manifested in at least three different diseases: XP, CS and the photosensitive form of trichothiodystrophy (TTD). Of these, only patients with XP are prone to sunlight-induced skin cancer, although patients with CS and the photosensitive form of trichothiodystrophy (TTD) are clearly UV-sensitive. For most cancers the causative agent is unknown but skin cancer is a notorious exception. In fact, XP is a paradigm for a causal link between defective DNA repair and exposure to an exogenous (environmental) component i.e. sunlight, as XP patients have a >1000-fold increased risk to develop skin tumors primarily at sun-exposed sites of their body. Mutation analysis of TTD revealed a complicated genotype as patients have been identified with mutations in the XPB, XPD and TTDA genes, all components of the TFIIH complex. Since TFIIH functions both in DNA repair and transcription it is assumed that photosensitive TTD patients have a defect in both processes; these patients are characterized by sulphur-deficient brittle hair and nails, ichthyosis, neurological/developmental abnormalities and short life span. Finally, patients exist that belong to the XP-B, XP-D or XP-G complementation group that display severe features of CS (early death and neurological/developmental abnormalities) and XP (skin lesions and skin cancer).

As mentioned above, the most overt phenotype of XP patients is their enhanced susceptibility to develop skin cancer including basal cell carcinomas (BCCs) and squamous cell carcinomas (SCCs) but also melanomas. Increased cancer susceptibility is not only seen at the sun-exposed parts of the body but is also evidenced by a low incidence of internal tumors. Since epidemiological data on the relationship between skin cancers and ambient solar UV radiation are very restricted, animal models i.e. (transgenic) mice, have been used to study the process of UV carcinogenesis in depth and to gain quantitative data on tumor development and dose, time and wavelength of the UV radiation and genetic make-up. Transgenic hairless mice (to facilitate UVB irradiation and the identification of tumors) mimicking the human XP phenotype have been extremely useful in studying the role of (exogenously-induced) DNA damage in mutagenesis, carcinogenesis and aging. The protective role of GG-NER and TC-NER against the acute (i.e. erythema, apoptosis, cell cycle arrest) and long term (i.e. skin cancer, aging) effects of genotoxic (UV-B light, bulky chemicals) exposure has been dissected in mouse models with defined mutations in NER genes, i.e. XPE (DDB2), XPA (defective in GG-NER and TC-NER), XPC (defective in GG-NER) or CSB (defective in TC-NER) deficient mice. DDB2$^{-/-}$ mice are deficient in GG-NER of CPD, but otherwise TC-NER proficient. XPA$^{-/-}$ and CSB$^{-/-}$ mice appeared to be 10-fold more sensitive to the acute toxic effects of UV-B light (erythema/edema of the skin) and to the polycyclic aromatic hydrocarbon DMBA (lethality) compared to normal, XPC$^{-/-}$ or DDB2$^{-/-}$ mice [6, 100]. The difference in UV-B sensitivity relates to enhanced apoptosis and severe cell cycle arrest of epidermal keratinocytes in XPA$^{-/-}$ and CSB$^{-/-}$ mice [95, 88]. These results highlight TC-NER as a profound survival pathway and identify TC-NER as the principal defense mechanism towards the deleterious effects of transcription blocking DNA lesions by counteracting apoptosis and cell cycle arrest. However, this increased survival occurs at expense of increased mutagenesis manifested by the fast appearance of epidermal patches expressing mutant p53 in UV-B irradiated XPC$^{-/-}$ mice [75] but also by increased spontaneous mutagenesis in lymphocytes [100]. Mutation spectrum analysis showed that almost all UV-B light induced mutations in rodent and human were at dipyrimidine positions with C$\rightarrow$T transition mutations being the most prominent. The latter is caused by three factors. Firstly, DNA polymerase η preferentially incorporates adenine residues opposite to non-instructional lesions. Secondly, 5-methylcytosines within CPD lesions display accelerated deamination rates, resulting in base changes to uracil. Moreover, CPDs are formed preferentially at dipyrimidines containing 5-methylcytosine when cells are irradiated with UV-B or sunlight. Finally, CC$\rightarrow$TT double transitions are caused in vivo exclusively by UV-induced pre-mutagenic lesions in XPA and XPC deficient mice and human XPC patients [86]. Also in tumors isolated from UV-B irradiated mice (with high frequencies of p53 mutations) defective GG-NER and TC-NER resulted in increased mutations in p53 through UV-targeted dipyrimidine sites but strikingly, only XPA$^{-/-}$ and CSB$^{-/-}$ mice developed benign papillomas before squamous cell carcinomas (SCC). These papillomas carried mutations in the 12th Hras codon with a dipyrimidine site in the transcribed strand; such mutations were not observed in the UV-induced SCCs. Evidently, proficient TC-NER prevents Hras mutagenesis and therefore prevents the development of papillomas.

Taken together, GG-NER and TC-NER protect against UV-B induced skin cancer in mice. Although the mouse cancer data reveals remarkable similarities with skin cancer susceptibility in human, striking differences exist as well. Most notably, the $XPA^{-/-}$ mice, in contrast to XP patients, do not develop melanoma whereas $CSB^{-/-}$ mice, but not CSB patients, are skin cancer prone. The latter is related to the poorly expressed GG-NER system in rodents. Unlike human cells, rodent epidermal cells express DDB2 at a low level. Mice ectopically expressing DDB2 display delayed onset of squamous cell carcinoma following chronic UV-B light exposure and at the cellular level enhanced repair of UV-photolesions [2, 69] whereas $DDB2^{-/-}$ mice were hypersensitive to UV-induced skin carcinogenesis. Ectopic expression of DDB2 in $CSB^{-/-}$ mice counteracts the cancer process of UVB exposed $CSB^{-/-}$ mice indicating that GG-NER serves as a back-up system for TC-NER deficiency (Pines et al., in preparation).

11.5 Perspectives

Molecular, cellular and animal studies over the last three decades have greatly improved our understanding of the interplay between cellular processes (DNA damage, NER and transcription) and human disease. However, much is to be learned about the exact functions of key players in NER and the mechanisms by which eukaryotic cells sense DNA damage in their genome and which signals activate and regulate NER. The mammalian genome is protected against genotoxic insults by a network of DNA damage response (DDR) mechanisms initiated by sensing of DNA damage or damage-induced chromatin alterations through specific sensors. The next stage in the process is to transmit the signal to transducers that are able to pass the signal to effectors that control various protective pathways i.e. different DNA repair pathways, cell cycle checkpoints, apoptosis, transcription and chromatin remodeling. Hence, full understanding of mammalian NER not only requires insights into the mechanisms of NER but also the DNA damage signaling cascade.

Faithful DNA damage processing in various chromatin environments requires process control at each individual step including regulation of the expression of NER factors, regulation of NER protein activity by post-translational modifications, remodeling of chromatin at sites of DNA damage, monitoring progress and completeness of repair and checking integrity of chromatin after damage removal. Presently, little is known about these regulatory processes and how NER is connected with DNA damage signaling pathways i.e. specific sensors of DNA damage (such as UV-induced photolesions) and transducers able to pass the signal to downstream effectors i.e. transcription, chromatin remodeling and protein modification. Which are the factors that control initiation, progression and completion of the NER process? Which factors enable GG-NER activity in different chromatin environments such as heterochromatic and euchromatic regions? The ultimate goal is to use this information to further improve the mathematical modeling of NER. The current model based on in vivo kinetic data [70] unveils that a sequential assembly mechanism appears remarkably advantageous in terms of repair efficiency and suggests that random assembly and preassembly are kinetically unfavorable.

Multiple gene products are implicated in TC-NER but we lack knowledge of the signals that regulate TC-NER and we do not know the precise function of key components in TC-NER. Most notably, it is not clear why chromatin remodeling would be required for TC-NER in addition to the structural changes that are needed to allow transcription of actively transcribed chromatin-embedded DNA substrates. Currently it is not well understood which factors or processes are required to resume transcription although it is clear that besides TC-NER other mechanisms play a role [77]. Of particular interest to resolve is the fate of stalled RNAPII when TC-NER fails to operate and to find out which types of oxidative DNA damage (induced by metabolic processes) can inhibit transcription in vivo. A stalled RNAPII transcription machinery senses DNA damage and leads to a strong signal for apoptosis. Moreover, there is increasing evidence that during S-phase collisions of replication forks with transcription complexes stalled at DNA lesions are a very mutagenic event [31]. Hence, it is of pivotal importance to dissect the contributions of impaired TC-NER and transcription defects in the aetiology of the progeroid, neurodevelopmental disorder of CS.

Knowledge of the NER pathway and repair proteins has lead to the identification of inherited polymorphisms of NER genes (SNPs). These SNPs may contribute to variations in DNA repair capacity and genetic susceptibility to cancer. Numerous published data provide emerging evidence that polymorphisms in NER genes may contribute to the genetic susceptibility to cancers in man. However, many of the studies are of limited value because of the limited size of the study populations. It is obvious that large and well-designed population-based studies are warranted to identify NER genes as biomarkers to screen high-risk populations for early detection of cancer. Knowledge of the NER pathway and repair proteins can also be applied as basis for enzyme therapy to counteract sunlight induced skin cancer. The bacterial DNA repair enzyme T4 endonuclease V packaged in an engineered delivery vehicle was shown to be capable of reversing the defective repair in xeroderma pigmentosum cells [102]. Moreover, expression of the CPD-photolyase in mouse epidermis is an effective tool to combat UV-B induced nonmelanoma skin cancer [37]. These findings directly proof that enhancement of repair activity can be used as a therapeutic tool to protect against UVB induced skin cancer although NER proteins have not been applied so far.

Acknowledgments This work was supported by grants from ZonMw (912-03-012 and 917-46-364) and from EU (MRTN-Ct-2003-503618).

References

1. Aboussekhra, A., Biggerstaff, M., Shivji, M.K., Vilpo, J.A., Moncollin, V., Podust, V.N., Protic, M., Hubscher, U., Egly, J.M., and Wood, R.D. (1995). Mammalian DNA nucleotide excision repair reconstituted with purified protein components. *Cell*, **80**, 859–868.

2. Alekseev, S., Kool, H., Rebel, H., Fousteri, M., Moser, J., Backendorf, C., de Gruijl, F.R., Vrieling, H., and Mullenders, L.H. (2005). Enhanced DDB2 expression protects mice from carcinogenic effects of chronic UV-B irradiation. *Cancer Res.*, **65**, 10298–10306.
3. Araki, M., Masutani, C., Takemura, M., Uchida, A., Sugasawa, K., Kondoh, J., Ohkuma, Y., and Hanaoka, F. (2001). Centrosome protein centrin 2/caltractin 1 is part of the xeroderma pigmentosum group C complex that initiates global genome nucleotide excision repair. *J. Biol. Chem.*, **276**, 18665–18672.
4. Auclair, Y., Rouget, R., Affar, E.B., and Drobetsky, E.A. (2008). ATR kinase is required for global genomic nucleotide excision repair exclusively during S phase in human cells. *Proc. Natl Acad. Sci. U.S.A*, **105**, 17896–17901.
5. Batty, D., Rapic'-Otrin, V., Levine, A.S., and Wood, R.D. (2000). Stable binding of human XPC complex to irradiated DNA confers strong discrimination for damaged sites. *J. Mol. Biol.*, **300**, 275–290.
6. Berg, R.J., Ruven, H.J., Sands, A.T., de Gruijl, F.R., and Mullenders, L.H. (1998). Defective global genome repair in XPC mice is associated with skin cancer susceptibility but not with sensitivity to UVB induced erythema and edema. *J. Invest Dermatol.*, **110**, 405–409.
7. Bergink, S., Salomons, F.A., Hoogstraten, D., Groothuis, T.A., de Waard, H., Wu, J., Yuan, L., Citterio, E., Houtsmuller, A.B., Neefjes, J., Hoeijmakers, J.H., Vermeulen, W., and Dantuma, N.P. (2006). DNA damage triggers nucleotide excision repair-dependent monoubiquitylation of histone H2A. *Genes Dev.*, **20**, 1343–1352.
8. Birger, Y., West, K.L., Postnikov, Y.V., Lim, J.H., Furusawa, T., Wagner, J.P., Laufer, C.S., Kraemer, K.H., and Bustin, M. (2003). Chromosomal protein HMGN1 enhances the rate of DNA repair in chromatin. *EMBO J.*, **22**, 1665–1675.
9. Camenisch, U., Dip, R., Schumacher, S.B., Schuler, B., and Naegeli, H. (2006). Recognition of helical kinks by xeroderma pigmentosum group A protein triggers DNA excision repair. *Nat. Struct. Mol. Biol.*, **13**, 278–284.
10. Camenisch, U., Dip, R., Vitanescu, M., and Naegeli, H. (2007). Xeroderma pigmentosum complementation group A protein is driven to nucleotide excision repair sites by the electrostatic potential of distorted DNA. *DNA Repair (Amst)*, **6**, 1819–1828.
11. Citterio, E., van den Boom, V., Schnitzler, G., Kanaar, R., Bonte, E., Kingston, R.E., Hoeijmakers, J.H., and Vermeulen, W. (2000). ATP-dependent chromatin remodeling by the Cockayne syndrome B DNA repair-transcription-coupling factor. *Mol. Cell Biol.*, **20**, 7643–7653.
12. Coin, F., Oksenych, V., and Egly, J.M. (2007). Distinct roles for the XPB/p52 and XPD/p44 subcomplexes of TFIIH in damaged DNA opening during nucleotide excision repair. *Mol. Cell*, **26**, 245–256.
13. Datta, A., Bagchi, S., Nag, A., Shiyanov, P., Adami, G.R., Yoon, T., and Raychaudhuri, P. (2001). The p48 subunit of the damaged-DNA binding protein DDB associates with the CBP/p300 family of histone acetyltransferase. *Mutat. Res.*, **486**, 89–97.
14. de Laat, W.L., Appeldoorn, E., Sugasawa, K., Weterings, E., Jaspers, N.G., and Hoeijmakers, J.H. (1998). DNA-binding polarity of human replication protein A positions nucleases in nucleotide excision repair. *Genes Dev.*, **12**, 2598–2609.
15. Delacroix, S., Wagner, J.M., Kobayashi, M., Yamamoto, K., and Karnitz, L.M. (2007). The Rad9-Hus1-Rad1 (9-1-1) clamp activates checkpoint signaling via TopBP1. *Genes Dev.*, **21**, 1472–1477.
16. Derheimer, F.A., O'Hagan, H.M., Krueger, H.M., Hanasoge, S., Paulsen, M.T., and Ljungman, M. (2007). RPA and ATR link transcriptional stress to p53. *Proc. Natl Acad. Sci. U.S.A*, **104**, 12778–12783.
17. Dunand-Sauthier, I., Hohl, M., Thorel, F., Jaquier-Gubler, P., Clarkson, S.G., and Scharer, O.D. (2005). The spacer region of XPG mediates recruitment to nucleotide excision repair complexes and determines substrate specificity. *J. Biol. Chem.*, **280**, 7030–7037.

18. Evans, E., Fellows, J., Coffer, A., and Wood, R.D. (1997). Open complex formation around a lesion during nucleotide excision repair provides a structure for cleavage by human XPG protein. *EMBO J.*, **16**, 625–638.
19. Ford, J.M. and Hanawalt, P.C. (1995). Li-Fraumeni syndrome fibroblasts homozygous for p53 mutations are deficient in global DNA repair but exhibit normal transcription-coupled repair and enhanced UV resistance. *Proc. Natl Acad. Sci. U.S.A.*, **92**, 8876–8880.
20. Fousteri, M., Vermeulen, W., van Zeeland, A.A., and Mullenders, L.H. (2006). Cockayne syndrome A and B proteins differentially regulate recruitment of chromatin remodeling and repair factors to stalled RNA polymerase II in vivo. *Mol. Cell*, **23**, 471–482.
21. Frit, P., Kwon, K., Coin, F., Auriol, J., Dubaele, S., Salles, B., and Egly, J.M. (2002). Transcriptional activators stimulate DNA repair. *mol. cell,* **10,** 1391–1401
22. Gaillard, H., Fitzgerald, D.J., Smith, C.L., Peterson, C.L., Richmond, T.J., and Thoma, F. (2003). Chromatin remodeling activities act on UV-damaged nucleosomes and modulate DNA damage accessibility to photolyase. *J. Biol. Chem.*, **278**, 17655–17663.
23. Gerard, A., Koundrioukoff, S., Ramillon, V., Sergere, J.C., Mailand, N., Quivy, J.P., and Almouzni, G. (2006). The replication kinase Cdc7-Dbf4 promotes the interaction of the p150 subunit of chromatin assembly factor 1 with proliferating cell nuclear antigen. *EMBO Rep.*, **7**, 817–823.
24. Gillet, L.C. and Scharer, O.D. (2006). Molecular mechanisms of mammalian global genome nucleotide excision repair. *Chem. Rev.*, **106**, 253–276.
25. Gong, F., Fahy, D., and Smerdon, M.J. (2006). Rad4-Rad23 interaction with SWI/SNF links ATP-dependent chromatin remodeling with nucleotide excision repair. *Nat. Struct. Mol. Biol.*, **13**, 902–907.
26. Green, C.M. and Almouzni, G. (2002). When repair meets chromatin. First in series on chromatin dynamics. *EMBO Rep.*, **3**, 28–33.
27. Groisman, R., Polanowska, J., Kuraoka, I., Sawada, J., Saijo, M., Drapkin, R., Kisselev, A.F., Tanaka, K., and Nakatani, Y. (2003). The ubiquitin ligase activity in the DDB2 and CSA complexes is differentially regulated by the COP9 signalosome in response to DNA damage. *Cell*, **113**, 357–367.
28. Hara, R., Mo, J., and Sancar, A. (2000). DNA damage in the nucleosome core is refractory to repair by human excision nuclease. *Mol. Cell Biol.*, **20**, 9173–9181.
29. Hara, R. and Sancar, A. (2003). Effect of damage type on stimulation of human excision nuclease by SWI/SNF chromatin remodeling factor. *Mol. Cell Biol.*, **23**, 4121–4125.
30. Hasan, S., Hassa, P.O., Imhof, R., and Hottiger, M.O. (2001). Transcription coactivator p300 binds PCNA and may have a role in DNA repair synthesis. *Nature*, **410**, 387–391.
31. Hendriks, G., Calleja, F., Vrieling, H., Mullenders, L.H., Jansen, J.G., and de Wind, N. (2008). Gene transcription increases DNA damage-induced mutagenesis in mammalian stem cells. *DNA Repair (Amst)*, **7**, 1330–1339.
32. Hoogstraten, D., Nigg, A.L., Heath, H., Mullenders, L.H., van Driel, R., Hoeijmakers, J.H., Vermeulen, W., and Houtsmuller, A.B. (2002). Rapid switching of TFIIH between RNA polymerase I and II transcription and DNA repair in vivo. *Mol. Cell*, **10**, 1163–1174.
33. Hoogstraten, D., Nigg, A.L., van Cappellen, W.A., Hoeijmakers, J.H., Houtsmuller, A.B., and Vermeulen, W. (2003). DNA-damage sensing in living cells by xeroderma pigmentosum group C. Thesis.
34. Hwang, B.J., Ford, J.M., Hanawalt, P.C., and Chu, G. (1999). Expression of the p48 xeroderma pigmentosum gene is p53-dependent and is involved in global genomic repair. *Proc. Natl Acad. Sci. U.S.A.*, **96**, 424–428.
35. Ito, S., Kuraoka, I., Chymkowitch, P., Compe, E., Takedachi, A., Ishigami, C., Coin, F., Egly, J.M., and Tanaka, K. (2007). XPG stabilizes TFIIH, allowing transactivation of nuclear receptors: implications for Cockayne syndrome in XP-G/CS patients. *Mol. Cell*, **26**, 231–243.

36. Janicijevic, A., Sugasawa, K., Shimizu, Y., Hanaoka, F., Wijgers, N., Djurica, M., Hoeijmakers, J.H., and Wyman, C. (2003). DNA bending by the human damage recognition complex XPC-HR23B. *DNA Repair (Amst)*, **2**, 325–336.
37. Jans, J., Schul, W., Sert, Y.G., Rijksen, Y., Rebel, H., Eker, A.P., Nakajima, S., van Steeg, H., de Gruijl, F.R., Yasui, A., Hoeijmakers, J.H., and van der Horst, G.T. (2005). Powerful skin cancer protection by a CPD-photolyase transgene. *Curr. Biol.*, **15**, 105–115.
38. Jaspers, N.G., Raams, A., Silengo, M.C., Wijgers, N., Niedernhofer, L.J., Robinson, A.R., Giglia-Mari, G., Hoogstraten, D., Kleijer, W.J., Hoeijmakers, J.H., and Vermeulen, W. (2007). First reported patient with human ERCC1 deficiency has cerebro-oculo-facio-skeletal syndrome with a mild defect in nucleotide excision repair and severe developmental failure. *Am. J. Hum. Genet.*, **80**, 457–466.
39. Kapetanaki, M.G., Guerrero-Santoro, J., Bisi, D.C., Hsieh, C.L., Rapic-Otrin, V., and Levine, A.S. (2006). The DDB1-CUL4ADDB2 ubiquitin ligase is deficient in xeroderma pigmentosum group E and targets histone H2A at UV-damaged DNA sites. *Proc. Natl Acad. Sci. U.S.A.*, **103**, 2588–2593.
40. Keeney, S., Chang, G.J., and Linn, S. (1993). Characterization of a human DNA damage binding protein implicated in xeroderma pigmentosum E. *J. Biol. Chem.*, **268**, 21293–21300.
41. Kumagai, A., Lee, J., Yoo, H.Y., and Dunphy, W.G. (2006). TopBP1 activates the ATR-ATRIP complex. *Cell*, **124**, 943–955.
42. Kuraoka, I., Ito, S., Wada, T., Hayashida, M., Lee, L., Saijo, M., Nakatsu, Y., Matsumoto, M., Matsunaga, T., Handa, H., Qin, J., Nakatani, Y., and Tanaka, K. (2008). Isolation of XAB2 complex involved in pre-mRNA splicing, transcription, and transcription-coupled repair. *J. Biol. Chem.*, **283**, 940–950.
43. Lange, S.S., Mitchell, D.L., and Vasquez, K.M. (2008). High mobility group protein B1 enhances DNA repair and chromatin modification after DNA damage. *Proc. Natl. Acad. Sci. U.S.A.*, **105**, 10320–10325.
44. Lim, J.H., Catez, F., Birger, Y., West, K.L., Prymakowska-Bosak, M., Postnikov, Y.V., and Bustin, M. (2004). Chromosomal protein HMGN1 modulates histone H3 phosphorylation. *Mol. Cell*, **15**, 573–584.
45. Lim, J.H., West, K.L., Rubinstein, Y., Bergel, M., Postnikov, Y.V., and Bustin, M. (2005). Chromosomal protein HMGN1 enhances the acetylation of lysine 14 in histone H3. *EMBO J.*, **24**, 3038–3048.
46. Ljungman, M., Zhang, F., Chen, F., Rainbow, A.J., and McKay, B.C. (1999). Inhibition of RNA polymerase II as a trigger for the p53 response. *Oncogene*, **18**, 583–592.
47. Maga, G. and Hubscher, U. (2003). Proliferating cell nuclear antigen (PCNA): a dancer with many partners. *J. Cell Sci.*, **116**, 3051–3060.
48. Marini, F., Nardo, T., Giannattasio, M., Minuzzo, M., Stefanini, M., Plevani, P., and Falconi, M.M. (2006). DNA nucleotide excision repair-dependent signaling to checkpoint activation. *Proc. Natl Acad. Sci. U.S.A.*, **103**, 17325–17330.
49. Marti, T.M., Hefner, E., Feeney, L., Natale, V., and Cleaver, J.E. (2006). H2AX phosphorylation within the G1 phase after UV irradiation depends on nucleotide excision repair and not DNA double-strand breaks. *Proc. Natl Acad. Sci. U.S.A.*, **103**, 9891–9896.
50. Masutani, C., Sugasawa, K., Yanagisawa, J., Sonoyama, T., Ui, M., Enomoto, T., Takio, K., Tanaka, K., van der Spek, P.J., and Bootsma, D., (1994). Purification and cloning of a nucleotide excision repair complex involving the xeroderma pigmentosum group C protein and a human homologue of yeast RAD23. *EMBO J.*, **13**, 1831–1843.
51. Matsumoto, M., Yaginuma, K., Igarashi, A., Imura, M., Hasegawa, M., Iwabuchi, K., Date, T., Mori, T., Ishizaki, K., Yamashita, K., Inobe, M., and Matsunaga, T. (2007). Perturbed gap-filling synthesis in nucleotide excision repair causes histone H2AX phosphorylation in human quiescent cells. *J. Cell Sci.*, **120**, 1104–1112.

52. Matsunaga, T., Mu, D., Park, C.H., Reardon, J.T., and Sancar, A. (1995). Human DNA repair excision nuclease. Analysis of the roles of the subunits involved in dual incisions by using anti-XPG and anti-ERCC1 antibodies. *J. Biol. Chem.*, **270**, 20862–20869.
53. Mayne, L.V. and Lehmann, A.R. (1982). Failure of RNA synthesis to recover after UV irradiation: an early defect in cells from individuals with Cockayne's syndrome and xeroderma pigmentosum. *Cancer Res.*, **42**, 1473–1478.
54. Min, J.H. and Pavletich, N.P. (2007). Recognition of DNA damage by the Rad4 nucleotide excision repair protein. *Nature*, **449**, 570–575.
55. Moggs, J.G., Yarema, K.J., Essigmann, J.M., and Wood, R.D. (1996). Analysis of incision sites produced by human cell extracts and purified proteins during nucleotide excision repair of a 1,3-intrastrand d(GpTpG)-cisplatin adduct. *J. Biol. Chem.*, **271**, 7177–7186.
56. Mone, M.J., Volker, M., Nikaido, O., Mullenders, L.H., van Zeeland, A.A., Verschure, P.J., Manders, E.M., and van Driel, R. (2001). Local UV-induced DNA damage in cell nuclei results in local transcription inhibition. *EMBO Rep.*, **2**, 1013–1017.
57. Moser, J., Kool, H., Giakzidis, I., Caldecott, K., Mullenders, L.H., and Fousteri, M.I. (2007). Sealing of chromosomal DNA nicks during nucleotide excision repair requires XRCC1 and DNA ligase III alpha in a cell-cycle-specific manner. *Mol. Cell*, **27**, 311–323.
58. Moser, J., Volker, M., Kool, H., Alekseev, S., Vrieling, H., Yasui, A., van Zeeland, A.A., and Mullenders, L.H. (2005). The UV-damaged DNA binding protein mediates efficient targeting of the nucleotide excision repair complex to UV-induced photo lesions. *DNA Repair (Amst)*, **4**, 571–582.
59. Mu, D. and Sancar, A. (1997). Model for XPC-independent transcription-coupled repair of pyrimidine dimers in humans. *J. Biol. Chem.*, **272**, 7570–7573.
60. Mullenders, L.H. (1998). Transcription response and nucleotide excision repair. *Mutat. Res.*, **409**, 59–64.
61. Mullenders, L.H., van Kesteren, A.C., Bussmann, C.J., van Zeeland, A.A., and Natarajan, A.T. (1986). Distribution of u.v.-induced repair events in higher-order chromatin loops in human and hamster fibroblasts. *Carcinogenesis*, **7**, 995–1002.
62. Mullenders, L.H., Vrieling, H., Venema, J., and van Zeeland, A.A. (1991). Hierarchies of DNA repair in mammalian cells: biological consequences. *Mutat. Res.*, **250**, 223–228.
63. Newman, J.C., Bailey, A.D., and Weiner, A.M. (2006). Cockayne syndrome group B protein (CSB) plays a general role in chromatin maintenance and remodeling. *Proc. Natl. Acad. Sci. U.S.A.*, **103**, 9613–9618.
64. Niedernhofer, L.J., Garinis, G.A., Raams, A., Lalai, A.S., Robinson, A.R., Appeldoorn, E., Odijk, H., Oostendorp, R., Ahmad, A., van Leeuwen, W., Theil, A.F., Vermeulen, W., van der Horst, G.T., Meinecke, P., Kleijer, W.J., Vijg, J., Jaspers, N.G., and Hoeijmakers, J.H. (2006). A new progeroid syndrome reveals that genotoxic stress suppresses the somatotroph axis. *Nature*, **444**, 1038–1043.
65. O'Donovan, A., Davies, A.A., Moggs, J.G., West, S.C., and Wood, R.D. (1994). XPG endonuclease makes the 3′ incision in human DNA nucleotide excision repair. *Nature*, **371**, 432–435.
66. O'Driscoll, M., Ruiz-Perez, V.L., Woods, C.G., Jeggo, P.A., and Goodship, J.A. (2003). A splicing mutation affecting expression of ataxia-telangiectasia and Rad3–related protein (ATR) results in Seckel syndrome. *Nat. Genet.*, **33**, 497–501.
67. Ogi, T. and Lehmann, A.R. (2006). The Y-family DNA polymerase kappa (pol kappa) functions in mammalian nucleotide-excision repair. *Nat. Cell Biol.*, **8**, 640–642.
68. Park, C.H., Mu, D., Reardon, J.T., and Sancar, A. (1995). The general transcription-repair factor TFIIH is recruited to the excision repair complex by the XPA protein independent of the TFIIE transcription factor. *J. Biol. Chem.*, **270**, 4896–4902.
69. Pines, A., Backendorf, C., Alekseev, S., Jansen, J.G., de Gruijl, F.R., Vrieling, H., and Mullenders, L.H. (2008). Differential activity of UV-DDB in mouse keratinocytes and fibroblasts: Impact on DNA repair and UV-induced skin cancer. *DNA Repair (Amst)*.

70. Politi, A., Mone, M.J., Houtsmuller, A.B., Hoogstraten, D., Vermeulen, W., Heinrich, R., and van Driel, R. (2005). Mathematical modeling of nucleotide excision repair reveals efficiency of sequential assembly strategies. *Mol. Cell*, **19**, 679–690.
71. Polo, S.E., Roche, D., and Almouzni, G. (2006). New histone incorporation marks sites of UV repair in human cells. *Cell*, **127**, 481–493.
72. Rademakers, S., Volker, M., Hoogstraten, D., Nigg, A.L., Mone, M.J., van Zeeland, A.A., Hoeijmakers, J.H., Houtsmuller, A.B., and Vermeulen, W. (2003). Xeroderma pigmentosum group A protein loads as a separate factor onto DNA lesions. *Mol. Cell Biol.*, **23**, 5755–5767.
73. Ramanathan, B. and Smerdon, M.J. (1989). Enhanced DNA repair synthesis in hyperacetylated nucleosomes. *J. Biol. Chem.*, **264**, 11026–11034.
74. Rapic-Otrin, V., McLenigan, M.P., Bisi, D.C., Gonzalez, M., and Levine, A.S. (2002). Sequential binding of UV DNA damage binding factor and degradation of the p48 subunit as early events after UV irradiation. *Nucleic Acids Res.*, **30**, 2588–2598.
75. Rebel, H., Kram, N., Westerman, A., Banus, S., van Kranen, H.J., and de Gruijl, F.R. (2005). Relationship between UV-induced mutant p53 patches and skin tumours, analysed by mutation spectra and by induction kinetics in various DNA-repair-deficient mice. *Carcinogenesis*, **26**, 2123–2130.
76. Riedl, T., Hanaoka, F., and Egly, J.M. (2003). The comings and goings of nucleotide excision repair factors on damaged DNA. *EMBO J.*, **22**, 5293–5303.
77. Rockx, D.A., Mason, R., van Hoffen, A., Barton, M.C., Citterio, E., Bregman, D.B., van Zeeland, A.A., Vrieling, H., and Mullenders, L.H. (2000). UV-induced inhibition of transcription involves repression of transcription initiation and phosphorylation of RNA polymerase II. *Proc. Natl. Acad. Sci. U.S.A.*, **97**, 10503–10508.
78. Sarker, A.H., Tsutakawa, S.E., Kostek, S., Ng, C., Shin, D.S., Peris, M., Campeau, E., Tainer, J.A., Nogales, E., and Cooper, P.K. (2005). Recognition of RNA polymerase II and transcription bubbles by XPG, CSB, and TFIIH: insights for transcription-coupled repair and Cockayne Syndrome. *Mol. Cell*, **20**, 187–198.
79. Savitsky, K., Bar-Shira, A., Gilad, S., Rotman, G., Ziv, Y., Vanagaite, L., Tagle, D.A., Smith, S., Uziel, T., Sfez, S., Ashkenazi, M., Pecker, I., Frydman, M., Harnik, R., Patanjali, S.R., Simmmons, A., Clines, G.A., Sartiel, A., Gatti, R.A., Chessa, L., Sanal, O., Lavin, M.F., Jaspers, N.G., Taylor, A.M.,Arlett, C.F., Miki, T., Weissman, S.M., Lovett, M., Collins, F.S., and Shiloh, Y.(1995). A single ataxia telangiectasia gene with a product simillar to PI-3 Kinase. *Science,* **268,** 1749–1753.
80. Selby, C.P. and Sancar, A. (1994). Mechanisms of transcription-repair coupling and mutation frequency decline. *Microbiol. Rev.*, **58**, 317–329.
81. Selby, C.P., Witkin, E.M., and Sancar, A. (1991). *Escherichia coli* mfd mutant deficient in "mutation frequency decline" lacks strand-specific repair: in vitro complementation with purified coupling factor. *Proc. Natl Acad. Sci. U.S.A.*, **88**, 11574–11578.
82. Shivji, K.K., Kenny, M.K., and Wood, R.D. (1992). Proliferating cell nuclear antigen is required for DNA excision repair. *Cell*, **69**, 367–374.
83. Shivji, M.K., Podust, V.N., Hubscher, U., and Wood, R.D. (1995). Nucleotide excision repair DNA synthesis by DNA polymerase epsilon in the presence of PCNA, RFC, and RPA. *Biochemistry*, **34**, 5011–5017.
84. Sijbers, A.M., de Laat, W.L., Ariza, R.R., Biggerstaff, M., Wei, Y.F., Moggs, J.G., Carter, K.C., Shell, B.K., Evans, E., de Jong, M.C., Rademakers, S., de Rooij, J., Jaspers, N.G., Hoeijmakers, J.H., and Wood, R.D. (1996). Xeroderma pigmentosum group F caused by a defect in a structure-specific DNA repair endonuclease. *Cell*, **86**, 811–822.
85. Smerdon, M.J., Lan, S.Y., Calza, R.E., and Reeves, R. (1982). Sodium butyrate stimulates DNA repair in UV-irradiated normal and xeroderma pigmentosum human fibroblasts. *J. Biol. Chem.*, **257**, 13441–13447.
86. Spatz, A., Giglia-Mari, G., Benhamou, S., and Sarasin, A. (2001). Association between DNA repair-deficiency and high level of p53 mutations in melanoma of Xeroderma pigmentosum. *cancer Res.*, **61,** 2480–2486.

87. Stiff, T., Cerosaletti, K., Concannon, P., O'Driscoll, M., and Jeggo, P.A. (2008). Replication independent ATR signalling leads to G2/M arrest requiring Nbs1, 53BP1 and MDC1. *Hum. Mol. Genet.*, **17**, 3247–3253.
88. Stout, G.J., Oosten, M., Acherrat, F.Z., Wit, J., Vermeij, W.P., Mullenders, L.H., Gruijl, F.R., and Backendorf, C. (2005). Selective DNA damage responses in murine Xpa–/–, Xpc–/– and Csb–/– keratinocyte cultures. *DNA Repair (Amst)*, **4**, 1337–1344.
89. Sugasawa, K. and Hanaoka, F. (2007). Sensing of DNA damage by XPC/Rad4: one protein for many lesions. *Nat. Struct. Mol. Biol.*, **14**, 887–888.
90. Sugasawa, K., Ng, J.M., Masutani, C., Iwai, S., van der Spek, P.J., Eker, A.P., Hanaoka, F., Bootsma, D., and Hoeijmakers, J.H. (1998). Xeroderma pigmentosum group C protein complex is the initiator of global genome nucleotide excision repair. *Mol. Cell*, **2**, 223–232.
91. Sugasawa, K., Okuda, Y., Saijo, M., Nishi, R., Matsuda, N., Chu, G., Mori, T., Iwai, S., Tanaka, K., Tanaka, K., and Hanaoka, F. (2005). UV-induced ubiquitylation of XPC protein mediated by UV-DDB-ubiquitin ligase complex. *Cell*, **121**, 387–400.
92. Tang, J.Y., Hwang, B.J., Ford, J.M., Hanawalt, P.C., and Chu, G. (2000). Xeroderma pigmentosum p48 gene enhances global genomic repair and suppresses UV-induced mutagenesis. *Mol. Cell*, **5**, 737–744.
93. Ura, K., Araki, M., Saeki, H., Masutani, C., Ito, T., Iwai, S., Mizukoshi, T., Kaneda, Y., and Hanaoka, F. (2001). ATP-dependent chromatin remodeling facilitates nucleotide excision repair of UV-induced DNA lesions in synthetic dinucleosomes. *EMBO J.*, **20**, 2004–2014.
94. van Gool, A.J., Citterio, E., Rademakers, S., van Os, R., Vermeulen, W., Constantinou, A., Egly, J.M., Bootsma, D., and Hoeijmakers, J.H. (1997). The Cockayne syndrome B protein, involved in transcription-coupled DNA repair, resides in an RNA polymerase II-containing complex. *EMBO J.*, **16**, 5955–5965.
95. van Oosten, M., Rebel, H., Friedberg, E.C., van Steeg, H., van der Horst, G.T., van Kranen, H.J., Westerman, A., van Zeeland, A.A., Mullenders, L.H., and de Gruijl, F.R. (2000). Differential role of transcription-coupled repair in UVB-induced G2 arrest and apoptosis in mouse epidermis. *Proc. Natl Acad. Sci. U.S.A.*, **97**, 11268–11273.
96. van Oosterwijk, M.F., Versteeg, A., Filon, R., van Zeeland, A.A., and Mullenders, L.H. (1996). The sensitivity of Cockayne's syndrome cells to DNA-damaging agents is not due to defective transcription-coupled repair of active genes. *Mol. Cell Biol.*, **16**, 4436–4444.
97. Volker, M., Mone, M.J., Karmakar, P., van Hoffen, A., Schul, W., Vermeulen, W., Hoeijmakers, J.H., van Driel, R., van Zeeland, A.A., and Mullenders, L.H. (2001). Sequential assembly of the nucleotide excision repair factors in vivo. *Mol. Cell*, **8**, 213–224.
98. Wakasugi, M., Reardon, J.T., and Sancar, A. (1997). The non-catalytic function of XPG protein during dual incision in human nucleotide excision repair. *J. Biol. Chem.*, **272**, 16030–16034.
99. Wang, H., Zhai, L., Xu, J., Joo, H.Y., Jackson, S., Erdjument-Bromage, H., Tempst, P., Xiong, Y., and Zhang, Y. (2006). Histone H3 and H4 ubiquitylation by the CUL4-DDB-ROC1 ubiquitin ligase facilitates cellular response to DNA damage. *Mol. Cell*, **22**, 383–394.
100. Wijnhoven, S.W., Kool, H.J., Mullenders, L.H., Slater, R., van Zeeland, A.A., and Vrieling, H. (2001). DMBA-induced toxic and mutagenic responses vary dramatically between NER-deficient Xpa, Xpc and Csb mice. *Carcinogenesis*, **22**, 1099–1106.
101. Yamaizumi, M. and Sugano, T. (1994). U.v.-induced nuclear accumulation of p53 is evoked through DNA damage of actively transcribed genes independent of the cell cycle. *Oncogene*, **9**, 2775–2784.
102. Yarosh, D.B. (2002). Enhanced DNA repair of cyclobutane pyrimidine dimers changes the biological response to UV-B radiation. *Mutat. Res.*, **509**, 221–226.
103. Yu, Y., Teng, Y., Liu, H., Reed, S.H., and Waters, R. (2005). UV irradiation stimulates histone acetylation and chromatin remodeling at a repressed yeast locus. *Proc. Natl Acad. Sci. U.S.A.*, **102**, 8650–8655.

104. Zotter, A., Luijsterburg, M.S., Warmerdam, D.O., Ibrahim, S., Nigg, A., van Cappellen, W.A., Hoeijmakers, J.H., van Driel, R., Vermeulen, W., and Houtsmuller, A.B. (2006). Recruitment of the nucleotide excision repair endonuclease XPG to sites of UV-induced DNA damage depends on functional TFIIH. *Mol. Cell Biol.*, **26**, 8868–8879.
105. Zou, L. and Elledge, S.J. (2003). Sensing DNA damage through ATRIP recognition of RPA-ssDNA complexes. *Science*, **300**, 1542–1548.

Chapter 12
Chromosomal Single-Strand Break Repair

Keith W. Caldecott

Abstract Tens of thousands of cellular single strand breaks (SSBs) arise in cells each day, from attack of deoxyribose and DNA bases by reactive oxygen species and other electrophilic molecules, and from the intrinsic instability of DNA. If not repaired, SSBs can block transcription and replication and can be converted into potentially clastogenic and/or lethal DNA double-strand breaks. SSBs can arise directly by disintegration of damaged deoxyribose, and indirectly as normal intermediates of DNA base excision repair (BER). Here, the molecular mechanism/s and organisation of the DNA repair pathways that remove single strand breaks are reviewed and the connection between defects in these pathways and hereditary neurodegenerative disease are discussed.

Keywords Strand break · XRCC1 · Neurodegeneration

12.1 The Source and Structure of Endogenous DNA Single-Strand Breakage

Single-strand breaks (SSBs) are discontinuities in one strand of the DNA double helix and are usually accompanied by loss of a single nucleotide and by damaged 5′- and/or 3′-termini at the site of the break. One of the commonest sources of SSBs is oxidative attack by endogenous reactive oxygen species (ROS). In the case of free radicals arising from hydrogen peroxide (H_2O_2), a physiologically relevant source of ROS, SSBs arise three orders of magnitude more frequently than DNA double-strand breaks (DSBs) [11]. SSBs can arise directly via disintegration of the oxidised sugar or indirectly during the DNA base excision repair (BER) of oxidised bases, abasic sites, or bases that are damaged or altered in other ways [31, 54, 110]. SSBs can also arise as a result of erroneous or abortive activity of cellular enzymes such

---

K.W. Caldecott (✉)
Genome Damage and Stability Centre, University of Sussex, Falmer, Brighton, BN1 9RQ, UK
e-mail: k.w.caldecott@sussex.ac.uk

K.K. Khanna, Y. Shiloh (eds.), *The DNA Damage Response: Implications on Cancer Formation and Treatment*, DOI 10.1007/978-90-481-2561-6_12,

as DNA topoisomerase 1 (Top1). Top1 creates a "cleavage complex" intermediate containing a DNA nick in order to relax DNA supercoils during transcription and DNA replication [42]. Cleavage complexes are normally very transient and rapidly resealed by Top1, but collision with RNA or DNA polymerases, or close proximity to other types of DNA lesion, can convert these intermediates into Top1-linked SSBs (Top1-SSBs) or DSBs (Top1-DSBs) in which Top1 peptide is covalently linked to the 3′-terminus of the DNA strand break [39, 112].

12.2 DNA Single-Strand Breaks and Cell Fate

Chromosomal SSBs can impact on cell fate in a number of ways if not repaired rapidly or appropriately (Fig. 12.1). The most likely consequence of un-repaired SSBs in proliferating cells is the blockage or collapse of DNA replication forks during S phase, likely leading to DSBs [72, 74]. Cells possess a remarkable ability to accurately repair this type of DSB, using homologous recombination (see below, *SSBR and the Cell Cycle*), but acute increases in cellular SSB levels might saturate this pathway, leading to genetic instability and/or cell death. In non-proliferating

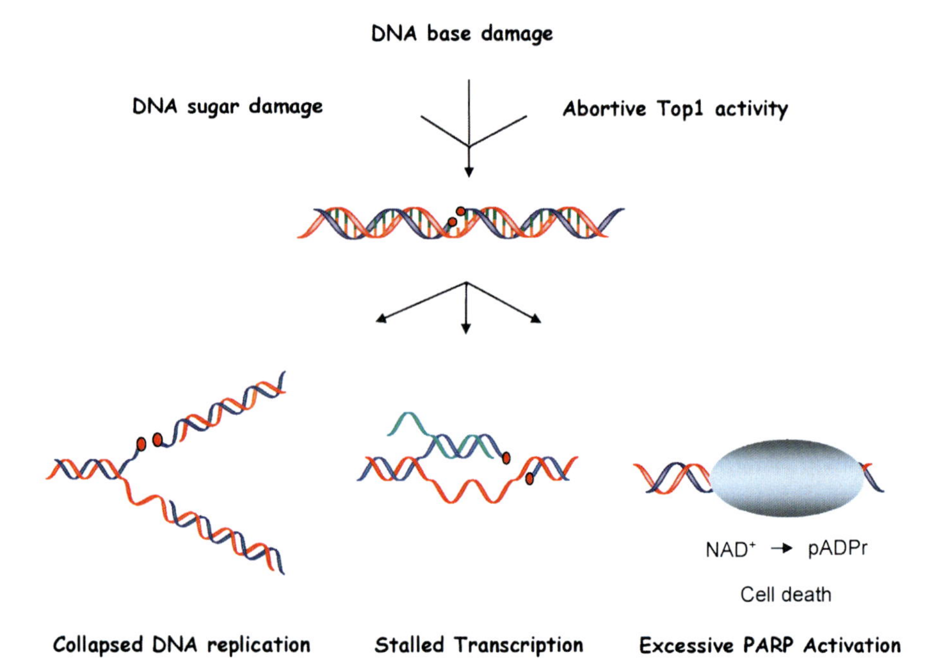

Fig. 1 SSBs and cell fate. SSBs can arise in a variety of ways including directly from disintegration of deoxyribose, indirectly as normal intermediates of BER, or as abortive intermediates of Top1 activity. If not repaired rapidly or appropriately SSBs can collapse replication forks, block transcription, or promote excessive activation of the SSB sensor protein PARP-1 (resulting in cell death). Red circles denote damaged DNA termini

cells, such as post-mitotic neurons, cell death induced by SSBs may involve stalling of RNA polymerases during transcription because SSBs can block RNA polymerase progression in vitro, particularly if they possess damaged termini [10, 66, 158, 159]. Alternatively, under some physiological situations, such as ischemia injury, high levels of SSBs may induce cell death through excessive activation of the SSB sensor protein poly ADP-ribose polymerase-1 (PARP-1) [53, 94]. Under these conditions, prolonged activation of PARP-1 leads to depletion of cellular NAD^+ and ATP and/or release of the apoptosis-inducing factor AIF from mitochondria. This type of cell death is relevant to a number of pathological conditions involving oxidative stress including diabetes, arthritis, and post-ischemic brain or heart damage resulting from stroke or heart attack, respectively.

12.3 Mechanisms of Chromosomal Single-Strand Break Repair (SSBR)

Whilst endogenous SSBs arise from a variety of different sources there is extensive overlap between the enzymes employed to remove these breaks. Consequently, the repair of SSBs from different sources (e.g. those arising during BER versus those arising directly from sugar disintegration) are considered to be sub-pathways of single-strand break repair (SSBR). Most SSBs are repaired by a very rapid global SSBR process that can be divided into four basic steps, which are SSB detection, DNA end processing, DNA gap filling, and DNA ligation (Fig. 12.2).

12.3.1 Detection of SSBs

SSBs arising directly from disintegration of oxidised deoxyribose are primarily detected by poly (ADP-ribose) polymerase-1 (PARP-1), although contributions from other members of PARP super-family are possible [3, 24, 69]. PARP-1 rapidly binds to and is activated by DNA strand breaks and subsequently modifies itself and other target proteins with branched chains of poly (ADP-ribose) (pADPr) of several hundred ADP-ribose units in length. The binding and activity of PARP-1 at DNA breaks is very transient, because poly ADP-ribosylated PARP-1 rapidly dissociates from DNA and pADPr is rapidly degraded by poly (ADP-ribose) glycohydrolase (PARG), thereby restoring PARP-1 (and other poly ADP-ribosylated proteins) to its de-ribosylated state for subsequent rounds of SSB detection and signalling [29]. PARP-1 is also activated at SSB arising indirectly, during BER [34, 93, 104], but it is less clear why such breaks require "detecting" since they arise as part of a co-ordinated repair process in which DNA intermediates might be passed from one enzyme to another in a molecular relay [90, 125, 153]. Perhaps some SSB intermediates become uncoupled from this relay and so require PARP-1 to re-engage the repair machinery. Alternatively, perhaps PARP-1 fulfils a different role during BER, downstream of SSB detection (see below). It is similarly unclear whether or

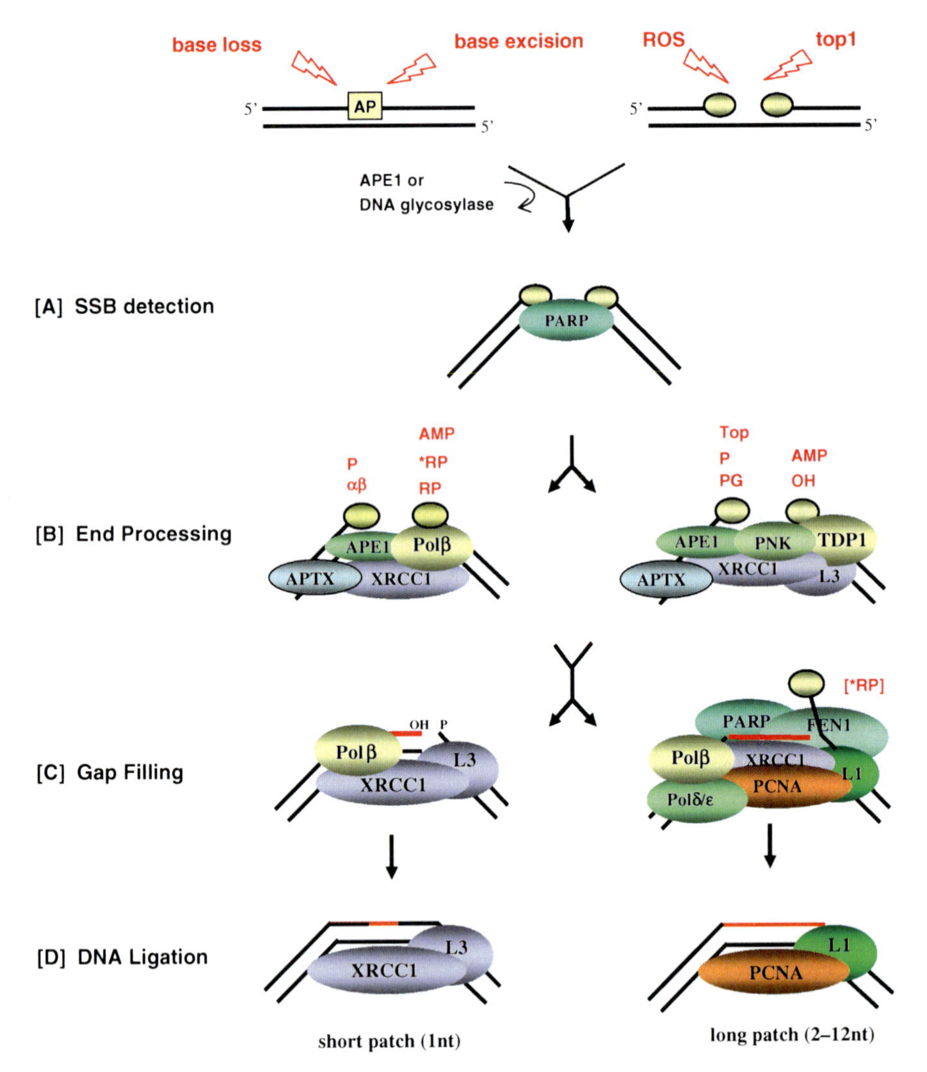

Fig. 2 A model for global SSBR. SSBs can arise indirectly during BER via enzymatic incision at an abasic (AP) site by AP endonuclease (APE1) or the lyase activity of a bifunctional DNA glycosylase (e.g. OGG1, NTH, NEIL1, NEIL2) (*top left*). Abasic sites arise via spontaneous or induced base loss or by enzymatic excision of damaged (e.g. alkylated, oxidized, deaminated) bases. SSBs can also arise directly by disintegration of oxidized deoxyribose or during abortive top1 activity (*top right*). (**A**) SSB detection; PARP-1 binds to an SSB and is activated. PARP activation may have multiple roles (see text for details), including chromatin modification and accumulation of repair factors at the SSBs. PARP then dissociates from SSBs allowing access to the break by other repair proteins. Note that PARG is also required during SSBR to restore PARP-1 to its pre-activated state in preparation for subsequent rounds of SSB detection. (**B**) End processing; *Left*, Indirect SSBs arising during BER possess 5′-RP termini, or oxidized RP (*RP) termini if they have arisen from an oxidized abasic site, if the AP site is cleaved by APE1. Alternatively, SSBs arising during BER possess either 3′-α,β unsaturated aldehyde (αβ) or 3′-phosphate (P) termini if the AP site is cleaved by a DNA glycosylase with β or β/δ elimination, respectively. The damaged 3′- and 5′-termini (*yellow ovals*) are converted to 3′-hydroxyl and 5′-phosphate moieties by APE1 and Pol β, respectively.

not PARP-1 is involved in detecting abortive Top1-SSBs, since such breaks might effectively be detected and signalled by collision with the transcription machinery.

A role for PARP activity in accelerating chromosomal SSBR is suggested both by inhibitor studies [34, 35, 62, 79, 93, 129] and by experiments employing cells selectively depleted or deleted of PARP-1 [33, 43, 77, 143]. PARG is also required for rapid rates of SSBR, suggesting that optimal SSBR rates require that levels of pADPr are dynamic and tightly regulated [43, 46]. PARP-1 may accelerate SSBR by promoting the focal accumulation or stability of SSBR protein complexes at chromosomal SSBs, via interaction with dedicated pADPr binding motifs [18, 38, 76, 87, 101, 108]. Arguably the most important of these is XRCC1 protein, which functions as a molecular scaffold that interacts with, stabilises, and stimulates multiple enzymatic components of the SSBR process [15]. Note that the impact of PARP-1 on focal accumulation of SSBR proteins at chromosomal SSBs does not necessarily reflect their initial recruitment, however, which mechanistically may be different and may require recognition of their cognate DNA substrates [32]. PARP-1 may also accelerate chromosomal SSBR by regulating chromatin structure, since histone proteins are targets for poly ADP-ribosylation and pADPr synthesis can disrupt nucleosomes and can regulate higher-order chromatin compaction [17, 88, 111, 144, 145]. Other possible roles include promoting DNA gap filling during long-patch repair [25, 121, 127], generating ATP for the final step of DNA ligation [100, 107], and inhibiting illegitimate recombination events and/or unwanted nucleolytic activity [80, 104]. Finally, since PARP-1 is also a transcriptional regulator, it remains possible that this enigmatic protein accelerates SSBR rates indirectly, by affecting the level of one or more SSBR proteins [127].

Fig. 2 Note that oxidized dRP termini (*RP) are most likely removed by FEN1 (see below) during long-patch repair. *Right*, SSBs possessing 3′-phosphate (P), 3′-phosphoglycolate (PG), 3′-top1 peptide (Top), or 5′-hydroxyl (OH) termini induced by oxidative damage to deoxyribose (3′-P or 3′-PG, and possibly 5′-OH) or by abortive top1 activity (3′-Top, 3′-P, and 5′-OH), are processed by APE1 (3′-PG and possibly 3′-P), PNK (3′-P and 5′-OH), or TDP1 (3′-Top). Note that some 5′-termini can become prematurely adenylated by DNA ligases resulting in 5′-AMP termini that are repaired by aprataxin (APTX). Also note that the scaffold protein XRCC1 interacts with many of the enzymatic components of end processing reactions, and may stimulate and/or facilitate accumulation of some such factors at SSBs, in vivo. (**C**) Gap filling; *Left*, direct and indirect SSBs can employ Pol β for single nucleotide gap filling (*short patch repair*). *Right*, under some circumstances (e.g. during BER if the 5′-terminus is an oxidized dRP that Pol β cannot remove) Pol β and/or Pol δ/ε may extend gap filling for ~2–12 nt at direct and indirect SSBs (*long patch repair*). Note that, in vivo, XRCC1 may promote the recruitment, accumulation, or stability of Pol β at direct SSBs, and Pol β may conduct this role for XRCC1 at indirect SSBs (during BER). Also note that long patch repair involves the removal of the 5′-terminus as a flap of two or more displaced nucleotides by FEN-1, and is a coordinated reaction that is stimulated by PARP-1 and PCNA. (**D**) DNA ligation; short patch and long patch repair patches are ligated by Lig3α and Lig1, respectively, though considerable redundancy may exist in vivo. Most if not all cellular Lig3α is constitutively bound and stabilized by XRCC1

12.3.2 DNA End Processing

The 3′- and/or 5′-termini of most, if not all, SSBs are "damaged" and must be restored to conventional 3′-hydroxyl (3′-OH) and 5′-phosphate (5′-P) moieties (Fig. 12.2B). End processing is a critical stage of SSBR since DNA strand breaks with abnormal DNA termini are particularly cytotoxic. For example, it is the role of DNA Polymerase β in removing 5′-deoxyribose phosphate (dRP) residues from 5′-termini during BER (see below) that accounts for the importance of this protein in maintaining cellular resistance to DNA base damage, rather than its role as a DNA polymerase during DNA gap filling [133]. In addition, over-expression of DNA glycosylase enzymes that initiate BER can actually increase the sensitivity of cells to DNA base damage, and their deletion decrease this sensitivity, due most likely to steady state increases in the level of SSB intermediates [126, 132, 142]. The cytotoxicity of damaged termini may in part reflect a greater ability to block progression of RNA polymerases, and/or difficulties encountered by cells in dealing with blocked termini at collapsed replication forks.

The variety of different types of damaged termini that can arise is reflected by the number of enzymes involved, including polynucleotide kinase (PNK), AP endonuclease-1 (APE1), DNA polymerase β (Pol β), tyrosyl phosphodiesterase 1 (TDP1), and flap endonuclease-1 (FEN-1). Indirect SSBs created by AP endonuclease (APE1) during BER possess a 5′-deoxyribose phosphate (dRP) that is usually removed by the 5′-dRP lyase activity of Pol β [89, 133]. Under some circumstances however (e.g. if the deoxyribose is oxidised) [31, 135], the damaged 5′ terminus can be displaced during long-patch gap filling and removed as a single-strand flap by flap endonuclease-1 (FEN-1) in a reaction likely stimulated by PCNA (see below and Fig. 12.2) [45, 48, 70, 81]. It is worth noting that SSBs arising during BER may alternatively possess a 3′-phosphate or 3′-α,β-unsaturated aldehyde terminus, rather than a 5′-dRP terminus, if created by cleavage of an abasic site by DNA glycosylase AP lyase activity rather than APE1. These termini are substrates for PNK and APE1, respectively [20, 61, 152]. 3′-phosphate and 3′-phosphoglycolate termini comprise the majority of direct SSBs induced by reactive oxygen species (ROS) and are major substrates for polynucleotide kinase/phosphatase (PNK) [63, 65, 151] and AP endonuclease 1 (APE1) [20, 61, 105, 154, 156], respectively.

Two other types of damaged terminus that warrant attention are 3′- and 5′-termini linked covalently to Topoisomerase 1 (Top1) and AMP, respectively. Top1 associated SSBs (Top1-SSBs) arise from abortive Top1 activity and are processed by tyrosyl phosphodiesterase 1 (TDP1) [114, 158]. TDP1 is strongly implicated in end processing at DSBs in budding yeast [113, 147], but the yeast and human enzyme can also process SSBs, in vitro [39, 113]. It is possible that other components of SSBR such as Lig3α, with which TDP1 interacts in a multi-protein SSBR complex [39, 109], or DNA bending at DNA gaps and nicks (see Fig. 12.1), stimulate TDP1 activity at chromosomal SSBs. The primary substrates for TDP1 are the top1-linked 3′-termini (top1-SSBs) that arise through abortive top1 activity [114, 158], which can be induced by the drug camptothecin (CPT) or by the close proximity of unusual DNA secondary structures or other types of DNA lesion ([16, 112]

and references therein). TDP1 may also process other types of termini to some extent, including 3′-sugar moieties, though these activities have so far only been demonstrated on single-stranded oligonucleotides and DSBs [58, 59, 160]. 3′-end processing by TDP1 creates a 3′-phosphate which, together with the 5′-hydroxyl terminus also present at top1-SSBs, can be processed by PNK [39, 122]. In contrast, 5′-AMP SSBs arise from abortive DNA ligase activity at existing SSBs and are processed by Aprataxin (APTX) [1, 123]. It should be noted, however, that 5′-AMP termini are hypothetical lesions, and their occurrence in vivo remains to be established. Nevertheless, it is intriguing that both TDP1 and Aprataxin appear to process SSBs that are products of failed or abortive activity of endogenous enzymatic activities, particularly since both are mutated in hereditary neurodegenerative diseases, which are discussed below.

Finally, it is important to note that the molecular scaffold protein XRCC1 plays a particularly important role during DNA end processing. XRCC1 directly interacts with PNK [83, 151], Aprataxin [22, 49, 84, 128], and Pol β [18, 73], and indirectly via DNA ligase IIIα with TDP1 [39, 109], and the repair of 5′-hydroxyl and/or 3′-phosphate termini is rate limiting for SSBR in XRCC1-mutant CHO cells and cell extracts [134, 151]. In the case of PNK it is clear that the interaction with XRCC1 stimulates both DNA kinase and DNA phosphatase activity, possibly by increasing damage-discrimination by PNK and by displacing the enzyme from its reaction product [83, 86, 151]. The interaction of XRCC1 with PNK, Aprataxin, and Pol β also promotes the accumulation of these enzymes at oxidative chromosomal breaks induced by H_2O_2 and UVA laser, in vivo [56, 76, 83].

12.3.3 DNA Gap Filling

Once damaged 3′-termini termini at SSB have been restored to their conventional hydroxyl configuration, gap filling can occur (Fig. 12.2C). This often involves insertion of the single nucleotide that is missing at most SSBs, but at some SSBs gap filling may continue for two or more nucleotides, with FEN-1 removing the displaced 5′-residue either one at a time or as a single-stranded flap (see [17] and references therein) (see Fig. 12.2). These two sub-pathways are termed short-patch and long-patch repair, respectively. Early studies employing permeabilized cells suggest that multiple DNA polymerases, including Pol β, are involved in the repair of single-strand breaks, with the choice of DNA polymerase influenced by the source and type of SSB. In vitro studies employing DNA molecules with defined types of break similarly implicate Pol β in gap filling, both at direct SSBs arising from oxidative damage and at indirect SSBs during BER, and show that DNA polymerase δ and/or DNA polymerase ε (Pol δ/ε) can also conduct this role (see [17] and references therein). Direct measurements of chromosomal repair rates in mouse embryonic fibroblasts confirm that Pol β is required for rapid repair MMS-induced SSBs during BER [44, 106]. In contrast, a requirement for Pol β for rapid SSBR rates at oxidative SSBs has not been observed [44]. This may reflect the availability of an alternative short-patch DNA polymerases such as Pol λ or Pol ι for repair of oxidative SSBs, or alternatively the use of long-patch DNA polymerases [7, 12, 47, 148]. A number of accessory proteins may also be important for gap filling during SSBR. For example, PARP-1 and FEN-1 stimulate long-patch gap filling by Pol β in vitro [120, 121], and the RFC/PCNA clamp-loader/clamp may promote long-patch gap filling by Pol δ/ε [45, 52, 70]. It is also noteworthy that XRCC1 interacts with PCNA [41, 146] and Pol β [18, 73] and that XRCC1 is important for the accumulation of RFC, PCNA, and Pol β at direct SSBs induced by UVA laser damage in cells [52, 76].

12.3.4 DNA Ligation

The final step of SSBR is DNA ligation (Fig. 12.2D). Three human DNA ligase genes have been identified (*LIG1*, *LIG3*, and *LIG4*), encoding five different polypeptides (see [40] for recent review). *LIG3* encodes three polypeptides, denoted DNA ligase IIIα (Lig3α), DNA ligase IIIβ, and a mitochondrial isoform (mtLig3). Of these Lig1, Lig3α, and mLig3 are implicated in SSBR. mtLig3 interacts with DNA polymerase γ and is required for integrity of the mitochondrial genome [30, 75], whereas Lig1 and Lig3α appear to be the enzymes of choice during nuclear long patch repair and short patch SSBR, respectively (see [17] and references therein). Lig3α is stable and active as a recombinant protein in vitro, but the nuclear enzyme requires constitutive interaction with XRCC1 for stability and for accumulation at sites of oxidative chromosome damage [19, 95]. In contrast, Lig1 requires interaction with PCNA for accumulation at such sites [95], although one might also expect a dependency on XRCC1, since the latter is required for PCNA accumulation at sites of oxidative damage [76]. It is important to note that our understanding concerning the requirement for specific DNA ligases during SSBR is derived largely from in vitro assays, and that the level of functional overlap between Lig1 and Lig3α during SSBR in living cells remains to be determined.

All Lig3 proteins possess an amino-terminal zinc finger (ZNF) homologous to the two present in PARP-1 [150]. The Lig3 ZNF is largely dispensable for binding and ligation of simple nicked DNA substrates in vitro, although a weak impact (~2-fold) on these activities is observed under some experimental conditions [18, 23, 85, 140, 141]. The dispensability of the ZNF for simple nick ligation most likely reflects the presence of an efficient nick-binding activity within the conserved catalytic core of the enzyme. Indeed, the ZNF has a major impact on binding and ligation of DNA strand breaks located near to unusual secondary structure or other intermediates of DNA repair, suggesting that one function of the ZNF is to broaden the range of substrates targeted by Lig3α [23, 140, 141]. The ZNF may thus be important for repair of chromosomal breaks in repetitive regions of the nuclear and/or mitochondrial genome, or at multiply-damaged sites or clustered lesions. Intriguingly, the ZNF is required for efficient DSB end joining even in the absence of proximal secondary structure, suggesting that this domain may play a major role in *alternative non-homologous end joining*, in which Lig3 is believed to participate [4, 149].

12.4 The Organisation of SSBR

As indicated in the text above, each of the four basic steps of SSBR employ multiple protein-protein interactions. Arguably the most important of these are the interactions between the scaffold protein XRCC1 and enzymatic components of the repair process. XRCC1 accelerates the overall process of SSBR, which it does by stabilising (e.g. Lig3α) [19, 157] and/or stimulating (e.g. PNK) [86, 151] its protein partners. XRCC1 also facilitates the accumulation of its protein partners at sites of chromosomal DNA damage, though whether or not this reflects the initial recruitment of these proteins to DNA breaks, which mechanistically may be very different, remains to be determined. For example, the accumulation of SSBR proteins mediated by XRCC1 may reflect an increased stability or persistence of XRCC1 protein partners at sites of DNA damage, whereas their initial recruitment may be mediated by direct binding to their cognate SSB substrate [32].

It is not obvious why higher eukaryotes employ XRCC1 to accelerate global rates of SSBR, since lower eukaryotes such as yeast cope perfectly well without these proteins. One possibility is that these factors have arisen in response to dramatic increases in genome size. Since the frequency of SSBs arising per cell per day is intrinsically related to DNA content it is possible that higher eukaryotes require faster global rates of SSBR to maintain their steady-state level of SSBs below a certain threshold. The organism with the smallest genome that contains XRCC1 is the amoeba Dictyostelium, which has a genome three times larger (34 Mb) than that of budding yeast, which lacks a recognisable XRCC1.

It is also not clear whether SSBR proteins are organised into one or more constitutive protein complexes or whether SSBR is organised into a sequential series of rapid but transient protein-protein interactions at SSBs. Most likely both are true. For example, the interactions between XRCC1 and PNK [83], aprataxin [22, 84], and APLF [9, 57] are largely constitutive because they are mediated by FHA domain-mediated interactions with threonine/serine residues that are phosphorylated by the constitutively active protein kinase, CK2. Note that these three proteins interact with the same general region of XRCC1, suggesting that they are unlikely to interact with XRCC1 simultaneously. Similarly, levels of Lig3α are reduced ~5-fold in cells lacking XRCC1 or expressing XRCC1 that cannot bind Lig3α, suggesting that at least 80% of cellular Lig3α is constitutively bound to XRCC1 [19]. The level of these complexes does not appear increase greatly, if at all, in response to acute increases in SSB levels. In contrast, the interaction between XRCC1 and PARP-1 clearly increases after DNA damage, consistent with the preferential interaction of XRCC1 with auto-modified PARP-1 [25, 87, 130]. The transient nature of this interaction makes perfect sense, since it enables the accumulation of XRCC1, and consequently those SSBR enzymes that interact with this scaffold protein, at chromosomal breaks.

12.5 SSBR and the Cell Cycle

The SSBR reactions described above represent a process that most likely operates throughout the genome and throughout the cell cycle to rapidly detect and remove the majority of chromosomal SSBs, denoted "global SSBR". However, it is possible that cellular SSBR capacity and mechanism is regulated to some extent depending on cell cycle status. For example, it has been suggested that XRCC1 expression is regulated by FoxM1 and E2F1, transcription factors that regulate a variety of genes required for DNA replication and proliferation [21, 138]. XRCC1 has also been reported to be down-regulated in terminally differentiated muscle cells, though how applicable this observation is to other non-proliferating cells remains to be determined [97]. There is also mounting evidence that S phase cells possess one or more additional SSBR processes for repairing SSBs, possibly once they are encountered by the DNA replication machinery [41, 91, 102, 103, 139] (Fig. 12.3). The molecular detail of this hypothetical process (which we have previously denoted

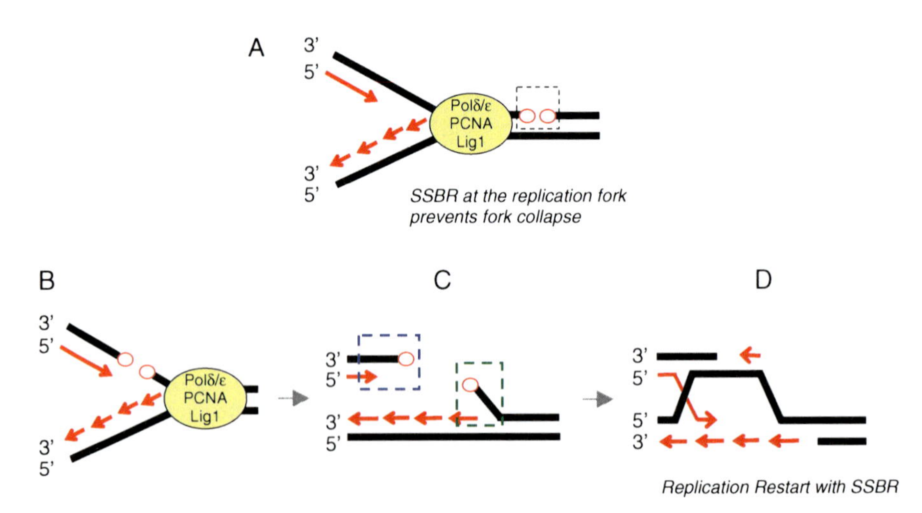

Fig. 3 Models for replication-coupled SSBR. Note that in this example an SSB is present in the leading strand template, but similar outcomes are applicable to a SSB in the lagging strand template. (**A**) An un-repaired SSB (*dotted black box*) is encountered by the DNA replication machinery (*yellow oval*) and is repaired ahead of the fork. (**B–D**) An un-repaired SSB collapses a replication fork (**B**and **C**) creating a "one-ended" DSB (*blue dotted box*) in one sister chromatid and a residual SSB (*green dotted box*) in the other. Note that the SSB shown here possesses a single-strand flap but alternatively might possess a nick or a single-strand gap, depending on the proximity of the terminus of the adjacent nascent strand (in the example depicted this is the nearest Okazaki fragment). The 5′-terminus of the DSB is subsequently repaired and resected (e.g. by CtIP/MRN) in preparation for Rad51-mediated template switching and reformation of the replication fork, (**D**). Repair of the SSB (*dotted green box*) is conducted by global SSBR factors (e.g. TDP1, PNK) and/or structure-specific nucleases (e.g. ERCC1/XPF at 3′-termini as depicted here). Note that PARP-1 and XRCC1 may fulfil an important but undefined role during RC-SSBR

RC-SSBR) [13, 15] is unclear, but is most likely similar to the long-patch sub-pathway of global SSBR described in Fig. 12.2D because many of the enzymes involved in the latter (e.g. Pol δ/ε, PCNA, Lig1, FEN1) are core components of the replication machinery. RC-SSBR most likely operates in conjunction with homologous recombination (HR) because SSBs that are encountered by a replication fork can be converted into DSBs [72, 74] (see Fig. 12.3). Consequently, RC-SSBR and HR may together function as a very effective "back-up" mechanism that ensures that SSBs that escape global repair do not irrevocably block DNA replication. RC-SSBR may also involve some global SSBR factors, such as XRCC1 and possibly components of the global SSBR end processing machinery. However, a key feature of this process may be the availability of structure-specific nucleases (e.g. CtIP & MRN complex, FEN-1, ERCC1/XPF) that can remove a variety of damaged termini. As proposed previously [14, 16], and discussed below, this feature of RC-SSBR/HR may in part account for the lack of genetic instability and cancer in human genetic diseases in which global SSBR proteins are mutated (e.g. SCAN1, AOA1). Thus, whereas global SSBR is most likely the primary determinant of genetic integrity and survival in non-cycling cells in response to SSBs, RC-SSBR (in conjunction with HR) may be the major determinant of these end points in proliferating cells.

12.6 SSBR and Hereditary Genetic Disease

12.6.1 Ataxia with Oculomotor Apraxia Type-1 (AOA1)

Ataxia oculomotor apraxia-1 (AOA1) is an autosomal recessive spinocerebellar ataxia syndrome that resembles Friedreich's ataxia and ataxia-telangiectasia (A-T) neurologically but which lacks extra-neurological features such as immunodeficiency and telangiectasia [2]. Initial studies also suggested that AOA lacks the overt cellular and chromosomal sensitivity to ionising radiation that accompanies A-T [51]. Although subsequent studies have revealed considerable clinical variation within AOA, characteristic features of AOA appear to be variable onset (1–16 years) cerebellar atrophy and ataxia (uncoordinated movement/gait), late axonal peripheral neuropathy, and oculomotor apraxia (limited eye movement on command). In addition to these features, a number of other features appear in AOA, to a greater or lesser extent, including cognitive impairment, hypercholesterolaemia, hypoalbuminaemia, and involuntary movements (choreoathetosis and dystonia). In a large study covering roughly half of the Portugese population, AOA accounted for ~21% (22 patients/11 families) of 107 individuals with recessive spinocerebellar ataxia, a frequency second only to Friedreich's ataxia (38%) [5].

Unlike A-T cells, AOA1 cells are only mildly sensitive, if at all, to ionising radiation and other genotoxins, and exhibit normal cell cycle checkpoint control and chromosome stability following ionising radiation [22]. Consistent with these observations, elevated cancer incidence has not been reported in AOA1 patients. In late 2001 the gene mutated in AOA1 was identified and designated *APTX* [28, 92]. The protein product of *APTX*, denoted aprataxin, was initially characterised as a polypeptide of 342 residues. However, the longest form of this polypeptide may include an extra 14 amino acids at the amino terminus, due to alternative splicing at the 5′-end of the gene [50]. The prevalence and function of the extra sequence is unclear, but since it is located immediately upstream of a putative mitochondrial targeting sequence (unpublished observation) it may influence the sub-cellular localisation of aprataxin.

Aprataxin contains a central histidine triad (HIT) domain and is a member of the HIT domain superfamily of nucleotide hydrolases/transferases. The amino terminus of aprataxin exhibits homology to PNK [92], and like PNK encodes a divergent forkhead-associated (FHA) domain [14]. Two splice variants of APTX were originally reported, one of which encoded a protein lacking the FHA domain. The FHA domains of both aprataxin and PNK facilitate constitutive interactions with CK2-phosphorylated XRCC1 [14, 22, 27, 49, 83, 84, 128]. It is noteworthy that a third member of this divergent FHA domain family has been identified (denoted APLF/PALF/Xip1) and shown to bind CK2-phosphorylated XRCC1 [9, 57, 64]. In fact, all three FHA domain proteins are also sequestered into the DSBR machinery, via FHA domain-mediated interaction with CK2-phosphorylated XRCC4 [22, 57, 71]. It is thus highly likely that aprataxin, PNK, and APLF play roles both in SSBR and DSBR. Aprataxin also associates with PARP-1 and p53, and additionally with the nucleolar proteins nucleolin, nucleophosmin, and UBF-1 [8, 49]. The association and partial co-localisation of aprataxin with nucleolar proteins is also mediated by the FHA domain, though whether these associations are direct or indirect (e.g. via another protein) remains to be determined.

Based on sequence comparisons and substrate specificity, aprataxin appears to represent a discrete branch of the HIT domain super-family [68]. Indeed, although aprataxin can hydrolyse substrates typical of either the Fhit or Hint branch of HIT domain proteins, releasing AMP from diadenosine tetraphosphate or AMP-lysine respectively, its catalytic activity is very low (Kcat <0.03 s-1) [68, 131]. Aprataxin has also been reported to process two types of damaged 3′-termini arising at oxidative DNA breaks; 3′-phosphate and 3′-phosphoglycolate termini, raising the possibility that it is an end processing factor. However, the activity of aprataxin on such substrates is also very low (Kcat $\sim$0.0003–0.003 s-1). A more likely physiological substrate for aprataxin are 5′-AMP termini, at which AMP is covalently linked to 5′-phosphate through a pyrophosphate bond [1, 123]. DNA strand breaks in which the 5′-terminus is linked to AMP are normal intermediates of DNA ligation, but can arise prematurely before 3′-DNA end processing has occurred. Aprataxin can remove AMP from the 5′-terminus of DNA breaks at such "abortive" DNA ligation events, effectively "proof-reading" the DNA ligase reaction (recently reviewed in [124]). In addition to the HIT domain, the C-terminal zinc finger is important for aprataxin activity on 5′-AMP, most likely to increase the affinity and/or specificity of aprataxin for 5′-AMP substrates [123].

12.6.2 Spinocerebellar Ataxia with Axonal Neuropathy-1 (SCAN-1)

Takashima et al. identified an autosomal recessive ataxia that they denoted *s*pinoc*e*rebellar *a*taxia with axonal *n*europathy (SCAN1) [137]. Like AOA1, SCAN1 patients lack chromosomal instability and cancer predisposition and exhibit cerebellar atrophy and peripheral neuropathy. Mild hypercholesterolaemia and hypoalbuminaemia are similarly present. There are differences between the two diseases, however. In particular, SCAN1 has a later age of onset (∼15 years) and does not involve cognitive decline. SCAN1 is associated with a mutation in *t*yrosyl *D*NA *p*hosphodiesterase *1* (TDP1), a DNA end processing protein that removes topoisomerase 1 (top1) peptide from the 3′-termini of Top1-SSBs [114] (see above). The level of top1-strand breaks is induced exogenously by treatment with camptothecin, which increases the half-life of Top1 cleavage complexes, and endogenously by "trapping" of cleavage complexes by other DNA lesions [78, 115–119]. The latter observation may explain why Top1-dependent strand breaks are also induced by H_2O_2 and ionising radiation [26, 37, 82, 99]. It is worth noting, however, that whilst top1-linked termini are likely to be the primary physiological substrate of TDP1, this protein may also process other types of 3′-termini, and possibly 5′-termini, particularly at DSBs [58, 59, 98, 160].

To date, SCAN1 is restricted nine patients from a single Saudi Arabian family, three of which have been examined in detail [137]. Affected individuals possess a common homozygous mutation (H493R) within the second HKD motif of the TDP1 active site. The ability of TDP1 to remove top1 peptide and create 3′-phosphate termini is reduced ∼25-fold and ∼100-fold in experiments employing recombinant mutated TDP1 or SCAN1 cell extracts, respectively [39, 60]. The greater defect in SCAN1 extracts most likely reflects the 2–3-fold reduction in TDP1 protein levels in these cells, due presumably to instability of the mutant protein. Intriguingly, the mutant TDP1 protein can partially process top1-linked termini, removing top1 but itself remaining trapped covalently on the 3′-terminus [55, 60]. The extent to which mutant TDP1 converts top1-breaks into TDP1-breaks is concentration dependent, and the relative proportion of unprocessed (and thus top1-associated) versus partially processed (and thus TDP1-associated) termini that accumulate in SCAN1 cells is unclear. Consequently, the relative contribution of these termini to the neurological defects in SCAN1 is unknown. It is worth noting, however, that in Tdp1$^{-/-}$ mice, which can accumulate Top1-linked, but not Tdp1-linked, termini, there is an age-dependent decrease in cerebellar size, consistent with an impact of Top1-linked termini on neurological function [67].

12.7 Do SSBs and/or DSBs Cause SCAN1 and AOA1?

Both SCAN1 lymphoblastoid cells and Tdp1$^{-/-}$primary mouse post-mitotic neurons display significantly reduced global rates of chromosomal SSBR [37, 39, 67]. This is true not only following camptothecin and IR but also following treatment with H_2O_2. Since H_2O_2 is a physiologically relevant oxidising agent these observations provide compelling support for the idea that SCAN1 neurons might accumulate SSBs or possess higher steady-state levels of this lesion. In contrast, neither non-replicating SCAN1 lymphoblastoid cells nor Tdp1$^{-/-}$ primary mouse post-mitotic neural cells display a measurable defect in DSBR following treatment with camptothecin, IR, or H_2O_2 [37, 39, 67]. This may indicate that the frequency of TDP1 DSB substrates is too rare to detect by existing cell based assays, or that the role of TDP1 is highly redundant during DSBR in non-cycling cells. The situation is more complex in AOA1, however. Aprataxin constitutively interacts with XRCC1 and XRCC4, and AOA1 cells exhibit mild to moderate hypersensitivity to genotoxins, consistent with a role for aprataxin in both SSBR and DSBR [22, 49]. However, AOA1 cells lack a measurable defect in DSBR, and the evidence for a defect in SSBR is conflicting [49, 56, 96]. Thus, whilst it is likely that aprataxin is involved in the repair of both SSBs and DSBs, the relationship between this disease and either or both of these lesions remains to be established.

12.8 SSBs and Cancer

One of the most striking features of SCAN1 and AOA1 is the absence of increased genetic instability and cancer. At first glance this seems surprising, given that unrepaired SSBs can result in potentially clastogenic DSBs during DNA replication. As we have proposed previously [14, 36, 39], and discussed above, this may in part reflect that proliferating cells possess alternative end processing factors (e.g. the structure specific nucleases ERCC1/XPF, FEN1, MRN) that operate during RC-SSBR and HR thereby providing an efficient and accurate mechanism for removing SSBs, most likely once they become DSBs, during DNA replication (see Fig. 12.3). Consistent with this idea, lymphoblastoid cells from SCAN1 individuals display a 3-fold increased frequency of "spontaneous" sister chromatid exchange, a hallmark of homologous recombination [39]. In short, it is likely that the roles of TDP1 and AOA1 are redundant in proliferating cells, at physiological levels of SSBs at least. Genetic instability in SCAN1 and AOA1 may be further limited by committing those few cells in which a DSB avoids repair, and/or is involved in a genetic rearrangement, to apoptosis. These arguments do not exclude an impact of SSBR on cancer frequency in other circumstances, however. Mutations in proteins with a role in RC-SSBR, or with a more extensive involvement in global SSBR such that the steady-state levels of SSBs is elevated above that which HR can tolerate, may impact on genetic instability. Whilst many such mutations are likely to be embryonic lethal, because they are incompatible with rapid cell proliferation during embryonic development, hypomorphic or somatic mutations in such genes could impact on cancer incidence. For example, the possible correlation between cancer incidence and a number of XRCC1 polymorphisms are the focus of intense epidemiological investigation, and mutations in Pol β have been identified in ~30% of human tumours, some of which can affect polymerase fidelity and/or activity during SSBR [136].

12.9 SSBs and Neurodegeneration

The existence of dedicated pathways for dealing with un-repaired SSBs during S phase may explain why SCAN1 and AOA1 are not associated with measurably increased genetic instability and cancer. However, this cannot explain why the impact of these diseases on post-mitotic cells is largely restricted to the nervous system. One possibility is that neurons may be more dependent on TDP1 and aprataxin for DNA end processing than are other post-mitotic cells, due to a more limited availability of alternative end processing factors. Another factor we have previously postulated is the high level of oxidative stress encountered by the nervous system, which consumes ~20% of inhaled oxygen and possesses low levels of antioxidant enzymes ([6, 14, 36] and references therein). In addition, there is a high transcriptional demand in post-mitotic neurons, which may further increase the dependency of these cells on SSBR. Finally, the limited regenerative capacity of neurons, compared to other non-cycling cell types that are more readily replaced by precursors, may render this tissue particularly sensitive to cell dysfunction or loss. Future analyses employing mouse models of these diseases will hopefully shed light on these critical issues.

Acknowledgments I thank the MRC, BBSRC, and EU for financial support, and members of my laboratory for comments and suggestions. I apologize to my colleagues for work I have was not able to reference in this article, due to restrictions in citation numbers.

References

1. Ahel, I., U. Rass, S. F. El-Khamisy, S. Katyal, P. M. Clements, P. J. McKinnon, K. W. Caldecott, and S. C. West. 2006. The neurodegenerative disease protein aprataxin resolves abortive DNA ligation intermediates. Nature **443**:713–6.
2. Aicardi, J., C. Barbosa, E. Andermann, F. Andermann, R. Morcos, Q. Ghanem, Y. Fukuyama, Y. Awaya, and P. Moe. 1988. Ataxia-ocular motor apraxia: a syndrome mimicking ataxia-telangiectasia. Ann Neurol **24**:497–502.
3. Ame, J. C., C. Spenlehauer, and G. de Murcia. 2004. The PARP superfamily. Bioessays **26**:882–93.
4. Audebert, M., B. Salles, and P. Calsou. 2004. Involvement of poly(ADP-ribose) polymerase-1 and XRCC1/DNA ligase III in an alternative route for DNA double-strand breaks rejoining. J Biol Chem **279**:55117–26.
5. Barbot, C., P. Coutinho, R. Chorao, C. Ferreira, J. Barros, I. Fineza, K. Dias, J. Monteiro, A. Guimaraes, P. Mendonca, M. do Ceu Moreira, and J. Sequeiros. 2001. Recessive ataxia with ocular apraxia: review of 22 Portuguese patients. Arch Neurol **58**:201–5.
6. Barzilai, A., G. Rotman, and Y. Shiloh. 2002. ATM deficiency and oxidative stress: a new dimension of defective response to DNA damage. DNA Repair **1**:3–25.
7. Bebenek, K., A. Tissier, E. G. Frank, J. P. McDonald, R. Prasad, S. H. Wilson, R. Woodgate, and T. A. Kunkel. 2001. 5′-Deoxyribose phosphate lyase activity of human DNA polymerase iota in vitro. Science **291**:2156–9.
8. Becherel, O. J., N. Gueven, G. W. Birrell, V. Schreiber, A. Suraweera, B. Jakob, G. Taucher-Scholz, and M. F. Lavin. 2006. Nucleolar localization of aprataxin is dependent on interaction with nucleolin and on active ribosomal DNA transcription. Hum Mol Genet **15**:2239–49.

9. Bekker-Jensen, S., K. Fugger, J. R. Danielsen, I. Gromova, M. Sehested, J. Celis, J. Bartek, J. Lukas, and N. Mailand. 2007. Human Xip1 (C2orf13) is a novel regulator of cellular responses to DNA strand breaks. J Biol Chem **282**:19638–43.
10. Bendixen, C., B. Thomsen, J. Alsner, and O. Westergaard. 1990. Camptothecin-stabilized topoisomerase I-DNA adducts cause premature termination of transcription. Biochemistry **29**:5613–9.
11. Bradley, M. O., and K. W. Kohn. 1979. X-ray induced DNA double-strand break production and repair in mammalian cells as measured by neutral filter elution. Nucleic Acids Res **7**:793–804.
12. Braithwaite, E. K., P. S. Kedar, L. Lan, Y. Y. Polosina, K. Asagoshi, V. P. Poltoratsky, J. K. Horton, H. Miller, G. W. Teebor, A. Yasui, and S. H. Wilson. 2005. DNA polymerase lambda protects mouse fibroblasts against oxidative DNA damage and is recruited to sites of DNA damage/repair. J Biol Chem **280**:31641–7.
13. Caldecott, K. W. 2001. Mammalian DNA single-strand break repair: an X-ra(y)ted affair. Bioessays **23**:447–55.
14. Caldecott, K. W. 2003. DNA single-strand break repair and spinocerebellar ataxia. Cell **112**:7–10.
15. Caldecott, K. W. 2003. XRCC1 and DNA strand break repair. DNA Repair (Amst) **2**:955–69.
16. Caldecott, K. W. 2004. DNA single-strand breaks and neurodegeneration. DNA Repair (Amst) **3**:875–82.
17. Caldecott, K. W. 2006. Mammalian single-strand break repair: Mechanisms and links with chromatin. DNA Repair (Amst).
18. Caldecott, K. W., S. Aoufouchi, P. Johnson, and S. Shall. 1996. XRCC1 polypeptide interacts with DNA polymerase beta and possibly poly (ADP-ribose) polymerase, and DNA ligase III is a novel molecular 'nick-sensor' in vitro. Nucleic Acids Res **24**:4387–94.
19. Caldecott, K. W., J. D. Tucker, L. H. Stanker, and L. H. Thompson. 1995. Characterization of the XRCC1-DNA ligase III complex in vitro and its absence from mutant hamster cells. Nucleic Acids Res **23**:4836–43.
20. Chen, D. S., T. Herman, and B. Demple. 1991. Two distinct human DNA diesterases that hydrolyze 3′-blocking deoxyribose fragments from oxidized DNA. Nucleic Acids Res **19**:5907–14.
21. Chen, D., Z. Yu, Z. Zhu, and C. D. Lopez. 2008. E2F1 regulates the base excision repair gene XRCC1 and promotes DNA repair. J Biol Chem **283**:15381–9
22. Clements, P. M., C. Breslin, E. D. Deeks, P. J. Byrd, L. Ju, P. Bieganowski, C. Brenner, M. C. Moreira, A. M. Taylor, and K. W. Caldecott. 2004. The ataxia-oculomotor apraxia 1 gene product has a role distinct from ATM and interacts with the DNA strand break repair proteins XRCC1 and XRCC4. DNA Repair (Amst) **3**:1493–502.
23. Cotner-Gohara, E., I. K. Kim, A. E. Tomkinson, and T. Ellenberger. 2008. Two DNA binding and nick recognition modules in human DNA ligase III. J Biol Chem **283**:10764–72.
24. D'Amours, D., S. Desnoyers, I. D'Silva, and G. G. Poirier. 1999. Poly(ADP-ribosyl)ation reactions in the regulation of nuclear functions. Biochem J **342(Pt 2)**:249–68.
25. Dantzer, F., R. G. de La, J. Menissier-de Murcia, Z. Hostomsky, G. de Murcia, and V. Schreiber. 2000. Base excision repair is impaired in mammalian cells lacking poly(ADP-ribose) polymerase-1. Biochemistry **39**:7559–7569.
26. Daroui, P., S. D. Desai, T. K. Li, A. A. Liu, and L. F. Liu. 2004. Hydrogen peroxide induces topoisomerase I-mediated DNA damage and cell death. J Biol Chem **279**: 14587–94.
27. Date, H., S. Igarashi, Y. Sano, T. Takahashi, T. Takahashi, H. Takano, S. Tsuji, M. Nishizawa, and O. Onodera. 2004. The FHA domain of aprataxin interacts with the C-terminal region of XRCC1. Biochem Biophys Res Commun **325**:1279–85.
28. Date, H., O. Onodera, H. Tanaka, K. Iwabuchi, K. Uekawa, S. Igarashi, R. Koike, T. Hiroi, T. Yuasa, Y. Awaya, T. Sakai, T. Takahashi, H. Nagatomo, Y. Sekijima, I. Kawachi, Y. Takiyama, M. Nishizawa, N. Fukuhara, K. Saito, S. Sugano, and S. Tsuji. 2001. Early-onset

ataxia with ocular motor apraxia and hypoalbuminemia is caused by mutations in a new HIT superfamily gene. Nat Genet **29**:184–8.
29. Davidovic, L., M. Vodenicharov, E. B. Affar, and G. G. Poirier. 2001. Importance of poly(ADP-ribose) glycohydrolase in the control of poly(ADP-ribose) metabolism. Exp Cell Res **268**:7–13.
30. De, A., and C. Campbell. 2007. A novel interaction between DNA ligase III and DNA polymerase gamma plays an essential role in mitochondrial DNA stability. Biochem J **402**:175–86.
31. Demple, B., and M. S. DeMott. 2002. Dynamics and diversions in base excision DNA repair of oxidized abasic lesions. Oncogene **21**:8926–34.
32. Dianov, G. L., and J. L. Parsons. 2007. Co-ordination of DNA single strand break repair. DNA Repair (Amst) **6**:454–60.
33. Ding, R., Y. Pommier, V. H. Kang, and M. Smulson. 1992. Depletion of poly(ADP-ribose) polymerase by antisense RNA expression results in a delay in DNA strand break rejoining. J Biol Chem **267**:12804–12.
34. Durkacz, B. W., O. Omidiji, D. A. Gray, and S. Shall. 1980. (ADP-ribose)n participates in DNA excision repair. Nature **283**:593–6.
35. Durkacz, B. W., S. Shall, and J. Irwin. 1981. The effect of inhibition of (ADP-ribose)n biosynthesis on DNA repair assayed by the nucleoid technique. Eur J Biochem **121**:65–9.
36. El-Khamisy, S. F., and K. W. Caldecott. 2006. TDP1-dependent DNA single-strand break repair and neurodegeneration. Mutagenesis **21**:219–24.
37. El-Khamisy, S. F., E. Hartsuiker, and K. W. Caldecott. 2007. TDP1 facilitates repair of ionizing radiation-induced DNA single-strand breaks. DNA Repair (Amst) **6**:1485–95.
38. El-Khamisy, S. F., M. Masutani, H. Suzuki, and K. W. Caldecott. 2003. A requirement for PARP-1 for the assembly or stability of XRCC1 nuclear foci at sites of oxidative DNA damage. Nucleic Acids Res **31**:5526–33.
39. El-Khamisy, S. F., G. M. Saifi, M. Weinfeld, F. Johansson, T. Helleday, J. R. Lupski, and K. W. Caldecott. 2005. Defective DNA single-strand break repair in spinocerebellar ataxia with axonal neuropathy-1. Nature **434**:108–113.
40. Ellenberger, T., and A. E. Tomkinson. 2008. Eukaryotic DNA ligases: structural and functional insights. Annu Rev Biochem **77**:313–38.
41. Fan, J., M. Otterlei, H. K. Wong, A. E. Tomkinson, and D. M. Wilson, 3rd. 2004. XRCC1 co-localizes and physically interacts with PCNA. Nucleic Acids Res **32**:2193–201.
42. Fischhaber, P. L., V. L. Gerlach, W. J. Feaver, Z. Hatahet, S. S. Wallace, and E. C. Friedberg. 2002. Human DNA polymerase kappa bypasses and extends beyond thymine glycols during translesion synthesis in vitro, preferentially incorporating correct nucleotides. J Biol Chem **277**:37604–37611.
43. Fisher, A., H. Hochegger, S. Takeda, and K. W. Caldecott. 2007. Poly (ADP-ribose) Polymerase-1 accelerates single-strand break repair in concert with poly (ADP-ribose) glycohydrolase. Mol Cell Biol **27**:5597–605.
44. Fortini, P., B. Pascucci, F. Belisario, and E. Dogliotti. 2000. DNA polymerase beta is required for efficient DNA strand break repair induced by methyl methanesulfonate but not by hydrogen peroxide. Nucleic Acids Res **28**:3040–6.
45. Frosina, G., P. Fortini, O. Rossi, F. Carrozzino, G. Raspaglio, L. S. Cox, D. P. Lane, A. Abbondandolo, and E. Dogliotti. 1996. Two pathways for base excision repair in mammalian cells. J Biol Chem **271**:9573–8.
46. Gao, H., D. L. Coyle, M. L. Meyer-Ficca, R. G. Meyer, E. L. Jacobson, Z. Q. Wang, and M. K. Jacobson. 2007. Altered poly(ADP-ribose) metabolism impairs cellular responses to genotoxic stress in a hypomorphic mutant of poly(ADP-ribose) glycohydrolase. Exp Cell Res **313**:984–96.
47. Garcia-Diaz, M., K. Bebenek, T. A. Kunkel, and L. Blanco. 2001. Identification of an intrinsic 5′-deoxyribose-5-phosphate lyase activity in human DNA polymerase lambda: a possible role in base excision repair. J Biol Chem **276**:34659–63.

48. Gary, R., K. Kim, H. L. Cornelius, M. S. Park, and Y. Matsumoto. 1999. Proliferating cell nuclear antigen facilitates excision in long-patch base excision repair. J Biol Chem **274**:4354–63.
49. Gueven, N., O. J. Becherel, A. W. Kijas, P. Chen, O. Howe, J. H. Rudolph, R. Gatti, H. Date, O. Onodera, G. Taucher-Scholz, and M. F. Lavin. 2004. Aprataxin, a novel protein that protects against genotoxic stress. Hum Mol Genet **13**:1081–93.
50. Habeck, M., C. Zuhlke, K. H. Bentele, S. Unkelbach, W. Kress, K. Burk, E. Schwinger, and Y. Hellenbroich. 2004. Aprataxin mutations are a rare cause of early onset ataxia in Germany. J Neurol **251**:591–4.
51. Hannan, M. A., D. Sigut, M. Waghray, and G. G. Gascon. 1994. Ataxia-ocular motor apraxia syndrome: an investigation of cellular radiosensitivity of patients and their families. J Med Genet **31**:953–6.
52. Hashiguchi, K., Y. Matsumoto, and A. Yasui. 2007. Recruitment of DNA repair synthesis machinery to sites of DNA damage/repair in living human cells. Nucleic Acids Res **35**: 2913–23.
53. Heeres, J. T., and P. J. Hergenrother. 2007. Poly(ADP-ribose) makes a date with death. Curr Opin Chem Biol **11**:644–53.
54. Hegde, M. L., T. K. Hazra, and S. Mitra. 2008. Early steps in the DNA base excision/single-strand interruption repair pathway in mammalian cells. Cell Res **18**:27–47.
55. Hirano, R., H. Interthal, C. Huang, T. Nakamura, K. Deguchi, K. Choi, M. B. Bhattacharjee, K. Arimura, F. Umehara, S. Izumo, J. L. Northrop, M. A. Salih, K. Inoue, D. L. Armstrong, J. J. Champoux, H. Takashima, and C. F. Boerkoel. 2007. Spinocerebellar ataxia with axonal neuropathy: consequence of a Tdp1 recessive neomorphic mutation? Embo J **26**: 4732–43.
56. Hirano, M., A. Yamamoto, T. Mori, L. Lan, T. A. Iwamoto, M. Aoki, K. Shimada, Y. Furiya, S. Kariya, H. Asai, A. Yasui, T. Nishiwaki, K. Imoto, N. Kobayashi, T. Kiriyama, T. Nagata, N. Konishi, Y. Itoyama, and S. Ueno. 2007. DNA single-strand break repair is impaired in aprataxin-related ataxia. Ann Neurol **61**:162–74.
57. Iles, N., S. Rulten, S. F. El-Khamisy, and K. W. Caldecott. 2007. APLF (C2orf13) is a novel human protein involved in the cellular response to chromosomal DNA strand breaks. Mol Cell Biol **27**:3793–803.
58. Inamdar, K. V., J. J. Pouliot, T. Zhou, S. P. Lees-Miller, A. Rasouli-Nia, and L. F. Povirk. 2002. Conversion of phosphoglycolate to phosphate termini on 3′ overhangs of DNA double strand breaks by the human tyrosyl-DNA phosphodiesterase hTdp1. J Biol Chem **277**:27162–8.
59. Interthal, H., H. J. Chen, and J. J. Champoux. 2005. Human Tdp1 cleaves a broad spectrum of substrates, including phosphoamide linkages. J Biol Chem **280**:36518–28.
60. Interthal, H., H. J. Chen, T. E. Kehl-Fie, J. Zotzmann, J. B. Leppard, and J. J. Champoux. 2005. SCAN1 mutant Tdp1 accumulates the enzyme – DNA intermediate and causes camptothecin hypersensitivity. Embo J **24**:2224–33.
61. Izumi, T., T. K. Hazra, I. Boldogh, A. E. Tomkinson, M. S. Park, S. Ikeda, and S. Mitra. 2000. Requirement for human AP endonuclease 1 for repair of 3′-blocking damage at DNA single-strand breaks induced by reactive oxygen species. Carcinogenesis **21**: 1329–34.
62. James, M. R., and A. R. Lehmann. 1982. Role of poly(adenosine diphosphate ribose) in deoxyribonucleic acid repair in human fibroblasts. Biochemistry **21**:4007–13.
63. Jilani, A., D. Ramotar, C. Slack, C. Ong, X. M. Yang, S. W. Scherer, and D. D. Lasko. 1999. Molecular cloning of the human gene, PNKP, encoding a polynucleotide kinase 3′-phosphatase and evidence for its role in repair of DNA strand breaks caused by oxidative damage. J Biol Chem **274**:24176–86.
64. Kanno, S., H. Kuzuoka, S. Sasao, Z. Hong, L. Lan, S. Nakajima, and A. Yasui. 2007. A novel human AP endonuclease with conserved zinc-finger-like motifs involved in DNA strand break responses. Embo J **26**:2094–103.

65. Karimi-Busheri, F., G. Daly, P. Robins, B. Canas, D. J. Pappin, J. Sgouros, G. G. Miller, H. Fakhrai, E. M. Davis, M. M. Le Beau, and M. Weinfeld. 1999. Molecular characterization of a human DNA kinase. J Biol Chem **274**:24187–94.
66. Kathe, S. D., G. P. Shen, and S. S. Wallace. 2004. Single-stranded breaks in DNA but not oxidative DNA base damages block transcriptional elongation by RNA polymerase II in HeLa cell nuclear extracts. J Biol Chem. **279**:18511–20.
67. Katyal, S., S. F. el-Khamisy, H. R. Russell, Y. Li, L. Ju, K. W. Caldecott, and P. J. McKinnon. 2007. TDP1 facilitates chromosomal single-strand break repair in neurons and is neuroprotective in vivo. Embo J **26**:4720–31.
68. Kijas, A. W., J. L. Harris, J. M. Harris, and M. F. Lavin. 2006. Aprataxin forms a discrete branch in the HIT (histidine triad) superfamily of proteins with both DNA/RNA binding and nucleotide hydrolase activities. J Biol Chem **281**:13939–48.
69. Kim, M. Y., T. Zhang, and W. L. Kraus. 2005. Poly(ADP-ribosyl)ation by PARP-1: 'PAR-laying' NAD+ into a nuclear signal. Genes Dev **19**:1951–67.
70. Klungland, A., and T. Lindahl. 1997. Second pathway for completion of human DNA base excision-repair: reconstitution with purified proteins and requirement for DNase IV (FEN1). Embo J **16**:3341–8.
71. Koch, C. A., R. Agyei, S. Galicia, P. Metalnikov, P. O'Donnell, A. Starostine, M. Weinfeld, and D. Durocher. 2004. Xrcc4 physically links DNA end processing by polynucleotide kinase to DNA ligation by DNA ligase IV. Embo J **23**:3874–85.
72. Kouzminova, E. A., and A. Kuzminov. 2006. Fragmentation of replicating chromosomes triggered by uracil in DNA. J Mol Biol **355**:20–33.
73. Kubota, Y., R. A. Nash, A. Klungland, P. Schar, D. E. Barnes, and T. Lindahl. 1996. Reconstitution of DNA base excision-repair with purified human proteins: interaction between DNA polymerase beta and the XRCC1 protein. EMBO J. **15**:6662–70.
74. Kuzminov, A. 2001. Single-strand interruptions in replicating chromosomes cause double-strand breaks. Proc Natl Acad Sci USA **98**:8241–6.
75. Lakshmipathy, U., and C. Campbell. 2001. Antisense-mediated decrease in DNA ligase III expression results in reduced mitochondrial DNA integrity. Nucleic Acids Res **29**: 668–76.
76. Lan, L., S. Nakajima, Y. Oohata, M. Takao, S. Okano, M. Masutani, S. H. Wilson, and A. Yasui. 2004. In situ analysis of repair processes for oxidative DNA damage in mammalian cells. Proc Natl Acad Sci USA **101**:13738–43.
77. Le Page, F., V. Schreiber, C. Dherin, G. De Murcia, and S. Boiteux. 2003. Poly(ADP-ribose) polymerase-1 (PARP-1) is required in murine cell lines for base excision repair of oxidative DNA damage in the absence of DNA polymerase beta. J Biol Chem **278**:18471–7.
78. Lebedeva, N., P. Auffret Vander Kemp, M. A. Bjornsti, O. Lavrik, and S. Boiteux. 2006. Trapping of DNA topoisomerase I on nick-containing DNA in cell free extracts of *Saccharomyces cerevisiae*. DNA Repair (Amst) **5**:799–809.
79. Lehmann, A. R., and B. C. Broughton. 1984. Poly(ADP-ribosylation) reduces the steady-state level of breaks in DNA following treatment of human cells with alkylating agents. Carcinogenesis **5**:117–9.
80. Lindahl, T., M. S. Satoh, G. G. Poirier, and A. Klungland. 1995. Post-translational modification of poly(ADP-ribose) polymerase induced by DNA strand breaks. Trends Biochem Sci **20**:405–11.
81. Liu, Y., W. A. Beard, D. D. Shock, R. Prasad, E. W. Hou, and S. H. Wilson. 2005. DNA polymerase beta and flap endonuclease 1 enzymatic specificities sustain DNA synthesis for long patch base excision repair. J Biol Chem **280**:3665–74.
82. Liu, C., J. J. Pouliot, and H. A. Nash. 2004. The role of TDP1 from budding yeast in the repair of DNA damage. DNA Repair (Amst) **3**:593–601.
83. Loizou, J. I., S. F. El-Khamisy, A. Zlatanou, D. J. Moore, D. W. Chan, J. Qin, S. Sarno, F. Meggio, L. A. Pinna, and K. W. Caldecott. 2004. The protein kinase CK2 facilitates repair of chromosomal DNA single-strand breaks. Cell **117**:17–28.

84. Luo, H., D. W. Chan, T. Yang, M. Rodriguez, B. P. Chen, M. Leng, J. J. Mu, D. Chen, Z. Songyang, Y. Wang, and J. Qin. 2004. A new XRCC1-containing complex and its role in cellular survival of methyl methanesulfonate treatment. Mol Cell Biol **24**:8356–65.
85. Mackey, Z. B., C. Niedergang, J. M. Murcia, J. Leppard, K. Au, J. Chen, G. de Murcia, and A. E. Tomkinson. 1999. DNA ligase III is recruited to DNA strand breaks by a zinc finger motif homologous to that of poly(ADP-ribose) polymerase. Identification of two functionally distinct DNA binding regions within DNA ligase III. J Biol Chem **274**:21679–87.
86. Mani, R. S., M. Fanta, F. Karimi-Busheri, E. Silver, C. A. Virgen, K. W. Caldecott, C. E. Cass, and M. Weinfeld. 2007. XRCC1 stimulates polynucleotide kinase by enhancing its damage discrimination and displacement from DNA repair intermediates. J Biol Chem **282**:28004–13.
87. Masson, M., C. Niedergang, V. Schreiber, S. Muller, J. Menissier-de Murcia, and G. de Murcia. 1998. XRCC1 is specifically associated with poly(ADP-ribose) polymerase and negatively regulates its activity following DNA damage. Mol Cell Biol **18**:3563–71.
88. Mathis, G., and F. R. Althaus. 1987. Release of core DNA from nucleosomal core particles following (ADP-ribose)n-modification in vitro. Biochem Biophys Res Commun **143**: 1049–54.
89. Matsumoto, Y., and K. Kim. 1995. Excision of deoxyribose phosphate residues by DNA polymerase beta during DNA repair. Science **269**:699–702.
90. Mol, C. D., T. Izumi, S. Mitra, and J. A. Tainer. 2000. DNA-bound structures and mutants reveal abasic DNA binding by APE1 DNA repair and coordination. Nature **403**:451–6.
91. Moore, D. J., R. M. Taylor, P. Clements, and K. W. Caldecott. 2000. Mutation of a BRCT domain selectively disrupts DNA single-strand break repair in noncycling Chinese hamster ovary cells. Proc Natl Acad Sci USA **97**:13649–54.
92. Moreira, M. C., C. Barbot, N. Tachi, N. Kozuka, E. Uchida, T. Gibson, P. Mendonca, M. Costa, J. Barros, T. Yanagisawa, M. Watanabe, Y. Ikeda, M. Aoki, T. Nagata, P. Coutinho, J. Sequeiros, and M. Koenig. 2001. The gene mutated in ataxia-ocular apraxia 1 encodes the new HIT/Zn-finger protein aprataxin. Nat Genet **29**:189–93.
93. Morgan, W. F., and J. E. Cleaver. 1983. Effect of 3-aminobenzamide on the rate of ligation during repair of alkylated DNA in human fibroblasts. Cancer Res **43**:3104–7.
94. Moroni, F. 2008. Poly(ADP-ribose)polymerase 1 (PARP-1) and postischemic brain damage. Curr Opin Pharmacol **8**:96–103.
95. Mortusewicz, O., U. Rothbauer, M. C. Cardoso, and H. Leonhardt. 2006. Differential recruitment of DNA ligase I and III to DNA repair sites. Nucleic Acids Res **34**: 3523–32.
96. Mosesso, P., M. Piane, F. Palitti, G. Pepe, S. Penna, and L. Chessa. 2005. The novel human gene aprataxin is directly involved in DNA single-strand-break repair. Cell Mol Life Sci **62**:485–91.
97. Narciso, L., P. Fortini, D. Pajalunga, A. Franchitto, P. Liu, P. Degan, M. Frechet, B. Demple, M. Crescenzi, and E. Dogliotti. 2007. Terminally differentiated muscle cells are defective in base excision DNA repair and hypersensitive to oxygen injury. Proc Natl Acad Sci USA **104**:17010–5.
98. Nitiss, K. C., M. Malik, X. He, S. W. White, and J. L. Nitiss. 2006. Tyrosyl-DNA phosphodiesterase (Tdp1) participates in the repair of Top2-mediated DNA damage. Proc Natl Acad Sci USA **103**:8953–8.
99. Nitiss, J. L., K. C. Nitiss, A. Rose, and J. L. Waltman. 2001. Overexpression of type I topoisomerases sensitizes yeast cells to DNA damage. J Biol Chem **276**:26708–14.
100. Oei, S. L., and M. Ziegler. 2000. ATP for the DNA ligation step in base excision repair is generated from poly(ADP-ribose). J Biol Chem **275**:23234–9.
101. Okano, S., L. Lan, K. W. Caldecott, T. Mori, and A. Yasui. 2003. Spatial and temporal cellular responses to single-strand breaks in human cells. Mol Cell Biol **23**:3974–81.

102. Otterlei, M., E. Warbrick, T. A. Nagelhus, T. Haug, G. Slupphaug, M. Akbari, P. A. Aas, K. Steinsbekk, O. Bakke, and H. E. Krokan. 1999. Post-replicative base excision repair in replication foci. EMBO J **18**:3834–44.
103. Parlanti, E., G. Locatelli, G. Maga, and E. Dogliotti. 2007. Human base excision repair complex is physically associated to DNA replication and cell cycle regulatory proteins. Nucleic Acids Res **35**:1569–77.
104. Parsons, J. L., I. I. Dianova, S. L. Allinson, and G. L. Dianov. 2005. Poly(ADP-ribose) polymerase-1 protects excessive DNA strand breaks from deterioration during repair in human cell extracts. Febs J **272**:2012–21.
105. Parsons, J. L., I. I. Dianova, and G. L. Dianov. 2004. APE1 is the major 3′-phosphoglycolate activity in human cell extracts. Nucleic Acids Res. **32**:3531–6.
106. Pascucci, B., M. T. Russo, M. Crescenzi, M. Bignami, and E. Dogliotti. 2005. The accumulation of MMS-induced single strand breaks in G1 phase is recombinogenic in DNA polymerase beta defective mammalian cells. Nucleic Acids Res **33**:280–8.
107. Petermann, E., M. Ziegler, and S. L. Oei. 2003. ATP-dependent selection between single nucleotide and long patch base excision repair. DNA Repair (Amst) **2**:1101–14.
108. Pleschke, J. M., H. E. Kleczkowska, M. Strohm, and F. R. Althaus. 2000. Poly(ADP-ribose) binds to specific domains in DNA damage checkpoint proteins. J Biol Chem **275**: 40974–80.
109. Plo, I., Z. Y. Liao, J. M. Barcelo, G. Kohlhagen, K. W. Caldecott, M. Weinfeld, and Y. Pommier. 2003. Association of XRCC1 and tyrosyl DNA phosphodiesterase (Tdp1) for the repair of topoisomerase I-mediated DNA lesions. DNA Repair (Amst) **2**: 1087–100.
110. Pogozelski, W. K., and T. D. Tullius. 1998. Oxidative strand scission of nucleic acids: routes initiated by hydrogen abstraction from the sugar moiety. Chem Rev **98**:1089–108.
111. Poirier, G. G., G. deMurcia, J. Jongstra-Bilen, C. Niedergang, and P. Mandel. 1982. Poly(ADP-ribosy)lation of polynucleosomes causes relaxation of chromatin structure. Proc Natl Acad Sci USA **79**:3423–7.
112. Pommier, Y., C. Redon, V. A. Rao, J. A. Seiler, O. Sordet, H. Takemura, S. Antony, L. Meng, Z. Liao, G. Kohlhagen, H. Zhang, and K. W. Kohn. 2003. Repair of and checkpoint response to topoisomerase I-mediated DNA damage. Mutat Res **532**:173–203.
113. Pouliot, J. J., C. A. Robertson, and H. A. Nash. 2001. Pathways for repair of topoisomerase I covalent complexes in *Saccharomyces cerevisiae*. Genes Cells **6**:677–87.
114. Pouliot, J. J., K. C. Yao, C. A. Robertson, and H. A. Nash. 1999. Yeast gene for a Tyr-DNA phosphodiesterase that repairs topoisomerase I complexes. Science **286**:552–5.
115. Pourquier, P., A. A. Pilon, G. Kohlhagen, A. Mazumder, A. Sharma, and Y. Pommier. 1997. Trapping of mammalian topoisomerase I and recombinations induced by damaged DNA containing nicks or gaps. Importance of DNA end phosphorylation and camptothecin effects. J Biol Chem **272**:26441–7.
116. Pourquier, P., and Y. Pommier. 2001. Topoisomerase I-mediated DNA damage. Adv Cancer Res **80**:189–216.
117. Pourquier, P., L. M. Ueng, J. Fertala, D. Wang, H. J. Park, J. M. Essigmann, M. A. Bjornsti, and Y. Pommier. 1999. Induction of reversible complexes between eukaryotic DNA topoisomerase I and DNA-containing oxidative base damages. 7,8-dihydro-8-oxoguanine and 5-hydroxycytosine. J Biol Chem **274**:8516–23.
118. Pourquier, P., L. M. Ueng, G. Kohlhagen, A. Mazumder, M. Gupta, K. W. Kohn, and Y. Pommier. 1997. Effects of uracil incorporation, DNA mismatches, and abasic sites on cleavage and religation activities of mammalian topoisomerase I. J Biol Chem **272**:7792–6.
119. Pourquier, P., J. L. Waltman, Y. Urasaki, N. A. Loktionova, A. E. Pegg, J. L. Nitiss, and Y. Pommier. 2001. Topoisomerase I-mediated cytotoxicity of *N*-methyl-*N*'-nitro-*N*-nitrosoguanidine: trapping of topoisomerase I by the O6-methylguanine. Cancer Res **61**:53–8.

120. Prasad, R., G. L. Dianov, V. A. Bohr, and S. H. Wilson. 2000. FEN1 stimulation of DNA polymerase beta mediates an excision step in mammalian long patch base excision repair. J Biol Chem **275**:4460–6.
121. Prasad, R., O. I. Lavrik, S. J. Kim, P. Kedar, X. P. Yang, B. J. Vande Berg, and S. H. Wilson. 2001. DNA polymerase beta-mediated long patch base excision repair. Poly(ADP-ribose)polymerase-1 stimulates strand displacement DNA synthesis. J Biol Chem **276**:32411–4.
122. Rasouli-Nia, A., F. Karimi-Busheri, and M. Weinfeld. 2004. Stable down-regulation of human polynucleotide kinase enhances spontaneous mutation frequency and sensitizes cells to genotoxic agents. Proc Natl Acad Sci USA **101**:6905–10.
123. Rass, U., I. Ahel, and S. C. West. 2007. Actions of aprataxin in multiple DNA repair pathways. J Biol Chem **282**:9469–74.
124. Rass, U., I. Ahel, and S. C. West. 2007. Defective DNA repair and neurodegenerative disease. Cell **130**:991–1004.
125. Rice, P. A. 1999. Holding damaged DNA together [news; comment]. Nat Struct Biol **6**: 805–6.
126. Roth, R. B., and L. D. Samson. 2002. 3-Methyladenine DNA glycosylase-deficient Aag null mice display unexpected bone marrow alkylation resistance. Cancer Res **62**: 656–60.
127. Sanderson, R. J., and T. Lindahl. 2002. Down-regulation of DNA repair synthesis at DNA single-strand interruptions in poly(ADP-ribose) polymerase-1 deficient murine cell extracts. DNA Repair (Amst) **1**:547–58.
128. Sano, Y., H. Date, S. Igarashi, O. Onodera, M. Oyake, T. Takahashi, S. Hayashi, M. Morimatsu, H. Takahashi, T. Makifuchi, N. Fukuhara, and S. Tsuji. 2004. Aprataxin, the causative protein for EAOH is a nuclear protein with a potential role as a DNA repair protein. Ann Neurol **55**:241–9.
129. Schraufstatter, I. U., P. A. Hyslop, D. B. Hinshaw, R. G. Spragg, L. A. Sklar, and C. G. Cochrane. 1986. Hydrogen peroxide-induced injury of cells and its prevention by inhibitors of poly(ADP-ribose) polymerase. Proc Natl Acad Sci USA **83**:4908–12.
130. Schreiber, V., J. C. Ame, P. Dolle, I. Schultz, B. Rinaldi, V. Fraulob, J. Menissier-de Murcia, and G. de Murcia. 2002. Poly(ADP-ribose) polymerase-2 (PARP-2) is required for efficient base excision DNA repair in association with PARP-1 and XRCC1. J Biol Chem **277**: 23028–36.
131. Seidle, H. F., P. Bieganowski, and C. Brenner. 2005. Disease-associated mutations inactivate AMP-lysine hydrolase activity of Aprataxin. J Biol Chem **280**:20927–31.
132. Sobol, R. W., M. Kartalou, K. H. Almeida, D. F. Joyce, B. P. Engelward, J. K. Horton, R. Prasad, L. D. Samson, and S. H. Wilson. 2003. Base excision repair intermediates induce p53-independent cytotoxic and genotoxic responses. J Biol Chem **278**:39951–9.
133. Sobol, R. W., R. Prasad, A. Evenski, A. Baker, X. P. Yang, J. K. Horton, and S. H. Wilson. 2000. The lyase activity of the DNA repair protein beta-polymerase protects from DNA-damage-induced cytotoxicity. Nature **405**:807–10.
134. Sossou, M., C. Flohr-Beckhaus, I. Schulz, F. Daboussi, B. Epe, and J. P. Radicella. 2005. APE1 overexpression in XRCC1-deficient cells complements the defective repair of oxidative single strand breaks but increases genomic instability. Nucleic Acids Res **33**:298–306.
135. Sung, J. S., and B. Demple. 2006. Roles of base excision repair subpathways in correcting oxidized abasic sites in DNA. Febs J **273**:1620–9.
136. Sweasy, J. B., T. Lang, and D. DiMaio. 2006. Is base excision repair a tumor suppressor mechanism? Cell Cycle **5**:250–9.
137. Takashima, H., C. F. Boerkoel, J. John, G. M. Saifi, M. A. Salih, D. Armstrong, Y. Mao, F. A. Quiocho, B. B. Roa, M. Nakagawa, D. W. Stockton, and J. R. Lupski. 2002. Mutation of TDP1, encoding a topoisomerase I-dependent DNA damage repair enzyme, in spinocerebellar ataxia with axonal neuropathy. Nat Genet **32**:267–72.

138. Tan, Y., P. Raychaudhuri, and R. H. Costa. 2007. Chk2 mediates stabilization of the FoxM1 transcription factor to stimulate expression of DNA repair genes. Mol Cell Biol **27**: 1007–16.
139. Taylor, R. M., D. J. Moore, J. Whitehouse, P. Johnson, and K. W. Caldecott. 2000. A cell cycle-specific requirement for the XRCC1 BRCT II domain during mammalian DNA strand break repair. Mol Cell Biol **20**:735–40.
140. Taylor, R. M., C. J. Whitehouse, and K. W. Caldecott. 2000. The DNA ligase III zinc finger stimulates binding to DNA secondary structure and promotes end joining. Nucleic Acids Res **28**:3558–63.
141. Taylor, R. M., J. Whitehouse, E. Cappelli, G. Frosina, and K. W. Caldecott. 1998. Role of the DNA ligase III zinc finger in polynucleotide binding and ligation. Nucleic Acids Res **26**:4804–10.
142. Trivedi R. N., X. H. Wang, E. Jelezcova, E. M. Goellner, J. Tang, and R. W. Sobol. 2008. Human methyl purine DNA glycosylase and DNA polymerase {beta} expression collectively predict sensitivity to temozolomide. Mol Pharmacol **74**:505–16.
143. Trucco, C., F. J. Oliver, G. de Murcia, and J. Menissier-de Murcia. 1998. DNA repair defect in poly(ADP-ribose) polymerase-deficient cell lines. Nucleic Acids Res **26**:2644–9.
144. Tulin, A., and A. Spradling. 2003. Chromatin loosening by poly(ADP)-ribose polymerase (PARP) at Drosophila puff loci. Science **299**:560–2.
145. Tulin, A., D. Stewart, and A. C. Spradling. 2002. The *Drosophila heterochromatic* gene encoding poly(ADP-ribose) polymerase (PARP) is required to modulate chromatin structure during development. Genes Dev **16**:2108–19.
146. Uchiyama, Y., Y. Suzuki, and K. Sakaguchi. 2008. Characterization of plant XRCC1 and its interaction with proliferating cell nuclear antigen. Planta **227**:1233–41.
147. Vance, J. R., and T. E. Wilson. 2002. Yeast Tdp1 and Rad1-Rad10 function as redundant pathways for repairing Top1 replicative damage. Proc Natl Acad Sci USA **99**:13669–74.
148. Vermeulen, C., M. Verwijs-Janssen, P. Cramers, A. C. Begg, and C. Vens. 2007. Role for DNA polymerase beta in response to ionizing radiation. DNA Repair (Amst) **6**:202–12.
149. Wang, H., B. Rosidi, R. Perrault, M. Wang, L. Zhang, F. Windhofer, and G. Iliakis. 2005. DNA ligase III as a candidate component of backup pathways of nonhomologous end joining. Cancer Res **65**:4020–30.
150. Wei, Y. F., P. Robins, K. Carter, K. Caldecott, D. J. Pappin, G. L. Yu, R. P. Wang, B. K. Shell, R. A. Nash, P. Schar, et al. 1995. Molecular cloning and expression of human cDNAs encoding a novel DNA ligase IV and DNA ligase III, an enzyme active in DNA repair and recombination. Mol Cell Biol **15**:3206–16.
151. Whitehouse, C. J., R. M. Taylor, A. Thistlethwaite, H. Zhang, F. Karimi-Busheri, D. D. Lasko, M. Weinfeld, and K. W. Caldecott. 2001. XRCC1 stimulates human polynucleotide kinase activity at damaged DNA termini and accelerates DNA single-strand break repair. Cell **104**:107–17.
152. Wiederhold, L., J. B. Leppard, P. Kedar, F. Karimi-Busheri, A. Rasouli-Nia, M. Weinfeld, A. E. Tomkinson, T. Izumi, R. Prasad, S. H. Wilson, S. Mitra, and T. K. Hazra. 2004. AP endonuclease-independent DNA base excision repair in human cells. Mol Cell **15**:209–20.
153. Wilson, S. H., and T. A. Kunkel. 2000. Passing the baton in base excision repair [news]. Nat Struct Biol **7**:176–8.
154. Winters, T. A., W. D. Henner, P. S. Russell, A. McCullough, and T. J. Jorgensen. 1994. Removal of 3′-phosphoglycolate from DNA strand-break damage in an oligonucleotide substrate by recombinant human apurinic/apyrimidinic endonuclease 1. Nucleic Acids Res **22**:1866–73.
155. Winters, T. A., M. Weinfeld, and T. J. Jorgensen. 1992. Human HeLa cell enzymes that remove phosphoglycolate 3′-end groups from DNA. Nucleic Acids Res **20**:2573–80.
156. Wong, H. K., D. Kim, B. A. Hogue, D. R. McNeill, and D. M. Wilson, 3rd. 2005. DNA damage levels and biochemical repair capacities associated with XRCC1 deficiency. Biochemistry **44**:14335–43.

157. Yang, S. W., A. B. Burgin, Jr., B. N. Huizenga, C. A. Robertson, K. C. Yao, and H. A. Nash. 1996. A eukaryotic enzyme that can disjoin dead-end covalent complexes between DNA and type I topoisomerases. Proc Natl Acad Sci USA **93**:11534–9.
158. Zhou, W., and P. W. Doetsch. 1993. Effects of abasic sites and DNA single-strand breaks on prokaryotic RNA polymerases. Proc Natl Acad Sci USA **90**:6601–5.
159. Zhou, W., and P. W. Doetsch. 1994. Transcription bypass or blockage at single-strand breaks on the DNA template strand: effect of different 3′ and 5′ flanking groups on the T7 RNA polymerase elongation complex. Biochemistry **33**:14926–34.
160. Zhou, T., J. W. Lee, H. Tatavarthi, J. R. Lupski, K. Valerie, and L. F. Povirk. 2005. Deficiency in 3′-phosphoglycolate processing in human cells with a hereditary mutation in tyrosyl-DNA phosphodiesterase (TDP1). Nucleic Acids Res **33**:289–97.

Chapter 13
Mouse Models of DNA Double Strand Break Repair Deficiency and Cancer

Sachin Katyal and Peter J. McKinnon

13.1 Overview

Understanding the physiological requirements for DNA repair is of paramount importance in establishing the links between genotoxic stress, development and disease. Defective DNA repair can lead to a variety of systemic pathology including cancer, neurodegeneration and immune defects. The genetically modified mouse is a critical resource for analysis of the effects of genotoxic stress and the tissue-specific requirements for DNA repair pathways. In the following we will outline how mouse models have aided our understanding of the connections between genotoxic stress and cancer and how they will contribute to therapeutic approaches for cancer treatment.

13.2 Introduction

A variety of exogenous and endogenous agents can affect cellular genomic integrity. DNA strand breaks can occur through exposure to ionizing radiation, radiomimetic drugs or from endogenous agents including free radicals that result from oxidative cellular metabolism. DNA double strand breaks (DSBs) also occur during normal physiological events such as meiotic recombination or V(D)J recombination during maturation of the immune system. The repair of DSBs is crucial for maintenance of genomic integrity and misrepair or translocation of the breaks can result in chromosomal rearrangements or the acquisition of tumorigenic mutations that can result in cancer.

The use of in vivo systems such as knockout mice has emerged as a critical tool for DNA repair research and cancer biology. The mouse genome is very similar to the human genome as it shares ~99% of all human genes [124]. Advanced mouse genomic technologies have provided a critical experimental biological system to

P.J. McKinnon (✉)
Department of Genetics and Tumor Cell Biology, St Jude Children's Research Hospital, Memphis, TN 38105, USA
e-mail: peter.mckinnon@stjude.org

K.K. Khanna, Y. Shiloh (eds.), *The DNA Damage Response: Implications on Cancer Formation and Treatment*, DOI 10.1007/978-90-481-2561-6_13,

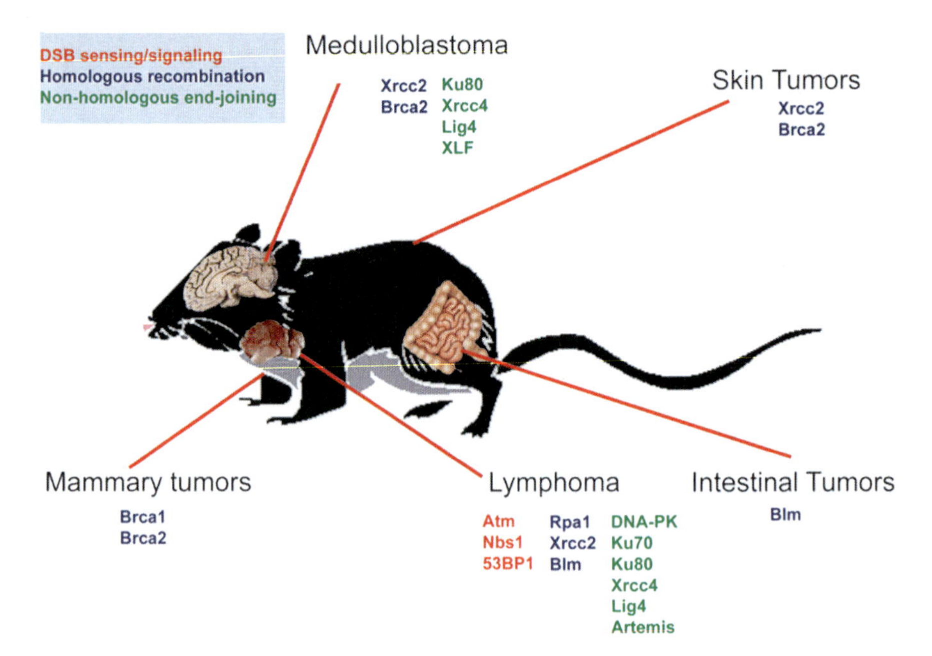

Fig. 13.1 Schematic of various tissues susceptible to tumorigenesis in DNA repair deficient mouse strains. Genes found to be involved in suppressing tumor formation in various tissues are indicated and are color-coded based on their involvement in specific DNA double strand break repair pathways

address key questions concerning DNA repair pathways. The use of germline or tissue-specific gene inactivation and transgenic techniques allows us to perform a systematic functional analysis of DNA repair factors during cellular function and tissue development and maintenance. Furthermore, the relatively short gestational time for mouse embryogenesis makes this system an ideal tool for investigating the biology of DNA repair pathways. Multiple mouse lines have been developed that mimic heritable human DNA repair syndromes, which have greatly facilitated our ability in developing biologically relevant treatment strategies for patients, afflicted with these syndromes.

In this chapter, we will illustrate mouse lines that have been developed to augment our understanding of DNA strand break repair biology and tumorigenesis. We will focus our discussion of mouse models that are deficient in DNA double strand break repair (DSBR), and how these have been important for understanding DNA repair and cancer (Fig. 13.1). While many of these lines result in embryonic lethality, some result in similar tumorigenic profiles as seen in human syndromes; while other lines, when crossed onto a p53-deficient background, rescue embryonic lethal phenotypes but increase tumorigenesis.

Mouse models resulting from germline gene inactivation were originally established in the early nineties and became widely used in the following decade [37]. Currently, a great deal of emphasis is now being placed upon the analysis of tissue-specific gene deletion to selectively inactivate DNA damage response genes and to

Table 13.1 Examples of tissue specific Cre lines

Cre Line (promoter)	Spatiotemporal expression of Cre recombinase	References
CamKII	Hippocampus and cortex	Minichiello et al. [84]
GFAP	Neuroprogenitors cells, all CNS except Purkinje neurons	Zhuo et al. [135]
L7/Pcp2	Retinal inner nuclear layer and cerebellar Purkinje neurons	Campsall et al. [14]
Nestin	Neuroprogenitor cells	Bates et al. [6], Isaka et al. [57]
Thy1	Post-mitotic neurons	Gelman et al. [44]
MBP	Oligodendrocytes	Hisahara et al. [55]
K14	Epithelial	Vasioukhin et al. [115]
WAP	Mammary gland	Wagner et al. [119]
CD19	Hematopoietic: pre-B cells	Rickert et al. [98]
Rag-1	Hematopoietic: B cells and T cells	McCormack et al. [81]
Tie1	Endothelial	Gustafsson et al. [51]
Tie2	Endothelial	Forde et al. [32]
Mx1	Hepatocyte	Rohlmann et al. [100]
Albumin	Hepatocyte	Postic and Magnuson [96]
Col1a1	Osteoblast	Dacquin et al. [23]
Ckmm (Mck)	Muscle	Bruning et al. [10]
Rip	Pancreatic beta cells	Postic et al. [97]

generate refined cancer models [42, 59]. This approach often relies on combining a Cre recombinase with an allele that contains recombination recognition sites (LoxP sites) so that in the presence of Cre the gene region is deleted, thereby either fully or partially inactivating the gene. A main utility of tissue-specific deletion is that this approach can generally bypass embryonic lethality that typically results from germline deletion, and is also a convenient way of generating compound mutants. Because conditional gene inactivation can be tissue-specific or inducible either in tissues or throughout development or postnatally, this approach to gene manipulation is extremely powerful and is continuing to provide an enormous amount of new insight into DNA repair processes. Table 13.1 lists a representative sample of various available mouse lines in which Cre recombinase is expressed in a tissue-specific manner.

13.3 DNA DSB Repair Pathways

Repair of DNA DSBs can occur via homologous recombination (HR) repair and/or non-homologous end joining (NHEJ). The choice of repair pathway that becomes activated is linked to the cell cycle, with HR being available in S- and G2-phases while NHEJ occurring throughout the cell cycle [29]. HR is an error-free DSBR mechanism that utilizes a sister chromatid as DNA template to achieve precise repair [52, 125]. HR-mediated repair requires initial DSB processing via the Mre11-Rad50-Nbs1 (MRN) complex to generate a 3′-single stranded DNA which is bound by the Rad51 recombinase complex (Xrcc2, Xrcc3, RAD51B, RAD51C, RAD51D), Rad52, Rad54, RPA, Brca1 and Brca2 to facilitate template homology and strand invasion.

In contrast to HR, NHEJ modifies the two broken DNA ends so that they are compatible for direct ligation and is the predominant pathway for repairing DSBs in non-cycling mammalian cells [4, 52, 114, 127]. NHEJ is an error-prone DSB repair mechanism whereby broken DNA ends are recognized by and bind the Ku70 and Ku80 heterodimeric proteins, which then mediates recruitment of the DNA-PK catalytic subunit (DNA-PKcs) [13]. DNA end-ligation ultimately occurs via the Xrcc4-Lig4 complex (X4/L4) with Cernunnos/XLF augmenting ligation efficiency [1, 12]. Some NHEJ-mediated repair may also involve additional factors such as the DNA nuclease Artemis [52, 71, 74, 114, 127].

Coincident with DNA DSB repair is the initiation of specific signaling events to activate cell cycle arrest and facilitate accurate DNA repair. DNA DSBs induce rapid phosphorylation of the histone H2A variant, H2AX, which allows for recruitment of a number of proteins including NBS1, MDC1 and 53BP1 in proximity to the breaksite [16, 103, 121]. This serves to amplify DNA DSB signaling by the ataxia-telangiectasia mutated (ATM) kinase to modulate the activity of a variety of downstream proteins involved in cell-cycle checkpoints and DNA repair. ATM is a key protein kinase that function at the top of a signaling pathway that coordinates cell cycle arrest after DNA damage [107]. ATM-dependent phosphorylation of cell-cycle checkpoint effectors such as p53, Chk2, SMC1, BRCA1, and NBS1 occurs to activate G1, intra-S and G2/M checkpoints. Detailed descriptions of ATM-signaling after DNA damage have recently been published [63, 70, 107, 108, 121].

A critical checkpoint protein in this process is the tumor suppressor protein p53 which functions primarily as a transcriptional regulator of genes involved in multiple response pathways [7, 117]. Cells or mice that are deficient in p53 may be unable to undergo cell-cycle arrest and/or apoptosis in response to DNA damage or perturbation of the cell cycle [7, 64, 85, 118]. As such, many lethal mouse phenotypes that arise from inactivation of DSB repair genes can be rescued in a p53-null background. However, this rescue is often associated with cancer, as the combined failure in cell cycle checkpoint control and apoptosis will allow genomically unstable cells to proliferate, thereby accumulating chromosomal rearrangements, which can lead to tumorigenesis. However, the use of $p53^{-/-}$ in combination with other DNA repair deficient lines has inherent pitfalls as long term studies are precluded by the fact that $p53^{-/-}$ mice will often develop tumors, such as thymic lymphoma, between 3 and 6 months of age. For a detailed description of $p53^{-/-}$ mouse lines, please refer to [78, 79, 93]. More refined approaches in utilizing p53-deficient lines for tumor formation studies have been developed, including the use of a conditional p53 allele, wherein a tissue-specific promoter driving Cre recombinase expression may allow for user-selected inactivation of p53 based on tissue type or spatiotemporal need [59, 60].

13.4 Mouse Models of DSBR Deficiency and Tumorigenesis

Murine inactivation of key DSBR proteins often leads to embryonic lethality, highlighting the critical importance of these factors for cellular proliferation in the maintenance of genomic integrity [37]. While the early embryonic lethality associated with these deficient lines often precludes significant insight into the role of DSBR during development and tumorigenesis, analysis of cell lines from these mutants are often useful as a source of experimental materials. However, intercrossing these mutant animals with *p53*$^{-/-}$ lines often rescues the observed phenotype to varying degrees, and in some cases, the impact that prolonged DSBR deficiency towards cellular homeostasis and tumor formation can be determined.

13.4.1 Inactivation of Homologous Recombination in the Mouse

Rad51: Inactivation of Rad51, a key component of the Rad51 recombinase complex, results in pre-implantation defects that fail to yield any *Rad51*$^{-/-}$ embryos [113]. Furthermore, *Rad51*$^{-/-}$ embryonic stem cells fail to establish, indicating that Rad51 is essential for cell proliferation. An additional Rad51-deficient mouse line, *Rad51*$^{M1-/-}$ (which harbors an out-of-frame deletion after amino acid 73), yields embryos that arrest by the egg cylinder stage (E5.5) whereby proliferation defects and widespread cellular apoptosis results in embryonic lethality [75]. Trophoblast-like cells isolated from *Rad51*$^{M1-/-}$ embryos also show hypersensitivity to gamma-radiation. *Rad51*$^{M1-/-}$ ES cells also fail to establish due to proliferation defects. Interestingly, *Rad51*$^{M1-/-}$ mice crossed onto a *p53*$^{-/-}$ background (*Rad51*$^{M-/-}$*p53*$^{-/-}$) extend embryonic development to E9.5. Although loss of p53 in this *Rad51*$^{-/-}$ background partially rescues the in vivo proliferation defect, *Rad51*$^{M-/-}$*p53*$^{-/-}$mouse embryonic fibroblasts (MEFs) fail to grow. The differences in these two published *Rad51*$^{-/-}$ knockout lines may relate to differing mouse backgrounds.

Inactivation of Rad51B (Rad51L1) also results in early embryonic lethality due to severe proliferation defects, with embryos progressing until E5.5 prior to maternal resorption [109]. A part of this lethality occurs via p53-dependent cell death as *Rad51B*$^{-/-}$*p53*$^{-/-}$ embryos are detected until E9.5.

Rad51C$^{-/-}$ mice show post-implantation lethality during embryogenesis [66]. However, generation of a Rad51C hemizygote mouse consisting of a *Rad51C* hypomorphic allele and a *Rad51C*-deficient allele (*Rad51C*$^{ko/neo}$) results in a mouse expressing only 5–30% of the normal levels of RAD51C protein. Although *Rad51C*$^{ko/neo}$ mice are viable, males and females show varied infertility due to meiotic recombination defects. Interestingly, almost half of all spermatocytes showed pronounced γH2AX at pachytene suggesting pronounced accumulation of unrepaired DNA breaks, while oocytes showed chromatid cohesion defects.

Rad51D$^{-/-}$ mice show mid-gestational embryonic lethality with death occurring by E8.5–E11.5. *Rad51D*$^{-/-}$ embryos are significantly smaller than their control counterparts thereby indicating proliferative defects [95]. Indeed, *Rad51D*$^{-/-}$ MEFs fail to propagate beyond one generation in culture. Combined loss of *Rad51D* and *p53* extends the embryonic phenotype by approximately 6 days, and these are sensitive to genotoxic agents such as inter-strand cross-links, and exhibit profound chromosome instability.

Replication Protein A (RPA): In contrast to complete inactivation, a knock-in approach was used to generate an RPA mutant line (*Rpa1*^{689C}), which results in an L230P amino acid substitution that disrupts one of three RPA DNA binding domains without affecting protein folding and stability [120]. *Rpa1*$^{689C/689C}$ mice show very early embryonic lethality and fail to develop past ~E3.5 due to failure in cell proliferation. Interestingly, RPA heterozygous mice (*Rpa1*$^{689C/+}$) were viable but showed premature morbidity between 4 and 14 months of age, compared to wild type controls. Histopathological analysis showed that *Rpa1*$^{689C/+}$ mice developed highly invasive lymphomas and altered bone marrow hematopoiesis. Loss of heterozygosity analysis of these tumors showed retention of the functional wild type copy of RPA consistent with its requirement in cellular proliferation. Array comparative genomic hybridization (CGH) analysis showed widespread chromosomal alterations including gains on chromosomes 6, 15 and 16, involving regions containing several oncogenes, most notably *Myc* (chr. 15) and *Raf1* (chr. 6) frequently involved in murine and human neoplasia [62]. Analysis of *Rpa1*$^{689C/+}$ MEFs also showed increased chromosomal instability evidenced by the accumulation of chromosomal breaks. Furthermore, *Rpa1*$^{689C/+}$ MEFs treated with the replication inhibitor aphidicolin show widespread induction of γH2AX foci, a marker of DNA DSBs that fail to repair upon subsequent treatment in drug-free media. These data demonstrate that RPA is essential for cell proliferation and that DNA repair and proliferation defects that arise from RPA dysfunction results in chromosomal instability, which can lead to tumorigenesis.

Brca1: Germline inactivation of *Brca1* also results in very early embryonic lethality (~E4.5–E6.5) [76]. Blastocysts from these *Brca1*$^{-/-}$ embryos are unable to grow in vitro. However, some *Brca1*mutant mice were obtained by inactivating exon 11 which supports embryonic development until E10–E13 [47, 128]. The majority of these mutant embryos exhibited neural tube development abnormalities with increased apoptosis being observed in the *Brca1*-deficient embryonic neuroepithelium. Spectral karyotyping of metaphase chromosomes revealed a multitude of chromosomal aberrations while Brca1-deficient embryos display hypersensitivity to ionizing radiation supporting an essential role for Brca1 in DNA repair and genome maintenance [106].

Analysis of *Brca1*$^{-/-}$*p53*$^{-/-}$ mice shows that some of the *Brca1*$^{-/-}$-mediated cell defects are in part p53-dependent as a few double knockout mice are born, however; they all show severe growth retardation and increased chromosomal aberrations than *Brca1*$^{-/-}$ single knockout mice. Of the few that are born, males are infertile due to meiotic recombination defects while females displayed reduced mammary epithelial growth [20]. Treatment of *Brca1*$^{+/-}$*p53*$^{-/-}$ mice with ionizing radiation results in mammary tumor formation [21]. Analysis of these tumors showed that expression of the remaining copy of the wild-type *Brca1* allele was lost (loss of heterozygosity; LOH), an observation reminiscent of heritable Brca1-dependent mammary tumorigenesis that occurs in human breast cancer patients. An especially relevant study to breast tumorigenesis was that of Xu and colleagues wherein, to overcome the embryonic lethality associated with germline *Brca1* inactivation, a conditional *Brca1* mutant that was inactivated in mammary epithelium using the whey-acidic protein promoter-driven Cre recombinase (*Wap-Cre*) [129]. *Brca1*$^{Ko/Co}$*Wap-Cre* mice, which harbors a Brca1-null allele (KO) and a conditionally inactivated allele (Co: loss of exon 11), displayed increased apoptosis and abnormal mammary duct formation. Analysis of 2–10 months old *Brca1*$^{Ko/Co}$*Wap-Cre* mice revealed hyperplasia in mammary tissues while analysis of older mice (10–13 months) revealed multi-variant mammary tumor formation. These tumors show profound genetic instability as multiple chromosomal translocations were detected. Of note were translocations involving chromosome 11 which houses several known tumor suppressor genes including *p53*, implying a need for additional genetic events for mammary tumorigenesis to occur. The addition of the *p53* heterozygote (*p53*$^{+/-}$) background to this line, thereby generating *Brca1*$^{Ko/Co}$*Wap-Cre p53*$^{+/-}$, confirmed that p53 inactivation is indeed involved in *Brca1*-mediated tumorigenesis as these double knockout mice displayed accelerated mammary tumorigenesis compared to the *Brca1*$^{Ko/Co}$*Wap-Cre* single knockout. Many *Brca1*$^{Ko/Co}$*Wap-Cre p53*$^{+/-}$ mice developed multi-focal mammary tumors and most tumors showed inactivation of the remaining wild type *p53* allele.

Brca2: Germline inactivation of *Brca2* also results in early embryonic lethality (E6.5–E8.5) due to developmental arrest [105, 111]. Two independently-derived *Brca2* hypomorphic lines harboring a C-terminal truncation mutation within exon 11 (encoding 45% of the wild-type BRCA2 protein) results in prenatal or perinatal lethality [19, 38]. This line is analogous to the inherited *BRCA2* truncating mutation in patients with breast and ovarian cancer [43]. Interestingly, the few *Brca2* hypomorphic mice that survive to adulthood develop thymic lymphomas while age-matched controls showed no signs of tumor development. Analysis of *Brca2*$^{-/-}$ hypomorphic cells revealed a progressive proliferation impediment that worsened with successive passages [94]. These cells showed increased expression of p53 and p21 indicating activation of cell cycle checkpoints that slowed cellular proliferation. Furthermore, these cells develop spontaneous chromosomal anomalies and are sensitive to exogenous genotoxic agents. Analysis of MEFs isolated from these *Brca2*$^{-/-}$ hypomorphs using the comet assay revealed substantial DNA repair deficiency upon exposure to ionizing radiation [19]. In an effort to circumvent the embryonic lethality associated with Brca2 loss, Jonkers and colleagues developed a conditional *Brca2* genetic line with loxP sites flanking exon 11 (*Brca2*^{F11}) [60]. Inactivation of *Brca2* in stratified epithelia, including mammary epithelia, driven by the K14*cre* line did not yield tumors of epithelial origin. However, combined simultaneous conditional inactivation of *Brca2* and *p53* (K14*cre*;$Brca2^{F11/F11}$;$Trp53^{F2-10/F2-10}$) resulted in a number skin and mammary epithelial tumors in female mice. Furthermore, conditional heterozygotes (K14*cre*;$Brca2^{F11/+}$;$Trp53^{F2-10/F2-10}$ and K14*cre*;$Brca2^{F11/F11}$;$Trp53^{F2-10/+}$) all developed these tumors after an extended tumor-free latency period. Many of these tumors demonstrated LOH of both of the remaining wild-type *Brca2* and *p53* alleles highlighting the requirement for inactivation of both *Brca2* and *p53* genes in epithelial and mammary tumor formation. A more recent study by Frappart and colleagues used the same conditional *Brca2* strain under control of the Nestin-*Cre*, resulting in inactivation of *Brca2* in the developing central nervous system (CNS) [35]. Although viable, $Brca2^{nes\text{-}Cre}$ mice developed microcephaly and showed impaired cerebellar development. These neurodevelopmental defects were primarily a result of increased DNA damage-induced apoptosis in proliferative cells and in early post-mitotic neurons, as evidenced by increased γH2AX formation and TUNEL signals. Furthermore, neuroprogenitor self-renewal and proliferation were defective. Inactivation of *p53* contributed to a partial rescue of the cerebellar phenotype by preventing cell loss associated with DNA damage-induced apoptosis. However, simultaneous inactivation of *p53* with *Brca2* in the developing CNS also led to rapid development of medulloblastoma, a pediatric tumor of the cerebellum [45]. The importance of BRCA2 in the nervous system is also underscored by Fanconi anemia (FA) which can result from disruption of proteins required for DNA crosslink repair, including BRCA2 [22]. FA individuals that result from bi-allelic *BRCA2* inactivation manifest neurodevelopmental abnormalities, and are prone to brain tumors [90]. Thus, these mouse studies have helped identify the requirement for BRCA2/HR-mediated DNA repair pathways in the maintenance of neural and epithelial homeostasis, which when perturbed, may result in cancer.

Xrcc2: Germline inactivation of *Xrcc2* typically results in mid-gestational lethality during embryogenesis (E9–E10) [24, 91]. Although no $Xrcc2^{-/-}$ mice are born alive, $Xrcc2^{-/-}$ embryos that progress further than E10 showed pronounced neurodevelopmental defects due to cell apoptosis in the proliferating region (ventricular zone) of the developing cortex. As such, these embryos die at birth due to respiratory failure. $Xrcc2^{-/-}$ cells show chromosomal instability and sensitivity to ionizing radiation [25, 91]. Combined simultaneous inactivation of Xrcc2 and p53 rescues the neuronal apoptosis in the developing nervous system, and all double mutant mice developed a multitude of tumor types including medulloblastoma, sarcomas, skin lesions and thymomas harboring many chromosomal rearrangements [91].

Bloom syndrome protein (Blm): Germline $Blm^{-/-}$ embryos are affected during development due to apoptosis in the epiblast during the early post-implantation stage [18]. These embryos die by E13.5 and show severe anemia. Red blood cells are heterogeneous and display increased macrocytes and micronuclei, pointing to inherent genomic instability and accumulation of chromosomal damage. *Blm* hypomorphic lines yield viable mice which show a slight reduction in postnatal growth rate [80]. However, like individuals afflicted with Bloom Syndrome, embryonic stem cells, lymphoblastoid cell lines and MEFs derived from these mice undergo a high rate of mitotic recombination resulting in a significant increase in sister chromatid exchanges (SCEs) which contributes to a gradual but widespread LOH events. It is proposed that this is the mechanistic basis for the widespread incidence and spectrum of tumor formation in these mice, a similar attribute also seen in Bloom syndrome patients. Blm-deficient mouse lines have been shown to develop lymphoma, various skin and intestinal carcinomas and sarcomas [46, 80].

Werner's Syndrome protein (Wrn): The Werner syndrome protein is a helicase, that has been implicated in HR [54]. *Wrn*$^{-/-}$ mice are viable and develop normally in the first year of life [67]. However, *Wrn*$^{-/-}$ ES cells display enhanced mutation rates and hypersensitivity to topoisomerase inhibitors. Furthermore, *Wrn*$^{-/-}$ MEFs develop premature proliferative defects.

Other HR factors: It is noteworthy that embryonic lethality is not featured in mouse lines deficient in other HR-associated factors, although *Rad52*$^{-/-}$ and *Rad54*$^{-/-}$ show reduced rates of HR activity while *Rad54*$^{-/-}$ cells show sensitivity to exogenous DNA damaging agents [30, 99].

13.4.2 Inactivation of Non-Homologous End-Joining in the Mouse

The core NHEJ components have all been subject to gene targeting in the mouse, and extensive phenotypic analysis has been undertaken. Notably, inactivation of either Lig4 or Xrcc4 results in an identical lethal phenotype around midgestation [3, 34, 41, 68, 91, 104]. In contrast, inactivation of either Ku70 or Ku80 results in viable, although growth retarded mice [49, 86, 92, 116], while inactivation of the DNA-PKcs results in viable mutant animals [39]. Artemis-null and Xlf-null animals are also viable [72, 102, 132]. These differences in viability illustrate the relative dispensability of the NHEJ core components, and point to the critical requirement for Lig4 in preventing the accumulation of DSBs.

DNA protein kinase (catalytic subunit) DNA-PK$_{cs}$: *DNA-PK$_{cs}$*$^{-/-}$ mice are viable and show normal size but show severe combined immunodeficiency [39, 65]. *DNA-PK$_{cs}$*$^{-/-}$ lung fibroblasts and MEFs are hypersensitive to ionizing radiation and defective in DNA DSB repair whereas *DNA-PK$_{cs}$*$^{-/-}$ ES cell do not show this defect [39]. Furthermore, *DNA-PK$_{cs}$*$^{-/-}$ T cells in the thymus show reduced frequency and diversity in V(D)J recombination. T and B lymphocyte development is arrested as a majority of thymocytes are blocked at the progenitor (CD4/CD8 double-negative) stage and bone marrow maturation is blocked at an early progenitor (B220+ CD43+) stage. *DNA-PK$_{cs}$*$^{-/-}$ also show a moderate increase in thymic lymphomagenesis compared to control mice. Furthermore, *DNA-PK$_{cs}$*$^{-/-}$ mice show hyperplasia and dysplasia of the intestinal mucosa and an abundance of aberrant crypt foci implicating a role for *DNA-PK$_{cs}$* in maintenance and tumor suppression of the intestinal mucosa. In contrast to this study, Jhappan and colleagues inactivated Prkdc/DNA-PK$_{cs}$ using a novel insertional mouse mutant and find that all resulting knockout mice develop thymic lymphomas [58]. These contradicting findings have yet to be properly resolved.

Ku70 and Ku80: Ku70 and Ku80 subunits act in concert to form a heteromeric complex with DNA-PK$_{CS}$ in order to bind DNA DSB ends, and their respective phenotypes after deletion are similar; in fact, mice containing dual inactivation of Ku70 and Ku80 have a phenotype like either single knockout [73].

Ku70$^{-/-}$ mice are viable but are dramatically smaller in comparison to control littermates [49, 92]. *Ku70*$^{-/-}$ mice display DNA repair deficiency and radiosensitivity, while ES cells derived from these mice lack DNA end-binding activity [48]. Furthermore, *Ku70*$^{-/-}$ MEFs undergo premature cell senescence. *Ku70*$^{-/-}$ mice display B cell developmental arrest, however; one *Ku70*$^{-/-}$ mutant line displays normal T cell maturation [92], while another line shows significant incidence of T cell lymphoma attributable to defects in the joining of V(D)J coding and recombination signal sequences [49]. Furthermore, *Ku70*$^{-/-}$ mice show neuronal death in the intermediate zone, albeit this cell loss is not as extensive as *Xrcc4*$^{-/-}$ or *Lig4*$^{-/-}$ deficient mice.

Ku80$^{-/-}$ mice are viable and can reproduce, but are approximately 50% reduced in size compared to control littermates [86]. *Ku80*$^{-/-}$ mice are hypersensitivity to ionizing radiation, which results in loss of hair pigmentation, severe gastrointestinal tract toxicity and enhanced mortality [87]. Exposure of postnatal *Ku80*$^{-/-}$ mice to sublethal doses of ionizing radiation extends the observed growth retardation while MEFs derived from these mice are defective in rejoining strand breaks and subsequently accumulate substantial chromosomal defects [27, 92]. *Ku80*$^{-/-}$ mice show arrest of T and B cell maturation at an early progenitor stage due to aberrant V(D)J recombination activity. *Ku80*$^{-/-}$ES and pre-B cells display enhanced gamma-ray induced spontaneous apoptosis. *Ku80*$^{-/-}$MEFs show early loss of proliferative capacity and prolonged doubling time, due to accumulation of unrepaired DNA damage thus triggering cell cycle checkpoints. Combined inactivation of *Ku80* and *p53* (*Ku80*$^{-/-}$*p53*$^{-/-}$), shows rescue of *Ku80*$^{-/-}$-associated growth defects; however, pro-B cell lymphomas occur early in these mice. SKY analysis of these tumors reveals translocations between chromosomes 12 and 15, which result in the juxtaposition and amplification of *IgH* and *c-Myc*, resembling the translocation that occurs in Burkitt's lymphoma [112]. These chromosomal translocations are typical for NHEJ mutants on a *p53*$^{-/-}$ background and reflect the need for this repair pathway during normal lymphocyte maturation involving V(D)J recombination.

Xrcc4: Germline inactivation of *Xrcc4* results in late-onset embryonic lethality due to extensive defects in lymphogenesis and neurogenesis [41].*Xrcc4*$^{-/-}$ ES cells are sensitive to ionizing radiation while *Xrcc4*$^{-/-}$ MEFs cannot repair DNA DSBs. *Xrcc4*$^{-/-}$MEF doubling time is two times longer than that of control cells. By E16.5, *Xrcc4*$^{-/-}$ thymi show ten-fold fewer cells than controls. Like *Ku70*$^{-/-}$, *Xrcc4*$^{-/-}$ thymocytes are arrested in an early progenitor (CD4/CD8 double-negative) stage. Furthermore, *Xrcc4*$^{-/-}$ B cells fail to progress out of the progenitor stage also indicating a block in B cell maturation. Finally, severe defects in neurogenesis were noted in the *Xrcc4*$^{-/-}$ embryo, accompanied by widespread neuronal apoptosis throughout the intermediate zone, where recently post-mitotic neurons are located. The coincident inactivation of p53 together with Xrcc4 rescued most defects observed in the *Xrcc4*$^{-/-}$ single mutant, including a complete rescue of neuronal apoptosis and reversal of cell proliferation and senescence defects, ultimately allowing survival of the mouse into postnatal stages [40]. However, lymphogenesis defects remained and these mice developed pro-B cell tumors [40]. SKY analysis of these lymphomas demonstrated non-reciprocal chromosomal translocation events occurring between chromosomes 12 and 15, while FISH and Southern analysis demonstrated *c-Myc* and *IgH* colocalization and amplification in these tumors, similar to pro-B cell tumors that develop in *Ku80*$^{-/-}$*p53*$^{-/-}$ mice.

DNA ligase IV (Lig4): Consistent with the need for effective repair of DNA DSBs by NHEJ, individuals with hypomorphic mutations in LIG4 exhibit immunodeficiency, developmental delay, growth retardation, and microcephaly, a disease that has been termed LIG4 syndrome [88, 89]. In view of the neuronal apoptosis in Lig4-deficient mice, it is possible that LIG4 syndrome patients also experience elevated neuronal apoptosis during development, which could underlie the reported microcephaly and developmental delay.

Inactivation of *Lig4* results in a similar phenotype as *Xrcc4*$^{-/-}$ mice including late onset embryonic lethality, defects in lymphopoeisis, failure in DNA DSBR, cellular radiosensitivity, proliferation defects and extensive apoptosis in the developing nervous system [3, 33, 91]. Simultaneous inactivation of *Lig4* and *p53* (*Lig4*$^{-/-}$*p53*$^{-/-}$), rescues embryonic lethality, proliferation and apoptosis defects including those in the developing nervous system [34]. However, like *Xrcc4*$^{-/-}$, p53-loss did not rescue *Lig4*$^{-/-}$-associated lymphopoeisis defects nor radiosensitivity. Rather, *Lig4*$^{-/-}$*p53*$^{-/-}$ mice also develop and succumb to pro-B cell lymphoma. Analysis of these tumors also showed *IgH/c-Myc* recombination and amplification. In addition, some of these mice were found to also develop medulloblastoma as early as 3 weeks of age [69]. Notably, loss of other NHEJ factors also lead to medulloblastoma when associated with p53 inactivation [56, 72, 91, 131].

Artemis and XLF: Like *DNA-PK*$_{cs}^{-/-}$ mice, *Artemis*$^{-/-}$ mice are of normal size, develop SCID, have V(D)J recombination defects and show cellular radiosensitivity [102]. Furthermore, *Artemis*$^{-/-}$ fibroblasts and ES cells display chromosomal instability [101]. Interestingly, *Artemis*$^{-/-}$*p53*$^{-/-}$ mice, like other NHEJ-deficient lines combined with p53 deficiency, results in pro-B cell lymphomagenesis. However, unlike these other models which harbor (12:15) chromosome translocations resulting in juxtaposition of *IgH* with *c-Myc*, *Artemis*$^{-/-}$*p53*$^{-/-}$-derived pro-B cell tumor cells harbor an interchromosome fusion between *IgH* and *N-Myc*. In contrast, XLF-deficient mice have a relatively subtle lymphoid defect, suggesting a more prominent role for XLF in general NHEJ than V(D)J recombination, but together with p53 loss, *Xlf*$^{-/-}$ mice develop medulloblastoma [72].

13.4.3 Inactivation of the DNA Damage Response

Ataxia telangiectasia, mutated (ATM): One of the most widely studied DNA damage response factors is ATM, a central kinase in orchestrating the DNA DSB response and the associated cell cycle events required for effective and efficient repair of DNA

DSBs. A heritable disease, A-T is a childhood affliction that is marked by severe neurodegeneration [82]. The DNA damage response defects in A-T strongly affect the brain as MRI and autopsy analysis reveal widespread loss of cerebellar Purkinje cell and granule neurons, that are subsequently accompanied by other neurological defects [36]. Individuals with A-T also develop immunodeficiency, radiosensitivity and a predisposition to malignancy including lymphoma and leukemia.

Many *Atm*-deficient mouse models have been generated; however, while these models mimic the extra-neurological symptoms found in humans they fail to model the neurological defects [2, 9, 28, 53, 110, 130]. $Atm^{-/-}$ mice display mild growth and immunological defects (reduced pre-B cells), sterility and succumb to thymic lymphoma around 4 months of age or later. Analysis of thymocyte chromosome structure reveals many complex translocations with many involving the T-cell receptor on chromosome 14 [2, 8, 77, 130]. $Atm^{-/-}$ mice show pronounced radiosensitivity marked by excessive intestinal and hematopoietic toxicity. $Atm^{-/-}$ MEFs show proliferation defects including abnormal G1 checkpoint function and early cell senescence.

NBS1: Nijmegen breakage syndrome (NBS) in humans results from mutations in *NBS1*. These individuals have microcephaly, immunological defects and lymphoid malignancy. Germline *Nbs1* deletion results in early embryonic lethality [134]; however, a variety of *Nbs1* hypomorphic mice models more closely reflecting NBS have subsequently been developed. Like $Atm^{-/-}$ mice, *Nbs1*-deficient mice recapitulate the extra-embryonic features of NBS (immunological defects and lymphomagenesis) but fail to model the neurological feature [26, 61, 126]. $Nbs1^{\Delta B/\Delta B}$ (lacking exons 1–5) fibroblasts exhibit numerous chromosomal abnormalities upon ionizing radiation, and these mice displayed an increased tumor incidence [126]. $Nbs1^{\Delta B/\Delta B}p53^{-/-}$ mice had reduced latency in the development of lymphoid tumors indicating that the increased genomic instability associated with the *Nbs1* mutation exacerbates the $p53^{-/-}$ tumor phenotype. Another line, $Nbs1^{m/m}$ (lacking exons 2–3) shows growth defects and defective lymphogenesis and immunodeficiency. $Nbs1^{m/m}$ ES cells displayed radiosensitivity after ionizing radiation while MEFs displayed defects in cell proliferation. $Nbs1^{m/m}$ mice also showed rapid onset of thymic lymphoma and PCR analysis showed enhanced interchromosonal translocations between the TCRβ and TCRν loci in $Nbs1^{m/m}$ thymocytes [61].

Histone H2AX: Mice lacking histone H2A variant, *H2AX* are viable but show growth retardation, immunological defects and male sterility and whole body irradiation experiments show a marked radiosensitivity [17]. $H2AX^{-/-}$ MEFs fail to recruit NBS1, 53BP1 and BRCA1 to radiation induced foci and thus showed a DNA DSBR defect. Similar to $ATM^{-/-}$, $Ku70^{-/-}$ and $Ku80^{-/-}$ lines, $H2AX^{-/-}$ MEFs show substantial proliferation defects and early cell senescence compared to controls. SKY analysis of $H2AX^{-/-}$ MEFs and activated T cells revealed marked accumulation of several random chromosomal translocations. H2AX loss was also found to significantly increase tumorigenesis in collaboration with p53 loss [5, 15] and showed a striking synergy with ATM inactivation that resulted in marked genomic instability and embryonic lethality [133].

53BP1: Similar to *Atm*$^{-/-}$ mice, *53BP1*$^{-/-}$ mice show mild growth defects, moderate radiation sensitivity and enhanced tumorigenesis [122]. However, unlike *Atm*$^{-/-}$ mice, there were no meiotic defects or chromosomal instability. *53BP1*$^{-/-}$ mice also show immunodeficiency as a result of defective class-switch recombination resulting in reduced numbers of pre-B cells, thymocytes, and peripheral T cells [123]. Furthermore, lymphomagenesis is apparent between 2 and 4 months of age characterized by increased numbers of CD4-CD8- double-negative progenitors and CD4+ mature lymphocytes.

13.5 Conclusions and Perspectives

As an experimental system to explore the biology of DNA damage responses, the mouse has been an extraordinary tool, and quite unprecedented in the opportunity it provides illuminating the physiological requirements for DNA repair processes. As these models accumulate and become more refined, such as the generation of conditional alleles, hypomorphic mutations and compound mutants, our understanding of the contribution of the DNA damage response in preventing cancer and other diseases will continue to expand. On the horizon will be enhanced understanding of the physiological interactions between various repair pathways. While in vitro analysis suggests relatively linear DNA repair pathways, the interaction between specific DNA damage response pathways is likely to be more complex and will relate to developmental stages and involve tissue specific effects of DNA damage. For example, genetic analysis of *Atm*$^{-/-}$ mice have revealed unanticipated synthetic lethality between ATM and DNA-PKcs or PARP-1 and ATM [50, 83]. Understanding the basis for these interactions may facilitate important insight into manipulation of repair pathways that can provide therapeutic benefit in cancer treatment. This concept underpins the elegant demonstration of the effectiveness of PARP inhibitors as a potent cytotoxic agent in the setting of HR deficiency in breast cancer [11, 31].

References

1. Ahnesorg P, Smith P, Jackson SP (2006) XLF interacts with the XRCC4-DNA ligase IV complex to promote DNA nonhomologous end-joining. *Cell* **124**(2): 301–313
2. Barlow C, Hirotsune S, Paylor R, Liyanage M, Eckhaus M, Collins F, Shiloh Y, Crawley JN, Ried T, Tagle D, Wynshaw-Boris A (1996) Atm-deficient mice: a paradigm of ataxia telangiectasia. *Cell* **86**(1): 159–171
3. Barnes DE, Stamp G, Rosewell I, Denzel A, Lindahl T (1998) Targeted disruption of the gene encoding DNA ligase IV leads to lethality in embryonic mice. *Curr Biol* **8**(25): 1395–1398
4. Bassing CH, Alt FW (2004) The cellular response to general and programmed DNA double strand breaks. *DNA Repair (Amst)* **3**, (8–9): 781–796
5. Bassing CH, Suh H, Ferguson DO, Chua KF, Manis J, Eckersdorff M, Gleason M, Bronson R, Lee C, Alt FW (2003) Histone H2AX: a dosage-dependent suppressor of oncogenic translocations and tumors. *Cell* **114**(3): 359–370

6. Bates B, Rios M, Trumpp A, Chen C, Fan G, Bishop JM, Jaenisch R (1999) Neurotrophin-3 is required for proper cerebellar development. *Nat Neurosci* **2**(2): 115–117
7. Bensaad K, Vousden KH (2005) Savior and slayer: the two faces of p53. *Nat Med* **11**(12): 1278–1279
8. Bishop AJ, Barlow C, Wynshaw-Boris AJ, Schiestl RH (2000) Atm deficiency causes an increased frequency of intrachromosomal homologous recombination in mice. *Cancer Res* **60**(2): 395–399
9. Borghesani PR, Alt FW, Bottaro A, Davidson L, Aksoy S, Rathbun GA, Roberts TM, Swat W, Segal RA, Gu Y (2000) Abnormal development of Purkinje cells and lymphocytes in Atm mutant mice. *Proc Natl Acad Sci USA* **97**(7): 3336–3341
10. Bruning JC, Michael MD, Winnay JN, Hayashi T, Horsch D, Accili D, Goodyear LJ, Kahn CR (1998) A muscle-specific insulin receptor knockout exhibits features of the metabolic syndrome of NIDDM without altering glucose tolerance. *Mol Cell* **2**(5): 559–569
11. Bryant HE, Schultz N, Thomas HD, Parker KM, Flower D, Lopez E, Kyle S, Meuth M, Curtin NJ, Helleday T (2005) Specific killing of BRCA2-deficient tumours with inhibitors of poly(ADP-ribose) polymerase. *Nature* **434**(7035): 913–917
12. Buck D, Malivert L, de Chasseval R, Barraud A, Fondaneche MC, Sanal O, Plebani A, Stephan JL, Hufnagel M, le Deist F, Fischer A, Durandy A, de Villartay JP, Revy P (2006) Cernunnos, a novel nonhomologous end-joining factor, is mutated in human immunodeficiency with microcephaly. *Cell* **124**(2): 287–299
13. Burma S, Chen DJ (2004) Role of DNA-PK in the cellular response to DNA double-strand breaks. *DNA Repair (Amst)* **3** (8–9): 909–918
14. Campsall KD, Mazerolle CJ, De Repentingy Y, Kothary R, Wallace VA (2002) Characterization of transgene expression and Cre recombinase activity in a panel of Thy-1 promoter-Cre transgenic mice. *Dev Dyn* **224**(2): 135–143
15. Celeste A, Difilippantonio S, Difilippantonio MJ, Fernandez-Capetillo O, Pilch DR, Sedelnikova OA, Eckhaus M, Ried T, Bonner WM, Nussenzweig A (2003a) H2AX haploinsufficiency modifies genomic stability and tumor susceptibility. *Cell* **114**(3): 371–383
16. Celeste A, Fernandez-Capetillo O, Kruhlak MJ, Pilch DR, Staudt DW, Lee A, Bonner RF, Bonner WM, Nussenzweig A (2003b) Histone H2AX phosphorylation is dispensable for the initial recognition of DNA breaks. *Nat Cell Biol* **5**(7): 675–679
17. Celeste A, Petersen S, Romanienko PJ, Fernandez-Capetillo O, Chen HT, Sedelnikova OA, Reina-San-Martin B, Coppola V, Meffre E, Difilippantonio MJ, Redon C, Pilch DR, Olaru A, Eckhaus M, Camerini-Otero RD, Tessarollo L, Livak F, Manova K, Bonner WM, Nussenzweig MC, Nussenzweig A (2002) Genomic instability in mice lacking histone H2AX. *Science* **296**(5569): 922–927
18. Chester N, Kuo F, Kozak C, O'Hara CD, Leder P (1998) Stage-specific apoptosis, developmental delay, and embryonic lethality in mice homozygous for a targeted disruption in the murine Bloom's syndrome gene. *Genes Dev* **12**(21): 3382–3393
19. Connor F, Bertwistle D, Mee PJ, Ross GM, Swift S, Grigorieva E, Tybulewicz VL, Ashworth A (1997) Tumorigenesis and a DNA repair defect in mice with a truncating Brca2 mutation. *Nat Genet* **17**(4): 423–430
20. Cressman VL, Backlund DC, Avrutskaya AV, Leadon SA, Godfrey V, Koller BH (1999a) Growth retardation, DNA repair defects, and lack of spermatogenesis in BRCA1-deficient mice. *Mol Cell Biol* **19**(10): 7061–7075
21. Cressman VL, Backlund DC, Hicks EM, Gowen LC, Godfrey V, Koller BH (1999b) Mammary tumor formation in p53- and BRCA1-deficient mice. *Cell Growth Differ* **10**(1): 1–10
22. D'Andrea AD, Grompe M (2003) The Fanconi anaemia/BRCA pathway. *Nat Rev Cancer* **3**(1): 23–34
23. Dacquin R, Starbuck M, Schinke T, Karsenty G (2002) Mouse alpha1(I)-collagen promoter is the best known promoter to drive efficient Cre recombinase expression in osteoblast. *Dev Dyn* **224**(2): 245–251

24. Deans B, Griffin CS, Maconochie M, Thacker J (2000) Xrcc2 is required for genetic stability, embryonic neurogenesis and viability in mice. *Embo J* **19**(24): 6675–6685
25. Deans B, Griffin CS, O'Regan P, Jasin M, Thacker J (2003) Homologous recombination deficiency leads to profound genetic instability in cells derived from Xrcc2-knockout mice. *Cancer Res* **63**(23): 8181–8187
26. Demuth I, Frappart PO, Hildebrand G, Melchers A, Lobitz S, Stockl L, Varon R, Herceg Z, Sperling K, Wang ZQ, Digweed M (2004) An inducible null mutant murine model of Nijmegen breakage syndrome proves the essential function of NBS1 in chromosomal stability and cell viability. *Hum Mol Genet* **13**(20): 2385–2397
27. Difilippantonio MJ, Zhu J, Chen HT, Meffre E, Nussenzweig MC, Max EE, Ried T, Nussenzweig A (2000) DNA repair protein Ku80 suppresses chromosomal aberrations and malignant transformation. *Nature* **404**(6777): 510–514
28. Elson A, Wang Y, Daugherty CJ, Morton CC, Zhou F, Campos-Torres J, Leder P (1996) Pleiotropic defects in ataxia-telangiectasia protein-deficient mice. *Proc Natl Acad Sci USA* **93**(23): 13084–13089
29. Esashi F, Christ N, Gannon J, Liu Y, Hunt T, Jasin M, West SC (2005) CDK-dependent phosphorylation of BRCA2 as a regulatory mechanism for recombinational repair. *Nature* **434**(7033): 598–604
30. Essers J, Hendriks RW, Swagemakers SM, Troelstra C, de Wit J, Bootsma D, Hoeijmakers JH, Kanaar R (1997) Disruption of mouse RAD54 reduces ionizing radiation resistance and homologous recombination. *Cell* **89**(2): 195–204
31. Farmer H, McCabe N, Lord CJ, Tutt AN, Johnson DA, Richardson TB, Santarosa M, Dillon KJ, Hickson I, Knights C, Martin NM, Jackson SP, Smith GC, Ashworth A (2005) Targeting the DNA repair defect in BRCA mutant cells as a therapeutic strategy. *Nature* **434**(7035): 917–921
32. Forde A, Constien R, Grone HJ, Hammerling G, Arnold B (2002) Temporal Cre-mediated recombination exclusively in endothelial cells using Tie2 regulatory elements. *Genesis* **33**(4): 191–197
33. Frank KM, Sekiguchi JM, Seidl KJ, Swat W, Rathbun GA, Cheng HL, Davidson L, Kangaloo L, Alt FW (1998) Late embryonic lethality and impaired V(D)J recombination in mice lacking DNA ligase IV. *Nature* **396**(6707): 173–177
34. Frank KM, Sharpless NE, Gao Y, Sekiguchi JM, Ferguson DO, Zhu C, Manis JP, Horner J, DePinho RA, Alt FW (2000) DNA ligase IV deficiency in mice leads to defective neurogenesis and embryonic lethality via the p53 pathway. *Mol Cell* **5**(6): 993–1002
35. Frappart PO, Lee Y, Lamont J, McKinnon PJ (2007) BRCA2 is required for neurogenesis and suppression of medulloblastoma. *Embo J* **26**(11): 2732–2742
36. Frappart PO, McKinnon PJ (2006) Ataxia-telangiectasia and related diseases. *Neuromolecular Med* **8**(4): 495–511
37. Friedberg EC, Meira LB (2006) Database of mouse strains carrying targeted mutations in genes affecting biological responses to DNA damage Version 7. *DNA Repair (Amst)* **5**(2): 189–209
38. Friedman LS, Thistlethwaite FC, Patel KJ, Yu VP, Lee H, Venkitaraman AR, Abel KJ, Carlton MB, Hunter SM, Colledge WH, Evans MJ, Ponder BA (1998) Thymic lymphomas in mice with a truncating mutation in Brca2. *Cancer Res* **58**(7): 1338–1343
39. Gao Y, Chaudhuri J, Zhu C, Davidson L, Weaver DT, Alt FW (1998) A targeted DNA-PKcs-null mutation reveals DNA-PK-independent functions for KU in V(D)J recombination. *Immunity* **9**(3): 367–376
40. Gao Y, Ferguson DO, Xie W, Manis JP, Sekiguchi J, Frank KM, Chaudhuri J, Horner J, DePinho RA, Alt FW (2000) Interplay of p53 and DNA-repair protein XRCC4 in tumorigenesis, genomic stability and development. *Nature* **404**(6780): 897–900
41. Gao Y, Sun Y, Frank KM, Dikkes P, Fujiwara Y, Seidl KJ, Sekiguchi JM, Rathbun GA, Swat W, Wang J, Bronson RT, Malynn BA, Bryans M, Zhu C, Chaudhuri J, Davidson L, Ferrini

R, Stamato T, Orkin SH, Greenberg ME, Alt FW (1998) A critical role for DNA end-joining proteins in both lymphogenesis and neurogenesis. *Cell* **95**(7): 891–902
42. Gaveriaux-Ruff C, Kieffer BL (2007) Conditional gene targeting in the mouse nervous system: Insights into brain function and diseases. *Pharmacol Ther* **113**(3): 619–634
43. Gayther SA, Mangion J, Russell P, Seal S, Barfoot R, Ponder BA, Stratton MR, Easton D (1997) Variation of risks of breast and ovarian cancer associated with different germline mutations of the BRCA2 gene. *Nat Genet* **15**(1): 103–105
44. Gelman DM, Noain D, Avale ME, Otero V, Low MJ, Rubinstein M (2003) Transgenic mice engineered to target Cre/loxP-mediated DNA recombination into catecholaminergic neurons. *Genesis* **36**(4): 196–202
45. Gilbertson RJ, Ellison DW (2008) The origins of medulloblastoma subtypes. *Annu Rev Pathol* **3**: 341–365
46. Goss KH, Risinger MA, Kordich JJ, Sanz MM, Straughen JE, Slovek LE, Capobianco AJ, German J, Boivin GP, Groden J (2002) Enhanced tumor formation in mice heterozygous for Blm mutation. *Science* **297**(5589): 2051–2053
47. Gowen LC, Johnson BL, Latour AM, Sulik KK, Koller BH (1996) Brca1 deficiency results in early embryonic lethality characterized by neuroepithelial abnormalities. *Nat Genet* **12**(2): 191–194
48. Gu Y, Jin S, Gao Y, Weaver DT, Alt FW (1997) Ku70-deficient embryonic stem cells have increased ionizing radiosensitivity, defective DNA end-binding activity, and inability to support V(D)J recombination. *Proc Natl Acad Sci USA* **94**(15): 8076–8081
49. Gu Y, Seidl KJ, Rathbun GA, Zhu C, Manis JP, van der Stoep N, Davidson L, Cheng HL, Sekiguchi JM, Frank K, Stanhope-Baker P, Schlissel MS, Roth DB, Alt FW (1997) Growth retardation and leaky SCID phenotype of Ku70-deficient mice. *Immunity* **7**(5): 653–665
50. Gurley KE, Kemp CJ (2001) Synthetic lethality between mutation in Atm and DNA-PK(cs) during murine embryogenesis. *Curr Biol* **11**(3): 191–194
51. Gustafsson E, Brakebusch C, Hietanen K, Fassler R (2001) Tie-1-directed expression of Cre recombinase in endothelial cells of embryoid bodies and transgenic mice. *J Cell Sci* **114** (Pt 4): 671–676
52. Helleday T, Lo J, van Gent DC, Engelward BP (2007) DNA double-strand break repair: from mechanistic understanding to cancer treatment. *DNA Repair (Amst)* **6**(7): 923–935
53. Herzog KH, Chong MJ, Kapsetaki M, Morgan JI, McKinnon PJ (1998) Requirement for Atm in ionizing radiation-induced cell death in the developing central nervous system. *Science* **280**(5366): 1089–1091
54. Hickson ID (2003) RecQ helicases: caretakers of the genome. *Nat Rev Cancer* **3**(3): 169–178
55. Hisahara S, Araki T, Sugiyama F, Yagami K, Suzuki M, Abe K, Yamamura K, Miyazaki J, Momoi T, Saruta T, Bernard CC, Okano H, Miura M (2000) Targeted expression of baculovirus p35 caspase inhibitor in oligodendrocytes protects mice against autoimmune-mediated demyelination. *EMBO J* **19**(3): 341–348
56. Holcomb VB, Vogel H, Marple T, Kornegay RW, Hasty P (2006) Ku80 and p53 suppress medulloblastoma that arise independent of Rag-1-induced DSBs. *Oncogene* **25**: 7159–7165
57. Isaka F, Ishibashi M, Taki W, Hashimoto N, Nakanishi S, Kageyama R (1999) Ectopic expression of the bHLH gene Math1 disturbs neural development. *Eur J Neurosci* **11**(7): 2582–2588
58. Jhappan C, Morse HC, 3rd, Fleischmann RD, Gottesman MM, Merlino G (1997) DNA-PKcs: a T-cell tumour suppressor encoded at the mouse scid locus. *Nat Genet* **17**(4): 483–486
59. Jonkers J, Berns A (2002) Conditional mouse models of sporadic cancer. *Nat Rev Cancer* **2**(4): 251–265
60. Jonkers J, Meuwissen R, van der Gulden H, Peterse H, van der Valk M, Berns A (2001) Synergistic tumor suppressor activity of BRCA2 and p53 in a conditional mouse model for breast cancer. *Nat Genet* **29**(4): 418–425
61. Kang J, Bronson RT, Xu Y (2002) Targeted disruption of NBS1 reveals its roles in mouse development and DNA repair. *Embo J* **21**(6): 1447–1455

62. Karlsson A, Deb-Basu D, Cherry A, Turner S, Ford J, Felsher DW (2003) Defective double-strand DNA break repair and chromosomal translocations by MYC overexpression. *Proc Natl Acad Sci USA* **100**(17): 9974–9979
63. Kastan MB, Bartek J (2004) Cell-cycle checkpoints and cancer. *Nature* **432**(7015): 316–323
64. Kastan MB, Onyekwere O, Sidransky D, Vogelstein B, Craig RW (1991) Participation of p53 protein in the cellular response to DNA damage. *Cancer Res* **51**(23 Pt 1): 6304–6311
65. Kurimasa A, Ouyang H, Dong LJ, Wang S, Li X, Cordon-Cardo C, Chen DJ, Li GC (1999) Catalytic subunit of DNA-dependent protein kinase: impact on lymphocyte development and tumorigenesis. *Proc Natl Acad Sci USA* **96**(4): 1403–1408
66. Kuznetsov S, Pellegrini M, Shuda K, Fernandez-Capetillo O, Liu Y, Martin BK, Burkett S, Southon E, Pati D, Tessarollo L, West SC, Donovan PJ, Nussenzweig A, Sharan SK (2007) RAD51C deficiency in mice results in early prophase I arrest in males and sister chromatid separation at metaphase II in females. *J Cell Biol* **176**(5): 581–592
67. Lebel M, Leder P (1998) A deletion within the murine Werner syndrome helicase induces sensitivity to inhibitors of topoisomerase and loss of cellular proliferative capacity. *Proc Natl Acad Sci USA* **95**(22): 13097–13102
68. Lee Y, Barnes DE, Lindahl T, McKinnon PJ (2000) Defective neurogenesis resulting from DNA ligase IV deficiency requires Atm. *Genes Dev* **14**(20): 2576–2580
69. Lee Y, McKinnon PJ (2002) DNA ligase IV suppresses medulloblastoma formation. *Cancer Res* **62**(22): 6395–6399
70. Lee Y, McKinnon PJ (2007) Responding to DNA double strand breaks in the nervous system. *Neuroscience* **145**(4): 1365–1374
71. Lees-Miller SP, Meek K (2003) Repair of DNA double strand breaks by non-homologous end joining. *Biochimie* **85**(11): 1161–1173
72. Li G, Alt FW, Cheng HL, Brush JW, Goff PH, Murphy MM, Franco S, Zhang Y, Zha S (2008) Lymphocyte-specific compensation for XLF/cernunnos end-joining functions in V(D)J recombination. *Mol Cell* **31**(5): 631–640
73. Li H, Vogel H, Holcomb VB, Gu Y, Hasty P (2007) Deletion of Ku70, Ku80, or both causes early aging without substantially increased cancer. *Mol Cell Biol* **27**(23): 8205–8214
74. Lieber MR, Ma Y, Pannicke U, Schwarz K (2003) Mechanism and regulation of human non-homologous DNA end-joining. *Nat Rev Mol Cell Biol* **4**(9): 712–720
75. Lim DS, Hasty P (1996) A mutation in mouse rad51 results in an early embryonic lethal that is suppressed by a mutation in p53. *Mol Cell Biol* **16**(12): 7133–7143
76. Liu CY, Flesken-Nikitin A, Li S, Zeng Y, Lee WH (1996) Inactivation of the mouse Brca1 gene leads to failure in the morphogenesis of the egg cylinder in early postimplantation development. *Genes Dev* **10**(14): 1835–1843
77. Liyanage M, Weaver Z, Barlow C, Coleman A, Pankratz DG, Anderson S, Wynshaw-Boris A, Ried T (2000) Abnormal rearrangement within the alpha/delta T-cell receptor locus in lymphomas from Atm-deficient mice. *Blood* **96**(5): 1940–1946
78. Lozano G, Liu G (1998) Mouse models dissect the role of p53 in cancer and development. *Semin Cancer Biol* **8**(5): 337–344
79. Lozano G, Zambetti GP (2005) What have animal models taught us about the p53 pathway? *J Pathol* **205**(2): 206–220
80. Luo G, Santoro IM, McDaniel LD, Nishijima I, Mills M, Youssoufian H, Vogel H, Schultz RA, Bradley A (2000) Cancer predisposition caused by elevated mitotic recombination in Bloom mice. *Nat Genet* **26**(4): 424–429
81. McCormack MP, Forster A, Drynan L, Pannell R, Rabbitts TH (2003) The LMO2 T-cell oncogene is activated via chromosomal translocations or retroviral insertion during gene therapy but has no mandatory role in normal T-cell development. *Mol Cell Biol* **23**(24): 9003–9013
82. McKinnon PJ (2004) ATM and ataxia telangiectasia. *EMBO Rep* **5**(8): 772–776

83. Menisser-de Murcia J, Mark M, Wendling O, Wynshaw-Boris A, de Murcia G (2001) Early embryonic lethality in PARP-1 Atm double-mutant mice suggests a functional synergy in cell proliferation during development. *Mol Cell Biol* **21**(5): 1828–1832
84. Minichiello L, Korte M, Wolfer D, Kuhn R, Unsicker K, Cestari V, Rossi-Arnaud C, Lipp HP, Bonhoeffer T, Klein R (1999) Essential role for TrkB receptors in hippocampus-mediated learning. *Neuron* **24**(2): 401–414
85. Minn AJ, Boise LH, Thompson CB (1996) Expression of Bcl-xL and loss of p53 can cooperate to overcome a cell cycle checkpoint induced by mitotic spindle damage. *Genes Dev* **10**(20): 2621–2631
86. Nussenzweig A, Chen C, da Costa Soares V, Sanchez M, Sokol K, Nussenzweig MC, Li GC (1996) Requirement for Ku80 in growth and immunoglobulin V(D)J recombination. *Nature* **382**(6591): 551–555
87. Nussenzweig A, Sokol K, Burgman P, Li L, Li GC (1997) Hypersensitivity of Ku80-deficient cell lines and mice to DNA damage: the effects of ionizing radiation on growth, survival, and development. *Proc Natl Acad Sci USA* **94**(25): 13588–13593
88. O'Driscoll M, Cerosaletti KM, Girard PM, Dai Y, Stumm M, Kysela B, Hirsch B, Gennery A, Palmer SE, Seidel J, Gatti RA, Varon R, Oettinger MA, Neitzel H, Jeggo PA, Concannon P (2001) DNA ligase IV mutations identified in patients exhibiting developmental delay and immunodeficiency. *Mol Cell* **8**(6): 1175–1185
89. O'Driscoll M, Gennery AR, Seidel J, Concannon P, Jeggo PA (2004) An overview of three new disorders associated with genetic instability: LIG4 syndrome, RS-SCID and ATR-Seckel syndrome. *DNA Repair (Amst)* **3**(8–9): 1227–1235
90. Offit K, Levran O, Mullaney B, Mah K, Nafa K, Batish SD, Diotti R, Schneider H, Deffenbaugh A, Scholl T, Proud VK, Robson M, Norton L, Ellis N, Hanenberg H, Auerbach AD (2003) Shared genetic susceptibility to breast cancer, brain tumors, and Fanconi anemia. *J Natl Cancer Inst* **95**(20): 1548–1551
91. Orii KE, Lee Y, Kondo N, McKinnon PJ (2006) Selective utilization of nonhomologous end-joining and homologous recombination DNA repair pathways during nervous system development. *Proc Natl Acad Sci USA* **103**(26): 10017–10022
92. Ouyang H, Nussenzweig A, Kurimasa A, Soares VC, Li X, Cordon-Cardo C, Li W, Cheong N, Nussenzweig M, Iliakis G, Chen DJ, Li GC (1997) Ku70 is required for DNA repair but not for T cell antigen receptor gene recombination In vivo. *J Exp Med* **186**(6): 921–929
93. Parant JM, Lozano G (2003) Disrupting TP53 in mouse models of human cancers. *Hum Mutat* **21**(3): 321–326
94. Patel KJ, Yu VP, Lee H, Corcoran A, Thistlethwaite FC, Evans MJ, Colledge WH, Friedman LS, Ponder BA, Venkitaraman AR (1998) Involvement of Brca2 in DNA repair. *Mol Cell* **1**(3): 347–357
95. Pittman DL, Schimenti JC (2000) Midgestation lethality in mice deficient for the RecA-related gene, Rad51d/Rad51l3. *Genesis* **26**(3): 167–173
96. Postic C, Magnuson MA (2000) DNA excision in liver by an albumin-Cre transgene occurs progressively with age. *Genesis* **26**(2): 149–150
97. Postic C, Shiota M, Niswender KD, Jetton TL, Chen Y, Moates JM, Shelton KD, Lindner J, Cherrington AD, Magnuson MA (1999) Dual roles for glucokinase in glucose homeostasis as determined by liver and pancreatic beta cell-specific gene knock-outs using Cre recombinase. *J Biol Chem* **274**(1): 305–315
98. Rickert RC, Roes J, Rajewsky K (1997) B lymphocyte-specific, Cre-mediated mutagenesis in mice. *Nucleic Acids Res* **25**(6): 1317–1318
99. Rijkers T, Van Den Ouweland J, Morolli B, Rolink AG, Baarends WM, Van Sloun PP, Lohman PH, Pastink A (1998) Targeted inactivation of mouse RAD52 reduces homologous recombination but not resistance to ionizing radiation. *Mol Cell Biol* **18**(11): 6423–6429
100. Rohlmann A, Gotthardt M, Hammer RE, Herz J (1998) Inducible inactivation of hepatic LRP gene by Cre-mediated recombination confirms role of LRP in clearance of chylomicron remnants. *J Clin Invest* **101**(3): 689–695

101. Rooney S, Alt FW, Lombard D, Whitlow S, Eckersdorff M, Fleming J, Fugmann S, Ferguson DO, Schatz DG, Sekiguchi J (2003) Defective DNA repair and increased genomic instability in Artemis-deficient murine cells. *J Exp Med* **197**(5): 553–565
102. Rooney S, Sekiguchi J, Zhu C, Cheng HL, Manis J, Whitlow S, DeVido J, Foy D, Chaudhuri J, Lombard D, Alt FW (2002) Leaky Scid phenotype associated with defective V(D)J coding end processing in Artemis-deficient mice. *Mol Cell* **10**(6): 1379–1390
103. Sedelnikova OA, Pilch DR, Redon C, Bonner WM (2003) Histone H2AX in DNA damage and repair. *Cancer Biol Ther* **2**(3): 233–235
104. Sekiguchi J, Ferguson DO, Chen HT, Yang EM, Earle J, Frank K, Whitlow S, Gu Y, Xu Y, Nussenzweig A, Alt FW (2001) Genetic interactions between ATM and the nonhomologous end-joining factors in genomic stability and development. *Proc Natl Acad Sci USA* **98**(6): 3243–3248
105. Sharan SK, Morimatsu M, Albrecht U, Lim DS, Regel E, Dinh C, Sands A, Eichele G, Hasty P, Bradley A (1997) Embryonic lethality and radiation hypersensitivity mediated by Rad51 in mice lacking Brca2. *Nature* **386**(6627): 804–810
106. Shen SX, Weaver Z, Xu X, Li C, Weinstein M, Chen L, Guan XY, Ried T, Deng CX (1998) A targeted disruption of the murine Brca1 gene causes gamma-irradiation hypersensitivity and genetic instability. *Oncogene* **17**(24): 3115–3124
107. Shiloh Y (2003) ATM and related protein kinases: safeguarding genome integrity. *Nat Rev Cancer* **3**(3): 155–168
108. Shiloh Y (2006) The ATM-mediated DNA-damage response: taking shape. *Trends Biochem Sci* **31**(7): 402–410
109. Shu Z, Smith S, Wang L, Rice MC, Kmiec EB (1999) Disruption of muREC2/RAD51L1 in mice results in early embryonic lethality which can Be partially rescued in a p53(–/–) background. *Mol Cell Biol* **19**(12): 8686–8693
110. Spring K, Cross S, Li C, Watters D, Ben-Senior L, Waring P, Ahangari F, Lu SL, Chen P, Misko I, Paterson C, Kay G, Smorodinsky NI, Shiloh Y, Lavin MF (2001) Atm knock-in mice harboring an in-frame deletion corresponding to the human ATM 7636del9 common mutation exhibit a variant phenotype. *Cancer Res* **61**(11): 4561–4568
111. Suzuki A, de la Pompa JL, Hakem R, Elia A, Yoshida R, Mo R, Nishina H, Chuang T, Wakeham A, Itie A, Koo W, Billia P, Ho A, Fukumoto M, Hui CC, Mak TW (1997) Brca2 is required for embryonic cellular proliferation in the mouse. *Genes Dev* **11**(10): 1242–1252
112. Taub R, Kirsch I, Morton C, Lenoir G, Swan D, Tronick S, Aaronson S, Leder P (1982) Translocation of the c-myc gene into the immunoglobulin heavy chain locus in human Burkitt lymphoma and murine plasmacytoma cells. *Proc Natl Acad Sci USA* **79**(24): 7837–7841
113. Tsuzuki T, Fujii Y, Sakumi K, Tominaga Y, Nakao K, Sekiguchi M, Matsushiro A, Yoshimura Y, Morita T (1996) Targeted disruption of the Rad51 gene leads to lethality in embryonic mice. *Proc Natl Acad Sci USA* **93**(13): 6236–6240
114. van Gent DC, van der Burg M (2007) Non-homologous end-joining, a sticky affair. *Oncogene* **26**(56): 7731–7740
115. Vasioukhin V, Degenstein L, Wise B, Fuchs E (1999) The magical touch: genome targeting in epidermal stem cells induced by tamoxifen application to mouse skin. *Proc Natl Acad Sci USA* **96**(15): 8551–8556
116. Vogel H, Lim DS, Karsenty G, Finegold M, Hasty P (1999) Deletion of Ku86 causes early onset of senescence in mice. *Proc Natl Acad Sci USA* **96**(19): 10770–10775
117. Vousden KH, Lane DP (2007) p53 in health and disease. *Nat Rev Mol Cell Biol* **8**(4): 275–283
118. Vousden KH, Lu X (2002) Live or let die: the cell's response to p53. *Nat Rev Cancer* **2**(8): 594–604
119. Wagner KU, Wall RJ, St-Onge L, Gruss P, Wynshaw-Boris A, Garrett L, Li M, Furth PA, Hennighausen L (1997) Cre-mediated gene deletion in the mammary gland. *Nucleic Acids Res* **25**(21): 4323–4330

120. Wang Y, Putnam CD, Kane MF, Zhang W, Edelmann L, Russell R, Carrion DV, Chin L, Kucherlapati R, Kolodner RD, Edelmann W (2005) Mutation in Rpa1 results in defective DNA double-strand break repair, chromosomal instability and cancer in mice. *Nat Genet* **37**(7): 750–755
121. Ward I, Chen J (2004) Early events in the DNA damage response. *Curr Top Dev Biol* **63**: 1–35
122. Ward IM, Minn K, van Deursen J, Chen J (2003) p53 Binding protein 53BP1 is required for DNA damage responses and tumor suppression in mice. *Mol Cell Biol* **23**(7): 2556–2563
123. Ward IM, Reina-San-Martin B, Olaru A, Minn K, Tamada K, Lau JS, Cascalho M, Chen L, Nussenzweig A, Livak F, Nussenzweig MC, Chen J (2004) 53BP1 is required for class switch recombination. *J Cell Biol* **165**(4): 459–464
124. Waterston RH, Lindblad-Toh K, Birney E, Rogers J, Abril JF, Agarwal P, Agarwala R, Ainscough R, Alexandersson M, An P, Antonarakis SE, Attwood J, Baertsch R, Bailey J, Barlow K, Beck S, Berry E, Birren B, Bloom T, Bork P, Botcherby M, Bray N, Brent MR, Brown DG, Brown SD, Bult C, Burton J, Butler J, Campbell RD, Carninci P, Cawley S, Chiaromonte F, Chinwalla AT, Church DM, Clamp M, Clee C, Collins FS, Cook LL, Copley RR, Coulson A, Couronne O, Cuff J, Curwen V, Cutts T, Daly M, David R, Davies J, Delehaunty KD, Deri J, Dermitzakis ET, Dewey C, Dickens NJ, Diekhans M, Dodge S, Dubchak I, Dunn DM, Eddy SR, Elnitski L, Emes RD, Eswara P, Eyras E, Felsenfeld A, Fewell GA, Flicek P, Foley K, Frankel WN, Fulton LA, Fulton RS, Furey TS, Gage D, Gibbs RA, Glusman G, Gnerre S, Goldman N, Goodstadt L, Grafham D, Graves TA, Green ED, Gregory S, Guigo R, Guyer M, Hardison RC, Haussler D, Hayashizaki Y, Hillier LW, Hinrichs A, Hlavina W, Holzer T, Hsu F, Hua A, Hubbard T, Hunt A, Jackson I, Jaffe DB, Johnson LS, Jones M, Jones TA, Joy A, Kamal M, Karlsson EK, Karolchik D, Kasprzyk A, Kawai J, Keibler E, Kells C, Kent WJ, Kirby A, Kolbe DL, Korf I, Kucherlapati RS, Kulbokas EJ, Kulp D, Landers T, Leger JP, Leonard S, Letunic I, Levine R, Li J, Li M, Lloyd C, Lucas S, Ma B, Maglott DR, Mardis ER, Matthews L, Mauceli E, Mayer JH, McCarthy M, McCombie WR, McLaren S, McLay K, McPherson JD, Meldrim J, Meredith B, Mesirov JP, Miller W, Miner TL, Mongin E, Montgomery KT, Morgan M, Mott R, Mullikin JC, Muzny DM, Nash WE, Nelson JO, Nhan MN, Nicol R, Ning Z, Nusbaum C, O'Connor MJ, Okazaki Y, Oliver K, Overton-Larty E, Pachter L, Parra G, Pepin KH, Peterson J, Pevzner P, Plumb R, Pohl CS, Poliakov A, Ponce TC, Ponting CP, Potter S, Quail M, Reymond A, Roe BA, Roskin KM, Rubin EM, Rust AG, Santos R, Sapojnikov V, Schultz B, Schultz J, Schwartz MS, Schwartz S, Scott C, Seaman S, Searle S, Sharpe T, Sheridan A, Shownkeen R, Sims S, Singer JB, Slater G, Smit A, Smith DR, Spencer B, Stabenau A, Stange-Thomann N, Sugnet C, Suyama M, Tesler G, Thompson J, Torrents D, Trevaskis E, Tromp J, Ucla C, Ureta-Vidal A, Vinson JP, Von Niederhausern AC, Wade CM, Wall M, Weber RJ, Weiss RB, Wendl MC, West AP, Wetterstrand K, Wheeler R, Whelan S, Wierzbowski J, Willey D, Williams S, Wilson RK, Winter E, Worley KC, Wyman D, Yang S, Yang SP, Zdobnov EM, Zody MC, Lander ES (2002) Initial sequencing and comparative analysis of the mouse genome. *Nature* **420**(6915): 520–562
125. West SC (2003) Molecular views of recombination proteins and their control. *Nat Rev Mol Cell Biol* **4**(6): 435–445
126. Williams BR, Mirzoeva OK, Morgan WF, Lin J, Dunnick W, Petrini JH (2002) A murine model of Nijmegen breakage syndrome. *Curr Biol* **12**(8): 648–653
127. Wyman C, Kanaar R (2006) DNA double-strand break repair: all's well that ends well. *Annu Rev Genet* **40**: 363–383
128. Xu X, Qiao W, Linke SP, Cao L, Li WM, Furth PA, Harris CC, Deng CX (2001) Genetic interactions between tumor suppressors Brca1 and p53 in apoptosis, cell cycle and tumorigenesis. *Nat Genet* **28**(3): 266–271
129. Xu X, Wagner KU, Larson D, Weaver Z, Li C, Ried T, Hennighausen L, Wynshaw-Boris A, Deng CX (1999) Conditional mutation of Brca1 in mammary epithelial cells results in blunted ductal morphogenesis and tumour formation. *Nat Genet* **22**(1): 37–43

130. Xu Y, Ashley T, Brainerd EE, Bronson RT, Meyn MS, Baltimore D (1996) Targeted disruption of ATM leads to growth retardation, chromosomal fragmentation during meiosis, immune defects, and thymic lymphoma. *Genes Dev* **10**(19): 2411–2422
131. Yan CT, Kaushal D, Murphy M, Zhang Y, Datta A, Chen C, Monroe B, Mostoslavsky G, Coakley K, Gao Y, Mills KD, Fazeli AP, Tepsuporn S, Hall G, Mulligan R, Fox E, Bronson R, De Girolami U, Lee C, Alt FW (2006) XRCC4 suppresses medulloblastomas with recurrent translocations in p53-deficient mice. *Proc Natl Acad Sci USA* **103**(19): 7378–7383
132. Zha S, Alt FW, Cheng HL, Brush JW, Li G (2007) Defective DNA repair and increased genomic instability in Cernunnos-XLF-deficient murine ES cells. *Proc Natl Acad Sci USA* **104**(11): 4518–4523
133. Zha S, Sekiguchi J, Brush JW, Bassing CH, Alt FW (2008) Complementary functions of ATM and H2AX in development and suppression of genomic instability. *Proc Natl Acad Sci USA* **105**(27): 9302–9306
134. Zhu J, Petersen S, Tessarollo L, Nussenzweig A (2001) Targeted disruption of the Nijmegen breakage syndrome gene NBS1 leads to early embryonic lethality in mice. *Curr Biol* **11**(2): 105–109
135. Zhuo L, Theis M, Alvarez-Maya I, Brenner M, Willecke K, Messing A (2001) hGFAP-Cre transgenic mice for manipulation of glial and neuronal function in vivo. *Genesis* **31**(2): 85–94

Chapter 14
Cancer Biomarkers Associated with Damage Response Genes

Anne E. Kiltie, Marie Fernet, and Janet Hall

Abstract The development and validation of prognostic and predictive cancers biomarkers associated with damage response genes is a rapidly moving field. Prognostic and predictive markers can be measured at the level of protein (by immunohistochemistry, proteomics, etc.), mRNA (mRNA expression, arrays, etc.) and DNA (DNA adducts, mutation and other sequence variant screens, epigenomics, including methylation studies, etc.). The challenge in the cancer biomarker field is in identifying which proteins play a critical limiting role in the cellular responses to DNA damage, the sequence variants, if any, that impact on the functionality of these responses and how these processes vary between normal and tumour cells. The understanding of these processes is essential in order that simple non-invasive tests that indicate cancer risk and allow early cancer detection and prognosis can be established. In addition biomarkers able to predict how an individual would respond to therapy could allow the personalisation and thus optimisation of therapeutic protocols.

Keywords Biomarkers · Damage response · Predictive · Prognostic

14.1 Introduction

A biomarker is any characteristic that is objectively measured and evaluated as an indicator of normal biological processes, pathogenic processes or pharmacological responses to a therapeutic intervention [10].

The goals of the cancer biomarker field have evolved over the last twenty years and differ between clinical and risk assessment settings. In the *clinical setting* the goals are to:

A.E. Kiltie (✉)
Section of Experimental Oncology, Leeds Institute of Molecular Medicine, St James's University Hospital, Leeds, LS9 7TF, UK
e-mail: a.e.kiltie@leeds.ac.uk

K.K. Khanna, Y. Shiloh (eds.), *The DNA Damage Response: Implications on Cancer Formation and Treatment*, DOI 10.1007/978-90-481-2561-6_14,

1. develop simple non-invasive tests that indicate cancer risk,
2. allow early cancer detection, and
3. classify tumours, in order for the patient to receive appropriate targeted therapy and for disease progression and therapeutic outcomes to be monitored [35].

As knowledge of cancer at a molecular level has increased, considerable efforts have been made to develop and validate prognostic, predictive and pharmacodynamic biomarkers that can be used in a clinical setting to fulfil these requirements. A subset of biomarkers can potentially substitute for clinical endpoints and in this scenario can be denoted as surrogate endpoints.

In the *risk assessment setting* biomarkers of exposure and of effect have also been developed and validated. Such biomarkers are being used to investigate the role of reactive molecules that produce DNA adducts thought to be involved in carcinogenesis, along with surrogates such as protein adducts and metabolites excreted in the urine, the consequences of their persistence in cellular macromolecules and mutation induction. Their use has greatly improved our understanding of species differences in metabolism and effects of chemical stability and DNA repair on tissue differences in molecular dose and the dose responses for both gene and chromosomal mutations (see for instance the recent review by [76]). A comprehensive review of all cancer biomarkers is outside the scope of this chapter and we will restrict our review to *prognostic* and *predictive* biomarkers and highlight the potential use of biomarkers associated with damage response genes.

14.2 The Cellular Damage Response

Through cellular metabolism, DNA is constantly being exposed to endogenous oxidative by-products that damage DNA, producing largely base damage and single strand breaks (SSBs). Exogenous sources of DNA damage include chemical carcinogens, ionizing radiation, ultraviolet light and chemotherapy agents. The cell has developed sophisticated mechanisms to identify and respond to such damage. The particular mechanism used depends on the type of damage, but essentially it involves recognition of the damage, signalling of the damage to effector molecules, and then activation of cell cycle checkpoints. If the damage is overwhelming the cell dies by apoptosis, but if less severe, cell cycle progression is halted in order to allow time for DNA repair processes to repair the damage before cellular replication. If repair is not accurate, the resulting cellular mutations or chromosomal aberrations can result in genomic instability and cancer formation (for review see [38]).

Base damage and single strand breaks, caused by oxidative damage, ionizing radiation and many alkylating agents, are repaired by the base excision repair (BER) pathway. The highly mutagenic methylated-DNA lesion O^6-methylguanine is dealt with in mammalian cells by the so called suicide enzyme O^6-methylguanine DNA-methyltransferase (MGMT), whilst bulky DNA adducts and intra-strand crosslinks, formed by carcinogenic metabolites, UV light, and platinum-based chemotherapy agents, are repaired by the nucleotide excision repair (NER) pathway. Replication errors result in mismatches and insertion or deletion loops, which are repaired by the mismatch repair (MMR) pathway. Double strand breaks (DSBs), the lethal lesions caused by ionizing radiation and topoisomerase II inhibitors, are repaired by homologous recombination (HR), non-homologous end-joining (NHEJ), and single strand annealing (SSA). The DNA DSB signalling cascade has been well characterized and involves recruitment of the MRE11-NBS1-RAD50 (MRN) complex to the site of the DSB, where NBS1 activates the ataxia telangiectasia (ATM) protein, which in turn phosphorylates various downstream substrates, including p53, CHK2, BRCA1 and NBS1 [41], thus resulting in checkpoint activation, and DNA repair.

Variations in expression or sequence variants of proteins involved in these processes may impact on the damage response and potentially modulate an individual's cancer risk and response to therapy. For instance radiotherapy is an effective cancer treatment, as ionizing radiation produces lethal double strand breaks in cancer cells. However, the ionizing radiation also damages cells in normal tissues. In fast-turnover tissues such as the epithelial lining of the gut and the hematopoietic system, damaged cells fail to reproduce so the cell population is depleted, resulting in acute radiation side effects. Cell death in slower turnover tissues accounts for some of the late side effects, but also gene transduction in response to radiation produces other late effects such as fibrosis. Inaccurate repair of DNA damage in surviving cells results in mutations and chromosome aberrations, which are responsible for the second malignancies sometimes seen after radiotherapy. The inter-individual variation in response to radiotherapy is large with some individuals tolerating standard protocols poorly. Similarly, chemotherapeutic agents damage DNA and inter-individual variation in responses is seen. Such agents also produce acute side effects as a result of cell death, particularly in the hematopoietic system and gut, and, whilst late side effects for chemotherapy are less well characterized than for ionizing radiation, second malignancies are observed in some individuals (see [75]).

The challenge in the cancer biomarker field is in identifying which proteins play a critical limiting role in the cellular responses to DNA damage, the sequence variants, if any, that impact on the functionality of these responses and how these processes vary between normal and tumour cells. The understanding of these processes is essential in order to develop simple non-invasive tests that indicate cancer risk, allow early cancer detection or are prognostic indicators. The identification of biomarkers predicting how an individual might respond to therapy would allow the personalisation and thus optimisation of treatment protocols.

14.3 Definitions of Prognostic and Predictive Factors

Prognostic biomarkers give a guide to the likely overall survival of an individual or group of patients and thus allow the natural course of an individual cancer to be predicted, distinguishing "good outcome" tumours from "poor outcome" tumours.

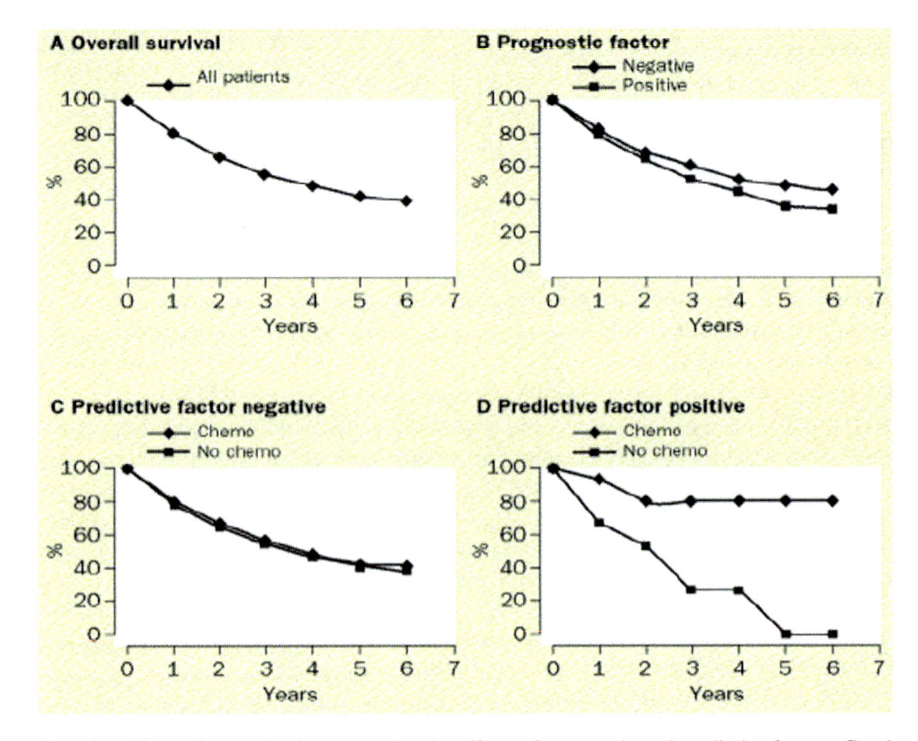

Fig. 14.1 Illustrative graphs demonstrating the effects of prognostic and predictive factors. Graph **A** shows a typical survival curve for a group of patients with Dukes' C carcinoma of the colon. Graph **B** shows that a prognostic factor will separate the patients into two groups with differing survival. In this case, patients with the prognostic factor have worse survival. Two graphs (**C** and **D**) are needed to show a predictive factor. For patients without the predictive factor (graph **C**), survival is similar whether or not chemotherapy is given. However, in graph **D**, possession of the predictive factor identifies a group of patients likely to benefit from chemotherapy (Adlard et al. [1] with permission)

Certain clinicopathological characteristics are proven prognostic biomarkers. For example, in resected colorectal cancer, Dukes' stage A is a good prognostic factor, with a 5-year survival of approximately 95%. *Predictive* markers differ in that they are used to assess the probability that a patient will benefit from a particular treatment. For instance they could be used to identify tumours likely to be responsive or resistant to treatment, allowing selection for treatment of only those patients likely to benefit, thereby sparing the others from unnecessary treatment-related side effects. The differences between prognostic and predictive factors are illustrated in Fig. 14.1, taken from [1]. There are a number of predictive markers that are routinely used for treatment decisions. For instance, patients with breast cancer in which the *ERBB2* gene (also known as *HER2* or *NEU*) is amplified, benefit from treatment with trastuzumab (Herceptin), whereas when the gene encoding the oestrogen receptor is expressed in a tumour, the patients respond to treatment with tamoxifen.

14.4 Biological Samples for Biomarker Measurements: Technical Considerations

Prognostic and predictive markers can be measured at the level of protein (by immunohistochemistry (IHC) and proteomics, etc.), mRNA (mRNA expression, arrays, etc.) and DNA (DNA adducts, mutation screens, epigenomics, including methylation studies, etc.). However the development and validation of biomarkers has been limited by access to appropriate tumour tissues. Thus, whilst predictive biomarkers are increasingly being used in clinical trials of chemotherapy agents in leukaemia, as large numbers of tumour cells are present in the peripheral blood and thus easily collectable, their incorporation into clinical trials for treating solid tumours has been more limited. Often the only time at which access to a solid tumour is guaranteed is at diagnosis, when the tumour is biopsied, or during surgery. In addition, phase I and II trials of experimental drugs are often carried out in patients with advanced or late stage disease, when additional tumour biopsies are not routinely taken.

Samples collected at biopsy and surgery for diagnostic purposes are usually fixed in formalin and embedded in paraffin before histological examination and long-term storage. This fixation process leads to structural changes in RNA, thus limiting the usefulness of such biological material for the assessment of RNA expression. Therefore, many cancer treatment centres are now also routinely banking frozen tissue samples for use in associated or future translational research studies, including those involving biomarkers.

The difficulties in obtaining and storing solid tumour tissue for use in biomarker studies have lead to the development of non-invasive approaches to detecting cancer cells. These include analyzing circulating tumor cells, carrying out mutation-specific PCR on circulating DNA, using proteomic approaches to study serum or plasma, imaging tumours at the molecular level and assessing auto-antibodies specific to tumour cells. These techniques also allow monitoring of tumour response during treatment. Some of these procedures have been hampered by problems with reproducibility, sensitivity of detection and the ability to make quantitative measurements. Goulart et al. [33] reviewed the trends in the use of biomarkers in published phase I clinical trials during the years 1991–2002 and found the proportion of trials that included biomarkers increased from 14 to 26% over this time.

14.5 Measurement of Biomarkers at the Protein Level

14.5.1 Protein Expression by Immunohistochemistry

The currently most widely applicable method of biomarker measurement is the evaluation of protein expression in tumours by IHC, due to the availability of formalin fixed paraffin embedded (FFPE) tissue. Both a quantitative assessment of the percentage of tumour cells staining positive for the antibody and a qualitative assessment of the intensity of staining can be made, and such scores are often combined to give a semi-quantitative score (for example [4]).

Whilst there are many single institution studies reporting the potential predictive and prognostic value of expression of various damage response proteins by IHC, these results are not often validated in further cohorts. The National Cancer Institute has therefore recommended a strategy delineating the full development of such tumour markers [20], based on clinical trial methodology. Phase I of this process involves pilot studies to develop a robust assay, where cut points are established, then retrospective phase II studies are performed to assess the potential clinical usefulness of the marker. Phase III prospective confirmatory analysis are then carried out using large cohorts of patients followed by phase IV validation studies in the context of a multi-centre clinical trial.

Two phase III confirmatory analyses have been performed in bladder cancer involving the transcription factor and cell cycle/apoptosis regulator, p53, which has been widely studied by IHC, and RB, the protein product of *RB*, a tumour suppressor gene involved in cell cycle regulation and apoptotic responses [17, 68]. p53 is the product of the tumour suppressor gene *TP53*, mutations in which have been extensively studied in terms of their predictive and prognostic value (see later). A multi-centre clinical trial of molecular-guided therapy based on p53 status in bladder cancer was opened in 1998 [74]. However, in this instance *TP53* mutation status was used rather than p53 expression level by IHC. Patients whose tumours were *TP53*-mutant were randomised to observation or three cycles of MVAC (methotrexate, vinblastine, adriamycin, and cisplatin) chemotherapy whilst those with *TP53*-wild-type tumours were observed. This trial has now closed to recruitment and results are awaited.

Expression of ATM and the MRN complex proteins has been found to be reduced in breast cancer compared to normal breast ductal tissue ([4, 72] and Fig. 14.2). In a clinical trial of 224 women with early breast cancer randomised to receive post-operative radiotherapy or adjuvant chemotherapy, Soderlund et al. [72] found that radiotherapy significantly reduced the risk of local recurrence compared with chemotherapy and it was those patients with moderate or strong expression of the MRN complex proteins that had the greatest benefit from radiotherapy compared to chemotherapy. Those with negative or weak MRN staining had no such benefit. This was counter to the hypothesis that down-regulated expression of the MRN complex proteins in tumours cells would lead to impaired repair of radiation-induced damage and hence radiosensitivity. A similar result has been found in bladder cancer, where in 190 patients, low tumour MRE11 expression was associated with a worse cancer-specific survival compared with high tumour MRE11 expression (Choudhury A, Nelson L, Chilka S, Johnston C, Bristow R, Bishop DT, Kiltie A, DNA double strand break signalling protein expression as a predictive factor in muscle invasive bladder cancer, NCRI meeting poster BOA13, NCRI meeting, Birmingham 2007 and Fig. 14.2). Low expression of MRN complex proteins may result in failure of induction of the ATM-mediated DNA damage response cascade and failure of the checkpoint activation and apoptosis responses, which normally result in permanent cell cycle arrest or cell death and radiosensitivity, which may be more important than the damage repair response.

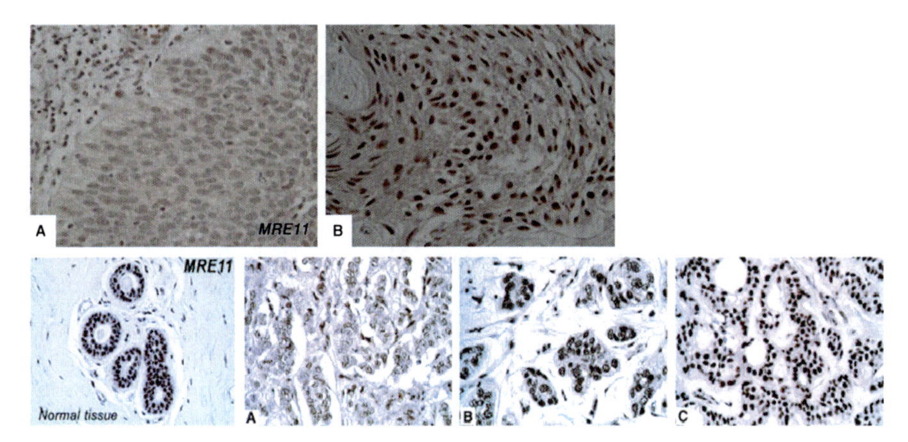

Fig. 14.2 Immunostaining of MRE11 in normal tissues and carcinomas of the bladder and breast. *Top panel*: the immunostaining is defined as weak (**A**) or strong (**B**) based on the intensity of staining in bladder tumour cells expressing MRE11 (moderate staining not shown). As normal urothelium was rarely seen on individual sections, the relative staining was based on intra-tumoral comparison rather than with normal tissue (unpublished data from Kiltie lab). *Lower panel*: the immunostaining is defined as low (**A**), moderate (**B**) and high (**C**) based on the percentage of breast epithelial cancer cells expressing MRE11 compared with the immunoreactivity seen in normal breast ducts (reproduced with permission from Angèle et al. [4])

Apoptosis has also been investigated for prognostic and predictive significance in radiotherapy studies. Callagy et al. [15] studied BCL2, in addition to p53, cyclin E and 10 other markers, by IHC in a tissue microarray (see below), in a developmental study of 930 breast cancer patients, which sought to evaluate the prognostic potential of this panel of markers in terms of improving on the prognostic power of the Nottingham Prognostic Index (NPI). BCL2 showed significant association with survival at 10 years on multivariate analysis (positive expression associated with improved survival), was independent of the NPI, and the use of BCL2 as a single marker was more powerful than using a panel of markers. BCL2 retained prognostic significance in an independent validation series of 1961 patients. In terms of effects on tumour cells, the BCL2 results would appear counterintuitive, with loss of an anti-apoptotic response being associated with worse patient outcome, as one would expect tumour cell death by apoptosis to be associated with improved prognosis. However, high levels of BCL2 can be associated with growth inhibition and so its prognostic role may be dependent on non-apoptotic functions of BCL2.

Kyndi et al. [48] also studied BCL2 expression, along with that of p53, in 1000 high-risk breast cancer patients randomly assigned to postmastectomy radiotherapy in the DBCG82 b&c studies. Whilst p53 accumulation was not found to have any prognostic value, negative BCL2 (an anti-apoptotic factor) expression was significantly associated with biological and clinical markers of poor prognosis and with increased overall mortality, development of distant metastases and locoregional recurrence. BCL2 was a stronger prognostic marker of locoregional recurrence than positive lymph node status. There was a significantly improved overall survival after postmastectomy radiotherapy (PMRT) in the BCL2 positive subgroup compared to the no radiotherapy group, whilst PMRT had virtually no impact on survival in the BCL2 negative subgroup, although the hazard ratio was of only borderline significance ($p=0.06$), thus BCL2 may be a predictive factor for outcome following radiotherapy.

Survivin is an inhibitor of apoptosis which is over-expressed in most solid tumours compared to normal tissues and high expression by IHC is associated with an aggressive phenotype and reduced disease-free and overall survival rates (see review by [16]). In terms of prediction of treatment outcome, high survivin is associated with increased risk of local tumour recurrences following pre-operative radiation or combined chemo-radiotherapy in rectal cancer, and with worse overall survival in cervical cancer patients treated with radiotherapy, although the opposite was found in oral squamous cell carcinoma.

ERCC1 is a key protein in the NER pathway involved, with its cofactor XPF, in dual incision of the damaged DNA strand during repair. In the International Adjuvant Lung Cancer Trial (IALT), in completely resected non-small cell lung cancer, 1867 patients were randomly assigned to treatment with adjuvant cisplatin-based chemotherapy or no further treatment. Those receiving chemotherapy had a 4.1% absolute benefit in terms of overall survival at 5 years [5]. The tumours from 761 patients were examined by IHC for ERCC1 and a benefit from chemotherapy was associated with the absence of ERCC1. Those patients whose tumours expressed ERCC1 did not benefit from chemotherapy and in patients not receiving chemotherapy, those with ERCC1 positive tumours survived longer [58]. This suggests that ERCC1 is a predictive factor for adjuvant chemotherapy outcome rather than a prognostic factor in lung cancer.

Another example of a large randomised controlled clinical trial where expression of damage response proteins by IHC has been investigated in terms of prediction of treatment outcome is the UK MRC FOCUS trial [11]. This trial compared fluouracil alone, with fluorouracil and irinotecan, and fluorouracil and oxaliplatin, in 2135 patients with advanced colorectal cancer. Formalin-fixed paraffin-embedded tumour specimens were available from 1628 patients. Tumour IHC was performed for the MLH1/MSH2, p53, topoisomerase-1 (Topo1), ERCC1, MGMT and COX2 proteins. These were screened in more than 750 patients for interaction with benefit from irinotecan or oxaliplatin and Topo1 and MLH1/MSH2 were taken forward for analysis in the full population. Patients with moderate to high Topo1 expression benefited from the addition of irinotecan (a topoisomerase-1 inhibitor) or oxaliplatin (a platinum-analogue) but patients with low Topo1 did not benefit. False-positive report probability analysis suggested however that the oxaliplatin result was probably a false positive. There was no significant interaction between MLH1/MSH2 expression with treatment, however, only 4.4% of patients had loss of these markers, so the study was underpowered to detect an effect for this biomarker. This study was an adequately powered training set, so independent validation is currently underway in the Dutch CAIRO trial and the FOCUS2 trial [11]. The FOCUS3 trial will incorporate Topo1 (along with K-ras testing) in a prospective comparison of molecularly-guided or standard therapy.

Around 15% of colorectal cancers have altered MMR gene function, and about one-third of these are associated with hereditary non-polyposis colorectal cancer (HNPCC) syndrome, whilst the remainder occur sporadically. Screening for HNPCC allows patients to undergo close colonoscopic surveillance and prophylactic surgery, thereby reducing morbidity and mortality rates. Approximately 30–70% of HNPCC patients have a germline MMR gene mutation, particularly in *MLH1* or *MSH2*, whilst sporadic cases usually arise due to abnormal methylation of the promoter region of *MLH1* [71]. MMR abnormalities are associated with microsatellite instability (MSI) reflecting the accumulation of insertions or deletions at the time of replication and failure of repair of the resulting mismatches. Whilst a tumour's MSI status can be assessed using a panel of microsatellite markers, the use of IHC to predict MSI status would obviate the need for such a labour-intensive, time consuming test. This is of potential clinical importance, as MSI tumours have different clinical outcomes and respond differently to some treatments. Shia et al. [71] reviewed the literature and reported an overall sensitivity of 90% and greater than 99% specificity of IHC in detecting MSI. IHC was more reliable in sporadic cases, where promoter hypermethylation was responsible for the gene defect, than in germline mutation carriers, where missense mutations or even protein truncating mutations can results in antigenically intact mutant proteins. Addition of IHC testing of PMS2 and MSH6 have made IHC as sensitive as MSI testing [70]. At present, IHC can be recommended to reduce the number of colorectal cancer cases suspicious for HNPCC that require MSI testing, but in the future IHC may be used routinely to identify MMR-abnormal colorectal cancers, in order for these to be given tailored treatment regimes, when such drugs have been developed.

Tissue microarray (TMA) technology is a relatively recent development, where cylindrical cores of representative tumours are taken from FFPE blocks and embedded in recipient paraffin blocks, each of which may contain up to 1000 tumour specimens. Sections can be taken and slides processed for IHC or in situ hybridisation [13]. This allows more rapid and direct comparisons to be made across a cohort of tumours, as experimental variation is minimised in terms of staining. Also, as only one slide is required per antibody, a large number of proteins can be studied. Despite a diameter of only 0.6 mm, the specimens are highly representative of their donor tissue. TMAs have been used for immunohistochemical studies of MMR repair proteins in colorectal cancer [37] and found to be reliable in order to simultaneously screen large numbers of tumours from suspected HNPCC patients for further mutation analysis of MMR genes. Studies in bladder cancer have also used TMAs to investigate panels of cell cycle proteins [67] and apoptosis proteins [42] in 101 and 226 patients respectively, and these have, in both cases, provided prognostic information. Interestingly, in the case of apoptosis markers, results were statistically significant only when all four markers (BCL-2, caspase-3, p53, survivin) were included in the analysis. This highlights the need for a pathway approach, with the associated requirement for high-throughput technologies, in this field. Such biomarker information, if validated in larger independent cohorts, could be added to normograms, which include standard histopathological features and staging data, to improve their accuracy.

14.5.2 Protein Expression in Serum and Plasma

Several tumour types secrete proteins that are detectable in serum and plasma, and such biomarkers are already in routine clinical use. For example prostatic adenocarcinomas generally secrete prostate specific antigen (PSA) and its measurement in serum allows early detection of prostate cancer, before it becomes symptomatic or even detectable on imaging. The PSA level can then be monitored to assess the rate of progression of disease if patients are on active monitoring, or it can be used to assess the response to radical prostatectomy or radiotherapy to the prostate. Patients treated with hormonal therapy usually have a good response to treatment, assessable by PSA measurement, and PSA is a particularly sensitive indicator of the development of resistance to these drugs, which allows early intervention with an alternative hormonal agent or chemotherapy. Serum carcinoembryonic antigen (CEA) measurements are used to monitor response to chemotherapy agents in colorectal cancer and CA125 levels are used similarly in ovarian cancer.

Proteomics is a rapidly developing area in the cancer biomarker field (for a recent review see [35]). This technology involves systematic searches for proteins which may act as biomarkers, either present in tumour tissues, in cell lines, or in plasma, from tumour tissues which contribute proteins to the plasma fraction, or alternatively in biological fluids containing tumour tissue proteins, such as ascites, breast ductal fluid, cerebrospinal fluid and pleural effusions. Modern techniques involve sample fractionation, using 2D gels or liquid chromatography, followed by mass spectrometry to identify and quantify proteins. It permits the study of multiple protein forms reflecting mRNA splicing and post-translational modifications [6]. There are a number of technical challenges to using this technique in plasma samples, including the low abundance of such proteins in plasma and false discovery rates due to sources of variability (see Table 14.1). However, such technology might be applied to early cancer detection, prediction of prognosis or for prediction of response to targeted therapies, hence allowing patient selection. There are also now early promising studies indicating that formalin-fixed tissue proteomics may be a powerful tool for biomarker-driven translational research in the future (reviewed by

Table 14.1 Sources of variability in proteomic analysis of plasma

Differences in the specificity and sensitivity of methodologies
Lack of standardised sample collection and storage
Differences between cases and controls in terms of sex, age and physiological states (e.g. fasting, weight gain or loss, hormonal status)
Differences in genetic make-up (e.g. ethnic differences between cases and controls, comparisons between different populations)
Changes in inflammation and acute-phase reactants
Changes in metabolic states
Other non-specific changes: for example, cell death and tissue necrosis
Changes reflecting underlying chronic disease: for example, those caused by smoking and chronic lung disease, in contrast to lung-cancer-specific changes

Adapted from Hanash et al. [35].

[55]). Due to the heterogeneity among cancers, it is unlikely that single biomarkers will provide the sensitivity and specificity required for most applications, and panels of biomarkers will probably be needed. These will need to be validated.

Recently, Moumen et al. [54] used differential proteomics (i.e. comparison in the presence and absence of a damaging agent) and found that the heterogeneous nuclear ribonucleoprotein K (hnRNP K) was rapidly induced by DNA damage caused by ionizing radiation or UV irradiation, in an ATM- or ATR-dependent manner respectively. This protein was found to act as a cofactor for p53, thus coordinating the transcriptional response to DNA damage. Similarly, a proteomics screen identified a subunit of the RNA polymerase II holoenzyme, RPB8, as a protein modified after epirubicin treatment in a BRCA1-dependent manner [82].

14.6 Measurement of Biomarkers at the mRNA Level

The introduction of microarray technology, such as the 3′-based expression Affymetrix human X3P microarray which includes 47,000 genes, or the Gene Chip Human Exon 1.0 ST Array which integrates 1 million exon clusters, allows the quantification of mRNA abundance and has enabled genome-wide gene expression data to be collected for large numbers of biological specimens (reviewed by [79]). The most unbiased approach to identify biomarkers using this technology is the *data-driven* approach, where no assumptions are made as to which genes are likely to be involved in the outcome of interest. However, *candidate-gene* and *model-driven* approaches are also used, the latter involving measurement of transcriptional responses of cells after exposure to a specific stimuli.

Hierarchical clustering (unsupervised classification) of breast cancer specimens has identified five naturally occurring subtypes of breast cancers [62, 73]. Some of these intrinsic subtypes differ in their aggressiveness, for instance a poor prognosis was found for individuals with the basal-like subtype. Gene expression signatures that assess the risk of recurrence of NSCLC and lung adenocarcinoma ([69] and references therein) and several other cancer types have also been established.

A major milestone in the use of molecular diagnostics has been inclusion of the option to use the Oncotype DX profile [60] in the recent National Comprehensive Cancer Network breast cancer treatment guidelines (www.nccn.org). This test system measures the expression of 16 cancer-related genes and 5 reference genes (developed from a prospectively chosen 250-candidate gene set) to generate a recurrence score that can help to predict the risk of recurrence in women with ER-positive, lymph node negative breast tumours. Interestingly, two of these genes, the anti-apoptotic gene *BCL2* and the cell cycle regulatory gene, *CCNB1* (cyclin B1), are involved in the cellular damage response. Marchionni et al. [52] recently reviewed three licensed gene expression-based prognostic breast cancer tests, including Oncotype DX, and reported that the body of evidence showed that this new generation of tests may improve prognostic and therapeutic prediction, although it was yet to be determined how best to incorporate the test results into treatment decision-making.

Table 14.2 Illustrative examples of DNA damage response genes that show potential as predictive biomarkers based on IHC or expression analysis

Tumour site	Predictive
Bladder	ERCC1 for platinum based therapy [9]
	TP53 for chemotherapy/radiotherapy [28]
Breast	ATM for chemotherapy/radiotherapy [72]
	BCL2 for radiotherapy [48]
Colon	MLHI for 5-FU treatment* [7]
Colorectal	Topoisomerase-1 for irinotecan or oxaliplatin therapy [11]
Glioma	MGMT for chemosensitivity [80]
HNSCC	Ku70 for chemotherapy/radiotherapy [61]
Gastric cancer	ERCC1 for 5-FU and oxaliplatin therapy [47]
Lung – non-small cell lung cancer	ERCC1 for platinum-based chemotherapy [5, 58]

*5-FU does not confer a survival benefit to those with MSI-H tumours.

The potential of damage response genes as prognostic and predictive biomarkers has been extensively investigated using the candidate-gene approach (Table 14.2). This can be illustrated by the search for predictive markers for normal tissue reaction after radiotherapy (see for instance recent reviews by [45, 57]). For instance, Hummerich et al. [39] generated expression profiles for 143 DNA repair and repair related genes in peripheral blood lymphocytes from prostate cancer patients. Cluster analysis identified 19 genes where high expression was associated with a lack of clinical radiation sensitivity thus indicating radioresistance. It is tempting to speculate that low expression of these same genes might be associated with an opposite effect but this remains to be confirmed.

Model-driven approaches have also been used to identify potential radiation sensitivity biomarkers. Overgaard, Børresen-Dale and colleagues have used microarray analysis to examine differential gene expression in subcutaneous fibroblasts from breast cancer patients after exposure to single and multiple doses of ionising radiation [3, 64, 63]. While many of the genes identified showing differential expression were those involved in known ionising radiation response pathways like cell cycle arrest, proliferation and detoxification, a substantial fraction of the genes were involved in processes not previously associated with the response to ionising radiation. For instance, transcriptional changes of genes involved in extracellular matrix remodelling, Wnt signalling, IGF signalling and ROS scavenging were found. Using a panel of fibroblast cultures the same approach was used to identify a minimum set of 18 genes that could differentiate patients with a high risk of radiation-induced fibrosis from those with a low risk [63]. This classifier of 18 genes has allowed the identification of genes that are involved in the fibrotic process and may provide a basis for a predictive assay for normal tissues reactions after radiotherapy and new candidate genes to be examined in single nucleotide polymorphism (SNP) association studies.

There are also reports that single DNA response gene transcript levels could be used as biomarkers for developing adverse reactions. For instance, Wiebalk et al.

[81] found that prostate cancer patients with *in vitro* radiation-induced levels of *XPC* mRNA above the 90th percentile compared to those with lower induction levels were at increased risk of suffering from adverse reactions during radiotherapy (odds ratio 5.3, 95% confidence interval 1.2–24.5; adjusted for smoking). The predictive value of such comparatively simple measurements as biomarkers for radiosensitivity requires confirmation and extension taking into account possible confounding factors: *XPC* induction levels were found to be strongly correlated with smoking in the population studied and smoking was associated with a non-significantly lower risk of acute toxicity.

Expression patterns may also be used for prognostic purposes. For example, Ye et al. [84] studied *ATM* gene expression in tumour and adjacent normal tissue, by quantitative real-time reverse transcription PCR, in a cohort of 471 breast cancer patients and found that *ATM* gene expression was down-regulated in breast cancer tissues (as seen by [4, 72] for ATM protein expression by IHC). High *ATM* expression levels were associated with improved disease-free survival and overall survival following surgery, and chemotherapy in 97% and radiotherapy in 45% of patients respectively.

In terms of prediction of response to chemotherapy, ERCC1 mRNA levels have been found to be associated with sensitivity to platinum-based chemotherapy regimens in a sizable number of studies, in lung, colorectal, ovarian, gastric, oesophageal and bladder cancer, where in general improved response/survival was associated with lower expression levels (reviewed by [32]). Such work has since been taken forward to customise chemotherapy in clinical trial settings based on ERCC1 mRNA levels, with encouraging results. For example, Cobo et al. [18] determined ERCC1 mRNA expression in pre-treatment biopsies from 444 stage IV non-small cell lung cancer patients and patients randomly assigned (1:2 ratio) to a control (docetaxel plus cisplatin) or genotypic arm, with the chemotherapy in the genotypic arm being determined by ERCC1 levels (docetaxel plus cisplatin for low expressors and docetaxel plus gemcitabine for high expressors respectively). In the 346 patients assessable for response, 39.3% of the control and 50.7% of the genotypic arm attained an objective response, implying that ERCC1 mRNA expression predicted response to docetaxel and cisplatin.

14.7 Measurement of Biomarkers at the DNA Level

14.7.1 DNA Adducts and Measurements of Oxidative Stress

DNA adducts have mainly been used as biomarkers of exposure, to such substances as aflatoxin, vinylchloride, and alkylating agents, their levels representing an integrated measure of exposure and repair (recently reviewed by [76]). The sensitivity and specificity of methods for studying DNA adducts have greatly advanced over the past 40 years such that technical considerations in terms of the ability to detect adducts are no longer a limiting factor for the integration of such measurements in studies: methodological issues relating to study design and statistical power do remain major concerns. As the damage caused by reactive oxygen species (ROS) has been implicated in a myriad of diseases, cancer and aging, much effort has been put into the development of techniques to measure ROS adducts. One of the most studied DNA oxidative damage biomarkers is 8-oxo-7,8-dihydroguanine (8-oxoGua) and its deoxyribonucleoside equivalent, 8-oxodG [19]. Whilst 8-oxoG is increasingly being regarded as a reliable biomarker of oxidative stress which may have a predictive value for cancer risk and its measurement has been particularly used is in dietary intervention studies on the ingestion of anti-oxidant containing foods or tablets (recently reviewed by Loft et al. [51]) it is not used clinically in the prognostic and predictive settings, and therefore shall not be discussed further here.

14.7.2 Germline Mutations as Biomarkers

Mutations in several DNA damage signalling (*ATM, NBS1*), cell cycle checkpoint (*TP53*) and DNA repair genes are seen in rare inherited disorders associated with a high risk of developing cancer (reviewed by [43]). For example, germline mutations in mismatch repair genes are associated with colorectal tumours, *BRCA1* and *BRCA2* mutations are associated with early onset breast and ovarian cancer, *NBS1* is mutated in Nijmegen breakage syndrome which is associated with lymphoma, and xeroderma pigmentosum, which carries with it a very high incidence of skin cancer, is a condition characterised by mutations in one of *XPA, B, C, E, F* or *G*. These genes are involved in the NER pathway, so skin cancers arise due to failure to repair UV-induced DNA damage.

The germline mutations in these inherited disorders result in disruption of DNA repair, with persistence of mutations and development of genomic instability through chromosomal breakage; for the tumour cells in affected individuals, this results in increased sensitivity to DNA damaging agents. For example, testicular germ cell tumours, which are extremely sensitive to cisplatin-based chemotherapy (whose damage is repaired by NER) exhibit loss of expression of *XPA*.

Identification of *BCRA1* or *BRCA2* mutations in women with a family history of early breast cancer (5–10% of the breast cancer population) allows these women to be counselled about their risk of developing breast and ovarian cancer and some may elect to undergo prophylactic mastectomy or ovariectomy. More recently, an exciting development is the finding that inhibitors of a BER enzyme, PARP-1, are highly selective for tumours cells with HR defects, particularly in the context of *BRCA1* and *BRCA2* mutations, a process known as synthetic lethality. Following DNA damage, SSBs persist as a result of PARP inhibition, and when a SSB meets a replication fork a DSB may be formed, which in normal cells is repair by HR. However, in *BRCA1* or *2* defective tumours, HR is inactivated, so the cells repair the DSB by NHEJ or SSA resulting in complex chromosomal rearrangements, and hence permanent cell cycle arrest or apoptosis [53]. Phase I and II trials of PARP inhibitors as monotherapy for breast and ovarian cancer in individuals carrying *BRCA1* or *2* mutations are currently underway, with the mutation itself being used as the predictive biomarker to direct molecularly-guided therapy.

The damage response pathways affected by germline mutations are often also those disrupted by somatic mutations or epigenetic changes in tumour cells of sporadic cancers. *TP53* is one of the most extensively investigated damage response genes in this regard, in terms of its potential as a prognostic or predictive biomarker. *TP53* germline mutations predispose to a wide spectrum of early onset cancers associated with the Li-Fraumeni and Li-Fraumeni-like syndromes, and somatic mutations are frequent in most human cancers.

Early reports investigating the prognostic value of *TP53* mutations for tumour responses and patient outcomes produced conflicting results, as many of these studies relied on IHC to assess p53 alterations. However the presence of a mutation has been correlated with shorter survival and a poorer response to treatment at several cancer sites. A number of studies have reported specific types of mutation that were associated with a worse prognosis compared with other mutations. For instance results from a large breast cancer cohort showed that *TP53* mutation was an independent factor of poor prognosis and that missense mutations and truncating mutations located within the DNA binding motifs were associated with worse outcome [59]. *TP53* mutation status has been incorporated into some expression array analysis of tumour subtypes and for instance for breast cancers, was found to cluster with the profiles associated with the worst outcomes [73].

TP53 mutational analysis (exons 4–8) was performed in a large international retrospective study of 3583 patients with colorectal cancer [65], and *TP53* mutations were not found to be of prognostic value in terms of overall survival. However, in the approximately one-third of Dukes' C patients treated with fluorouracil-based chemotherapy wild-type *TP53* status was associated with improved survival following chemotherapy compared to no chemotherapy in the rectal and proximal colon tumours and mutated *TP53* was associated with improved survival in the proximal colon group.

14.7.3 Detection of Circulating Free Mutant DNA (ctDNA)

It has been found that tumour-derived mutant DNA can be detected in plasma or serum of individuals with cancer [31]. This circulating tumour DNA (ctDNA) is difficult to detect and quantify as the number of circulating mutated gene fragments is small compared to the number of normal circulating DNA fragments (in the region of <0.01%). Diehl et al. [22] studied 162 plasma samples from 18 patients receiving surgery for colorectal cancer and determined the mutant DNA concentration in plasma using BEAMing (bead emulsion amplification magnetics) and real-time PCR, having previously identified the somatic mutations in the patients' tumours by direct sequencing. Whilst ctDNA levels were generally reduced after surgery, they were still detectable in most cases. They were prognostic for relapse-free survival and appeared to be a more reliable and sensitive indicator of relapse than the CEA measurements currently used. The authors proposed that ctDNA measurements could be used for non-invasive monitoring of many types of cancer.

14.7.4 Gene Promoter Methylation as a Predictive Factor

Around 15% of sporadic cases of colorectal cancer display microinstability (MSI), which in this situation is generally caused by abnormal promoter methylation of the MMR gene, *MLH1* [71]. Whilst MSI is associated with better prognosis, these tumours are resistant to fluorouracil-based chemotherapy [56]. As promoter methylation prevents the normal activation of this MMR gene, the development of a drug targeting this DNA repair pathway could have an impact on treatment of colorectal cancer, and *MLH1* promoter methylation status could be used as a predictive test for case selection.

The MGMT gene is responsible for removing alkyl groups from the O^6-position of guanine and this gene can also be epigenetically silenced by promoter methylation. Such silencing is associated with loss of MGMT expression and diminished DNA repair activity. In glioblastoma, promoter methylation of MGMT was associated with a longer survival in patients treated with radiotherapy and the alkylating agent temozolomide. In a subsequent phase III randomized trial, where patients received radiotherapy plus or minus temozolamide, Hegi et al. [36] found that in 206 glioblastoma specimens, where MGMT methylation status was determined by methylation-specific polymerase chain reaction, patients whose tumours had MGMT promoter methylation had a significantly improved overall survival irrespective of treatment assignment, with a hazard ratio for death of 0.45 (95% confidence interval 0.32–0.61). The patients with the longest survival were those who received temozolamide and radiotherapy with promoter methylation, with a 46% 2 year survival compared to 22.7% in those with promoter methylation assigned to radiotherapy alone. However, in patients whose tumours were not methylated at the MGMT promoter, there was only a marginally significant difference in overall survival between the two treatment arms. These data suggested that MGMT promoter methylation may have both prognostic value and also predictive value in terms of response to temozolamide.

14.7.5 Single Nucleotide Polymorphisms and Genome Wide Association Studies: Cancer Risk and Pharmacogenetics

The most frequent form of genetic variation is the single nucleotide polymorphism (SNP), involving the alteration of a single DNA base at polymorphic frequency, defined as greater than 1%. It is estimated that there are around 10 million SNPs in the human genome. SNPs account for most of the known genetic variation between individuals and may fall within coding sequences of genes, non-coding regions of genes such as the promoter region, introns or 3′untranslated region, or the regions between genes. Those within a coding sequence may or may not change the amino acid sequence of the protein that is produced (non-synonymous and synonymous SNPs respectively), due to degeneracy of the genetic code. SNPs outside the protein-coding regions can modify transcriptional or translational control. SNP association studies have been undertaken on the assumption that patients with a common, complex disease such as cancer are likely to share some common, low penetrance alleles that increase their disease susceptibility [8].

Many early studies, where only a few SNPs were investigated, were small and failed to be replicated, as false-positive reporting rates were high. However, in terms of larger studies in single tumour sites, several studies have been undertaken in bladder cancer for example, where known risk factors include exposure to cigarette smoke, which produces oxidative DNA damage, and to chemical agents whose metabolites produce bulky DNA adducts. These lesions are repaired by the BER and NER pathways respectively. Studies have found associations between BER and NER SNP variants and increased risk of bladder cancer, and in a pathway approach, Wu et al. [83] found significant gene dosage effects [27, 29, 66, 83]. Not only are the bulky adducts formed by the chemical carcinogens repaired by DNA repair mechanisms, but the carcinogens themselves are metabolised and detoxified by several metabolising enzymes, including glutathione S-transferases (GSTs), N-acetyltransferases (NATs) and cytochrome p450 (CYP) 1B1, so an individual's ability to detoxify carcinogens is also an important risk factor in bladder cancer formation. The GSTM1 null genotype and NAT2 slow acetylation status are associated with increased risk of bladder cancer [30].

Dong et al. [23] recently reviewed 161 meta-analyses and pooled analyses for genetic polymorphisms and cancer risk, which included 18 cancer sites and 99 genes, in studies of at least 500 cases. They took into account false-positive report probabilities (FPRP), and for a prior probability of 0.001, considered appropriate for a candidate gene with known function and biologically plausible effects, and a statistical power to detect an odds ratio of 1.5, they identified 13 gene variants associated with cancer a FPRP of <0.2. These included the DNA repair genes *XPD* and *XRCC1*, the cell cycle genes *MDM2* and *CHK2* and the apoptosis gene *CASP8*. For a very low prior probability of 0.000001 (as for genome-wide association studies, see below) and a statistical power to detect an odds ratio of 1.5, they identified four gene-variant cancer associations with a FPRP of <0.2, which included GSTM1 null status and bladder cancer and NAT2 slow acetylator status and bladder cancer.

A number of genome-wide association studies have recently been reported in prostate, breast, colon and bladder cancer, where hundreds of thousands of variants have been genotyped across the whole genome, allowing capture of 65–75% of common genetic variation (for recent review of genome wide association studies see [49]). These have shown various SNPs, particularly in the gene desert region of chromosome 8q24, to be associated with increased cancer risk, but effects sizes are modest, with odds ratios in the region of 1.1–1.5 [12, 34, 40, 44, 77, 78, 85]. None of the statistically significant variants had been previously identified as they mostly did not reside in "interesting" candidate regions. Genome wide statistical significance levels are necessarily very stringent, in the region of $p < 5 \times 10^{-7}$ (0.05/number of SNPs tested) to allow for multiple testing. This means that many disease-causing variants with p values greater than 5×10^{-7} will be missed by this approach.

Whilst the above association studies are concerned with determining cancer risk, there is also interest in using SNPs to predict therapeutic outcomes following radiotherapy or chemotherapy treatments. There is a wide variation in the efficacy and toxicity of drug and radiation treatment across the population of individuals treated, and toxicities can be life-threatening or severely impair quality of life, thus an ability to predict and avoid such effects would be highly valuable. Up to 95% of inter-individual variability in responses to drugs and radiation is determined by genetic factors [25]. The evolving field of pharmacogenetics uses genetic profiling to predict the response of tumour and normal tissues to therapy. Gene products of SNPs can act upstream of drug targets (drug transporters, and phase I and II metabolising enzymes), or downstream (apoptosis-regulating proteins and chemokines), or can be proteins which interact with anticancer drugs, including DNA biosynthesis and metabolism enzymes, DNA repair enzymes, and mitotic spindle proteins (for review see [24]).

The cytotoxicity of both chemotherapy and radiotherapy is to a large extent related to their ability to induce DNA damage and thus the ability of a cell to activate the necessary signalling pathways that slow cell cycle progression, activate apoptotic mechanisms of cell death and allow damage repair. There is a small but growing body of evidence that sequence variants within DNA damage response genes may contribute to therapeutic responses to chemotherapy and radiotherapy and normal tissue response to radiotherapy (see for instance recent reviews by [2, 26, 32]). To date attempts to identify sequence variants associated with therapeutic responses have almost exclusively used a candidate gene approach assessing the frequency of common variants in different groups of patients. Over the past few years there has been a clear trend away from studies where only single SNPs in a gene were genotyped and more groups are reporting genotyping data for multiple SNPs in the same gene allowing associations based on the presence of inferred haplotypes to be assessed. Certain groups have put together genotyping panels for assessing responses to cancer chemotherapy [21]. In such studies the selection of genes and polymorphisms is critical.

A comprehensive search for genetic prognostic and predictive biomarkers would ideally examine all the genetic differences in a large number of affected individuals and controls. As discussed above, genome wide association studies have successfully identified variants associated with increased risk of developing cancer and reports on the identification of sequence variants associated with modified therapeutic responses are beginning to appear in the literature in other diseases. For example, Byun et al. [14] studied DNA from 206 multiple sclerosis patients using Affymetrix 100 K GeneChips and compared allelic frequencies between good responders and non-responders to interferon beta therapy. They found that the two groups had significantly different genotype frequencies for SNPs located in many genes, particularly those associated with ion channels and signal transduction pathways. Interestingly, genes from previous pharmacogenetic studies were not confirmed. Liu et al. [50], using the Illumina HapMap300 SNP chip on 89 patients with rheumatoid arthritis, prospectively followed after starting anti-TNF therapy, found multiple significant associations between SNP markers and response but cautioned that independent replications studies were required. It can only be a matter of time before such reports emerge in the cancer field.

14.8 Conclusions

Biomarker research in the clinical setting ultimately strives to identify markers for diagnosis, prognosis and therapy. Success of a new marker can be defined as its showing statistically significant associations independent of parameters that are currently being used. Combining routinely used, often pathological-based techniques with more molecular prognostic factors is becoming more frequent but most remain to be validated in randomized trials and the option to use, for example, gene expression profiles in the treatment decision-making process is far from standard practice for many cancer sites. Traditionally, diagnostic tests for cancers are relatively cost-effective and have been carried out in local hospitals. However the introduction of DNA-microarray based tests, for example, which are more expensive and technically challenging, into therapy decision-making and the possibility that bodies such as the FDA and the European Medicines Agency will regulate molecular diagnostics will impact on their integration into medical practice.

In order to develop and validate new biomarkers integrated approaches are needed. For instance, investigating DNA, RNA, and proteins in the same biological samples and examining whether loss of expression of a protein is linked to the presence of DNA mutations, loss of heterozygosity or gene copy numbers, promoter methylation, aneuploidy, microsatellite instability or RNA expression profiles, including that of microRNAs, will allow a better understanding of the underlying molecular mechanisms that are altered during carcinogenesis. There is clearly a need to embed biomarker discovery and validation into randomised clinical trials and prospective studies that minimise biases and a need for multi-centre cooperative efforts so that studies are not statistically underpowered and give robust outcome data. There are instances where it is useful to monitor tumour response during treatment but obtaining additional tissue samples is often challenging, and here the development of non-invasive approaches to biomarker discovery will be particularly useful.

There are potentially several endpoints of the DNA damage response pathway that might be useable as predictive biomarkers in the context of cancer treatment and therapy. For instance the formation of DNA damage foci, such as gamma-H2AX foci, is extensively used in experimental settings as a marker for DSB formation and persistence (recently reviewed by [46]). However foci detection requires the use of antibodies and thus appropriate biological samples for ex-vivo measurements and whilst such measurement in hair follicles are a useful means of assessing PARP-1 inactivation by PARP inhibitors they are not yet routinely applied to clinical practice.

Biomarkers associated with DNA damage response genes clearly have potential in a clinical setting. An ability to predict an individual's outcome to several alternative treatment regimens, including radiotherapy and various DNA damaging chemotherapeutic agents, would allow patients to be directed to the treatment combination with the highest likelihood of cure and the lowest likelihood of unacceptable toxicity. The challenge for the future will be to identify and validate those biomarkers that can add information on prognosis and outcome independently of previously known and established clinical and biomarkers parameters.

Acknowledgments This review was prepared during Anne Kiltie's Collaborator's visit to the Institute Curie, Orsay, funded by a Cancer Research UK Collaborator Bursary as part of her Cancer Research UK Clinician Scientist Fellowship.

References

1. Adlard JW, Richman SD, Seymour MT et al. (2002) Prediction of the response of colorectal cancer to systemic therapy. Lancet Oncol 3: 75–82.
2. Alsner J, Andreassen CN and Overgaard J (2008) Genetic markers for prediction of normal tissue toxicity after radiotherapy. Semin Radiat Oncol 18: 126–35.
3. Alsner J, Rodningen OK and Overgaard J (2007) Differential gene expression before and after ionizing radiation of subcutaneous fibroblasts identifies breast cancer patients resistant to radiation-induced fibrosis. Radiother Oncol 83: 261–6.
4. Angèle S, Treilleux I, Bremond A et al. (2003) Altered expression of DNA double-strand break detection and repair proteins in breast carcinomas. Histopathology 43: 347–53.
5. Arriagada R, Bergman B, Dunant A et al. (2004) Cisplatin-based adjuvant chemotherapy in patients with completely resected non-small-cell lung cancer. N Engl J Med 350: 351–60.
6. Banks RE, Craven RA, Harnden P et al. (2007) Key clinical issues in renal cancer: a challenge for proteomics. World J Urol 25: 537–56.
7. Barrow E, McMahon R, Evans DG et al. (2008) Cost analysis of biomarker testing for mismatch repair deficiency in node-positive colorectal cancer. Br J Surg 95: 868–75.
8. Baynes C, Healey CS, Pooley KA et al. (2007) Common variants in the ATM, BRCA1, BRCA2, CHEK2 and TP53 cancer susceptibility genes are unlikely to increase breast cancer risk. Breast Cancer Res 9: R27.
9. Bellmunt J, Paz-Ares L, Cuello M et al. (2007) Gene expression of ERCC1 as a novel prognostic marker in advanced bladder cancer patients receiving cisplatin-based chemotherapy. Ann Oncol 18: 522–8.
10. Biomarkers Definitions Working Group (2001) Biomarkers and surrogate endpoints: preferred definitions and conceptual framework. Clin Pharmacol Ther 69: 89–95.
11. Braun MS, Richman SD, Quirke P et al. (2008) Predictive biomarkers of chemotherapy efficacy in colorectal cancer: results from the UK MRC FOCUS trial. J Clin Oncol 26: 2690–8.
12. Broderick P, Carvajal-Carmona L, Pittman AM et al. (2007) A genome-wide association study shows that common alleles of SMAD7 influence colorectal cancer risk. Nat Genet 39: 1315–7.
13. Bubendorf L, Nocito A, Moch H et al. (2001) Tissue microarray (TMA) technology: miniaturized pathology archives for high-throughput in situ studies. J Pathol 195: 72–9.
14. Byun E, Caillier SJ, Montalban X et al. (2008) Genome-wide pharmacogenomic analysis of the response to interferon beta therapy in multiple sclerosis. Arch Neurol 65: 337–44.
15. Callagy GM, Pharoah PD, Pinder SE et al. (2006) Bcl-2 is a prognostic marker in breast cancer independently of the Nottingham Prognostic Index. Clin Cancer Res 12: 2468–75.
16. Capalbo G, Rodel C, Stauber RH et al. (2007) The role of survivin for radiation therapy. Prognostic and predictive factor and therapeutic target. Strahlenther Onkol 183: 593–9.

17. Chatterjee SJ, Datar R, Youssefzadeh D et al. (2004) Combined effects of p53, p21, and pRb expression in the progression of bladder transitional cell carcinoma. J Clin Oncol 22: 1007–13.
18. Cobo M, Isla D, Massuti B et al. (2007) Customizing cisplatin based on quantitative excision repair cross-complementing 1 mRNA expression: a phase III trial in non-small-cell lung cancer. J Clin Oncol 25: 2747–54.
19. Collins AR, Cadet J, Moller L et al. (2004) Are we sure we know how to measure 8-oxo-7,8-dihydroguanine in DNA from human cells? Arch Biochem Biophys 423: 57–65.
20. Cordon-Cardo C (2004) p53 and RB: simple interesting correlates or tumor markers of critical predictive nature? J Clin Oncol 22: 975–7.
21. Dai Z, Papp AC, Wang D et al. (2008) Genotyping panel for assessing response to cancer chemotherapy. BMC Med Genomics 1: 24.
22. Diehl F, Schmidt K, Choti MA et al. (2008) Circulating mutant DNA to assess tumor dynamics. Nat Med 14: 985–90.
23. Dong LM, Potter JD, White E et al. (2008) Genetic susceptibility to cancer: the role of polymorphisms in candidate genes. JAMA 299: 2423–36.
24. Efferth T and Volm M (2005) Pharmacogenetics for individualized cancer chemotherapy. Pharmacol Ther 107: 155–76.
25. Evans WE and McLeod HL (2003) Pharmacogenomics – drug disposition, drug targets, and side effects. N Engl J Med 348: 538–49.
26. Fernet M and Hall J (2008) Predictive markers for normal tissue reactions: fantasy or reality? Cancer Radiother 12: 614–18.
27. Figueroa JD, Malats N, Real FX et al. (2007) Genetic variation in the base excision repair pathway and bladder cancer risk. Hum Genet 121: 233–42.
28. Garcia del Muro X, Condom E, Vigues F et al. (2004) p53 and p21 Expression levels predict organ preservation and survival in invasive bladder carcinoma treated with a combined-modality approach. Cancer 100: 1859–67.
29. Garcia-Closas M, Malats N, Real FX et al. (2006) Genetic variation in the nucleotide excision repair pathway and bladder cancer risk. Cancer Epidemiol Biomarkers Prev 15: 536–42.
30. Garcia-Closas M, Malats N, Silverman D et al. (2005) NAT2 slow acetylation, GSTM1 null genotype, and risk of bladder cancer: results from the Spanish Bladder Cancer Study and meta-analyses. Lancet 366: 649–59.
31. Gormally E, Caboux E, Vineis P et al. (2007) Circulating free DNA in plasma or serum as biomarker of carcinogenesis: practical aspects and biological significance. Mutat Res 635: 105–17.
32. Gossage L and Madhusudan S (2007) Current status of excision repair cross complementing-group 1 (ERCC1) in cancer. Cancer Treat Rev 33: 565–77.
33. Goulart BH, Clark JW, Pien HH et al. (2007) Trends in the use and role of biomarkers in phase I oncology trials. Clin Cancer Res 13: 6719–26.
34. Gudmundsson J, Sulem P, Manolescu A et al. (2007) Genome-wide association study identifies a second prostate cancer susceptibility variant at 8q24. Nat Genet 39: 631–7.
35. Hanash SM, Pitteri SJ and Faca VM (2008) Mining the plasma proteome for cancer biomarkers. Nature 452: 571–9.
36. Hegi ME, Diserens AC, Gorlia T et al. (2005) MGMT gene silencing and benefit from temozolomide in glioblastoma. N Engl J Med 352: 997–1003.
37. Hendriks Y, Franken P, Dierssen JW et al. (2003) Conventional and tissue microarray immunohistochemical expression analysis of mismatch repair in hereditary colorectal tumors. Am J Pathol 162: 469–77.
38. Hoeijmakers JH (2001) Genome maintenance mechanisms for preventing cancer. Nature 411: 366–74.
39. Hummerich J, Werle-Schneider G, Popanda O et al. (2006) Constitutive mRNA expression of DNA repair-related genes as a biomarker for clinical radio-resistance: a pilot study in prostate cancer patients receiving radiotherapy. Int J Radiat Biol 82: 593–604.

40. Hunter DJ, Kraft P, Jacobs KB et al. (2007) A genome-wide association study identifies alleles in FGFR2 associated with risk of sporadic postmenopausal breast cancer. Nat Genet 39: 870–4.
41. Jackson SP (2002) Sensing and repairing DNA double-strand breaks. Carcinogenesis 23: 687–96.
42. Karam JA, Lotan Y, Karakiewicz PI et al. (2007) Use of combined apoptosis biomarkers for prediction of bladder cancer recurrence and mortality after radical cystectomy. Lancet Oncol 8: 128–36.
43. Kennedy RD and D'Andrea AD (2006) DNA repair pathways in clinical practice: lessons from pediatric cancer susceptibility syndromes. J Clin Oncol 24: 3799–808.
44. Kiemeney LA, Thorlacius S, Sulem P et al. (2008) Sequence variant on 8q24 confers susceptibility to urinary bladder cancer. Nat Genet 40: 1307–12.
45. Kruse JJ and Stewart FA (2007) Gene expression arrays as a tool to unravel mechanisms of normal tissue radiation injury and prediction of response. World J Gastroenterol 13: 2669–74.
46. Kuo LJ and Yang LX (2008) Gamma-H2AX – a novel biomarker for DNA double-strand breaks. In Vivo 22: 305–9.
47. Kwon HC, Roh MS, Oh SY et al. (2007) Prognostic value of expression of ERCC1, thymidylate synthase, and glutathione S-transferase P1 for 5-fluorouracil/oxaliplatin chemotherapy in advanced gastric cancer. Ann Oncol 18: 504–9.
48. Kyndi M, Sorensen FB, Knudsen H et al. (2008) Impact of BCL2 and p53 on postmastectomy radiotherapy response in high-risk breast cancer. A subgroup analysis of DBCG82 b&c. Acta Oncol 47: 608–17.
49. Lango H and Weedon MN (2008) What will whole genome searches for susceptibility genes for common complex disease offer to clinical practice? J Intern Med 263: 16–27.
50. Liu C, Batliwalla F, Li W et al. (2008) Genome-wide association scan identifies candidate polymorphisms associated with differential response to anti-TNF treatment in Rheumatoid Arthritis. Mol Med 14: 575–81.
51. Loft S, Moller P, Cooke MS et al. (2008) Antioxidant vitamins and cancer risk: is oxidative damage to DNA a relevant biomarker? Eur J Nutr 47 Suppl 2: 19–28.
52. Marchionni L, Wilson RF, Wolff AC et al. (2008) Systematic review: gene expression profiling assays in early-stage breast cancer. Ann Intern Med 148: 358–69.
53. Martin SA, Lord CJ and Ashworth A (2008) DNA repair deficiency as a therapeutic target in cancer. Curr Opin Genet Dev 18: 80–6.
54. Moumen A, Masterson P, O'Connor MJ et al. (2005) hnRNP K: an HDM2 target and transcriptional coactivator of p53 in response to DNA damage. Cell 123: 1065–78.
55. Nirmalan NJ, Harnden P, Selby PJ et al. (2008) Mining the archival formalin-fixed paraffin-embedded tissue proteome: opportunities and challenges. Mol Biosyst 4: 712–20.
56. Niv Y (2007) Microsatellite instability and MLH1 promoter hypermethylation in colorectal cancer. World J Gastroenterol 13: 1767–9.
57. Nuyten DS and van de Vijver MJ (2008) Using microarray analysis as a prognostic and predictive tool in oncology: focus on breast cancer and normal tissue toxicity. Semin Radiat Oncol 18: 105–14.
58. Olaussen KA, Dunant A, Fouret P et al. (2006) DNA repair by ERCC1 in non-small-cell lung cancer and cisplatin-based adjuvant chemotherapy. N Engl J Med 355: 983–91.
59. Olivier M, Langerod A, Carrieri P et al. (2006) The clinical value of somatic TP53 gene mutations in 1,794 patients with breast cancer. Clin Cancer Res 12: 1157–67.
60. Paik S, Shak S, Tang G et al. (2004) A multigene assay to predict recurrence of tamoxifen-treated, node-negative breast cancer. N Engl J Med 351: 2817–26.
61. Pavon MA, Parreno M, Leon X et al. (2008) Ku70 predicts response and primary tumor recurrence after therapy in locally advanced head and neck cancer. Int J Cancer 123: 1068–79.

62. Perou CM, Sorlie T, Eisen MB et al. (2000) Molecular portraits of human breast tumours. Nature 406: 747–52.
63. Rodningen OK, Borresen-Dale AL, Alsner J et al. (2008) Radiation-induced gene expression in human subcutaneous fibroblasts is predictive of radiation-induced fibrosis. Radiother Oncol 86: 314–20.
64. Rodningen OK, Overgaard J, Alsner J et al. (2005) Microarray analysis of the transcriptional response to single or multiple doses of ionizing radiation in human subcutaneous fibroblasts. Radiother Oncol 77: 231–40.
65. Russo A, Bazan V, Iacopetta B et al. (2005) The TP53 colorectal cancer international collaborative study on the prognostic and predictive significance of p53 mutation: influence of tumor site, type of mutation, and adjuvant treatment. J Clin Oncol 23: 7518–28.
66. Sak SC, Barrett JH, Paul AB et al. (2006) Comprehensive analysis of 22 XPC polymorphisms and bladder cancer risk. Cancer Epidemiol Biomarkers Prev 15: 2537–41.
67. Shariat SF, Karakiewicz PI, Ashfaq R et al. (2008) Multiple biomarkers improve prediction of bladder cancer recurrence and mortality in patients undergoing cystectomy. Cancer 112: 315–25.
68. Shariat SF, Tokunaga H, Zhou J et al. (2004) p53, p21, pRB, and p16 expression predict clinical outcome in cystectomy with bladder cancer. J Clin Oncol 22: 1014–24.
69. Shedden K, Taylor JM, Enkemann SA et al. (2008) Gene expression-based survival prediction in lung adenocarcinoma: a multi-site, blinded validation study. Nat Med 14: 822–7.
70. Shia J (2008) Immunohistochemistry versus microsatellite instability testing for screening colorectal cancer patients at risk for hereditary nonpolyposis colorectal cancer syndrome. Part I. The utility of immunohistochemistry. J Mol Diagn 10: 293–300.
71. Shia J, Ellis NA and Klimstra DS (2004) The utility of immunohistochemical detection of DNA mismatch repair gene proteins. Virchows Arch 445: 431–41.
72. Soderlund K, Stal O, Skoog L et al. (2007) Intact Mre11/Rad50/Nbs1 complex predicts good response to radiotherapy in early breast cancer. Int J Radiat Oncol Biol Phys 68: 50–8.
73. Sorlie T, Perou CM, Tibshirani R et al. (2001) Gene expression patterns of breast carcinomas distinguish tumor subclasses with clinical implications. Proc Natl Acad Sci USA 98: 10869–74.
74. Stadler WM, Cote R, Learner S et al. (2004) Phase III trial assessing 3 cycles of MVAC in organ-confined bladder cancer with p53. Oncology Times 26: 4.
75. Steel GG (eds) (2002) Basic Clinical Radiobiology. Arnold, London.
76. Swenberg JA, Fryar-Tita E, Jeong YC et al. (2008) Biomarkers in toxicology and risk assessment: informing critical dose-response relationships. Chem Res Toxicol 21: 253–65.
77. Thomas G, Jacobs KB, Yeager M et al. (2008) Multiple loci identified in a genome-wide association study of prostate cancer. Nat Genet 40: 310–5.
78. Tomlinson I, Webb E, Carvajal-Carmona L et al. (2007) A genome-wide association scan of tag SNPs identifies a susceptibility variant for colorectal cancer at 8q24.21. Nat Genet 39: 984–8.
79. van't Veer LJ and Bernards R (2008) Enabling personalized cancer medicine through analysis of gene-expression patterns. Nature 452: 564–70.
80. Wager M, Menei P, Guilhot J et al. (2008) Prognostic molecular markers with no impact on decision-making: the paradox of gliomas based on a prospective study. Br J Cancer 98: 1830–8.
81. Wiebalk K, Schmezer P, Kropp S et al. (2007) In vitro radiation-induced expression of XPC mRNA as a possible biomarker for developing adverse reactions during radiotherapy. Int J Cancer 121: 2340–5.
82. Wu W, Nishikawa H, Hayami R et al. (2007) BRCA1 ubiquitinates RPB8 in response to DNA damage. Cancer Res 67: 951–8.

83. Wu X, Gu J, Grossman HB et al. (2006) Bladder cancer predisposition: a multigenic approach to DNA-repair and cell-cycle-control genes. Am J Hum Genet 78: 464–79.
84. Ye C, Cai Q, Dai Q et al. (2007) Expression patterns of the ATM gene in mammary tissues and their associations with breast cancer survival. Cancer 109: 1729–35.
85. Yeager M, Orr N, Hayes RB et al. (2007) Genome-wide association study of prostate cancer identifies a second risk locus at 8q24. Nat Genet 39: 645–9.

Chapter 15
Linking Human RecQ Helicases to DNA Damage Response and Aging

Wen-Hsing Cheng, Byungchan Ahn, and Vilhelm A. Bohr

Abstract Maintenance of genome integrity is crucial for the survival of an organism. Increasing evidence suggests that defective DNA damage response promotes genome instability and accelerates age-related degeneration. The RecQ family of DNA helicases plays important roles in DNA replication and the DNA damage response. There are five human RecQ helicases, amongst which mutations in *BLM, WRN,* and *RECQ4* have been linked to human diseases displaying symptoms of accelerated aging to various degrees. Recent studies have provided new insights into the mechanisms that control early stages of the DNA damage response, and shed light on common roles of the RecQ helicases in DNA replication and the DNA damage response. Here, we review the evidence for defective DNA damage response being a feature linking mutations in RecQ helicases to premature aging diseases. Data are accumulating to suggest that RecQ helicases impart signaling and repair functions in the response to DNA damage.

Keywords Aging · RecQ · Helicase · DNA damage response

15.1 Introduction: Genome Instability Syndromes and Aging

The genome is the fundamental component for the existence of all living creatures. The DNA in each chromosome intrinsically experiences chemical modifications and breaks during energy production and DNA replication. Moreover, exogenous stress

W. Cheng (✉)
Department of Nutrition and Food Science, University of Maryland, College Park, MD 20742, USA
e-mail: whcheng@umd.edu

B. Ahn (✉)
Department of Life Sciences, University of Ulsan, Ulsan 680-749, Korea
e-mail: bbccahn@mail.ulsan.ac.kr

V.A. Bohr (✉)
Laboratory of Molecular Gerontology, Gerontology Research Center, National Institute on Aging, NIH, Baltimore, MD 21224, USA
e-mail: vbohr@nih.gov

K.K. Khanna, Y. Shiloh (eds.), *The DNA Damage Response: Implications on Cancer Formation and Treatment*, DOI 10.1007/978-90-481-2561-6_15,

including sunlight exposure, environmental clastogens and many other toxins, can invoke DNA damage.

To restore genetic information and avoid mutations, strict regulation is required to recognize and repair damaged DNA, a process known as the DNA damage response (DDR). The DDR includes checkpoint response, DNA repair, and apoptosis. In the double strand break (DSB) branch of DDR, the upstream kinases DNA-dependent protein kinase (DNA-PK), ataxia-telangiectasia mutated (ATM) and ataxia-telangiectasia and RAD3-related (ATR) are activated by the key protein partners (Ku, MRN, and ATRIP, respectively) that bridge a DSB and the kinases (detailed below). In the case of ATM-mediated pathway of DSB response (Fig. 15.1), the inactive ATM dimers are converted to active monomers via direct interaction with MRN [4, 64]. Additionally, recent results link ATM activation to chromatin factors. It is known that ATM monomerization requires acetylation of ATM on lysine 3016 by the Tip60 histone acetyltransferase [104]. Moreover, ATM can mediate chromatin decondensation by phosphorylating KAP-1 (KRAB-associated protein) [115], allowing the genome at heterochromatin region to be preferentially expanded for efficient DSB repair [6, 39]. Then, activated ATM phosphorylates H2AX, known as γH2AX, at the vicinity of the DSB, where MDC1 is recruited and binds to γH2AX for DSB signal transduction [103]. ATM also phosphorylates several downstream substrates (MRN, CHK2, and BRCA1) to initiate checkpoint responses that allow DNA repair, senescence and/or apoptosis to act on the DSB. Once repaired, phospatases such as PP2A and Wip 1 are activated

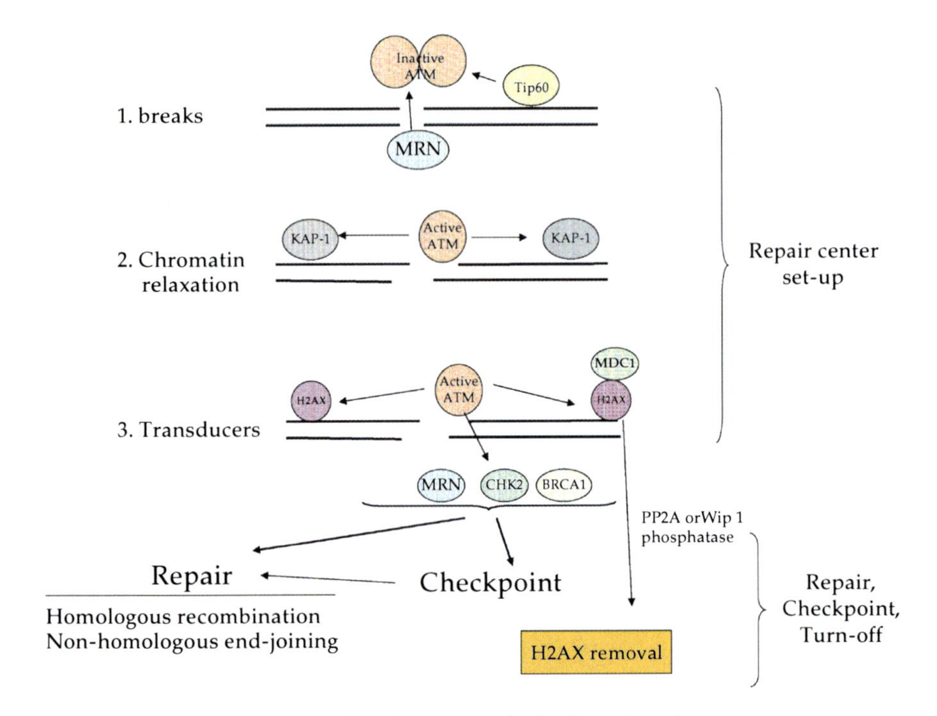

Fig. 15.1 The DNA double strand break response: checkpoint and repair

and dephosphorylates γH2AX, followed by H2AX removal and checkpoint signals being turned off. Pathways of DSB repair include non-homologous end-joining (NHEJ), which is predominant in G1-phase and homologous recombination in S- and G2/M-phases of the cell cycle (for details, see [74]).

The level of somatic mutations increases with age, and this causes defective DDR, which is considered a major factor driving age-related chromosome alternations. This notion is consistent with the known clinical and cellular features of many human premature aging syndromes, where mutations in genes encoding DNA damage signaling or repair proteins are frequently observed.

A strong body of evidence has linked human RecQ helicases to DDR and premature aging syndromes. Human RecQ helicases deal with diverse processes of DNA metabolism including DDR, replication and telomere maintenance (detailed below). There are five human RecQ helicases identified to date, and mutations in three of these lead to chromosome instability syndromes. In particular, inactivation mutations of WRN cause the hallmark premature aging Werner syndrome, and of RECQ4 lead to a subset of Rothmund-Thomson syndrome cases, both of which display multiple symptoms of age-related disorders including early graying and hair loss, cataracts, and osteosarcomas. In contrast, inactivation mutations of BLM lead to Bloom syndrome that is generally considered to predispose to cancer instead of aging [43]. To date, there are no genetic disorders associated with RECQ1 or RECQ5 mutations.

Upstream DNA damage signaling proteins are also crucial in preventing accelerated aging. Ataxia-telangiectasia and Nijmegen breakage syndrome are human genomic instability syndromes caused by inactivating mutations of ATM and NBS1, respectively. These two hereditary disorders show many similarities in clinical and cellular phenotypes including hypersensitivity to cellular exposure to γ-irradiation, defective intra-S-phase checkpoint regulation, cancer predisposition (lymphoma), and immunodeficiency (for details, see [96, 27, 63]). In response to DSBs and other types of DNA damage, the ATM kinase is rapidly activated in the presence of the *M*RE11-*R*AD50-*N*BS1 (MRN) complex. Human ATM activation involves ATM autophosphorylation on specific residues including Ser-1981, Ser-367, and Ser-1893 [4, 58], leading to phosphorylation of key proteins in the various branches of the DDR pathways that mediate checkpoint functions.

The link between a defective ATR pathway and aging has received increasing appreciation. ATR is generally thought to be an upstream kinase, which is activated in response to single-stranded regions of DNA coated with the specific ATR-interacting protein (ATRIP) during replication fork stalling or the repair of bulky lesions such as UV irradiation [77]. Hypomorphic mutations of ATR are causative of a subset of patients with Seckel syndrome, displaying overlapping clinical and cellular features of genomic instability reminiscent of ataxia-telangiectasia and Nijmegen breakage syndrome [74]. ATR is activated in a manner similar to that of ATM [32, 21], and also phosphorylates many of the ATM phosphorylation substrates, leading to cell cycle checkpoint activation [86, 32]. Interestingly, in addition to an important role in activating ATM after γ-irradiation, NBS1 has also been reported to activate ATR pathways after replication stress [101]. Although ATR null mice are embryonic lethal [9, 30], conditional deletion of ATR in adult mice display

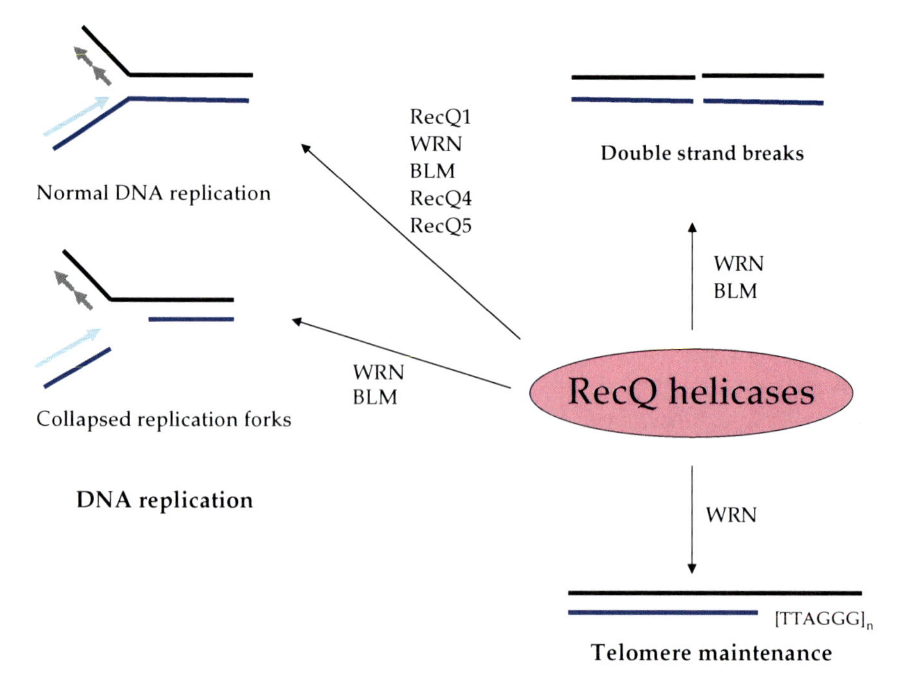

Fig. 15.2 Proposed roles of human RecQ helicases in DNA replication, DNA damage response, and telomere maintenance

premature appearance of age-related phenotypes together with reduced regeneration of stem cells [87].

As chromosome instability syndromes are also associated with accelerated aging processes and age-related disorders, it is conceivable that DDR genes are important in attenuating age-related phenotypes. This review provides information about physical and functional interactions of RecQ helicases with other DDR proteins, have identified specific and distinct functions of human RecQ helicases in DDR, and have helped explained the overlapping but non-redundant activities of human RecQ helicases in DDR pathways. In particular, this review focuses on human RecQ helicases in the DDR mechanisms, and how it relates to accelerated aging in genomic instability syndromes (Fig. 15.2).

15.2 Human RecQ Helicases and DNA Double Strand Break Response

DSBs are severe lesions that compromise the integrity of cells. To restore genomic stability, proteins are dynamically and rapidly recruited to the site for coordinated actions of signal transduction, cell-cycle checkpoint response, and DSB repair. DSBs are induced in response to various DNA-damaging agents including γ-irradiation, radiomimetic chemicals, or reactive oxygen species (ROS). γ-irradiation can directly generate DSBs by high energy input to the chromosome.

Oxidants and γ-irradiation can generate ROS that lead to damaged bases, sugar phosphates, and single-stranded breaks (SSB), some of which can be converted into DSBs following DNA replication.

Central to the early DSB response is the ATM kinase (Fig. 15.1). ATM can be activated by DSBs and other conditions including changes in chromatin structures [20, 4, 96]. Activated ATM then initiates the DSB response network by phosphorylating key players in its numerous branches [64, 96]. Activated ATM can phosphorylate the transcriptional corepressor KAP-1 to regulate the heterochromatin relaxation and facilitate DNA repair [115, 39]. Structural analysis demonstrates that the 53BP1 protein binds to Lysine-79 methylated histone H3 or to lysine-20 dimethylated histone H4 in the response to DSBs [49, 8]. However, whether and how 53BP1 and MRN interact for ATM activation is not fully understood. Biochemical and cellular analyses have shown that NBS1 is required for ATM activation after exposure to γ-irradiation or in the presence of a model DNA structure representing processed DSBs [106, 64]. Using a site-specific chromatin immunoprecipitation assay, it was observed that the recruitment of human ATM to sites of DSBs requires NBS1, and this interaction can facilitate DSB repair [6]. In yeast, an area of about 50-kb of DNA surrounding a DSB is loaded with an array of proteins [56], suggesting a complicated nature of DDR in that it may require cross-talks of multiple checkpoint and repair pathways. H2AX phosphorylation by ATM is a hallmark event after the induction of DSBs. Analysis of a site-specific DSB in yeast suggests that, during DNA damage checkpoint recovery following completion of DSB repair, H2AX falls off the damaged site and is dephosphorylated by the Pph3 phosphatase [56], and chromatin restoration is driven by acetylation of histone H3 at lysine 56 [13].

Inactivation mutations of WRN lead to the genomically instable Werner syndrome, which is characterized by the early onset of aging phenotypes. A cDNA microarray analysis revealed that the gene expression profiles are similar in Werner syndrome patients and normal aging individuals, suggesting that Werner syndrome may be used as a model to study normative aging [60]. WRN is a conserved DNA helicase of the RecQ family that is essential for genomic stability [114]. WRN possesses $3'$–$5'$ helicase and $3'$–$5'$ exonuclease activities [41, 48]. Recent results suggest that WRN plays a role in the response to DSBs. We first identified that the DSB sensor NBS1 interacts with and stimulates WRN helicase activity in the response to DNA damage [16]. Furthermore, WRN re-localization to DNA damage foci after exposure to γ-irradiation or association with γH2AX requires NBS1 [15]. However, WRN may not act as a DNA damage mediator (such as MDC1 and 53BP1), since these proteins contain the BRCT phosphoprotein interaction domain that might facilitate the transient binding to γH2AX [68, 38, 91]. WRN does not contain the BRCT domain and does not bind directly to γH2AX [15]. Rather, WRN localization to sites of DSBs requires the presence of its DNA-binding HRDC domain and follows similar kinetics as NBS1 [61]. Thus, WRN may act as an upstream DNA damage response protein in an NBS1-dependent manner.

Roles of BLM in the response to DSBs have also been documented. A hallmark cellular phenotype of Bloom syndrome is hyper-recombination manifested with an elevated frequency of sister chromatid exchanges and interchromosomal recombination. Biochemical analysis suggests that BLM plays an important role in disassembling the double Holliday junction, preventing crossing-over events during homologous recombination of DSB repair [112]. BLM also may function in the NHEJ pathway of DSB repair. Genetic and cellular analyses suggest interactions of BLM with NHEJ components including the catalytic subunit of DNA-PK, Ku70/80, and Ligase IV [75, 99]. In agreement with the results, there is a Ku70/80-dependent increase in the repair of DSBs by NHEJ in Bloom syndrome cells [36]. Also, BLM was found to localize to sites of DSBs [55].

Mutations in RECQ4 lead to a subset of cases of Rothmund-Thomson syndrome [57], also known as poikiloderma congenitale. Although the role of RECQ4 in unwinding duplex DNA is still unclear, biochemical analysis demonstrated that RECQ4 possesses ATPase activity and is activated by ssDNA [70]. RECQ4 also interacts with Rad51, a key component in the strand invasion event of homologous recombination [78]. Moreover, cellular analysis has demonstrated that RECQ4 interacts with the XPA nucleotide excision repair protein in the response to UV-induced DNA damage [33]. Although RECQ1 or RECQ5 mutations have not been linked to human genetic diseases, their roles in DDR have been suggested. Knockdown of RECQ1 results in the formation of spontaneous γH2AX foci and elevated sister chromatid exchanges [94]. Biochemical analysis suggests that RECQ1 can process Holliday junctions and migrate D-loops [66, 10].

15.3 Human RecQ Helicases and DNA Replication Stress

It is crucial to maintain the integrity of the genome from one generation to the next. During replication, chromosomes are unpacked and vulnerable to damage induction. To avoid genomic instability, accurate replication and effective removal of damaged DNA are required. When DNA replication happens at chromosome ends, telomeres must dissociate from their binding partners, leading to a structure reminiscent of DSBs. To resolve this dilemma, cells have evolved protective mechanisms to avoid induction of spurious DDR [108, 37]. In addition, DNA can be damaged by factors such as ROS or irradiation that can interfere with the replication machinery. When replication is not appropriately managed or DNA lesions are not repaired, genomic instability and accelerated onset of aging can occur.

Replicating chromosomes form a forked structure that encompasses the discontinuous lagging strand DNA synthesis, leaving temporary single-stranded regions of DNA vulnerable for induction of DNA lesions. In particular, a SSB in the vicinity of stalled DNA replication forks can lead to DSB formation, a process known as replication fork collapse. RecQ helicases have been proposed to safeguard the integrity of the replication fork [3]. Here, we focus on recent advances in the roles of human RecQ helicases in the response to replication stress (Fig. 15.2).

BLM plays prominent roles in various aspects of the DNA replication process. Using a fluorescence-based technique to visualize stages of DNA replication, it was observed that an ATR-dependent BLM phosphorylation event is required to prevent replication fork restart and new origin firing under replication stress [28]. By employing oligonucleotide- or plasmid-based assays to assess fork regression activities, BLM has been found to convert a stalled replication fork into structures such as a Holliday junction or a "chicken foot" structure to prevent DSB formation [69, 84]. Fork regression is considered an early event in response to replication fork blockage. The fork regression activity of BLM requires its DNA unwinding activity [69], suggesting that BLM applies a recombinational role for the maintenance of fork stability. Indeed, BLM is suggested to impart functions in resolving replication fork blockage after replication fork collapse or in response to DNA interstrand crosslinks [79]. A recent study showed that BLM, together with ATR and the Mus81 nuclease, is required to generate DSBs at stalled forks and this process is required to resolve mild replication stress that appears to be independent of the DNA replication checkpoint [97, 35]. Moreover, BLM has been reported to be recruited to sites of stalled replication via the DNA damage mediator 53BP1 [93].

Rqh1 and Sgs1 are the sole members of the RecQ helicase family in fission and budding yeasts, respectively. Rqh1 is not essential for a hydroxyurea (HU)-induced S-phase checkpoint activation, but prevents aberrant recombination events during checkpoint recovery [100]. In contrast, Sgs1 has been implicated in checkpoint activation and in preventing the accumulation of aberrant recombinational intermediates in response to replication stress [34, 67]. Although studies in yeast suggest a role of RECQ helicases in S-phase checkpoint activation, there is no substantial evidence to support a role for BLM in S-phase checkpoint response. BLM is not required for either HU-induced S-phase checkpoint or for the recovery from S-phase arrest [28, 1].

WRN may play multifaceted roles during DNA replication. Werner syndrome fibroblasts display a reduced number of replication initiation sites, a prolonged S phase, and defective sister chromatid exchange at telomeres in unstressed cells [105, 25, 62, 82]. Interestingly, Werner syndrome cells with DNA interstrand crosslinks undergo apoptosis in S-phase, which is likely due to accumulation of cytotoxic DSBs during DNA replication [83, 7, 81, 85]. Consistent with this notion, available evidence indicates a role for WRN in the S-phase checkpoint response after replication stress. Biochemical and electron microscopic analyses support that WRN binds preferentially to DNA substrates that resemble replication structures [109, 22]. In humans, earlier studies suggested that WRN and RPA interact to resolve blocked oligonucleotide substrates resembling replication fork structures [19], and prevent aberrant recombination intermediates in HU-arrested S-phase cells [88, 23]. Under certain conditions, WRN is required for ATM pathway activation, and this checkpoint function is specific to conditions that induce replication fork collapse [14]. Similarly, it has been demonstrated that there is an increased rate of replication fork elongation after replication stress in Werner syndrome cells [98]. Moreover, chromosome fragile sites are considered as regions sensitive to replication stress, and WRN has been shown to collaborate with ATR in the maintenance of fragile site stability [80]. Although ATM acts upstream of ATR in the S-phase checkpoint response to γ-irradiation, ATR acts upstream of ATM in the response to UV damage and stalled replication forks [26, 52, 102]. Altogether, it is likely that ATM and ATR collaborate in the response to collapsed replication forks. Under such circumstances, WRN is required for ATM pathway activation that confers an intra-S checkpoint function [14]. RecQ in *E. coli* also processes replication fork structures, thus signaling and stabilizing stalled replication forks in vivo [44]. Collectively, we propose that WRN may recognize DNA damage during replication, yet WRN's helicase or exonuclease activities may reveal secondary DNA structures necessary to generate checkpoint signals, leading to ATM activation and suppression of tumorigenesis. Future experiments are required to delineate the intricacies between ATM, ATR, and WRN in the cellular response to replication fork collapse.

Other RecQ helicases have also been implicated in DNA replication. Although a role of human RECQ4 protein in DNA replication is not clear, data generated from studying Xenopus RECQ4 reveals a function in DNA replication. Xenopus RECQ4 can interact with polymerase α for replication initiation [71, 59, 89]. In addition to RECQ4, recent biochemical analysis demonstrated that RECQL5β possesses fork regression activities [54], and RECQ1 has been shown to exhibit branch migration activity [10]. Clearly, future studies are required to elucidate the mechanism by which RECQ1, RECQ4 and RECQ5 participate in DDR pathways and replication stress.

15.4 Mouse Models Associated with RecQ Helicase Deficiency

Studies on knockout and transgenic mice manipulating expression of RecQ helicases have provided useful models for age-related and genome instability syndromes. Interestingly, *Wrn*$^{-/-}$ mice do not show premature aging phenotypes observed in human Werner syndrome unless mouse Wrn is deficient in a Terc-null background [12, 31]. These results suggest that the existence of long telomeres in mice prohibit the appearance of aging phenotypes in *Wrn*$^{-/-}$ mice, and that WRN is crucial to limit genome instability and senescence at telomeres. Indeed, mice have longer telomeres than humans (25–40 vs. 4–15 kb). By titrating out the telomere factor, *Wrn*$^{-/-}$*Terc*$^{-/-}$ mice are useful for the dissection of physiological roles of WRN during the aging process. Consistent with this concept, some premature senescence and genomic instability characteristics of Werner syndrome cells can be prevented by ectopic expression of telomerase [113, 24]. In the absence of telomerase, primary human fibroblasts exhibit telomere shortening after each round of cell division [111], resulting in the onset of replicative senescence. As originally described by Hayflick [42], there is a finite replicative lifespan of human fibroblasts in culture. Additionally, recent results have shown that dysfunctional telomeres can activate DDR pathways as cells may fail to distinguish a shortened telomere from a DNA break [51, 11, 29]. Functional telomeres are presumably protected by creating a "D-loop" structure comprising telomeres and associating proteins; however, critically short telomeres in senescing cells may not be able to form such structure, and thus chromosome ends are exposed and recognized as DNA breaks. Interesting, functional telomeres can also be recognized as DNA damage during the G2 phase of the cell cycle prior to completion of a cell cycle division [107]. It is clear that results using the $Wrn^{-/-}Terc^{-/-}$ mice indicate that WRN maintains telomere stability and attenuates the onset of aging pathologies.

Targeted disruption of BLM in mice is embryonic lethal [18]. Conditional knockout of BLM yields viable mice and exhibits increased cancer susceptibility after 1 year [17]. However, there are no symptoms of premature aging observed in the BLM conditional knockout mice. The results obtained from mouse studies are consistent with the phenotypes of Bloom syndrome in which the patients exhibit cancer predisposition, but not symptoms of accelerated aging.

Rothmund-Thomson syndrome is considered a premature aging syndrome. Similar to BLM targeted disruption, RECQ4 null mice are embryonic lethal [45]. However, deletion of the RECQ4 helicase domain results in viable mice, and these mice show symptoms of severe growth retardation and skin abnormalities similar to what is observed in a subset of Rothmund-Thomson syndrome patients [45]. Although a helicase activity of RECQ4 has not been demonstrated, results from the RECQ4 helicase domain deficient mice suggest an important role of RECQ4 helicase to prevent the onset of aging phenotypes.

Although cells from RECQ5 or RECQ1 knockout mice show elevated rates of sister chromatid exchange and defective homologous recombination [46, 47, 94, 95], there has been no report yet that describes age-related pathologies in these mice. Since mutations of RECQ1 and RECQ5 have not been linked to any genetic disorders, mouse models of RECQ1 and RECQ5 will be valuable to assess their physiological roles during the process of aging and DDR.

15.5 Other RecQ Helicases

In *Drosophila melanogaster* there are three RecQ helicases, and in *Caenorhabditis elegans*, there are four. The suitability of two organisms for extensive and sophisticated genetic manipulation permits us to investigate regulatory pathways and the impacts of RecQ helicase loss at organismal, cellular, and molecular levels.

In Drosophila, a WRN-like exonuclease and BLM have been studied genetically and only RecQ5 has been characterized biochemically. Loss of the *Drosophila* ortholog of BLM (DmBlm) causes hypersensitivity to DNA-damaging agents, female sterility, and defects in repairing DSBs [5], suggesting that DmBLM may function in DSB repair. Sekelsky and his colleagues showed that DmBlm null mutants have shorter repair products from DSBs, resulting in large deletions and impairment in their ability to repair DNA synthesis during synthesis-dependent strand annealing [2]. Further, the Sekelsky group examined the mechanims responsible for deletion formation in the absence of DmBlm using repair after excision of the P[w(a)] element in various Dm mutants [73]. It was found that loss of DmRad51 suppresses deletion formation in DmBlm mutants. Thus, Sekelsky and his colleagues proposed that DmBlm acts downstream of strand invasion to unwind a D-loop intermediate to free the newly synthesized strand [73]. In the absence of

DmBlm, alternative pathways of D-loop disassembly result in short repair synthesis tract or flanking deletions.

Recently, it was reported that DmBlm function in multiple cellular contexts to promote genome stability [72]. Mutations at the N terminus of DmBlm have anaphase bridges, other mitotic defects, and decreased crossovers in meiotic recombination [72]. In addition, DmBlm null mutants showed increased spontaneous mitotic crossovers by several orders of magnitude [72].

Studies for *Drosophila* homologs of WRN are less active. Recently, a *Drosophila* ortholog of WRN exonuclease was identified and characterized [90]. DmWRNexo homozygous mutant flies showed a significant loss of viability when exposed to CPT (topoisomerase I inhibitor) and showed high levels of recombination in the developing *Drosophila* wing resulting from excessive reciprocal exchange, suggesting an important role for WRN exonuclecase in maintaining genome stability.

RECQ5 orthologs have been identified in *Drosophila melanogaster* and alternative splicing of the RECQ5 transcript has been reported in *Drosophila* [92]. Although differential functions of the three isoforms are not known, a small isoform of the *Drosophila* RECQ5 protein (DmRECQ5) was purified and characterized biochemically [76]. The DmRECQ5 is a structure-specific DNA helicase like other RecQ helicases. This RECQ5 is capable of unwinding 3′ flaps, three-way junctions, forks and three-strand junction substrates at lower protein concentrations [76]. This study speculated that the DmRECQ5 plays a role in the processing of stalled replication forks that is complementary to the roles played by WRN and BLM and has similar roles in recombinational repair in the cell.

In *C. elegans*, BLM and RecQ5 have been investigated genetically and WRN has been studied genetically and biochemically. Loss of function mutations in the BLM helicase ortholog HIM-6 result in an enhanced irradiation sensitivity, a partially defective cell cycle arrest phenotype in response to HU, and in reduced levels of DNA-damage induced apoptosis [110]. Moreover, the *him-6* mutants show genomic instability signs including a mutator phenotype such as increased genomic insertions, deletions, and germ line apoptosis [40]. As shown in human RecQ defective syndromes, defects in the *him-6* mutant show a shortened life span [40], supporting that genomic instability can adversely affect longevity. When the number of DSBs was measured in *him-6* mutants, *top-3* mutants, and *top-3* (RNAi); *him-6* mutants, the combined depletion of the two genes leads to a massive increase in the level of DSBs compared to the single mutant. However, this phenotype is completely suppressed by *rad-51*, suggesting that *him-6* and *top-3* act on partially redundant pathways downstream of *rad-51* to prevent the accumulation of recombination intermediates [110]. Although no biochemical activity for CeBLM has been reported, Ahn et al. recently purified recombinant CeBLM from *E. coli* and detected its helicase activity (personal communication).

A WRN homolog of *C. elegans* (WRN-1), lacking the exonuclease domain, was identified and *wrn-1*(RNAi) strains showed a shortened life span, signs of premature aging, and various developmental defects [65]. When the *wrn-1*(RNAi) worms were treated with HU to block DNA replication pre-meiotic germ cells had an abnormal checkpoint response, suggesting that WRN-1 protein may be required to activate the S-phase checkpoint in germ cells [65]. This idea is supported by the observation that S-phase is accelerated in the early stages of development. Furthermore, *wrn-1*(RNAi) worms show accelerated larval growth and the acceleration is irrespective of IR, suggesting that WRN-1 acts as a checkpoint protein for DNA damage [65]. These observations indirectly propose a role for WRN-1 helicase in checkpoint activation. However, it remains to be investigated how WRN-1 protein acts on two checkpoints distinctly.

Recently, a study by Hyun et al. reports the biochemical characterization of *C. elegans* WRN-1 protein [50]. The authors found that CeWRN-1 is an ATP-dependent 3′-5′ helicase capable of unwinding a variety of DNA structures, similar to human WRN helicase. Furthermore, CeWRN-1 interacts with CeRPA (replication protein A) functionally and physically. The authors proposed that CeWRN-1 helicase activity improves the access for DNA repair and replication proteins to prevent the accumulation of abnormal structures.

A *C. elegans* homolog of human RecQ5 is CeRCQ5. *rcq5* (RNAi) worms show reduce life span by 37% and are hypersensitive to IR, suggesting an involvement of CeRCQ5 in a cellular response to DNA damage, consequently its deficiency may cause genomic instability [53].

15.6 Perspectives

Our understanding of RecQ helicase function in DNA damage response and the aging process is still inadequate. Several studies have provided clues linking critical roles of the RecQ family of DNA helicases with DDR, DNA replication and telomere maintenance, all linked to the aging process. As DDR pathways are potential targets for selective killing of senescent cells, it will be interesting to determine how RecQ helicases exert dual roles in signaling and repair events in the context of cellular and organism aging. In particular, our understanding of cellular and physiological functions of RECQ1, RECQ4 and RECQ5 lags behind other helicases in the family. Moreover, it is important to compare and contrast roles of all five RECQ helicases in DDR, and the aging phenotypes at the cellular and physiological levels, as well as their common and unique biochemical properties. One question for the future will be to determine whether RECQ1 and RECQ5 deficient mice exhibit symptoms of premature aging.

Acknowledgments This work was partially supported by the Intramural Research Program of the NIH, National Institute on Aging, and also by a Korean Research Foundation Grant of the Korean Government (MOEHRD) (KRF-2007-412-J00303, KRF-2007-521-C00211) to B. Ahn.

References

1. Ababou, M., Dumaire, V., Lecluse, Y., and mor-Gueret, M. (2002). Bloom's syndrome protein response to ultraviolet-C radiation and hydroxyurea-mediated DNA synthesis inhibition. Oncogene *21*, 2079–2088.

2. Adams, M.D., McVey, M., and Sekelsky, J.J. (2003). *Drosophila* BLM in double-strand break repair by synthesis-dependent strand annealing. Science *299*, 265–267.
3. Bachrati, C.Z. and Hickson, I.D. (2008). RecQ helicases: guardian angels of the DNA replication fork. Chromosoma *117*, 219–233.
4. Bakkenist, C.J. and Kastan, M.B. (2003). DNA damage activates ATM through intermolecular autophosphorylation and dimer dissociation. Nature *421*, 499–506.
5. Beall, E.L. and Rio, D.C. (1996). *Drosophila* IRBP/Ku p70 corresponds to the mutagen-sensitive mus309 gene and is involved in P-element excision in vivo. Genes Dev. *10*, 921–933.
6. Berkovich, E., Monnat, R.J., Jr., and Kastan, M.B. (2007). Roles of ATM and NBS1 in chromatin structure modulation and DNA double-strand break repair. Nat. Cell Biol. *9*, 683–690.
7. Bohr, V.A., Souza, P.N., Nyaga, S.G., Dianov, G., Kraemer, K., Seidman, M.M., and Brosh, R.M., Jr. (2001). DNA repair and mutagenesis in Werner syndrome. Environ. Mol. Mutagen. *38*, 227–234.
8. Botuyan, MV., Lee, J., Ward, IM., Kim, JE., Thompson, JR., Chen, J., and Mer, G. (2006). Structural basis for the methylation state-specific recognition of histone H4-K20 by 53BP1 and Crb2 in DNA repair. Cell *127,* 1361–1373.
9. Brown, E.J. and Baltimore, D. (2000). ATR disruption leads to chromosomal fragmentation and early embryonic lethality. Genes Dev. *14*, 397–402.
10. Bugreev, D.V., Brosh, R.M., Jr., and Mazin, A.V. (2008). RECQ1 possesses DNA branch migration activity. J. Biol. Chem. *283*, 20231–20242.
11. Campisi, J. (2005). Senescent cells, tumor suppression, and organismal aging: good citizens, bad neighbors. Cell *120*, 513–522.
12. Chang, S., Multani, A.S., Cabrera, N.G., Naylor, M.L., Laud, P., Lombard, D., Pathak, S., Guarente, L., and DePinho, R.A. (2004). Essential role of limiting telomeres in the pathogenesis of Werner syndrome. Nat. Genet. *36*, 877–882.
13. Chen, C.C., Carson, J.J., Feser, J., Tamburini, B., Zabaronick, S., Linger, J., and Tyler, J.K. (2008). Acetylated lysine 56 on histone H3 drives chromatin assembly after repair and signals for the completion of repair. Cell *134*, 231–243.
14. Cheng, W.H., Muftic, D., Muftuoglu, M., Dawut, L., Morris, C., Helleday, T., Shiloh, Y., and Bohr, V.A. (2008). WRN is required for ATM activation and the S-phase checkpoint in response to interstrand crosslink-induced DNA double strand breaks. Mol. Biol. Cell *19*, 3923–3933.
15. Cheng, W.H., Sakamoto, S., Fox, J.T., Komatsu, K., Carney, J., and Bohr, V.A. (2005). Werner syndrome protein associates with gamma H2AX in a manner that depends upon Nbs1. FEBS Lett. *579*, 1350–1356.
16. Cheng, W.H., von Kobbe, K.C., Opresko, P.L., Arthur, L.M., Komatsu, K., Seidman, M.M., Carney, J.P., and Bohr, V.A. (2004). Linkage between Werner syndrome protein and the Mre11 complex via Nbs1. J. Biol. Chem. *279*, 21169–21176.
17. Chester, N., Babbe, H., Pinkas, J., Manning, C., and Leder, P. (2006). Mutation of the murine Bloom's syndrome gene produces global genome destabilization. Mol. Cell Biol. *26*, 6713–6726.
18. Chester, N., Kuo, F., Kozak, C., O'Hara, C.D., and Leder, P. (1998). Stage-specific apoptosis, developmental delay, and embryonic lethality in mice homozygous for a targeted disruption in the murine Bloom's syndrome gene. Genes Dev. *12*, 3382–3393.
19. Choudhary, S., Doherty, K.M., Handy, C.J., Sayer, J.M., Yagi, H., Jerina, D.M., and Brosh, R.M., Jr. (2006). Inhibition of Werner syndrome helicase activity by benzo[a]pyrene diol epoxide adducts can be overcome by replication protein A. J. Biol. Chem. *281*, 6000–6009.
20. Chun, H.H. and Gatti, R.A. (2004). Ataxia-telangiectasia, an evolving phenotype. DNA Repair (Amst) *3*, 1187–1196.
21. Cimprich, K.A. and Cortez, D. (2008). ATR: an essential regulator of genome integrity. Nat. Rev. Mol. Cell Biol. *9*, 616–627.

22. Compton, S.A., Tolun, G., Kamath-Loeb, A.S., Loeb, L.A., and Griffith, J.D. (2008). The Werner syndrome protein binds replication fork and holliday junction DNAs as an oligomer. J. Biol. Chem. *283*, 24478–24483.
23. Constantinou, A., Tarsounas, M., Karow, J.K., Brosh, R.M., Bohr, V.A., Hickson, I.D., and West, S.C. (2000). Werner's syndrome protein (WRN) migrates Holliday junctions and co-localizes with RPA upon replication arrest. EMBO Rep. *1*, 80–84.
24. Crabbe, L., Jauch, A., Naeger, C.M., Holtgreve-Grez, H., and Karlseder, J. (2007). Telomere dysfunction as a cause of genomic instability in Werner syndrome. Proc. Natl. Acad. Sci. U.S.A. *104*, 2205–2210.
25. Crabbe, L., Verdun, R.E., Haggblom, C.I., and Karlseder, J. (2004). Defective telomere lagging strand synthesis in cells lacking WRN helicase activity. Science *306*, 1951–1953.
26. Cuadrado, M., Martinez-Pastor, B., Murga, M., Toledo, L.I., Gutierrez-Martinez,P., Lopez, E., and Fernandez-Capetillo, O. (2006). ATM regulates ATR chromatin loading in response to DNA double-strand breaks. J. Exp. Med. *203*, 297–303.
27. D'Amours, D. and Jackson, S.P. (2002). The Mre11 complex: at the crossroads of dna repair and checkpoint signalling. Nat. Rev. Mol. Cell Biol. *3*, 317–327.
28. Davies, S.L., North, P.S., and Hickson, I.D. (2007). Role for BLM in replication-fork restart and suppression of origin firing after replicative stress. Nat. Struct. Mol. Biol. *14*, 677–679.
29. d'Adda di Fagagna, F., Teo, S.H., and Jackson, S.P. (2004). Functional links between telomeres and proteins of the DNA-damage response. Genes Dev. *18*, 1781–1799.
30. De Klein, A., Muijtjens, M., van Os, R., Verhoeven, Y., Smit, B., Carr, A.M., Lehmann, A.R., and Hoeijmakers, J.H. (2000). Targeted disruption of the cell-cycle checkpoint gene ATR leads to early embryonic lethality in mice. Curr. Biol. *10*, 479–482.
31. Du, X., Shen, J., Kugan, N., Furth, E.E., Lombard, D.B., Cheung, C., Pak, S., Luo, G., Pignolo, R.J., DePinho, R.A., Guarente, L., and Johnson, F.B. (2004). Telomere shortening exposes functions for the mouse Werner and Bloom syndrome genes. Mol. Cell Biol. *24*, 8437–8446.
32. Falck, J., Coates, J., and Jackson, S.P. (2005). Conserved modes of recruitment of ATM, ATR and DNA-PKcs to sites of DNA damage. Nature *434*, 605–611.
33. Fan, W. and Luo, J. (2008). RecQ4 facilitates UV-induced DNA damage repair through interaction with nucleotide excision repair factor XPA. J. Biol. Chem *283*, 29037–29044.
34. Frei, C. and Gasser, S.M. (2000). The yeast Sgs1p helicase acts upstream of Rad53p in the DNA replication checkpoint and colocalizes with Rad53p in S-phase-specific foci. Genes Dev. *14*, 81–96.
35. Froget, B., Blaisonneau, J., Lambert, S., and Baldacci, G. (2008). Cleavage of stalled forks by fission yeast mus81/eme1 in absence of DNA replication checkpoint. Mol. Biol. Cell *19*, 445–456.
36. Gaymes, T.J., North, P.S., Brady, N., Hickson, I.D., Mufti, G.J., and Rassool, F.V. (2002). Increased error-prone non homologous DNA end-joining – a proposed mechanism of chromosomal instability in Bloom's syndrome. Oncogene *21*, 2525–2533.
37. Gilson, E. and Geli, V. (2007). How telomeres are replicated. Nat. Rev. Mol. Cell Biol. *8*, 825–838.
38. Goldberg, M., Stucki, M., Falck, J., D'Amours, D., Rahman, D., Pappin, D., Bartek, J., and Jackson, S.P. (2003). MDC1 is required for the intra-S-phase DNA damage checkpoint. Nature *421*, 952–956.
39. Goodarzi, A.A., Noon, A.T., Deckbar, D., Ziv, Y., Shiloh, Y., Lobrich, M., and Jeggo, P.A. (2008). ATM signaling facilitates repair of DNA double-strand breaks associated with heterochromatin. Mol. Cell *31*, 167–177.
40. Grabowski, M.M., Svrzikapa, N., and Tissenbaum, H.A. (2005). Bloom syndrome ortholog HIM-6 maintains genomic stability in *C. elegans*. Mech. Ageing Dev. *126*, 1314–1321.

41. Gray, M.D., Shen, J.C., Kamath-Loeb, A.S., Blank, A., Sopher, B.L., Martin, G.M., Oshima, J., and Loeb, L.A. (1997). The Werner syndrome protein is a DNA helicase. Nat. Genet. *17*, 100–103.
42. Hayflick, L. (1965). The limited in vitro lifetime of human diploid cell strains. Exp. Cell Res. *37*, 614–636.
43. Hickson, I.D. (2003). RecQ helicases: caretakers of the genome. Nat. Rev. Cancer *3*, 169–178.
44. Hishida, T., Han, Y.W., Shibata, T., Kubota, Y., Ishino, Y., Iwasaki, H., and Shinagawa, H. (2004). Role of the *Escherichia coli* RecQ DNA helicase in SOS signaling and genome stabilization at stalled replication forks. Genes Dev. *18*, 1886–1897.
45. Hoki, Y., Araki, R., Fujimori, A., Ohhata, T., Koseki, H., Fukumura, R., Nakamura, M., Takahashi, H., Noda, Y., Kito, S., and Abe, M. (2003). Growth retardation and skin abnormalities of the Recql4-deficient mouse. Hum. Mol. Genet. *12*, 2293–2299.
46. Hu, Y., Lu, X., Barnes, E., Yan, M., Lou, H., and Luo, G. (2005). Recql5 and Blm RecQ DNA helicases have nonredundant roles in suppressing crossovers. Mol. Cell Biol. *25*, 3431–3442.
47. Hu, Y., Raynard, S., Sehorn, M.G., Lu, X., Bussen, W., Zheng, L., Stark, J.M., Barnes, E.L., Chi, P., Janscak, P., Jasin, M., Vogel, H., Sung, P., and Luo, G. (2007). RECQL5/Recql5 helicase regulates homologous recombination and suppresses tumor formation via disruption of Rad51 presynaptic filaments. Genes Dev. *21*, 3073–3084.
48. Huang, S., Li, B., Gray, M.D., Oshima, J., Mian, I.S., and Campisi, J. (1998). The premature ageing syndrome protein, WRN, is a 3′ 5′ exonuclease. Nat. Genet. *20*, 114–116.
49. Huyen, Y., Zgheib, O., Ditullio, R.A., Jr., Gorgoulis, V.G., Zacharatos, P., Petty, T.J., Sheston, E.A., Mellert, H.S., Stavridi, E.S., and Halazonetis, T.D. (2004). Methylated lysine 79 of histone H3 targets 53BP1 to DNA double-strand breaks. Nature *432*, 406–411.
50. Hyun, M., Bohr, V.A., and Ahn, B. (2008). Biochemical characterization of the WRN-1 RecQ helicase of *Caenorhabditis elegans*. Biochemistry *47*, 7583–7593.
51. Jacobs, J.J. and de, L.T. (2004). Significant role for p16INK4a in p53-independent telomere-directed senescence. Curr. Biol. *14*, 2302–2308.
52. Jazayeri, A., Falck, J., Lukas, C., Bartek, J., Smith, G.C., Lukas, J., and Jackson, S.P. (2006). ATM- and cell cycle-dependent regulation of ATR in response to DNA double-strand breaks. Nat. Cell Biol. *8*, 37–45.
53. Jeong, Y.S., Kang, Y., Lim, K.H., Lee, M.H., Lee, J., and Koo, H.S. (2003). Deficiency of *Caenorhabditis elegans* RecQ5 homologue reduces life span and increases sensitivity to ionizing radiation. DNA Repair (Amst) *2*, 1309–1319.
54. Kanagaraj, R., Saydam, N., Garcia, P.L., Zheng, L., and Janscak, P. (2006). Human RECQ5beta helicase promotes strand exchange on synthetic DNA structures resembling a stalled replication fork. Nucleic Acids Res. *34*, 5217–5231.
55. Karmakar, P., Seki, M., Kanamori, M., Hashiguchi, K., Ohtsuki, M., Murata, E., Inoue, E., Tada, S., Lan, L., Yasui, A., and Enomoto, T. (2006). BLM is an early responder to DNA double-strand breaks. Biochem. Biophys. Res. Commun. *348*, 62–69.
56. Keogh, M.C., Kim, J.A., Downey, M., Fillingham, J., Chowdhury, D., Harrison, J.C., Onishi, M., Datta, N., Galicia, S., Emili, A., Lieberman, J., Shen, X., Buratowski, S., Haber, J.E., Durocher, D., Greenblatt, J.F., and Krogan, N.J. (2006). A phosphatase complex that dephosphorylates gammaH2AX regulates DNA damage checkpoint recovery. Nature *439*, 497–501.
57. Kitao, S., Shimamoto, A., Goto, M., Miller, R.W., Smithson, W.A., Lindor, N.M., and Furuichi, Y. (1999). Mutations in RECQL4 cause a subset of cases of Rothmund-Thomson syndrome.Nat. Genet. 22, 82–84.
58. Kozlov, S.V., Graham, M.E., Peng, C., Chen, P., Robinson, P.J., and Lavin, M.F. (2006). Involvement of novel autophosphorylation sites in ATM activation. EMBO J. *25*, 3504–3514.

59. Kubota, Y., Takase, Y., Komori, Y., Hashimoto, Y., Arata, T., Kamimura, Y., Araki, H., and Takisawa, H. (2003). A novel ring-like complex of Xenopus proteins essential for the initiation of DNA replication. Genes Dev. *17*, 1141–1152.
60. Kyng, K.J., May, A., Kolvraa, S., and Bohr, V.A. (2003). Gene expression profiling in Werner syndrome closely resembles that of normal aging. Proc. Natl. Acad. Sci. U.S.A. *100*, 12259–12264.
61. Lan, L., Nakajima, S., Komatsu, K., Nussenzweig, A., Shimamoto, A., Oshima, J., and Yasui, A. (2005). Accumulation of Werner protein at DNA double-strand breaks in human cells. J. Cell Sci. *118*, 4153–4162.
62. Laud, P.R., Multani, A.S., Bailey, S.M., Wu, L., Ma, J., Kingsley, C., Lebel, M., Pathak, S., DePinho, R.A., and Chang, S. (2005). Elevated telomere-telomere recombination in WRN-deficient, telomere dysfunctional cells promotes escape from senescence and engagement of the ALT pathway. Genes Dev. *19*, 2560–2570.
63. Lee, J.H. and Paull, T.T. (2007). Activation and regulation of ATM kinase activity in response to DNA double-strand breaks. Oncogene *26*, 7741–7748.
64. Lee, J.H. and Paull, T.T. (2005). ATM activation by DNA double-strand breaks through the Mre11-Rad50-Nbs1 complex. Science *308*, 551–554.
65. Lee, S.J., Yook, J.S., Han, S.M., and Koo, H.S. (2004). A Werner syndrome protein homolog affects *C. elegans* development, growth rate, life span and sensitivity to DNA damage by acting at a DNA damage checkpoint. Development *131*, 2565–2575.
66. LeRoy, G., Carroll, R., Kyin, S., Seki, M., and Cole, M.D. (2005). Identification of RecQL1 as a Holliday junction processing enzyme in human cell lines. Nucleic Acids Res. *33*, 6251–6257.
67. Liberi, G., Maffioletti, G., Lucca, C., Chiolo, I., Baryshnikova, A., Cotta-Ramusino, C., Lopes, M., Pellicioli, A., Haber, J.E., and Foiani, M. (2005). Rad51-dependent DNA structures accumulate at damaged replication forks in sgs1 mutants defective in the yeast ortholog of BLM RecQ helicase. Genes Dev. *19*, 339–350.
68. Lou, Z., Minter-Dykhouse, K., Franco, S., Gostissa, M., Rivera, M.A., Celeste, A., Manis, J.P., van, D.J., Nussenzweig, A., Paull, T.T., Alt, F.W., and Chen, J. (2006). MDC1 maintains genomic stability by participating in the amplification of ATM-dependent DNA damage signals. Mol. Cell *21*, 187–200.
69. Machwe, A., Xiao, L., Groden, J., and Orren, D.K. (2006). The Werner and Bloom syndrome proteins catalyze regression of a model replication fork. Biochemistry *45*, 13939–13946.
70. Macris, M.A., Krejci, L., Bussen, W., Shimamoto, A., and Sung, P. (2006). Biochemical characterization of the RECQ4 protein, mutated in Rothmund-Thomson syndrome. DNA Repair (Amst) *5*, 172–180.
71. Matsuno, K., Kumano, M., Kubota, Y., Hashimoto, Y., and Takisawa, H. (2006). The N-terminal noncatalytic region of Xenopus RecQ4 is required for chromatin binding of DNA polymerase alpha in the initiation of DNA replication. Mol. Cell Biol. *26*, 4843–4852.
72. McVey, M., Andersen, S.L., Broze, Y., and Sekelsky, J. (2007). Multiple functions of *Drosophila* BLM helicase in maintenance of genome stability. Genetics *176*, 1979–1992.
73. McVey, M., Larocque, J.R., Adams, M.D., and Sekelsky, J.J. (2004). Formation of deletions during double-strand break repair in *Drosophila* DmBlm mutants occurs after strand invasion. Proc. Natl. Acad. Sci. U.S.A. *101*, 15694–15699.
74. O'Driscoll, M. and Jeggo, P.A. (2006). The role of double-strand break repair – insights from human genetics. Nat. Rev. Genet. *7*, 45–54.
75. Onclercq-Delic, R., Calsou, P., Delteil, C., Salles, B., Papadopoulo, D., and mor-Gueret, M. (2003). Possible anti-recombinogenic role of Bloom's syndrome helicase in double-strand break processing. Nucleic Acids Res. *31*, 6272–6282.
76. Ozsoy, A.Z., Ragonese, H.M., and Matson, S.W. (2003). Analysis of helicase activity and substrate specificity of *Drosophila* RECQ5. Nucleic Acids Res. *31*, 1554–1564.
77. Petermann, E. and Caldecott, K.W. (2006). Evidence that the ATR/Chk1 pathway maintains normal replication fork progression during unperturbed S phase. Cell Cycle *5*, 2203–2209.

78. Petkovic, M., Dietschy, T., Freire, R., Jiao, R., and Stagljar, I. (2005). The human Rothmund-Thomson syndrome gene product, RECQL4, localizes to distinct nuclear foci that coincide with proteins involved in the maintenance of genome stability. J. Cell Sci. *118*, 4261–4269.
79. Pichierri, P., Franchitto, A., and Rosselli, F. (2004). BLM and the FANC proteins collaborate in a common pathway in response to stalled replication forks. EMBO J. *23*, 3154–3163.
80. Pirzio, L.M., Pichierri, P., Bignami, M., and Franchitto, A. (2008). Werner syndrome helicase activity is essential in maintaining fragile site stability. J. Cell Biol. *180*, 305–314.
81. Poot, M., Gollahon, K.A., Emond, M.J., Silber, J.R., and Rabinovitch, P.S. (2002). Werner syndrome diploid fibroblasts are sensitive to 4-nitroquinoline-*N*-oxide and 8-methoxypsoralen: implications for the disease phenotype. FASEB J. *16*, 757–758.
82. Poot, M., Hoehn, H., Runger, T.M., and Martin, G.M. (1992). Impaired S-phase transit of Werner syndrome cells expressed in lymphoblastoid cell lines. Exp. Cell Res. *202*, 267–273.
83. Poot, M., Yom, J.S., Whang, S.H., Kato, J.T., Gollahon, K.A., and Rabinovitch, P.S. (2001). Werner syndrome cells are sensitive to DNA cross-linking drugs. FASEB J. *15*, 1224–1226.
84. Ralf, C., Hickson, I.D., and Wu, L. (2006). The Bloom's syndrome helicase can promote the regression of a model replication fork. J. Biol. Chem. *281*, 22839–22846.
85. Rothfuss, A. and Grompe, M. (2004). Repair kinetics of genomic interstrand DNA cross-links: evidence for DNA double-strand break-dependent activation of the Fanconi anemia/BRCA pathway. Mol. Cell Biol. *24*, 123–134.
86. Rouse, J. and Jackson, S.P. (2002). Interfaces between the detection, signaling, and repair of DNA damage. Science *297*, 547–551.
87. Ruzankina, Y., Pinzon-Guzman, C., Asare, A., Ong, T., Pontano, L., Cotsarelis, G., Zediak, V.P., Velez, M., Bhandoola, A., and Brown, E.J. (2007). Deletion of the developmentally essential gene ATR in adult mice leads to age-related phenotypes and stem cell loss. Cell Stem Cell *1*, 113–126.
88. Saintigny, Y., Makienko, K., Swanson, C., Emond, M.J., and Monnat, R.J., Jr. (2002). Homologous recombination resolution defect in werner syndrome. Mol. Cell Biol. *22*, 6971–6978.
89. Sangrithi, M.N., Bernal, J.A., Madine, M., Philpott, A., Lee, J., Dunphy, W.G., and Venkitaraman, A.R. (2005). Initiation of DNA replication requires the RECQL4 protein mutated in Rothmund-Thomson syndrome. Cell *121*, 887–898.
90. Saunders, R.D., Boubriak, I., Clancy, D.J., and Cox, L.S. (2008). Identification and characterization of a *Drosophila* ortholog of WRN exonuclease that is required to maintain genome integrity. Aging Cell *7*, 418–425.
91. Schultz, L.B., Chehab, N.H., Malikzay, A., and Halazonetis, T.D. (2000). p53 binding protein 1 (53BP1) is an early participant in the cellular response to DNA double-strand breaks. J. Cell Biol. *151*, 1381–1390.
92. Sekelsky, J.J., Brodsky, M.H., Rubin, G.M., and Hawley, R.S. (1999). *Drosophila* and human RecQ5 exist in different isoforms generated by alternative splicing. Nucleic Acids Res. *27*, 3762–3769.
93. Sengupta, S., Robles, A.I., Linke, S.P., Sinogeeva, N.I., Zhang, R., Pedeux, R., Ward, I.M., Celeste, A., Nussenzweig, A., Chen, J., Halazonetis, T.D., and Harris, C.C. (2004). Functional interaction between BLM helicase and 53BP1 in a Chk1-mediated pathway during S-phase arrest. J. Cell Biol. *166*, 801–813.
94. Sharma, S. and Brosh, R.M., Jr. (2007). Human RECQ1 is a DNA damage responsive protein required for genotoxic stress resistance and suppression of sister chromatid exchanges. PLoS. ONE. *2*, e1297.
95. Sharma, S. and Brosh, R.M., Jr. (2008). Unique and important consequences of RECQ1 deficiency in mammalian cells. Cell Cycle *7*, 989–1000.
96. Shiloh, Y. (2006). The ATM-mediated DNA-damage response: taking shape. Trends Biochem. Sci. *31*, 402–410.
97. Shimura, T., Torres, M.J., Martin, M.M., Rao, V.A., Pommier, Y., Katsura, M., Miyagawa, K., and Aladjem, M.I. (2008). Bloom's syndrome helicase and Mus81 are required to induce

transient double-strand DNA breaks in response to DNA replication stress. J. Mol. Biol. *375*, 1152–1164.

98. Sidorova, J.M., Li, N., Folch, A., and Monnat, R.J., Jr. (2008). The RecQ helicase WRN is required for normal replication fork progression after DNA damage or replication fork arrest. Cell Cycle *7*, 796–807.
99. So, S., Adachi, N., Lieber, M.R., and Koyama, H. (2004). Genetic interactions between BLM and DNA ligase IV in human cells. J. Biol. Chem. *279*, 55433–55442.
100. Stewart, E., Chapman, C.R., Al-Khodairy, F., Carr, A.M., and Enoch, T. (1997). rqh1+, a fission yeast gene related to the Bloom's and Werner's syndrome genes, is required for reversible S phase arrest. EMBO J. *16*, 2682–2692.
101. Stiff, T., Reis, C., Alderton, G.K., Woodbine, L., O'Driscoll, M., and Jeggo, P.A. (2005). Nbs1 is required for ATR-dependent phosphorylation events. EMBO J. *24*, 199–208.
102. Stiff, T., Walker, S.A., Cerosaletti, K., Goodarzi, A.A., Petermann, E., Concannon, P., O'Driscoll, M., and Jeggo, P.A. (2006). ATR-dependent phosphorylation and activation of ATM in response to UV treatment or replication fork stalling. EMBO J. *25*, 5775–5782.
103. Stucki, M., Clapperton, J.A., Mohammad, D., Yaffe, M.B., Smerdon, S.J., and Jackson, S.P. (2005). MDC1 directly binds phosphorylated histone H2AX to regulate cellular responses to DNA double-strand breaks. Cell 123, 1213–1226.
104. Sun, Y., Jiang, X., Chen, S., Fernandes, N., and Price, B.D. (2005). A role for the Tip60 histone acetyltransferase in the acetylation and activation of ATM. Proc. Natl. Acad. Sci. U.S.A. *102*, 13182–13287.
105. Takeuchi, F., Hanaoka, F., Goto, M., Akaoka, I., ori, T., Yamada, M., and Miyamoto, T. (1982). Altered frequency of initiation sites of DNA replication in Werner's syndrome cells. Hum. Genet. *60*, 365–368.
106. Uziel, T., Lerenthal, Y., Moyal, L., Andegeko, Y., Mittelman, L., and Shiloh, Y. (2003). Requirement of the MRN complex for ATM activation by DNA damage. EMBO J. *22*, 5612–5621.
107. Verdun, R.E., Crabbe, L., Haggblom, C., and Karlseder, J. (2005). Functional human telomeres are recognized as DNA damage in G2 of the cell cycle. Mol. Cell *20*, 551–561.
108. Verdun, R.E. and Karlseder, J. (2007). Replication and protection of telomeres. Nature *447*, 924–931.
109. von Kobbe, K.C., Thoma, N.H., Czyzewski, B.K., Pavletich, N.P., and Bohr, V.A. (2003). Werner syndrome protein contains three structure-specific DNA binding domains. J. Biol. Chem. *278*, 52997–53006.
110. Wicky, C., Alpi, A., Passannante, M., Rose, A., Gartner, A., and Muller, F. (2004). Multiple genetic pathways involving the *Caenorhabditis elegans* Bloom's syndrome genes him-6, rad-51, and top-3 are needed to maintain genome stability in the germ line. Mol. Cell Biol. *24*, 5016–5027.
111. Wright, W.E. and Shay, J.W. (1992). The two-stage mechanism controlling cellular senescence and immortalization. Exp. Gerontol. *27*, 383–389.
112. Wu, L. and Hickson, I.D. (2003). The Bloom's syndrome helicase suppresses crossing over during homologous recombination. Nature *426*, 870–874.
113. Wyllie, F.S., Jones, C.J., Skinner, J.W., Haughton, M.F., Wallis, C., Wynford-Thomas, D., Faragher, R.G., and Kipling, D. (2000). Telomerase prevents the accelerated cell ageing of Werner syndrome fibroblasts. Nat. Genet. *24*, 16–17.
114. Yu, C.E., Oshima, J., Fu, Y.H., Wijsman, E.M., Hisama, F., Alisch, R., Matthews, S., Nakura, J., Miki, T., Ouais, S., Martin, G.M., Mulligan, J., and Schellenberg, G.D. (1996). Positional cloning of the Werner's syndrome gene. Science *272*, 258–262.
115. Ziv, Y., Bielopolski, D., Galanty, Y., Lukas, C., Taya, Y., Schultz, D.C., Lukas, J., Bekker-Jensen, S., Bartek, J., and Shiloh, Y. (2006). Chromatin relaxation in response to DNA double-strand breaks is modulated by a novel ATM- and KAP-1 dependent pathway. Nat. Cell Biol. *8*, 870–876.

Chapter 16
Single-Stranded DNA Binding Proteins Involved in Genome Maintenance

Derek J. Richard and Kum Kum Khanna

Abstract The Single Stranded DNA Binding (SSB) family of proteins are ubiquitous to life. Structurally they are characterised by their oligonucleotide oligosaccharide-binding fold (OB fold), which binds to the single stranded DNA (ssDNA) substrate. Functionally they act in a number of cellular processes where ssDNA is exposed, such as DNA replication and DNA repair. They act by binding the exposed ssDNA, protecting it from nucleolytic degradation and attack from reactive chemical species. SSBs also function to stop the formation of secondary structures, prevent DNA re-annealing until appropriate and in the recruitment of protein partners. In humans two structurally distinct classes of SSBs exist. Replication Protein A (RPA), is a heterotrimeric polypeptide, widely believed to be a central component of both DNA replication and DNA repair pathways. Mammalian cells have long been thought to rely exclusively on one SSB, RPA, to perform repair function, however recent discovery of two other members of the SSB family in humans (hSSB1 and hSSB2) challenges many of the established models of DNA transactions involving ssDNA. Theses proteins are much more closely related to the bacterial and archaeal SSB families than RPA. hSSB1 has recently being described to have a central function in the repair of double strand DNA breaks (DSB) by homology directed repair (HDR). Unlike RPA however, hSSB1 appears to be dispensable for normal DNA replication. This chapter aims primarily to review the function of RPA in the DNA DSB repair process. It will also examine to some degree the implications of the recent discovery of hSSB1 and hSSB2 as well as look at the potential of these proteins as anticancer therapeutic targets.

Keywords SSB · DNA · RPA · ATM · Replication · Repair

K.K. Khanna (✉)
Cancer and Cell Biology Division, The Queensland Institute of Medical Research, 300 Herston Road, Herston, QLD 4006, Australia
e-mail: kumkumK@qimr.edu.au

K.K. Khanna, Y. Shiloh (eds.), *The DNA Damage Response: Implications on Cancer Formation and Treatment*, DOI 10.1007/978-90-481-2561-6_16,

16.1 Single Stranded DNA

Although our genomic DNA exists as an annealed double stranded helix, there are many occasions when single stranded DNA regions are present. In fact, to release the information contained within the genetic code it is a requirement that the DNA is melted (breaking of the Watson-Crick base pairs), exposing the base code and allowing it to be replicated or transcribed. DNA damage is also a major cause of duplex DNA destabilisation. DNA damage can result in a loss of correct hydrogen bond formation between DNA strands caused by bulky adducts, mis-matched base pairs or inter-strand cross links, it may also result in single or double strand breaks which destabilise the surrounding DNA duplex. All these damage types serve not necessarily to result in permanent ssDNA exposure but act to increase the breathing rate of the duplex DNA. The breathing of DNA refers to the constant equilibrium that exists between an annealed duplex and the non-anneal ssDNA. Breathing is generally contained within regions of tens of base pairs, forming transient bubbles in the DNA strand. The rates of breathing and duration of the bubble however are markedly increased if Watson-Crick hydrogen bonding is lost or if the phosphate backbone of DNA is broken. This breathing DNA is a potential substrate to which SSBs can bind.

16.2 Evolution of SSBs

All life that exists presently on earth originated from a common ancestor, which has aptly been named LUCA (Last Universal Common Ancestor). LUCA is believed to have been a RNA organism and its divergence into the three domains of life that exist today (Archaea, Bacteria and Eukarya), was initially catalysed by the evolution of UDNA (uracil containing DNA). SSBs exists in all three domains of life and are also represented in viruses, their sequence preservation suggests that they originated from a common ancestoral protein, which was present in LUCA. During evolution through the eukaryote branch of life, SSBs have structurally evolved from relatively simple structural configurations to higher order heterotrimeric structures. This has been a necessity, allowing SSBs to interact differentially with multiple partners as well as accommodating modulation of function by post-translational modifications.

16.3 Structural Organisation

The SSB family can be subdivided based on their structural organization. The bacteria have what can be described as a "simple SSB", this is characterised by a N-terminally located OB fold followed by an unstructured proline and glycine rich tail region (Fig. 16.1). The c-terminal of this flexible unstructured tail is often characterised by acidic residues. Predominantly in solution the bacterial SSBs form dimers, however on binding DNA they organise into a homotetrameric structure. The *E. coli* SSB is unusual in that it is a tetramer in solution and has two distinct binding modes based on the relative ratio of SSB to substrate [86]. The archaea, have examples of both the simple SSB and a more complex RPA like SSB (Fig. 16.1).

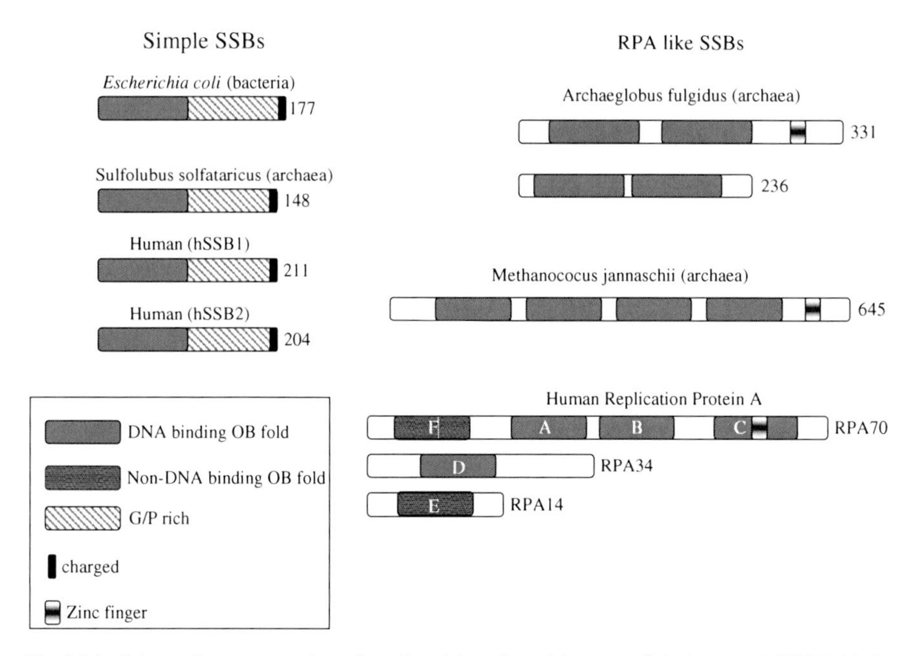

Fig. 16.1 Schematic representation of predicted domain architecture of single stranded DNA binding proteins in three domains life including bacteria, archaea and eukaryote. The letters within the *boxes* represent individual OB folds. The domains are not drawn to scale

Sulfolobus solfataricus has a single SSB encoding gene, its SSB has a structural organization very similar to bacterial SSB. Methanococus, by contrast, expresses a single polypeptide containing four OB folds while Archaeoglobus, expresses two distinct polypeptides, both containing two OB folds. In eukaryotes it was believed that the simple SSBs had been lost in evolution and had been replaced functionally by the heterotrimeric RPA. The RPA polypeptide has six OB folds, four of which bind to ssDNA. These four DNA binding OB folds are arranged over two of the three polypeptides, which comprise RPA (RPA70 and 34). However recently it has been discovered that eukaryotes also encode two simple SSBs [88]. Many lineages of eukaryotes appear to have preserved at least one of these simple SSBs, however surprisingly they are absent from the yeast genome, a factor, which may explain why they have remained undiscovered until recently.

16.4 *E. coli* SSB

The best studied simple SSB is from *E. coli*. This protein is an essential cellular component, which binds ssDNA, protecting it from further damage or incorrect processing [66, 69]. It functions secondarily to attract appropriate repair enzymes to the sites of ssDNA processing and in many cases it stimulates the enzymatic activities of these proteins. The *E. coli* SSB has been shown to interact with multiple partners, with its protein-protein interaction domain being at its c-terminal tail. This c-terminal tail is an amphipathic sequence, composed of acidic residues and a terminal hydrophobic tripeptide. SSB has been shown to interact with the RecG, RecJ, RecQ and RecO components of the *E. coli* HR pathway [17, 39, 57, 95, 98, 109] as well as other repair proteins such as uracil DNA glycosylase [40] and DNA polymerases II, III and V [4, 17, 47, 73, 113, 124]. When the correct substrate becomes available, such as stalled DNA replication forks, then the appropriate SSB bound repair enzyme is brought into action [57].

16.5 An Introduction to Replication Protein A

RPA was originally identified as an essential component required for in vitro replication of the Simian virus 40 (SV40) DNA [30, 34, 114, 116]. Its purification from cell lysates allowed the identification of its three subunits, RPA70, 34 and 14. To date all eukaryotes with a sequenced genome have been found to encode the three subunits of RPA, and this along with biochemical evidence pointed towards RPA being the eukaryotic equivalent of bacterial SSB. Like the bacterial SSBs, RPA, has a high affinity for DNA substrates with exposed regions of ssDNA, has little affinity to double stranded DNA and reduced affinity to RNA [34, 49, 114, 116, 115]. Although the RPA heterotrimer contains the classical four DNA binding OB folds it also contained a further two OB folds which are involved in Protein-Protein interactions. The structure of RPA also introduced a new DNA binding motif into the SSB family. The Zinc finger motif, although present in many other DNA binding proteins was absent from the bacterial SSBs. However this Zn finger motif does exist in some archaea which encode RPA like SSBs. Its role at first was unclear as deletion of the zinc finger appeared to have little effect on RPAs ability to bind ssDNA [51]. However it was the discovery by Park et al. that RPAs binding to ssDNA was in fact regulated by reduction-oxidation (redox) that eventually gave clues to its function [82]. It was demonstrated that RPAs binding to ssDNA was enhanced by up to 10 fold in the presence of reducing agents and that removal of the Zinc finger motif removed RPAs sensitivity to oxidation. This remarkable observation gives rise to the possibility that RPA senses and reacts to oxidative stress.

RPA is now understood to have multiple functions within the cell. Although originally identified from its role in viral DNA replication, RPA has since been shown to be required for the progression of the DNA replication fork in the host cell as well as in origin firing. RPAs function is also central to a number of DNA repair processes such as homologous recombination (HR), base excision repair (BER), mismatch repair and nucleotide excision repair. This multifaceted nature of RPA is probably due to differential protein:protein interactions as well as post-translational modifications that modify its cellular activities.

RPAs role in DNA damage repair is essential for the cell to maintain genomic stability. In yeast, a missense mutation, L221P, within the RPA70 subunit results in a reduced ability of RPA to bind DNA, this results in gross chromosomal rearrangements within the yeast, similar to those frequently found in human cancers [22, 21].

The same mutation within the mouse RPA70 subunit is embryonic lethal in homozygous animals. Heterozygous mutant mice however develop lymphoid tumors, with the tumors showing large-scale chromosomal changes, segmental gains and losses. Embryonic fibroblasts from these mice showed a defective ability to repair double-strand DNA breaks and frequent chromosomal rearrangements [111]. This implies that loss of RPA functionality may be a causally linked to some cancers, although there are no documented reports of cancer-associated mutations in RPA.

16.6 RPA Structure and DNA Binding

The RPA heterotrimer has six OB folds located within the three polypeptides. The RPA OB folds share a similar structure to the bacterial SSB OB folds, being composed of five β-strands arranged in a β-barrel [13, 38]. RPA70 is the major ssDNA-binding component of RPA. It contains four OB folds, three of which are involved in binding to the DNA substrate. The major binding activity however is found within the A and B OB folds of RPA70, these two OB folds are separated by a short flexible linker region. It is believed that the A-OB fold binds initially as it has the highest affinity for ssDNA, this result in a higher localised concentration of the B-OB fold which subsequently binds the ssDNA. This binding occludes a region of approximately 10 nt. The binding of the A and B-OB folds results in a conformational change to RPA70s structure, two finger like loops from each domain enclose around the ssDNA. The AB binding mode is then followed by the binding of the C-OB fold and lastly the D-OB fold of RPA34, these binding modes occlude 12–23 nt and 28–30 nt respectively [6, 15, 29, 42, 53, 120]. RPA binds ssDNA with a specific 5′ to 3′ polarity with the RPA70 subunit binding 5′ and RPA34 subunit 3′. The affinity of RPA for ssDNA depends on which OB folds are bound and can have association constants ranging from $10^8$to 10^{11} M^{-1} [48, 50, 65]. Microscopic analysis of RPA binding to ssDNA revealed multiple modes of binding as visualised by different shapes, these included globular, elongated contracted and elongated extended [11]. How and what regulates these binding modes remains unknown. Interestingly unlike the bacterial SSB, which wrap DNA around their homotetrameric structure [86, 87], RPA alters its structural conformation and elongates along the DNA. Evidence also exists that would support the possibility that RPA can bind to more than one DNA substrate at any given time, this gives RPA the potential to bridge DNA molecules or to hold primers [15, 85]. One big question, which remains unanswered, is how RPA is displaced from DNA once its role is complete. This could possibly involve post-translational modifications of RPA or interaction of RPA with other proteins, which might induce a conformational change in RPA. Altered RPA functionality could reverse the binding reaction with dissociation of OB folds in turn, in a reverse order. The function of the other non-DNA binding OB folds of RPA is not completely clear. E-OB fold in RPA14 is required to maintain a stable trimeric RPA while the F-OB fold of RPA70 is involved in protein-protein interactions.

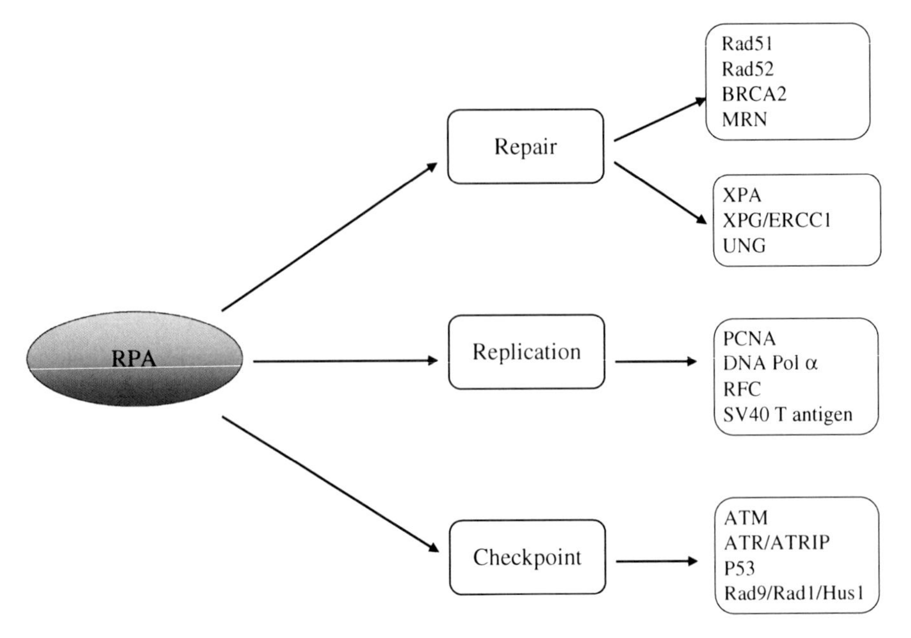

Fig. 16.2 RPA interacting proteins. RPA is involved in multiple cellular functions including DNA replication, repair and checkpoint control. RPA interacts with multiple proteins in each pathway (interactors are not limited to listed factors)

16.7 RPA Interacting Proteins

Involvement of RPA in multiple cellular functions reveals a requirement for it to interact with multiple partners. The majority of the interactions characterised to date occur through the RPA70 and RPA32 subunits and are predominantly with DNA repair proteins. RPA however has been shown to interact with a number of DNA replication proteins, the best characterised interaction is with the SV40 T antigen, an essential component for SV40 DNA replication [7]. RPA has been shown to interact with many DNA repair proteins (Fig. 16.2), including components of the nucleotide excision repair pathway (NER), base excision repair (BER) and both double strand DNA repair pathways [non homologous end joining (NHEJ) and homology directed repair (HDR)].

One major site of protein interaction within RPA is the winged helix C-terminal tail of the RPA32 subunit. The c-terminal tail of RPA32 has been shown to interact with a common α-helix found in uracil DNA glycosylase 2 (BER), XPA (NER) and Rad52 (HDR) [68]. RPA 70 has multiple interaction domains. Perhaps the best-described interaction is with the tumor suppressor P53. Acidic transactivation domain of p53 interacts directly with the F-OB fold present in N-terminus RPA70 [14, 32, 41, 59]. This interaction is disrupted after exposure of cells to DNA damaging agents. DNA damage induces hyperphosphorylation of RPA32 and a peptide mimetic of phosphorylated RPA32N and ssDNA, both consequences of DNA damage, are able to dissociate a p53 peptide from its OB binding site on RPA70N, suggesting that this OB-fold is subjected to competition between binding p53 and ssDNA [14]. Thus, P53 interaction with RPA may be displaced by binding of RPA to ssDNA or phosphorylation of the RPA itself. RPA70 also binds to the DNA strand exchange protein Rad51. The interaction between Rad51 and RPA70 is also interesting. An acidic region of Rad51 binds to the basic A-OB fold of RPA70 and as such would compete with ssDNA for binding this site. Indeed mutation of Rad51 shows a requirement for this interaction for the displacement of RPA prior to the formation of the presynaptic Rad51 nucleofilament [101]. Other interactions occur with both the RPA A- and B-OB fold domains. These include the interaction with both the Werner and Bloom Syndrome helicases. The interaction of RPA with a number of other DNA repair and signalling proteins has not yet been fully mapped, these include the MRN complex, 53BP1, BRCA1, BRCA2 and ATRIP [8, 26, 75, 89, 118, 122, 126]. The further characterisation of these interactions and how these interactions are controlled will clearly lead to a better understanding of the multiple functions of RPA and how it regulates different DNA repair processes.

16.8 Phosphorylation of RPA

Phosphoryaltion is a key mechanism by which a proteins function, localisation and abundance can be regulated in eukaryotic cells. SSBs from bacteria have been shown to be phosphorylated on tyrosines, this phosphorylation event increase their affinity for ssDNA [71]. RPA itself is phosphorylated at multiple sites. These phosphorylation events are stimulated by a number of factors including cell cycle stage [1, 31, 33, 78] and exposure to DNA damaging agents including UV and IR [9, 64].

During the cell cycle, the N terminus of RPA34 is phosphorylated [1, 31, 33, 35, 76, 79, 125]. The primary sites of phosphorylation are the cyclin dependent kinase (Cdk) sites Ser23 and Ser29. Phosphorylation on these sites has been shown to inhibit the binding of RPA to a number of proteins including ATM and DNA-PK [78, 84] as well as DNA alpha-primase. While the activity of Cdk in yeast, Cdc2, has been shown to be directly involved in the selection of the HDR pathway for repair of DSBs [20, 36, 44]. In mammalian cells, Cdk2 plays an important role in end resection and treatment of cells with Cdk inhibitor, roscovitine, suppresses the generation of ssDNA after IR [46]. These data clearly show Cdk2 has an important role to play in the control of HDR pathway however the mechanism by which it acts is yet unknown.

DNA damage results in what is referred to as hyperphosphorylation of RPA. A total of ten sites have been identified, Serines 4, 8, 11, 12, 13, 23, 29 and 33 as well as Threonines 21 and 98 [76, 77, 125]. However it is not clear if all sites can be phosphorylated simultaneously or in what combinations they are phosphorylated. The phosphatidyl inositol 3 kinase like kinase (PIKK) family members including ATM, ATR and DNAPK are required for the initiation of hyperphosphorylation of RPA. ATM, ATR and DNAPK phosphorylate Threonine 21 and Serine 33 in vivo in response to a variety of DNA damaging agents [12, 92]. The exact role of hyperphosphorylation is still contentious, however growing evidence suggests that damage-induced hyperphosphorylation is designed to switch off RPAs function in DNA replication and to modulate its DNA repair activity. It has been shown that hyperphosphorylated RPA cannot efficiently support in vitro DNA replication of SV40 [19, 43, 63, 84, 112]. Consistent with this hyperphosphorylation mimicking mutants of RPA, unlike native RPA, cannot associate with replication centres in vivo [110] although they are still capable of associating with DNA damage foci. Hyperphosphorylation has no effect on the ability of RPA to stimulate NER [5, 80, 84] but it does stimulate its interaction with HDR factors Rad51, Rad52 and the ATR kinase in response to UV or camptothecin treatment [119]. This may suggest that RPA is directed to the repair of cytotoxic DNA breaks preferentially over other damage-induced lesions.

16.9 RPA and the Link with HDR Repair

HDR involves a multitude of proteins which initially recognise the break, resect one strand of the break to expose the bases, locate a sister strand, invade, replicate the missing DNA, end capture the second strand and finally resolving the chromosomal bridges (for details refer to Chapter 8). HDR can also repair a number of other substrates including ssDNA gaps, interstrand cross links, as well as being required for the recovery of stalled and collapsed DNA replication forks [70, 81]. However in order to prevent loss of heterozygosity, a cell will only utilise HDR when it contains greater than 2 N copies of its chromosomes, so HDR occurs in S phase through to G2 [25, 90, 100, 108]. Consistent with this, the strand invasion protein Rad51 is only observed to form foci in S and G2 of the cell cycle consistent with its role in HR [10, 52]. The formation of IR induced RPA foci only occur in cyclin A positive cells indicating that RPA foci like Rad51 form during the S and G2 stages of the cell cycle [46].

One of the central complexes required for the initial stages of HDR is the MRN complex [28, 61, 96, 102]. The MRN complex is composed of three central proteins; Mre11, Rad50 and Nbs1. This complex functions to process the double strand DNA break, exposing a ssDNA length that will be the substrate for strand invasion of sister chromatids. Interestingly the exonuclease activity of the MRN nuclease (Mre11) has a polarity of $3'$ to $5'$ inconsistent with the generated $5'$ recessed end. This may just reflect the artificial nature of the in vitro assays [55], or that there are still missing components or it is even possible that polarity switching may occur as is observed for the bacteria RecBCD [2] or alternatively, as shown recently the endonuclease rather than exonuclease activity of MRN is required for resection [46]. RPA, like the MRN complex localises rapidly to sites of DSBs, however it does not precede the arrival of the MRN complex or its interacting partner CtIP (C-terminal region of Adenovirus E1A binding protein (CtBP)-interacting protein), which stimulate endonuclease activity and end resection by MRN. Consistent with this, activity of both CtIP and MRE11 are required for the formation of RPA foci [23, 46, 74, 93]. Interestingly the retention of RPA to sites of DSBs also requires the presence of BRCA1 [23]. This observation requires further investigation as it either implies that resection of the DSB by the MRN complex does not occur in the absence of BRCA1 or that RPA requires BRCA1 for loading or maintenance. Once loaded RPA is believed to protect the naked ssDNA from attack by nucleases and free radicals as well as preventing formation of secondary structures and binding by inappropriate proteins. It is not yet clear at what point resection of the DSB stops in HDR. It is known however that recombination can occur between two episomal plasmids with approximately 200 bp of homology [62, 91].

The RPA:ssDNA fibre formed as a result of resection in S and G2 cells now becomes a substrate for the ATR:ATRIP complex. Binding of ATR:ATRIP to the RPA coated ssDNA results in the activation of ATR which then phosphoryates downstream targets including Chk1 and Rad17 [46, 126]. This process is dependent on, and is down stream of ATM [46]. Interestingly however, although RPA coated ssDNA is required for the localisation of ATRIP:ATR to DNA breaks, an ATRIP mutant which does not bind to RPA or form foci at DSBs still allows ATR to phosphorylate Chk1 [8], suggesting a recruitment independent mechanism of activation of ATR (for details refer to Chapter 2). The recruitment of the ATR to ssDNA is dependant on RPA but also requires the cofactors claspin and Cut5 [56, 58, 60, 83]. Therefore RPA now exerts its role in the signalling of DSBs.

RPA must be removed from the ssDNA strand to allow the loading of Rad51 recombinase. The exact mechanism by which this happens is not clear. It is known that Rad51 can bind the A-OB fold of RPA70 [101]. This interaction could compete for binding with ssDNA, and as such may be the mechanism by which RPA is displaced from DNA. Mutational analysis of Rad51 confirmed the importance of Rad51 interaction with RPA in HDR pathway. Loss of this interaction prevents the formation of the Rad51 nucleofilament [101]. A proposed mechanism of RPA displacement suggests that the Rad51 N-terminus captures RPA that has been displaced from a 3′ ssDNA overhang. The elongation of the Rad51 nucleofilament proceeds in this manner assisted by ATP hydrolysis, displacing RPA as it progresses [101]. Rad51 alone is not sufficient for displacement of RPA from ssDNA. Addition of saturating amounts of RPA prior to or at the same time as Rad51 is inhibitory to strand invasion [54, 104, 106, 107]. However, RPA itself can facilitate the loading of Rad51 onto the ssDNA filament under certain conditions [99, 103, 107]. Another protein, Rad52, also acts to facilitate the loading of Rad51 and displacement of RPA. Rad52 can bind to both the RPA32 and RPA70 subunits [27, 45, 68], although it is not clear if it can interact with both subunits simultaneously. Similarly in yeast, Rad52 through its interaction with RPA facilitates the loading of Rad51 onto DNA [99, 103, 107]. This occurs by accelerating the displacement of RPA from the DNA and facilitating presynaptic complex formation [103]. Rad52 and Rad51 have been shown to interact directly although through different domains to RPA [72, 94]. In higher eukaryotes it has been shown that Rad52 can function in vitro to stimulate Rad51 loading, however the in vivo function of Rad52 is not required for Rad51 nucleo-filament formation. Rad51 loading, is however mediated in vivo through Rad51 paralogs [18, 67, 97, 99, 103, 105].

RPA displacement may also be mediated by BRCA2. The crystal structure of the BRCA2 c-terminal domain reveals five domains [121]. Three of these domains are homologous to each other, containing a single OB fold. These BRCA2 OB folds bind to ssDNA in a similar manner to RPA [121] and are arranged so that their binding grooves are aligned. The forth domain contains a structure referred to as a tower domain. This structure has previously been reported in bacterial recombinases as well as Myb and homeodomain proteins. The tower domain consists of a pair of long α-helices supporting a three helix bundle. The tower domains from other proteins have been shown to bind the major groove of double stranded DNA. Although, this has not been observed for BRCA2, the importance of this domain is clear as four of the seven most common missense mutations of BRCA2 occur in this c-terminal domain. The BRCA2 OB folds therefore may compete with and displace RPA from ssDNA. The involvement of BRCA2 in Rad51 loading is also backed by evidence that cells defective for BRCA2 do not form Rad51 foci [37, 123]. Rad51 has also been shown to interact with six of the eight BRC repeats of BRCA2 [16, 24, 117]. RPA has also been shown to interact with BRCA2 and that a common cancer predisposing mutant of BRCA2 fails to interact correctly with BRCA2 [118]. Together these studies suggest that BRCA2 functions in the displacement of RPA and in the loading of Rad51.

16.10 hSSB1 and hSSB2

Until recently it was believed that RPA was the only functional SSB homologue in humans, however, recently we have reported the identification of two additional SSBs, which have been named hSSB1 and hSSB2 [88]. hSSB1 is stabilized in response to DSBs by ATM-catalyzed phosphorylation of hSSB1. hSSB1 also accumulates at the sites of DSBs and co-localizes with other repair proteins. In contrast to RPA, hSSB1 does not localize to replication foci in s-phase cells and hSSB1-deficiency does not influence S-phase progression. Notably, hSSB1 is absolutely required for repair of DSBs by HDR pathway.

hSSB1 functions both at early and late stages of HDR where it is required for efficient signalling responses to double strand DNA breaks as well as, like RPA, functioning in vitro to stimulate the Rad51 mediated D loop formation [88] and Rad52 and polymerase eta mediated second end capture and DNA synthesis reaction. Consistent with this, loss of hSSB1 results in spontaneous chromosomal aberrations and a significant increase in the radiation-induced chromosomal aberrations [88]. Taken together these data suggest that, like RPA, hSSB1 plays a crucial role in HDR. It is unclear at present what distinguishes the roles of these two proteins. High resolution wide field miscroscopy has demonstrated that RPA and hSSB1 do not colocalise directly but are proximal to each other, forming touching foci. This may suggest that there functions are distinct but they act within the same repair centres. Future study of these two proteins will define more clearly their cellular roles. It will be also of interest to determine whether hSSB1 and RPA co-operate together or do they have distinct functions in HR?

Biochemical analysis reveals that hSSB1 is dimeric in solution and binds to the classical ssDNA substrates. The function of hSSB2 however is not clear at this point. Expression data suggests that hSSB2 is primarily expressed in lymphocytes and testes, suggesting it may have a role in class switch and meiotic recombination.

16.11 SSBs as Drug Targets

RPAs cellular function in DNA replication and DNA damage repair and signalling makes it a very attractive cancer therapeutic target. Cancer cells are known to replicate rapidly, a process requiring DNA replication to which RPA is central. Cancer cells also have much elevated levels of DNA damage. This is the result of higher metabolic rates increasing endogenous radicals as well as chromosomal rearrangements, DNA replication induced breaks and incorrect checkpoint control. As a result cancer cells have highly active repair systems, RPA inhibition therefore could result in massive chromosomal instability and cell death. One study sought to identify potential chemicals that would inhibit the interaction of RPA with DNA. Using a fluorescent based reporter assay they identified a number of inhibitory chemicals with possible therapeutic potential [3]. hSSB1, like RPA is central to the repair of double strand DNA breaks. These cytotoxic lesions must be repaired to prevent chromosomal fragmentation, an event, which is ultimately lethal. hSSB1 itself is the centre of a drug discovery program which aims to find potential therapeutic inhibitors of its function.

16.12 Summary

SSBs are clearly an evolutionarily important family of proteins, which have roles in numerous cellular processes where ssDNA is exposed. RPA not only functions to sequester the exposed ssDNA, protecting it from incorrect processing or degradation, but also to recruit and coordinate its correct processing. This review has outlined in part our understanding of RPA function. RPA is clearly a central component of many processes and as such it has the ability to alter many aspects of its function. This includes its modular conformation, which sets it aside from the bacterial SSBs, its ability to alter the mode by which it binds ssDNA, its functional modulation by post-translational modifications and its ability to interact with its protein partners. Although the field has progressed remarkably towards the understanding the functions of RPA, however, many questions have been raised by the discovery of two new simple SSB in humans (hSSB1 and hSSB2). Future investigations should address why different SSB's are needed, what are the functional differences between them, what do they interact with, do they have some overlapping and unique functions. It can be implied for now that hSSB1 and hSSB2 must have important cellular functions due to the level of conservation through the evolution. Their discovery challenges existing models of DNA transactions involving ssDNA. Deciphering the role of these two new hSSB's would allow us to understand cellular functions of RPA more clearly but also help in the development of new models of DNA damage repair and may ultimately result in the development of novel cancer therapeutics.

References

1. Anantha, R.W., E. Sokolova, and J.A. Borowiec. 2008. RPA phosphorylation facilitates mitotic exit in response to mitotic DNA damage. *Proc Natl Acad Sci USA*. 105:12903–8.
2. Anderson, D.G., and S.C. Kowalczykowski. 1997. The recombination hot spot chi is a regulatory element that switches the polarity of DNA degradation by the RecBCD enzyme. *Genes Dev*. 11:571–81.
3. Andrews, B.J., and J.J. Turchi. 2004. Development of a high-throughput screen for inhibitors of replication protein A and its role in nucleotide excision repair. *Mol Cancer Ther*. 3:385–91.
4. Arad, G., A. Hendel, C. Urbanke, U. Curth, and Z. Livneh. 2008. Single-stranded DNA-binding protein recruits DNA polymerase V to primer termini on RecA-coated DNA. *J Biol Chem*. 283:8274–82.
5. Ariza, R.R., S.M. Keyse, J.G. Moggs, and R.D. Wood. 1996. Reversible protein phosphorylation modulates nucleotide excision repair of damaged DNA by human cell extracts. *Nucleic Acids Res*. 24:433–40.
6. Arunkumar, A.I., M.E. Stauffer, E. Bochkareva, A. Bochkarev, and W.J. Chazin. 2003. Independent and coordinated functions of replication protein A tandem high affinity single-stranded DNA binding domains. *J Biol Chem*. 278:41077–82.
7. Arunkumar, A.I., V. Klimovich, X. Jiang, R.D. Ott, L. Mizoue, E. Fanning, and W.J. Chazin. 2005. Insights into hRPA32 C-terminal domain – mediated assembly of the simian virus 40 replisome. *Nat Struct Mol Biol*. 12:332–9.
8. Ball, H.L., J.S. Myers, and D. Cortez. 2005. ATRIP binding to replication protein A-single-stranded DNA promotes ATR-ATRIP localization but is dispensable for Chk1 phosphorylation. *Mol Biol Cell*. 16:2372–81.
9. Binz, S.K., A.M. Sheehan, and M.S. Wold. 2004. Replication protein A phosphorylation and the cellular response to DNA damage. *DNA Repair (Amst)*. 3:1015–24.
10. Bishop, D.K., U. Ear, A. Bhattacharyya, C. Calderone, M. Beckett, R.R. Weichselbaum, and A. Shinohara. 1998. Xrcc3 is required for assembly of Rad51 complexes in vivo. *J Biol Chem*. 273:21482–8.
11. Blackwell, L.J., J.A. Borowiec, and I.A. Masrangelo. 1996. Single-stranded-DNA binding alters human replication protein A structure and facilitates interaction with DNA-dependent protein kinase. *Mol Cell Biol*. 16:4798–807.
12. Block, W.D., Y. Yu, and S.P. Lees-Miller. 2004. Phosphatidyl inositol 3-kinase-like serine/threonine protein kinases (PIKKs) are required for DNA damage-induced phosphorylation of the 32 kDa subunit of replication protein A at threonine 21. *Nucleic Acids Res*. 32:997–1005.
13. Bochkarev, A., and E. Bochkareva. 2004. From RPA to BRCA2: lessons from single-stranded DNA binding by the OB-fold. *Curr Opin Struct Biol*. 14:36–42.
14. Bochkareva, E., L. Kaustov, A. Ayed, G.S. Yi, Y. Lu, A. Pineda-Lucena, J.C. Liao, A.L. Okorokov, J. Milner, C.H. Arrowsmith, and A. Bochkarev. 2005. Single-stranded DNA mimicry in the p53 transactivation domain interaction with replication protein A. *Proc Natl Acad Sci USA*. 102:15412–7.

15. Bochkareva, E., S. Korolev, S.P. Lees-Miller, and A. Bochkarev. 2002. Structure of the RPA trimerization core and its role in the multistep DNA-binding mechanism of RPA. *Embo J.* 21:1855–63.
16. Bork, P., N. Blomberg, and M. Nilges. 1996. Internal repeats in the BRCA2 protein sequence. *Nat Genet.* 13:22–3.
17. Butland, G., J.M. Peregrin-Alvarez, J. Li, W. Yang, X. Yang, V. Canadien, A. Starostine, D. Richards, B. Beattie, N. Krogan, M. Davey, J. Parkinson, J. Greenblatt, and A. Emili. 2005. Interaction network containing conserved and essential protein complexes in *Escherichia coli. Nature.* 433:531–7.
18. Cahill, D., B. Connor, and J.P. Carney. 2006. Mechanisms of eukaryotic DNA double strand break repair. *Front Biosci.* 11:1958–76.
19. Carty, M.P., M. Zernik-Kobak, S. McGrath, and K. Dixon. 1994. UV light-induced DNA synthesis arrest in HeLa cells is associated with changes in phosphorylation of human single-stranded DNA-binding protein. *Embo J.* 13:2114–23.
20. Caspari, T., J.M. Murray, and A.M. Carr. 2002. Cdc2-cyclin B kinase activity links Crb2 and Rqh1-topoisomerase III. *Genes Dev.* 16:1195–208.
21. Chen, C., K. Umezu, and R.D. Kolodner. 1998. Chromosomal rearrangements occur in S. cerevisiae rfa1 mutator mutants due to mutagenic lesions processed by double-strand-break repair. *Mol Cell.* 2:9–22.
22. Chen, C., and R.D. Kolodner. 1999. Gross chromosomal rearrangements in *Saccharomyces cerevisiae* replication and recombination defective mutants. *Nat Genet.* 23:81–5.
23. Chen, L., C.J. Nievera, A.Y. Lee, and X. Wu. 2008. Cell cycle-dependent complex formation of BRCA1.CtIP.MRN is important for DNA double-strand break repair. *J Biol Chem.* 283:7713–20.
24. Chen, P.L., C.F. Chen, Y. Chen, J. Xiao, Z.D. Sharp, and W.H. Lee. 1998b. The BRC repeats in BRCA2 are critical for RAD51 binding and resistance to methyl methanesulfonate treatment. *Proc Natl Acad Sci USA.* 95:5287–92.
25. Cheong, N., X. Wang, Y. Wang, and G. Iliakis. 1994. Loss of S-phase-dependent radioresistance in irs-1 cells exposed to X-rays. *Mutat Res.* 314:77–85.
26. Choudhary, S.K., and R. Li. 2002. BRCA1 modulates ionizing radiation-induced nuclear focus formation by the replication protein A p34 subunit. *J Cell Biochem.* 84:666–74.
27. Davis, A.P., and L.S. Symington. 2003. The Rad52-Rad59 complex interacts with Rad51 and replication protein A. *DNA Repair (Amst).* 2:1127–34.
28. de Jager, M., J. van Noort, D.C. van Gent, C. Dekker, R. Kanaar, and C. Wyman. 2001. Human Rad50/Mre11 is a flexible complex that can tether DNA ends. *Mol Cell.* 8:1129–35.
29. de Laat, W.L., E. Appeldoorn, K. Sugasawa, E. Weterings, N.G. Jaspers, and J.H. Hoeijmakers. 1998. DNA-binding polarity of human replication protein A positions nucleases in nucleotide excision repair. *Genes Dev.* 12:2598–609.
30. Dean, F.B., P. Bullock, Y. Murakami, C.R. Wobbe, L. Weissbach, and J. Hurwitz. 1987. Simian virus 40 (SV40) DNA replication: SV40 large T antigen unwinds DNA containing the SV40 origin of replication. *Proc Natl Acad Sci USA.* 84:16–20.
31. Din, S., S.J. Brill, M.P. Fairman, and B. Stillman. 1990. Cell-cycle-regulated phosphorylation of DNA replication factor A from human and yeast cells. *Genes Dev.* 4:968–77.
32. Dutta, A., J.M. Ruppert, J.C. Aster, and E. Winchester. 1993. Inhibition of DNA replication factor RPA by p 53. *Nature.* 365:79–82.
33. Dutta, A., and B. Stillman. 1992. cdc2 family kinases phosphorylate a human cell DNA replication factor, RPA, and activate DNA replication. *Embo J.* 11:2189–99.
34. Fairman, M.P., and B. Stillman. 1988. Cellular factors required for multiple stages of SV40 DNA replication in vitro. *Embo J.* 7:1211–8.
35. Fang, F., and J.W. Newport. 1993. Distinct roles of cdk2 and cdc2 in RP-A phosphorylation during the cell cycle. *J Cell Sci.* 106 (Pt 3):983–94.
36. Ferreira, M.G., and J.P. Cooper. 2001. The fission yeast Taz1 protein protects chromosomes from Ku-dependent end-to-end fusions. *Mol Cell.* 7:55–63.

37. Godthelp, B.C., F. Artwert, H. Joenje, and M.Z. Zdzienicka. 2002. Impaired DNA damage-induced nuclear Rad51 foci formation uniquely characterizes Fanconi anemia group D 1. *Oncogene*. 21:5002–5.
38. Gomes, X.V., L.A. Henricksen, and M.S. Wold. 1996. Proteolytic mapping of human replication protein A: evidence for multiple structural domains and a conformational change upon interaction with single-stranded DNA. *Biochemistry*. 35:5586–95.
39. Han, E.S., D.L. Cooper, N.S. Persky, V.A. Sutera, Jr., R.D. Whitaker, M.L. Montello, and S.T. Lovett. 2006. RecJ exonuclease: substrates, products and interaction with SSB. *Nucleic Acids Res*. 34:1084–91.
40. Handa, P., N. Acharya, and U. Varshney. 2001. Chimeras between single-stranded DNA-binding proteins from *Escherichia coli* and Mycobacterium tuberculosis reveal that their C-terminal domains interact with uracil DNA glycosylases. *J Biol Chem*. 276: 16992–7.
41. He, Z., B.T. Brinton, J. Greenblatt, J.A. Hassell, and C.J. Ingles. 1993. The transactivator proteins VP16 and GAL4 bind replication factor A. *Cell*. 73:1223–32.
42. Iftode, C., and J.A. Borowiec. 2000. $5'$ $3'$ molecular polarity of human replication protein A (hRPA) binding to pseudo-origin DNA substrates. *Biochemistry*. 39:11970–81.
43. Iftode, C., Y. Daniely, and J.A. Borowiec. 1999. Replication protein A (RPA): the eukaryotic SSB. *Crit Rev Biochem Mol Biol*. 34:141–80.
44. Ira, G., A. Pellicioli, A. Balijja, X. Wang, S. Fiorani, W. Carotenuto, G. Liberi, D. Bressan, L. Wan, N.M. Hollingsworth, J.E. Haber, and M. Foiani. 2004. DNA end resection, homologous recombination and DNA damage checkpoint activation require CDK1. *Nature*. 431: 1011–7.
45. Jackson, D., K. Dhar, J.K. Wahl, M.S. Wold, and G.E. Borgstahl. 2002. Analysis of the human replication protein A:Rad52 complex: evidence for crosstalk between RPA32, RPA70, Rad52 and DNA. *J Mol Biol*. 321:133–48.
46. Jazayeri, A., J. Falck, C. Lukas, J. Bartek, G.C. Smith, J. Lukas, and S.P. Jackson. 2006. ATM- and cell cycle-dependent regulation of ATR in response to DNA double-strand breaks. *Nat Cell Biol*. 8:37–45.
47. Kelman, Z., A. Yuzhakov, J. Andjelkovic, and M. O'Donnell. 1998. Devoted to the lagging strand-the subunit of DNA polymerase III holoenzyme contacts SSB to promote processive elongation and sliding clamp assembly. *Embo J*. 17:2436–49.
48. Kim, C., B.F. Paulus, and M.S. Wold. 1994. Interactions of human replication protein A with oligonucleotides. *Biochemistry*. 33:14197–206.
49. Kim, C., R.O. Snyder, and M.S. Wold. 1992. Binding properties of replication protein A from human and yeast cells. *Mol Cell Biol*. 12:3050–9.
50. Kim, C., and M.S. Wold. 1995. Recombinant human replication protein A binds to polynucleotides with low cooperativity. *Biochemistry*. 34:2058–64.
51. Kim, D.K., E. Stigger, and S.H. Lee. 1996. Role of the 70-kDa subunit of human replication protein A (I). Single-stranded dna binding activity, but not polymerase stimulatory activity, is required for DNA replication. *J Biol Chem*. 271:15124–9.
52. Kim, J.S., T.B. Krasieva, H. Kurumizaka, D.J. Chen, A.M. Taylor, and K. Yokomori. 2005. Independent and sequential recruitment of NHEJ and HR factors to DNA damage sites in mammalian cells. *J Cell Biol*. 170:341–7.
53. Kolpashchikov, D.M., S.N. Khodyreva, D.Y. Khlimankov, M.S. Wold, A. Favre, and O.I. Lavrik. 2001. Polarity of human replication protein A binding to DNA. *Nucleic Acids Res*. 29:373–9.
54. Krejci, L., S. Van Komen, Y. Li, J. Villemain, M.S. Reddy, H. Klein, T. Ellenberger, and P. Sung. 2003. DNA helicase Srs2 disrupts the Rad51 presynaptic filament. *Nature*. 423:305–9.
55. Krogh, B.O., and L.S. Symington. 2004. Recombination proteins in yeast. *Annu Rev Genet*. 38:233–71.

56. Kumagai, A., S.M. Kim, and W.G. Dunphy. 2004. Claspin and the activated form of ATR-ATRIP collaborate in the activation of Chk1. *J Biol Chem*. 279:49599–608.
57. Lecointe, F., C. Serena, M. Velten, A. Costes, S. McGovern, J.C. Meile, J. Errington, S.D. Ehrlich, P. Noirot, and P. Polard. 2007. Anticipating chromosomal replication fork arrest: SSB targets repair DNA helicases to active forks. *Embo J*. 26:4239–51.
58. Lee, J., A. Kumagai, and W.G. Dunphy. 2003. Claspin, a Chk1-regulatory protein, monitors DNA replication on chromatin independently of RPA, ATR, and Rad17. *Mol Cell*. 11:329–40.
59. Li, R., and M.R. Botchan. 1993. The acidic transcriptional activation domains of VP16 and p53 bind the cellular replication protein A and stimulate in vitro BPV-1 DNA replication. *Cell*. 73:1207–21.
60. Lin, S.Y., K. Li, G.S. Stewart, and S.J. Elledge. 2004. Human Claspin works with BRCA1 to both positively and negatively regulate cell proliferation. *Proc Natl Acad Sci USA*. 101:6484–9.
61. Lisby, M., J.H. Barlow, R.C. Burgess, and R. Rothstein. 2004. Choreography of the DNA damage response: spatiotemporal relationships among checkpoint and repair proteins. *Cell*. 118:699–713.
62. Liskay, R.M., A. Letsou, and J.L. Stachelek. 1987. Homology requirement for efficient gene conversion between duplicated chromosomal sequences in mammalian cells. *Genetics*. 115:161–7.
63. Liu, J.S., S.R. Kuo, M.M. McHugh, T.A. Beerman, and T. Melendy. 2000. Adozelesin triggers DNA damage response pathways and arrests SV40 DNA replication through replication protein A inactivation. *J Biol Chem*. 275:1391–7.
64. Liu, V.F., and D.T. Weaver. 1993. The ionizing radiation-induced replication protein A phosphorylation response differs between ataxia telangiectasia and normal human cells. *Mol Cell Biol*. 13:7222–31.
65. Liu, Y., Z. Yang, C.D. Utzat, Y. Liu, N.E. Geacintov, A.K. Basu, and Y. Zou. 2005. Interactions of human replication protein A with single-stranded DNA adducts. *Biochem J*. 385:519–26.
66. Lohman, T.M., and M.E. Ferrari. 1994. *Escherichia coli* single-stranded DNA-binding protein: multiple DNA-binding modes and cooperativities. *Annu Rev Biochem*. 63:527–70.
67. McIlwraith, M.J., E. Van Dyck, J.Y. Masson, A.Z. Stasiak, A. Stasiak, and S.C. West. 2000. Reconstitution of the strand invasion step of double-strand break repair using human Rad51 Rad52 and RPA proteins. *J Mol Biol*. 304:151–64.
68. Mer, G., A. Bochkarev, R. Gupta, E. Bochkareva, L. Frappier, C.J. Ingles, A.M. Edwards, and W.J. Chazin. 2000. Structural basis for the recognition of DNA repair proteins UNG2, XPA, and RAD52 by replication factor RPA. *Cell*. 103:449–56.
69. Meyer, R.R., and P.S. Laine. 1990. The single-stranded DNA-binding protein of *Escherichia coli*. *Microbiol Rev*. 54:342–80.
70. Michel, B., G. Grompone, M.J. Flores, and V. Bidnenko. 2004. Multiple pathways process stalled replication forks. *Proc Natl Acad Sci USA*. 101:12783–8.
71. Mijakovic, I., D. Petranovic, B. Macek, T. Cepo, M. Mann, J. Davies, P.R. Jensen, and D. Vujaklija. 2006. Bacterial single-stranded DNA-binding proteins are phosphorylated on tyrosine. *Nucleic Acids Res*. 34:1588–96.
72. Milne, G.T., and D.T. Weaver. 1993. Dominant negative alleles of RAD52 reveal a DNA repair/recombination complex including Rad51 and Rad52. *Genes Dev*. 7:1755–65.
73. Molineux, I.J., and M.L. Gefter. 1974. Properties of the *Escherichia coli* in DNA binding (unwinding) protein: interaction with DNA polymerase and DNA. *Proc Natl Acad Sci USA*. 71:3858–62.
74. Myers, J.S., and D. Cortez. 2006. Rapid activation of ATR by ionizing radiation requires ATM and Mre11. *J Biol Chem*. 281:9346–50.
75. Namiki, Y., and L. Zou. 2006. ATRIP associates with replication protein A-coated ssDNA through multiple interactions. *Proc Natl Acad Sci USA*. 103:580–5.

76. Niu, H., H. Erdjument-Bromage, Z.Q. Pan, S.H. Lee, P. Tempst, and J. Hurwitz. 1997. Mapping of amino acid residues in the p34 subunit of human single-stranded DNA-binding protein phosphorylated by DNA-dependent protein kinase and Cdc2 kinase in vitro. *J Biol Chem*. 272:12634–41.
77. Nuss, J.E., S.M. Patrick, G.G. Oakley, G.M. Alter, J.G. Robison, K. Dixon, and J.J. Turchi. 2005. DNA damage induced hyperphosphorylation of replication protein A. 1. Identification of novel sites of phosphorylation in response to DNA damage. *Biochemistry*. 44:8428–37.
78. Oakley, G.G., S.M. Patrick, J. Yao, M.P. Carty, J.J. Turchi, and K. Dixon. 2003. RPA phosphorylation in mitosis alters DNA binding and protein-protein interactions. *Biochemistry*. 42:3255–64.
79. Pan, Z.Q., A.A. Amin, E. Gibbs, H. Niu, and J. Hurwitz. 1994. Phosphorylation of the p34 subunit of human single-stranded-DNA-binding protein in cyclin A-activated G1 extracts is catalyzed by cdk-cyclin A complex and DNA-dependent protein kinase. *Proc Natl Acad Sci USA*. 91:8343–7.
80. Pan, Z.Q., C.H. Park, A.A. Amin, J. Hurwitz, and A. Sancar. 1995. Phosphorylated and unphosphorylated forms of human single-stranded DNA-binding protein are equally active in simian virus 40 DNA replication and in nucleotide excision repair. *Proc Natl Acad Sci USA*. 92:4636–40.
81. Paques, F., and J.E. Haber. 1999. Multiple pathways of recombination induced by double-strand breaks in *Saccharomyces cerevisiae*. *Microbiol Mol Biol Rev*. 63:349–404.
82. Park, J.S., M. Wang, S.J. Park, and S.H. Lee. 1999. Zinc finger of replication protein A, a non-DNA binding element, regulates its DNA binding activity through redox. *J Biol Chem*. 274:29075–80.
83. Parrilla-Castellar, E.R., and L.M. Karnitz. 2003. Cut5 is required for the binding of Atr and DNA polymerase alpha to genotoxin-damaged chromatin. *J Biol Chem*. 278:45507–11.
84. Patrick, S.M., G.G. Oakley, K. Dixon, and J.J. Turchi. 2005. DNA damage induced hyperphosphorylation of replication protein A. 2. Characterization of DNA binding activity, protein interactions, and activity in DNA replication and repair. *Biochemistry*. 44: 8438–48.
85. Pestryakov, P.E., D.Y. Khlimankov, E. Bochkareva, A. Bochkarev, and O.I. Lavrik. 2004. Human replication protein A (RPA) binds a primer-template junction in the absence of its major ssDNA-binding domains. *Nucleic Acids Res*. 32:1894–903.
86. Raghunathan, S., A.G. Kozlov, T.M. Lohman, and G. Waksman. 2000. Structure of the DNA binding domain of *E. coli* SSB bound to ssDNA. *Nat Struct Biol*. 7:648–52.
87. Raghunathan, S., C.S. Ricard, T.M. Lohman, and G. Waksman. 1997. Crystal structure of the homo-tetrameric DNA binding domain of *Escherichia coli* single-stranded DNA-binding protein determined by multiwavelength x-ray diffraction on the selenomethionyl protein at 2.9-A resolution. *Proc Natl Acad Sci USA*. 94:6652–7.
88. Richard, D.J., E. Bolderson, L. Cubeddu, R.I. Wadsworth, K. Savage, G.G. Sharma, M.L. Nicolette, S. Tsvetanov, M.J. McIlwraith, R.K. Pandita, S. Takeda, R.T. Hay, J. Gautier, S.C. West, T.T. Paull, T.K. Pandita, M.F. White, and K.K. Khanna. 2008. Single-stranded DNA-binding protein hSSB1 is critical for genomic stability. *Nature*. 453:677–81.
89. Robison, J.G., J. Elliott, K. Dixon, and G.G. Oakley. 2004. Replication protein A and the Mre11.Rad50.Nbs1 complex co-localize and interact at sites of stalled replication forks. *J Biol Chem*. 279:34802–10.
90. Rothkamm, K., I. Kruger, L.H. Thompson, and M. Lobrich. 2003. Pathways of DNA double-strand break repair during the mammalian cell cycle. *Mol Cell Biol*. 23:5706–15.
91. Rubnitz, J., and S. Subramani. 1984. The minimum amount of homology required for homologous recombination in mammalian cells. *Mol Cell Biol*. 4:2253–8.
92. Sakasai, R., K. Shinohe, Y. Ichijima, N. Okita, A. Shibata, K. Asahina, and H. Teraoka. 2006. Differential involvement of phosphatidylinositol 3-kinase-related protein kinases in hyperphosphorylation of replication protein A2 in response to replication-mediated DNA double-strand breaks. *Genes Cells*. 11:237–46.

93. Sartori, A.A., C. Lukas, J. Coates, M. Mistrik, S. Fu, J. Bartek, R. Baer, J. Lukas, and S.P. Jackson. 2007. Human CtIP promotes DNA end resection. *Nature*. 450:509–14.
94. Shen, Z., K.G. Cloud, D.J. Chen, and M.S. Park. 1996. Specific interactions between the human RAD51 and RAD52 proteins. *J Biol Chem*. 271:148–52.
95. Shereda, R.D., D.A. Bernstein, and J.L. Keck. 2007. A central role for SSB in *Escherichia coli* RecQ DNA helicase function. *J Biol Chem*. 282:19247–58.
96. Shroff, R., A. Arbel-Eden, D. Pilch, G. Ira, W.M. Bonner, J.H. Petrini, J.E. Haber, and M. Lichten. 2004. Distribution and dynamics of chromatin modification induced by a defined DNA double-strand break. *Curr Biol*. 14:1703–11.
97. Sigurdsson, S., S. Van Komen, W. Bussen, D. Schild, J.S. Albala, and P. Sung. 2001. Mediator function of the human Rad51B-Rad51C complex in Rad51/RPA-catalyzed DNA strand exchange. *Genes Dev*. 15:3308–18.
98. Slocum, S.L., J.A. Buss, Y. Kimura, and P.R. Bianco. 2007. Characterization of the ATPase activity of the *Escherichia coli* RecG protein reveals that the preferred cofactor is negatively supercoiled DNA. *J Mol Biol*. 367:647–64.
99. Song, B., and P. Sung. 2000. Functional interactions among yeast Rad51 recombinase, Rad52 mediator, and replication protein A in DNA strand exchange. *J Biol Chem*. 275:15895–904.
100. Sonoda, E., H. Hochegger, A. Saberi, Y. Taniguchi, and S. Takeda. 2006. Differential usage of non-homologous end-joining and homologous recombination in double strand break repair. *DNA Repair (Amst)*. 5:1021–9.
101. Stauffer, M.E., and W.J. Chazin. 2004. Physical interaction between replication protein A and Rad51 promotes exchange on single-stranded DNA. *J Biol Chem*. 279:25638–45.
102. Stracker, T.H., J.W. Theunissen, M. Morales, and J.H. Petrini. 2004. The Mre11 complex and the metabolism of chromosome breaks: the importance of communicating and holding things together. *DNA Repair (Amst)*. 3:845–54.
103. Sugiyama, T., and S.C. Kowalczykowski. 2002. Rad52 protein associates with replication protein A (RPA)-single-stranded DNA to accelerate Rad51-mediated displacement of RPA and presynaptic complex formation. *J Biol Chem*. 277:31663–72.
104. Sugiyama, T., E.M. Zaitseva, and S.C. Kowalczykowski. 1997. A single-stranded DNA-binding protein is needed for efficient presynaptic complex formation by the *Saccharomyces cerevisiae* Rad51 protein. *J Biol Chem*. 272:7940–5.
105. Sung, P. 1997. Function of yeast Rad52 protein as a mediator between replication protein A and the Rad51 recombinase. *J Biol Chem*. 272:28194–7.
106. Sung, P. 1997. Yeast Rad55 and Rad57 proteins form a heterodimer that functions with replication protein A to promote DNA strand exchange by Rad51 recombinase. *Genes Dev*. 11:1111–21.
107. Symington, L.S. 2002. Role of RAD52 epistasis group genes in homologous recombination and double-strand break repair. *Microbiol Mol Biol Rev*. 66:630–70, table of contents.
108. Takata, M., M.S. Sasaki, E. Sonoda, C. Morrison, M. Hashimoto, H. Utsumi, Y. Yamaguchi-Iwai, A. Shinohara, and S. Takeda. 1998. Homologous recombination and non-homologous end-joining pathways of DNA double-strand break repair have overlapping roles in the maintenance of chromosomal integrity in vertebrate cells. *Embo J*. 17:5497–508.
109. Umezu, K., N.W. Chi, and R.D. Kolodner. 1993. Biochemical interaction of the *Escherichia coli* RecF, RecO, and RecR proteins with RecA protein and single-stranded DNA binding protein. *Proc Natl Acad Sci USA*. 90:3875–9.
110. Vassin, V.M., M.S. Wold, and J.A. Borowiec. 2004. Replication protein A (RPA) phosphorylation prevents RPA association with replication centers. *Mol Cell Biol*. 24:1930–43.
111. Wang, Y., C.D. Putnam, M.F. Kane, W. Zhang, L. Edelmann, R. Russell, D.V. Carrion, L. Chin, R. Kucherlapati, R.D. Kolodner, and W. Edelmann. 2005. Mutation in Rpa1 results in defective DNA double-strand break repair, chromosomal instability and cancer in mice. *Nat Genet*. 37:750–5.

112. Wang, Y., X.Y. Zhou, H. Wang, M.S. Huq, and G. Iliakis. 1999. Roles of replication protein A and DNA-dependent protein kinase in the regulation of DNA replication following DNA damage. *J Biol Chem*. 274:22060–4.
113. Witte, G., C. Urbanke, and U. Curth. 2003. DNA polymerase III chi subunit ties single-stranded DNA binding protein to the bacterial replication machinery. *Nucleic Acids Res*. 31:4434–40.
114. Wobbe, C.R., L. Weissbach, J.A. Borowiec, F.B. Dean, Y. Murakami, P. Bullock, and J. Hurwitz. 1987. Replication of simian virus 40 origin-containing DNA in vitro with purified proteins. *Proc Natl Acad Sci USA*. 84:1834–8.
115. Wold, M.S., D.H. Weinberg, D.M. Virshup, J.J. Li, and T.J. Kelly. 1989. Identification of cellular proteins required for simian virus 40 DNA replication. *J Biol Chem*. 264:2801–9.
116. Wold, M.S., and T. Kelly. 1988. Purification and characterization of replication protein A, a cellular protein required for in vitro replication of simian virus 40 DNA. *Proc Natl Acad Sci USA*. 85:2523–7.
117. Wong, A.K., R. Pero, P.A. Ormonde, S.V. Tavtigian, and P.L. Bartel. 1997. RAD51 interacts with the evolutionarily conserved BRC motifs in the human breast cancer susceptibility gene brca2. *J Biol Chem*. 272:31941–4.
118. Wong, J.M., D. Ionescu, and C.J. Ingles. 2003. Interaction between BRCA2 and replication protein A is compromised by a cancer-predisposing mutation in BRCA2. *Oncogene*. 22: 28–33.
119. Wu, X., Z. Yang, Y. Liu, and Y. Zou. 2005. Preferential localization of hyperphosphorylated replication protein A to double-strand break repair and checkpoint complexes upon DNA damage. *Biochem J*. 391:473–80.
120. Wyka, I.M., K. Dhar, S.K. Binz, and M.S. Wold. 2003. Replication protein A interactions with DNA: differential binding of the core domains and analysis of the DNA interaction surface. *Biochemistry*. 42:12909–18.
121. Yang, H., P.D. Jeffrey, J. Miller, E. Kinnucan, Y. Sun, N.H. Thoma, N. Zheng, P.L. Chen, W.H. Lee, and N.P. Pavletich. 2002. BRCA2 function in DNA binding and recombination from a BRCA2-DSS1-ssDNA structure. *Science*. 297:1837–48.
122. Yoo, E., B.U. Kim, S.Y. Lee, C.H. Cho, J.H. Chung, and C.H. Lee. 2005. 53BP1 is associated with replication protein A and is required for RPA2 hyperphosphorylation following DNA damage. *Oncogene*. 24:5423–30.
123. Yuan, S.S., S.Y. Lee, G. Chen, M. Song, G.E. Tomlinson, and E.Y. Lee. 1999. BRCA2 is required for ionizing radiation-induced assembly of Rad51 complex in vivo. *Cancer Res*. 59:3547–51.
124. Yuzhakov, A., Z. Kelman, and M. O'Donnell. 1999. Trading places on DNA – a three-point switch underlies primer handoff from primase to the replicative DNA polymerase. *Cell*. 96:153–63.
125. Zernik-Kobak, M., K. Vasunia, M. Connelly, C.W. Anderson, and K. Dixon. 1997. Sites of UV-induced phosphorylation of the p34 subunit of replication protein A from HeLa cells. *J Biol Chem*. 272:23896–904.
126. Zou, L., and S.J. Elledge. 2003. Sensing DNA damage through ATRIP recognition of RPA-ssDNA complexes. *Science*. 300:1542–8.

Chapter 17
The Fanconi anemia-BRCA Pathway and Cancer

Toshiyasu Taniguchi

Abstract The Fanconi anemia (FA)-BRCA pathway has emerged as an important pathway in cancer biology. Fanconi anemia (FA) is a rare genetic disease characterized by aplastic anemia, developmental defects, cancer susceptibility and cellular hypersensitivity to interstrand DNA crosslinking agents. Thirteen FA genes have been identified (*FANCA*, *FANCB*, *FANCC*, *FANCD1/BRCA2*, *FANCD2*, *FANCE*, *FANCF*, *FANCG*, *FANCI*, *FANCJ/BRIP1*, *FANCL*, *FANCM*, and *FANCN/PALB2*). The FA proteins and breast/ovarian cancer susceptibility proteins (BRCA1 and BRCA2) cooperate in a common DNA damage-activated signaling pathway (the FA-BRCA pathway) which regulates DNA repair and is required for cellular resistance to DNA crosslinking agents. Inactivation of this pathway has been found in a wide variety of human cancers and is implicated in the sensitivity of cancer cells to anti-cancer DNA crosslinking agents, such as cisplatin and mitomycin C. Reactivation of this pathway is implicated in acquired resistance to DNA crosslinking agents. Therefore, inhibition of the FA-BRCA pathway is an attractive therapeutic strategy to overcome DNA-crosslinker resistance of tumor cells.

Keywords BRCA1 · BRCA2 · Cisplatin · Fanconi anemia · PARP inhibitors · Chemotherapy

17.1 Introduction

Defects of DNA repair and DNA damage response have at least two important implications in cancer biology. First, they cause genomic instability, which promotes malignant transformation of cells. Second, they lead to cellular hypersensitivity to DNA damaging agents, which can contribute to effective chemo/radio-therapy. The recently emerged concept of "the Fanconi anemia (FA)-BRCA pathway" [52, 257] (also called "the FA-BRCA network" [279, 282], or "the FA pathway"), which is a

T. Taniguchi (✉)
Divisions of Human Biology and Public Health Sciences, Fred Hutchinson Cancer Research Center, Seattle, WA 98109-1024, USA
e-mail: ttaniguc@fhcrc.org

K.K. Khanna, Y. Shiloh (eds.), *The DNA Damage Response: Implications on Cancer Formation and Treatment*, DOI 10.1007/978-90-481-2561-6_17,

DNA damage-activated signaling pathway regulating DNA repair, provides a good model for these implications [52].

Fanconi anemia (FA), a rare genetic disease and cancer susceptibility syndrome, has attracted attention from cancer biologists for many years because its understanding can provide insights into the pathogenesis of cancer in the general population [118, 257]. Recent identification of all of the 13 responsible genes for FA has changed our view of the disease. Now, FA attracts broader attention, because a connection between the FA genes and the breast/ovarian cancer susceptibility genes, BRCA1 and BRCA2, has been demonstrated, and thus the concept of the Fanconi anemia-BRCA pathway has been proposed. It has also become clearer that inactivation of this pathway in cancer cells can lead to sensitivity to anti-cancer DNA crosslinking agents, such as cisplatin and mitomycin C (MMC) [118]. Moreover, interaction among FA proteins and numerous proteins involved in DNA repair and damage response has been reported. FA now attracts attention from researchers who works on DNA repair as well.

In this chapter, we will discuss what the FA-BRCA pathway is and how the pathway is implicated in cancer biology.

17.2 Fanconi anemia

FA is a rare genetic disease characterized by childhood-onset bone marrow failure (aplastic anemia), cancer/leukemia susceptibility, multiple congenital abnormalities and cellular hypersensitivity to interstrand DNA crosslinking (ICL) agents, such as MMC, diepoxybutane (DEB), melphalan, cisplatin and cyclophosphamide (reviewed in [257, 282]). It is an autosomal (all complementation groups except FA-B group) or X-linked (FA-B group) recessive disease. FA is a genetically heterogeneous disease. FA can be divided into at least 13 complementation groups (A, B, C, D1, D2, E, F, G, I, J, L, M and N) defined by cell fusion studies and all of the 13 responsible FA genes have been identified (Table 17.1) (Fig. 17.1) [62, 148–150, 176, 221, 236, 237, 297].

FA is a very rare disease (1–5 per million) and its heterozygous carrier frequency is estimated to be 1/300 [13, 121]. The clinical course of FA and the therapy for FA have been extensively reviewed elsewhere [7, 13, 66, 264], therefore I will not go into great detail. The median survival age of patients with FA has improved to older than 30 years, according to a recent report [7]. The common congenital defects seen in FA patients includes short statue, abnormalities of the thumb or arm, skin, head, eyes, kidneys, and ears as well as developmental disability [7].

FA patients develop aplastic anemia (a condition where bone marrow does not produce sufficient new blood cells), typically during the first 10 years of life. In spite of recent improvements of treatment, the most common cause of mortality in FA is still bone marrow failure, followed by leukemia and solid tumors. For the treatment of bone marrow failure, androgens and hematopoietic growth factors are used. Hematopoietic stem cell transplantation is often performed, if a donor is available [66, 225, 264].

Table 17.1 Thirteen complementation groups and responsible genes for Fanconi anemia

Subtype	Responsible gene	FA patients, estimated, %*	Chromosome location	Protein kDa	Requirement for FANCD2 monoubiquitination	Main function of the protein	Comments
A	*FANCA*	57	16q24.3	163	+	FA core complex	
B	*FANCB (FAAP95)*	0.3	Xp22.31	95	+	FA core complex	
C	*FANCC*	15	9q22.3	63	+	FA core complex	Cytoplasmic functions
D1	*FANCD1/BRCA2*	4	13q12-13	380	-	RAD51 recruitment	
D2	*FANCD2*	3	3p25.3	155,162	+	partner of FANCI	Monoubiquitinated protein
E	*FANCE*	1	6p21-22	60	+	FA core complex	Direct binding to FANCD2
F	*FANCF*	2	11p15	42	+	FA core complex	
G	*FANCG/XRCC9*	9	9p13	68	+	FA core complex	
I	*FANCI*	Rare	15q25-26	140,147	+	partner of FANCD2	Monoubiquitinated protein
J	*FANCJ/BACH1/BRIP1*	1.6	17q22-q24	130	-	5'->3' DNA helicase/ATPase	Binding to BRCA1
L	*FANCL/PHF9/POG (FAAP43)*	0.1	2p16.1	43	+	FA core complex, ubiquitin ligase	PHD/RING finger motif
M	*FANCM/Hef (FAAP250)*	Rare	14q21.3	250	+	FA core complex, ATPase/translocase	Helicase/nuclease motif
N	*FANCN/PALB2*	Rare	16p12.1	140	-	Required for BRCA2 stability and nuclear localization	

Adapted from Taniguchi et al. [257] with modifications.

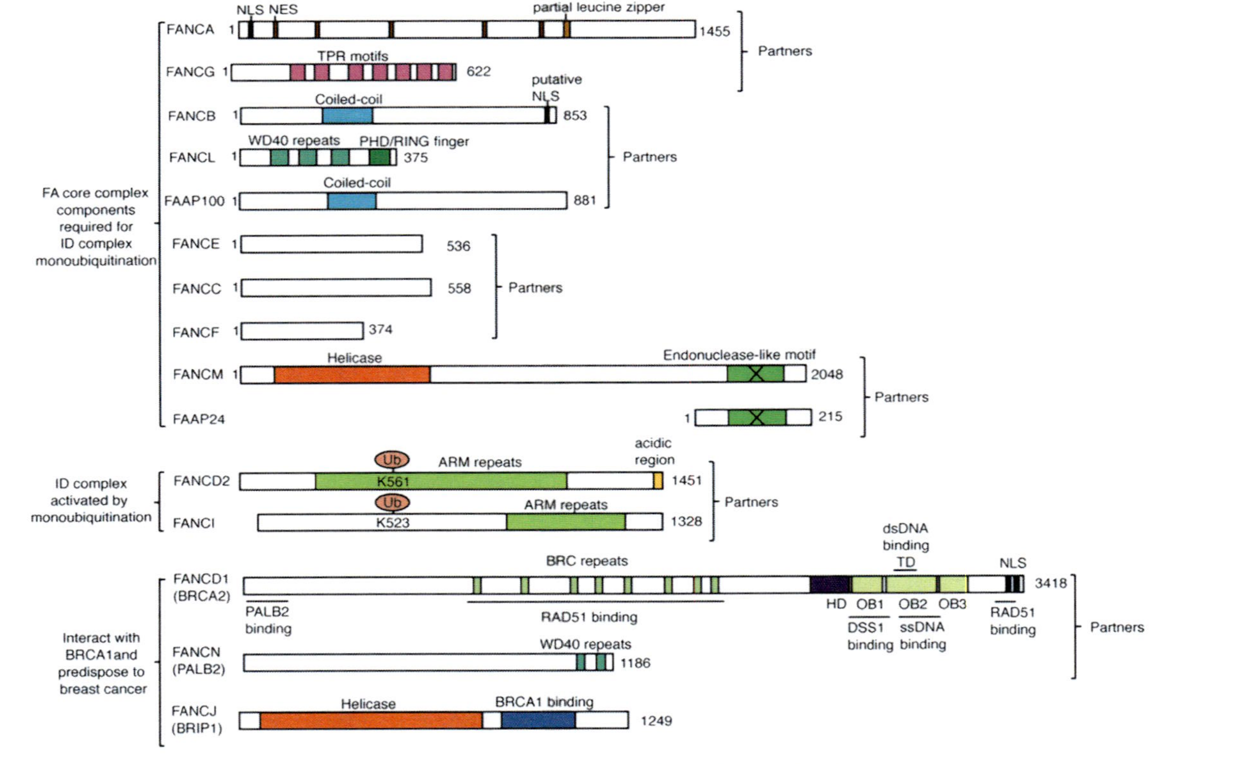

Fig. 17.1 Schematic representation of the thirteen human Fanconi anemia proteins, FAAP24, and FAAP100. (Adapted with modifications from Fig. 2 in Weidong Wang [282]). The FA proteins are classified into three groups on the basis of their roles in the monoubiquitination of FANCD2 and FANCI. The only FA proteins with known enzymatic activity are FANCJ (helicase), FANCM (DNA translocase), and FANCL (E3 ubiquitin ligase). FANCD2 and FANCD1/BRCA2 have been shown to have direct DNA binding activity. ARM, armadillo repeat; BRC, the internal repeat domains of BRCA2; dsDNA, double strand DNA; FAAP, FA-associated protein; HD, helical domain; NES, nuclear export sequences; NLS, nuclear localization signals; OB, oligonucleotide/oligosaccharide-binding folds; PALB2, partner and localizer of BRCA2; ssDNA, single strand DNA; TD, tower domain; TPR, tetratricopeptide repeat motifs; Ub, ubiquitin

According to recent studies, at least 20% of FA patients develop some kind of malignancy, and the actuarial risks of developing hematological malignancies (mainly, acute myelogenous leukemia (AML) and myelodysplastic syndrome) and solid tumors (head and neck squamous cell carcinoma, gynecological squamous cell carcinoma, esophageal carcinoma, liver tumors, brain tumors, skin tumors, renal tumors and other tumors) by 40 years of age are 33 and 28%, respectively [6, 140, 224]. FA subtypes FA-D1 and FA-N are associated with increased risk to develop medulloblastoma, Wilms' tumor and acute leukemia in very early childhood [104, 203, 221, 280, 297].

In some of the tumors that develop in FA patients, involvement of certain etiologic agents has been suggested. For example, some FA patients receiving androgens for bone marrow failure develop liver tumors, suggesting that androgens promote liver tumor formation [6]. In anogenital or head and neck squamous cell carcinoma, a very high incidence of human papilloma virus (HPV) DNA positivity was reported [141], although another group reported a conflicting result (lack of HPV positivity) [277]. Interestingly, it has been reported that FA pathway activation is triggered by the HPV type 16 (HPV-16) E7 oncoprotein, which may suggest that the FA pathway works as an early host cell response to high-risk HPV infection [242].

Whether heterozygous carriers of the FA gene mutation have increased risk of developing cancer is a clinically important issue. Over all, there is no clear evidence of increased cancer susceptibility in heterozygous carriers of mutation in FA genes, except for *FANCD1/BRCA2*, *FANCN/PALB2*, and *FANCJ/BRIP1/BACH1*. *BRCA2/FANCD1* is a cancer susceptibility gene, whose heterozygous mutation carriers have increased susceptibility to breast/ovarian and other cancers [270, 280]. Inherited mono-allelic mutations in *BACH1/BRIP1/FANCJ* and *FANCN/PALB2* have been implicated in breast cancer predisposition (familial breast cancer) [221, 231, 297], although germline mutations in *FANCJ* are rare in breast/ovarian cancer families [125]. There is some evidence that *FANCC* mutations are possibly breast cancer susceptibility alleles [17]. Mutations of some FA genes (*FANCA*, *FANCC*, *FANCD2*, *FANCE*, *FANCF* and *FANCG*) have been tested in *BRCA1/2* mutation-negative families with inherited breast cancer and have been reported to be rare [230].

Cellular hypersensitivity to ICL agents is the most prominent cellular phenotype of FA. Treatment of FA cells with ICL agents causes increased chromosome breakage and marked late S phase (or G2 phase) accumulation [3]. The accumulation of FA cells treated with ICL agents in late S phase (or G2 phase) is mediated by activation of ATR and CHK1 kinases [93]. The DEB-induced chromosome breakage assay (DEB test) is widely used as a diagnostic test for FA [12].

Determining the complementation group of each FA patient has become important clinically in order to confirm the diagnosis, to distinguish FA from other chromosome breakage disorders, and to manage each FA patient and family better [8, 234]. As described above, patients in FA-D1 and FA-N complementation groups have a much severer disease compared to patients in other complementation groups [104, 203, 221, 280, 297]. Family members of FA-D1, FA-N, and FA-J patients may be carriers of mutation in *BRCA2/FANCD1*, *FANCN/PALB2*, and

FANCJ/BRIP1/BACH1, respectively and may therefore be predisposed to cancers [221, 231, 280, 297]. FA-D2 patients tend to have a relatively severe form of FA [122]. In one report, FA patients in complementation group C (FA-C) had a significantly poorer survival than patients in groups A and G [140]. Some patients with other rare chromosome breakage syndrome, such as Nijmegen breakage syndrome (NBS) [129], AT-like disorder (ATLD) [246], and Seckel syndrome can have a positive DEB test [9, 188]. Therefore, to distinguish FA from other chromosome breakage disorders and to manage each FA patient and family better, FA subtyping and confirmation of the mutations should be performed routinely in the near future.

17.3 The Fanconi anemia-BRCA Pathway

17.3.1 The Fanconi anemia Genes

Since the identification of the first FA gene, *FANCC*,in 1992 [247], a total of 13 FA genes have been identified. Some of them (*FANCA* [158], *FANCC* [247], *FANCE* [57], *FANCF* [58], and *FANCG* [56]) have been identified by a functional complementation expression cloning method. *FANCA* [259] and *FANCD2* [262] were identified by positional cloning. *FANCB* [175], *FANCL* [173], and *FANCM*

Fig. 17.2 (continued) A simplified model of the Fanconi anemia-BRCA pathway. The FA proteins are depicted in the normal cell nucleus. Eight FA proteins (FANCA, FANCB, FANCC, FANCE, FANCF, FANCG, FANCL, and FANCM) along with FAAP100 and FAAP24 form a nuclear protein complex (FA core complex), which is considered to be a multi-subunit ubiquitin ligase (E3) for FANCD2 and FANCI. FANCL is the catalytic subunit of the FA core complex and directly interacts with the E2 ubiquitin conjugating enzyme UBE2T. UBE2T can be inactivated by auto-monoubiquitination on lysine 91 (K91). The FA core complex, BLM, RPA, topoisomerase IIIα, BLAP75 and BLAP250 form a super-complex called BRAFT. FANCI and FANCD2 form another complex called the ID complex. In response to DNA damage, or during S phase, FANCD2 and FANCI are monoubiquitinated on specific lysine residues (lysine 561 (K561) for FANCD2, lysine 523 (K523) for FANCI) in an FA core complex-, UBE2T-, and ID complex-dependent manner. DNA damage-induced monoubiquitination of FANCD2 also requires ATR, RPA and HCLK2. ATR directly phosphorylates and activates CHK1 kinase. CHK1 phosphorylates FANCE. Monoubiquitinated FANCD2 and monoubiquitinated FANCI are translocated into nuclear foci and colocalizes with BRCA1, FANCD1/BRCA2, FANCN/PALB2, RAD51, FANCJ/BRIP1, and other proteins. BRCA1 is required for FANCD2 foci formation in response to DNA damage. FANCC, FANCE, and FANCG also form nuclear foci and colocalize with FANCD2. All of these factors are required for cellular resistance to interstrand DNA crosslinking agents, through homologous recombination, translesion synthesis and some other unknown mechanisms. Monoubiquitination of PCNA on lysine 164 (K164) requires RAD6 (E2) and RAD18 (E3). Monoubiquitination of PCNA causes recruitment of translesion synthesis DNA polymerases at the site of stalled replication forks. USP1 deubiquitinates both PCNA and FANCD2, and possibly FANCI. USP1 forms complex with UAF1. The FA core complex itself can interact with DNA and displays some important functions (required for cellular resistance to DNA crosslinking agents) outside of FANCD2 monoubiquitination

[176] were identified by a proteomic approach. *FANCD1* [110], *FANCN* [297], and *FANCJ* [149, 150, 155] turned out to be identical to known genes, *BRCA2*, *PALB2* and *BRIP1/BACH1*, respectively. In 2007, the 13th FA gene, *FANCI*, was finally identified [62, 236, 237].

The 13 FA complementation groups and the responsible genes and proteins are summarized in Table 17.1 and Fig. 17.1. Products of all of the 13 FA genes are required for cellular resistance to ICL agents, such as MMC, DEB, melphalan, and cisplatin and are presumed to interact in a common network (the FA-BRCA network, the FA-BRCA pathway or the FA pathway), which regulates DNA repair (Fig. 17.2).

In the activation of this pathway, monoubiquitination of FANCD2 and FANCI is a critical step. Monoubiquitination (conjugation of one ubiquitin molecule onto

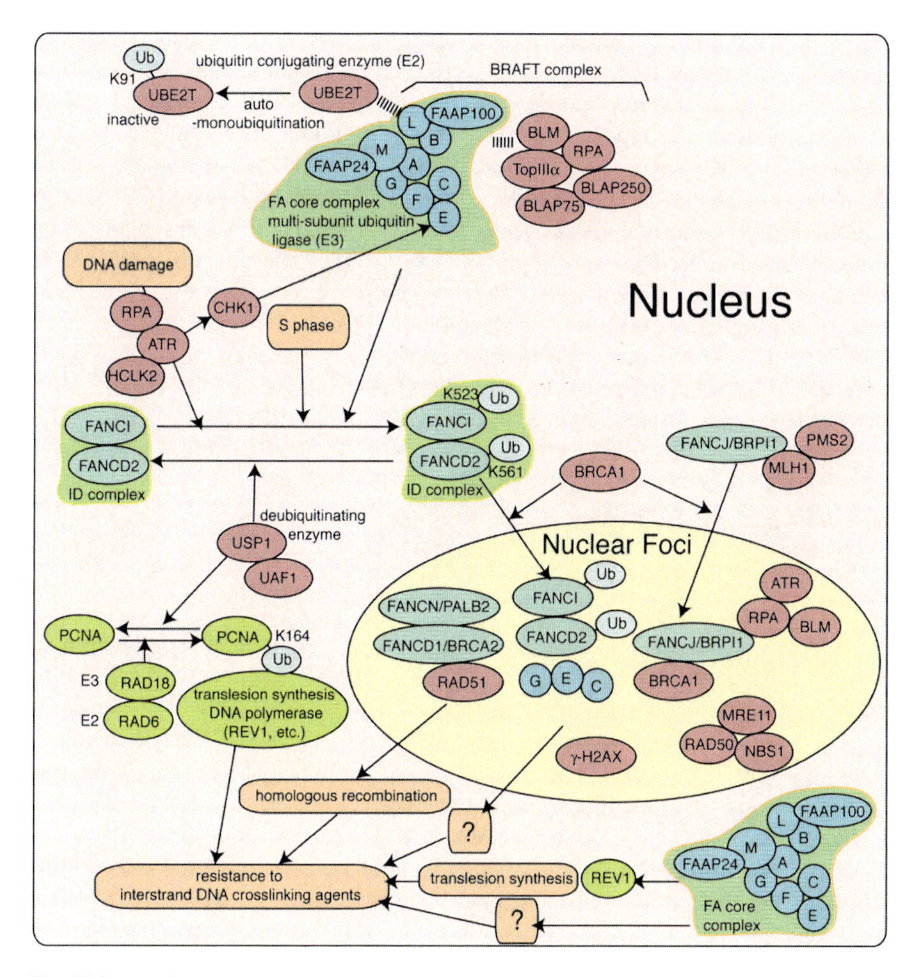

Fig. 17.2 (continued)

a protein) is a posttranslational modification in which a 76-residue small protein, ubiquitin, is covalently attached to a target protein and acts as a reversible signal regulating either protein targeting, membrane trafficking, histone function, transcription regulation, or DNA repair [235].

Using monoubiquitination of FANCD2 and FANCI as an indicator, the 13 FA proteins can be divided into 3 groups (Table 17.1 and Fig. 17.1): (1) FA core complex components required for FANCD2 and FANCI monoubiquitination (FANCA, FANCB, FANCC, FANCE, FANCF, FANCG, FANCL, FANCM), (2) FA 'ID' complex components (FANCD2 and FANCI), which are also required for monoubiquitination of FANCD2 and FANCI, and (3) FA proteins not required for monoubiquitination of FANCD2 or FANCI (FANCD1/BRCA2, FANCN/PALB2, and FANCJ/BRIP1/BACH1) [282].

FANCA is a phosphoprotein, and contains two overlapping bipartite nuclear localization signals (NLS), five functional leucine-rich nuclear export sequences (NESs) and a partial leucine zipper sequence [76, 158, 259]. The nuclear export of FANCA is reported to be regulated in a CRM1-dependent manner [76]. FANCA protein is one of the clients of HSP90 [201].

FANCB (previously reported as FAAP95) (FAAP stands for *F*ANC*A*-*a*ssociated *p*olypeptide) contains a putative bipartite NLS [175]. *FANCB* gene is located on the X chromosome, therefore, FA-B subtype of FA is an X-linked disease [175].

FANCC [247] is a component of the nuclear FA core complex, but some fraction of FANCC protein localizes in cytoplasm as well [303]. Other functions of FANCC outside of the FA core complex have also been proposed. For example, involvement in Jak/STAT signaling and apoptotic signaling [73] (reviewed in [14, 21]).

FANCE [57] is a nuclear protein required for nuclear accumulation of FANCC [206, 253]. The nuclear accumulation of FANCE also depends on FANCC [147]. FANCE interacts with monoubiquitinated FANCD2 and BRCA2 in the chromatin fraction [283]. FANCE can directly bind to FANCC and FANCD2 [88]. It mediates interaction between FANCC and FANCF [88]. Thus FANCE is a key mediator of protein interactions, both in the FA core complex and in connection with complex components to FANCD2. The structure of FANCE protein has been solved [199]. FANCE is directly phosphorylated on threonine 346 and serine 374 by CHK1 kinase in response to DNA damage [284].

FANCF acts as a flexible adaptor protein required for the assembly of the FA core complex [146]. Structure of C-terminal domain of FANCF protein has been solved and revealed a helical repeat structure similar to the Cand1 regulator of the Cul1-Rbx1-Skp1-F box (Skp2) ubiquitin ligase complex [134]. FANCF has a region of homology with the prokaryotic RNA-binding protein ROM [58], but this homology did not turn out to be functionally important [57].

FANCG is identical to *XRCC9*gene, which was originally cloned as a cDNA that partially corrects the MMC sensitivity of a Chinese hamster mutant UV40 cell line [56, 156]. FANCG has seven tetratricopeptide repeat motifs (TPRs) [20], which are degenerate 34-amino acid repeat motifs mediating protein-protein interactions. Serines 383 and 387 on FANCG are phosphorylated in M phase, presumably by cdc2 kinase [178]. These serines are reported to be important for exclusion of

FANCG from chromatin in mitosis. Phosphorylation of serine 7 is upregulated after MMC treatment [219] and is ATR-dependent [289]. In addition to its role as a subunit of the FA core complex, FANCG promotes formation of a protein complex containing BRCA2, FANCD2 and XRCC3 [289]. Phosphorylation of serine 7 is required for this interaction [289]. The role of this FANCG-BRCA2-FANCD2-XRCC3 complex in DNA repair is currently unclear.

FANCL, also known as PHF9, POG, or FAAP43, has three WD40 repeats and a PHD (plant homeodomain) finger motif (variant RING finger motif) [173]. WD40 repeats mediate protein-protein interactions. FANCL is presumed to be the catalytic subunit of the FA core complex as a ubiquitin ligase for FANCD2 and FANCI because it has an autoubiquitin ligase activity in vitro and is required for FANCD2 monoubiquitination in vivo. The mouse homolog of *FANCL* was previously identified as a gene responsible for the phenotype of gcd (*g*erm *c*ell-*d*eficient) mice with reduced fertility due to defective proliferation of germ cells [2].

FANCM (also called FAAP250) and its partner FAAP24 are the only known components of the FA core complex which have been shown to have direct DNA binding activity. FAAP24 shares homology with the XPF family of flap/fork endonucleases and forms a heterodimer with FANCM [43]. FANCM contains seven helicase-specific motifs and one degenerate endonuclease domain [176], but the helicase activity or endonuclease activity of FANCM has not been shown. FANCM has been shown to have ssDNA and dsDNA-stimulated ATPase activity, DNA translocase activity, a specific ATP-independent DNA-binding affinity to fork and four-way junction DNA, and an ATP-hydrolysis-dependent branch-migration activity for four-way junction DNA in both directions [86, 176, 298]. FANCM is related to the archeal protein Hef (*h*elicase-associated *e*ndonuclease for *f*ork-structured DNA), which has a functional helicase domain, a functional endonuclease domain, and resolves stalled replication forks [131]. Human XPF/ERCC4 endonuclease and yeast MPH1 helicase also have high homology to archeal Hef. These proteins may share a common ancestor and diverged [176]. FANCM is phosphorylated in response to DNA damage [176], but the biological significance of this phosphorylation has not been shown.

FANCD2 is an important substrate of the FA core complex as an E3 ubiquitin ligase. Normal cells express two isoforms of the FANCD2 protein, a non-ubiquitinated form (155 kDa) and a monoubiquitinated form (162 kDa) [85]. Monoubiquitination occurs on a single residue, lysine 561, in an FA core complex dependent manner. Not only monoubiquitination, but also phosphorylation of FANCD2 has an important role. FANCD2 can be directly phosphorylated by ATM (*a*taxia *t*elangiectasia *m*utated) and possibly by ATR [106, 255]. FANCD2 protein has a region (amino acids 267–1163) predicted by the SCOP database [187] to contain ARM repeats which represent alpha-alpha superhelix folds [237]. Except for this, FANCD2 does not have any known functional motifs [262].

FANCI is the most recently cloned FA gene [62, 236, 237]. FANCI protein has regions predicted as ARM repeats (amino acids 985–1207) and a lipocalin fold (amino acids 612–650), which is predicted to bind hydrophobic ligands in its interior [237]. FANCI and FANCD2 share a modest 13% identity and 20% similarity across the entire protein, and are considered to be paralogues [237]. FANCI is also monoubiquitinated on lysine 523 in an FA core complex dependent manner. FANCI can be phosphorylated in response to treatment with ionizing radiation. It has several SQ/TQ sites, which may be phosphorylated by ATM/ATR kinases [237].

FANCD1 is identical to a famous breast/ovarian cancer susceptibility gene, *BRCA2* [110]. Biallelic germline mutations cause FA-D1 subtype of FA, while mono-allelic germline mutation of *FANCD1/BRCA2* leads to breast/ovarian cancer susceptibility. The main function of BRCA2 is regulation of homologous recombination (HR) through interaction with RAD51 recombinase (eukaryotic homologue of bacterial RecA protein) [54, 184]. BRCA2 also has some other functions, such as stabilization of stalled replication forks [160] and regulation of cytokinesis [53]. BRCA2 protein has 8 BRC repeats, which directly bind to RAD51 [211]. The C-terminal region of BRCA2 also has an additional RAD51 binding domain [55, 71]. The C-terminal domain (CTD) has five domains; a helical domain (HD), three oligonucleotide/oligosaccharide-binding folds (OB1, OB2, and OB3) and a tower domain (TD) [304]. The OB folds bind to ssDNA and the TD interacts with double stranded DNA (dsDNA), suggesting that BRCA2 may bind to ssDNA-dsDNA junction of resected DNA ends of double strand DNA breaks and help RAD51 loading onto ssDNA [290]. Interestingly, a component of the lid subcomplex of the regulatory subunit of the 26S proteasome [137], DSS1 (*d*eleted in *s*plit-hand/*s*plit foot syndrome), interacts with BRCA2 through the HD, OB1, and OB2 domains [304]. DSS1 is required for stability of BRCA2 [151].

FANCN is identical to *PALB2* (*p*artner *a*nd *l*ocalizer of *B*RCA*2*) [221, 297]. PALB2 directly binds to N-terminal region of BRCA2 and regulates stability and localization of BRCA2 [296]. The *PALB2* gene is a breast cancer susceptibility gene. Monoallelic germline mutations of *FANCN/PALB2* lead to breast cancer susceptibility [69, 220, 265], while biallelic germline mutations of *FANCN/PALB2* cause FA-N subtype of FA.

The *FANCJ/BRIP1* gene is also a breast cancer susceptibility gene. Monoallelic germline mutations of *FANCJ/BRIP1* lead to breast cancer susceptibility [231], while biallelic germline mutations of *FANCJ/BRIP1* cause FA-J subtype of FA [149, 150, 155]. *FANCJ/BRIP1* was originally cloned as the helicase BACH1(*B*RCA1-*a*ssociated *C*-terminal *h*elicase 1) which can directly bind to the BRCT domain of BRCA1 protein [29] and later was identified as the responsible gene for FA-J subtype of FA [149, 150, 155]. FANCJ/BRIP1 is a DNA-dependent ATPase and a 5'-to-3' DNA helicase (DEAH helicase) [29]. The BRCT domain is a phosphoprotein-binding domain and phosphorylation of serine 990 of FANCJ/BRIP1 is required for BRCA1-FANCJ/BRIP1 interaction [309]. However, the FANCJ-S990A mutant (or FANCJ lacking the BRCT-interaction domain) can correct ICL sensitivity of FANCJ-deficient cells. Additionally, serine 990 of FANCJ is not conserved in chicken or *Caenorhabditis elegans*, further indicating that BRCA1-FANCJ/BRIP1 interaction is not required for cellular resistance to ICL agents [26, 212, 308]. Consistently, *C elegans dog-1* (*FANCJ*) is nonepistatic with *brc-1* (*BRCA1*) in terms of ICL sensitivity, while *dog-1* is epistatic with *fcd-2* (*FANCD2*) [308].

FANCJ-MLH1 interaction appears to be important for cellular resistance to ICL, at least in human cells. FANCJ can interact with the mismatch repair complex composed of PMS2 and MLH1 [212]. FANCJ directly interacts with MLH1 through its helicase domain (lysines 141 and 142 of FANCJ are required for the interaction) and this interaction is required for cellular resistance to ICL. FANCJ helicase activity is also required for cellular resistance to ICL. Other FA proteins (FANCA, FANCD2 and FANCD1/BRCA2) are not required for FANCJ-MLH1 interaction and FANCJ or MLH1 are not required for FANCD2 monoubiquitination. Currently, how FANCJ-mismatch repair proteins interaction functions in the ICL response/repair remains unknown.

FANCJ directly binds to RPA70 subunit of replication protein A (RPA), a single stranded DNA binding protein complex implicated in DNA replication and repair [94]. FANCJ co-localizes with RPA in nuclear foci after DNA damage, and RPA stimulates FANCJ helicase to unwind duplex DNA substrates.

In addition to its role in ICL resistance, FANCJ has a role in unwinding G4 DNA substrates [295]. Guanine (G)-rich DNA readily forms G4 DNA (G-quadruplex DNA), in which guanines joined in G-quartets stabilize interactions between four DNA strands [163, 164]. *C elegans DOG-1* (*d*eletion *o*f *g*uanine-rich regions) gene is a homolog of *FANCJ* [308] and has been shown to be necessary for the stability of poly(G)/poly(C) tracts [40]. Especially, G4 DNA sequences that have the potential to fold into replication blocking quadruplex structures are intrinsically mutagenic in *C elegans*, and DOG-1 (FANCJ) is required to prevent deletions at G4 DNA sequences [138]. In contrast, G/C tracts remain stable in the absence of ATL-1 (ATR homolog), CLK-2 (HCLK2 homolog), FCD-2 (FANCD2 homolog), BRC-2 (FANCD1/BRCA2 homolog) and MLH-1, indicating that DOG-1 (FANCJ homolog) is the sole FA component required for G/C tract maintenance in a wild-type background [308]. In the absence of DOG-1, FCD-2 is required for deletion-free repair at G/C tracts [308]. Human FANCJ protein can unwind G4 DNA substrates in vitro in an ATP hydrolysis-dependent manner, and RPA stimulates FANCJ unwinding activity [295]. Taken together, FANCJ has a critical role in unwinding G4 DNA structures that potentially impede replication and cause genomic instability.

17.3.2 The FA Core Complex

Eight FA proteins (FANCA, FANCB, FANCC, FANCE, FANCF, FANCG, FANCL, and FANCM), along with FAAP100 and FAAP24 form a nuclear protein complex (the FA core complex) with E3 ubiquitin ligase activity, which is required for monoubiquitination of two other FA proteins, FANCD2 and FANCI (Fig. 17.2) [43, 62, 85, 154, 173, 174, 176, 236, 237].

The FA core complex, BLM (responsible protein for another rare genetic disease Bloom syndrome), RPA (single strand DNA binding protein complex), BLAP75, BLAP250 and topoisomerase III α form a larger complex called BRAFT (for *B*LM, *R*PA, *F*A, and *t*opoisomerase III α) [174, 282, 307].

In the FA core complex, there seem to be some sub-complexes [172]. FANCA and FANCG bind to each other and stabilize each other [84]. FANCB, FANCL and FAAP100 interact with each other directly [154, 172]. The nuclear accumulation of FANCE depends on FANCC [147] and FANCE promotes the nuclear accumulation of FANCC [253].

Cells deficient in any one of the 10 components of the FA core complex lack monoubiquitination of FANCD2. One of the components of the FA core complex, FANCL/PHF9/POG (FAAP43) contains a PHD (plant homeodomain) finger motif [173], which is a variant of RING finger, a well-known motif with ubiquitin ligase activity. FANCL associates through its PHD/RING finger domain with UBE2T, a ubiquitin conjugating enzyme (E2). UBE2T is also required for in vivo FANCD2 monoubiquitination [162]. FANCL has autoubiquitin ligase activity in vitro and is considered to be the catalytic subunit of the FA core complex as an ubiquitin ligase for FANCD2. Although in vitro reconstitution of monoubiquitination of FANCD2 by FANCL and UBE2T has not been demonstrated, the FA core complex is considered to be a multi-subunit E3 ubiquitin ligase complex for FANCD2 (and FANCI), in which FANCL is the catalytic E3 ubiquitin ligase subunit working together with the E2, UBE2T.

The FA core complex has functions other than a ubiquitin ligase for FANCD2 and FANCI. The FA core complex is required for the translocation into chromatin of monoubiquitinated FANCD2 and for cellular resistance to ICL even if FANCD2 is localized in chromatin [169], suggesting that the FA core complex must have other functions than monoubiquitinating FANCD2.

Among the proteins in the FA core complex, two proteins, FANCM and FAAP24, have DNA-interacting domains and activities. FANCM contains DNA helicase motifs, a degenerate ERCC4/XPF-like endonuclease motif and in vitro DNA-stimulated ATPase and translocase activities [176]. FAAP24 protein, which directly interacts with FANCM and forms a FANCM-FAAP24 heterodimer, has been shown to bind to ssDNA and branched DNA structures [43].

FANCM has a specific ATP-independent DNA-binding affinity to fork and four-way junction DNA [86, 176, 298]. This binding specificity in FANCM is derived from the helicase domain [298]. The helicase domain of FANCM is required for both FANCD2 and FANCI monoubiquitination. FANCM also has an ATP-hydrolysis-dependent branch-migration activity for four-way junction DNA in both directions. This ATP-dependent activity of FANCM is not required for FANCD2 and FANCI monoubiquitination, but is required for cellular resistance to a DNA crosslinking agent, MMC. This indicates that FANCM does have functions outside of FANCD2 and FANCI monoubiquitination.

FANCM DNA translocase activity could play an important role in displacing the FA core complex along the DNA thereby allowing DNA damage recognition or FAAP24 specificity for ssDNA structures may target FANCM and the FA core complex to abnormal, branched DNA structures. In addition, ATP-dependent DNA-processing activities of FANCM seem to be required for repairing DNA crosslinks after the FANCD2 monoubiquitination step [298].

In summary, the FA core complex itself can interact with DNA and displays some important functions outside of FANCD2 monoubiquitination.

17.3.3 Monoubiquitination of FANCD2 and FANCI

Monoubiquitination of FANCD2 is a critical event in the activation of the FA-BRCA pathway. FANCI is reported to share sequence similarity with FANCD2 and they form a protein complex (ID complex) [237]. Monoubiquitination of FANCD2 or FANCI depends on the monoubiquitination of the other. In response to DNA damage or in S phase of the cell cycle, the FA core complex activates the monoubiquitination of FANCD2 (lysine 561 (K561)) and FANCI (lysine 523 (K523)). Monoubiquitinated FANCD2 and monoubiquitinated FANCI are translocated to chromatin and form nuclear foci containing BRCA1, BRCA2/FANCD1, PALB2/FANCN, RAD51 and other proteins involved in DNA repair [85, 181, 237]. The foci formation of FANCD2 and FANCI may reflect accumulation of these proteins at the sites of DNA damage and repair.

Monoubiquitination of FANCD2 is required for nuclear foci formation, chromatin localization, and is critical for cellular resistance to ICL [85, 283]. However, the precise mechanism of resistance to ICL by FANCD2 protein has not been elucidated.

17.3.4 Activation of the FA-BRCA Pathway

In response to DNA damage, the FA pathway becomes activated. In response to treatment with DNA damaging agents, such as ionizing radiation (IR), ultraviolet, DNA crosslinking agents and hydroxyurea, FANCD2 and FANCI become monoubiquitinated and are targeted to BRCA1/BRCA2/RAD51-containing nuclear foci, presumably at the sites of DNA damage and repair (Fig. 17.2) [85, 114, 237, 283]. In these DNA damage-induced nuclear foci, FANCD2 also colocalizes with many other factors involved in DNA damage response and DNA repair, which include γ-H2AX [22], ATR, RPA [9], FANCE [206], FANCC [169], NBS1 [188], BLM [103, 214], FANCJ [155], and FANCG [113]. All of these factors are required for cellular resistance to ICL agents.

For the activation of the FA pathway in response to DNA damage, a DNA damage-activated signaling kinase, ATR, plays a critical role. ATR, but not ATM, is required for the DNA damage-induced monoubiquitination and nuclear foci formation of FANCD2 [9, 255]. In ATR siRNA-treated cells or ATR-deficient Seckel syndrome cells [200], DNA-damage induced monoubiquitination and nuclear foci formation of FANCD2 is impaired [9]. A single strand DNA binding protein complex, RPA, is also required for DNA-damage induced monoubiquitination and nuclear foci formation of FANCD2 [9], consistent with a role of RPA in activation of ATR [315, 316]. Consistently, an ATR-interacting protein, HCLK2, is also required for DNA damage-induced FANCD2 monoubiquitination and FANCD2 and RAD51 foci formation in response to DNA damage [48]. The requirement of ATR, RPA and HCLK2 for DNA damage-induced nuclear foci formation of FANCD2 is conserved even in *C elegans* [48]. RAD17 and RAD9, which help ATR activation, are also required for DNA damage-induced monoubiquitination of FANCD2 [93].

However, how ATR activates DNA damage-induced FANCD2 monoubiquitination has not been elucidated. Phosphorylation of substrates sometimes enhances ubiquitination [124, 130], and therefore it has been hypothesized that phosphorylation of FANCD2 or FANCI itself may trigger monoubiquitination of FANCD2 or FANCI. FANCD2 has several SQ/TQ sites which are phosphorylated in response to DNA damage [106, 255]. FANCD2 can be directly phosphorylated by ATR in in vitro ATR kinase assay [9] and phosphorylation of FANCD2 in response to ICL is reported to be ATR dependent [215]. Furthermore, the partner of FANCD2, FANCI, also has several SQ/TQ sites which are possible ATR phosphorylation sites. Therefore, one possible explanation is that ATR-mediated phosphorylation of FANCD2 and FANCI may enhance monoubiquitination of these proteins.

On the other hand, ATR may phosphorylate some of the components of FA core complex and may enhance ubiquitin ligase activity of FA core complex. FANCM is reported to be phosphorylated in response to DNA damage [176]. Phosphorylation of serine 7 of FANCG is upregulated after MMC treatment [219] and is reported to be ATR-dependent [289], although which kinase phosphorylates this site is not known. The FANCG mutant of serine 7 (S7A) only partially correct MMC sensitivity of FA-G cells, suggesting a role of this phosphorylation in MMC resistance. However, this phosphorylation may not be involved in FANCD2 monoubiquitination, because S7A mutant can correct lack of FANCD2 monoubiquitination in FA-G cells [219].

It is possible that some other substrates of ATR kinase (for example, CHK1) modulate the FA pathway. In response to DNA damage, especially UV-induced DNA damage, Chk1 directly phosphorylates FANCE on threonine 346 and serine 374 [284]. Phosphorylated FANCE colocalizes with FANCD2 in nuclear foci. These phosphorylations are not required for FANCD2 monoubiquitination or FANCD2 foci formation, because a nonphosphorylated mutant form of FANCE (FANCE-T346A/S374A) allows FANCD2 monoubiquitination and foci assembly in FANCE-deficient cells. However, FANCE-T346A/S374A fails to complement the MMC hypersensitivity of FANCE-deficient cells. Therefore, Chk1-mediated phosphorylation of FANCE is required for a function independent of FANCD2 monoubiquitination [284].

CHK1 itself is not required for monoubiquitination of FANCD2 or nuclear foci formation of FANCD2 [48, 284]. In CHK1-depleted cells, even in the absence of exogenous DNA damage, basal level of DNA damage is increased, ATR kinase is activated [93, 251] and FANCD2 monoubiquitination and FANCD2 foci formation are upregulated [48, 93, 284]. In the CHK1-depleted cells, monoubiquitination and nuclear foci formation of FANCD2 are not further increased following DNA damage

[93, 284]. From these observations, some people concluded that CHK1 is required specifically for exogenous DNA damage-inducible upregulation of monoubiquitination/foci formation of FANCD2 [93], while others concluded that CHK1 is not involved in basal and DNA damage-induced monoubiquitination of FANCD2 [48, 284].

Claspin is an adaptor protein required for optimal CHK1 activation [41]. Claspin-depleted cells are quite similar to CHK1-depleted cells in terms of FANCD2 monoubiquitination and foci formation [93]. In claspin-depleted cells, even in the absence of exogenous DNA damage, FANCD2 monoubiquitination and foci formation is upregulated and is not further increased following DNA damage [93].

Another factor that seems to be important for DNA damage-induced activation of the FA pathway is γ-H2AX. γ-H2AX is a phosphorylated form of a histone H2A variant, H2AX. In response to DNA damage, histone H2AX surrounding the sites of DNA damage become phosphorylated. γ-H2AX is required for recruiting FANCD2 to chromatin at stalled replication forks [22], although the precise mechanism of this requirement is yet to be elucidated.

BRCA1 is involved in the activation of the FA pathway. BRCA1 is a famous breast/ovarian cancer susceptibility gene, regulating DNA damage response and DNA repair [91, 229]. BRCA1 is a RING finger protein with ubiquitin ligase activity and is required for efficient DNA damage-inducible increase of monoubiquitinated FANCD2 in a breast cancer cell line, HCC1937 [85]. Therefore, initially, BRCA1 was thought to be a candidate for the ubiquitin ligase for FANCD2. However, this phenomenon was not reproducible in HeLa cells treated with BRCA1 siRNA, nor in BRCA1 knockout chicken DT40 cells [278]. Therefore, BRCA1 is no longer considered to be the ubiquitin ligase for monoubiquitination of FANCD2, although BRCA1 together with BARD1 can monoubiquitinate FANCD2 in vitro [278]. BRCA1 is required for DNA damage-inducible nuclear foci formation of FANCD2 in these cells [85, 278] and is still considered to be a critical factor regulating the FA pathway. Similarly, BRCA1 also facilitates FANCJ/BRIP1 [29] and RAD51 [18] nuclear foci formation after DNA damage.

In the absence of exogenous DNA damage, FANCD2 is monoubiquitinated and co-localizes with BRCA1 and RAD51 in nuclear foci in S phase of the cell cycle [254]. For this activation, ATR is not required [9]. The functional significance and mechanism of S phase activation of the FA pathway is yet to be determined.

The proteasome controls the destruction of damaged and abnormally folded proteins as well as protein turnover, hence tightly regulating multiple cellular processes [216]. The proteasome function is required for monoubiquitination of FANCD2, foci formation of FANCD2, BRCA1, RAD51 and some other proteins involved in DNA damage signaling (53BP1, phospho-ATM, and NBS1) [119], but mechanisms behind this phenomenon are still unknown.

Monoubiquitination of FANCD2 seems to work as a signal for translocation of FANCD2 protein to chromatin fraction, because monoubiquitinated FANCD2 is preferentially found in the chromatin fraction [181, 283]. In the chromatin fraction, monoubiquitinated FANCD2 is co-immunoprecipitated with BRCA2 and FANCE [283]. How monoubiquitinated FANCD2 regulates DNA repair is still unclear. In one report, monoubiquitinated FANCD2 is required for the increase of nuclear foci of BRCA2 and RAD51 in response to DNA damage [283], suggesting a role of BRCA2 and RAD51 downstream of monoubiquitinated FANCD2. However, whether FA proteins (proteins in the FA core complex and ID complex) other than FANCD1/BRCA2 and FANCN/PALB2 increase the assembly of RAD51 foci is still controversial. Some reports claim that FA core complex proteins and FANCD2 are required for efficient RAD51 foci formation in response to certain types of DNA damage [61, 213, 283], while other reports claim that these FA proteins are not required for DNA-damage inducible RAD51 foci formation [87, 155, 204, 301, 302].

17.3.5 Deubiquitination of FANCD2 by USP1

Monoubiquitination of FANCD2 and FANCI is a regulated and reversible process. A deubiquitinating enzyme, USP1, removes monoubiquitin from FANCD2 and FANCI and negatively regulates the FA pathway [196, 236, 237]. Knockdown of USP1 using siRNA causes increased monoubiquitination of FANCD2 and FANCI [196, 236, 237]. USP1 and a WD40 repeat-containing protein, UAF1 (*U*SP1 *a*ssociated *f*actor *1*) form a protein complex where UAF1 functions as an activator of USP1 [45]. USP1/UAF1 complex can deubiquitinate monoubiquitinated FANCD2 in vitro.

PCNA (*p*roliferating *c*ell *n*uclear *a*ntigen) is a processivity factor for DNA polymerase. Monoubiquitination of PCNA on lysine 164 (K164) is mediated by RAD6 (E2 ubiquitin conjugating enzyme) and RAD18 (E3 ubiquitin ligase) and is implicated in DNA polymerase switch from a replicative polymerase to translesion synthesis (TLS) polymerases, such as DNA polymerase η (Rad30, XPV), at the site of blocked replication forks [123]. These factors are required for cellular resistance to ICL [64]. Interestingly, USP1 deubiquitinates monoubiquitinated PCNA as well as monoubiquitinated FANCD2 [112]. Therefore, these two pathways (the FA pathway and TLS activation through PCNA monoubiquitination) have different activation mechanisms, but share a common shutoff mechanism.

USP1 mRNA level is regulated in a cell cycle dependent manner; peaking during S phase and decreasing during G1 phase [196]. This may partly explain cell cycle regulated monoubiquitination of FANCD2 [254].

USP1 seems to be inactivated by autocleavage in response to DNA damage (UV irradiation) in human cells, which may explain the mechanism of activation of FANCD2 monoubiquitination and PCNA monoubiquitination in response to UV irradiation [112]. However, in chicken DT40 cells, UV irradiation leads to FANCD2 monoubiquitination before USP1 is degraded and USP1 degradation after UV treatment seems to be more related to apoptosis [202]. Furthermore, in DT40 cells, self-cleavage of chicken USP1 is not induced by DNA damage and is not strictly required for USP1 to improve cellular DNA repair efficiency [202].

siRNA knockdown of USP1 in human cells (HEK293 cells) seems to protect cells from MMC-induced chromosomal aberrations [196]. In contrast, USP1 knock-out chicken DT40 cells show hypersensitivity to DNA crosslinking agents (cisplatin and MMC) and persistent monoubiquitinated FANCD2 (not PCNA) is responsible for this DNA crosslinker sensitivity [202]. These results suggest that monoubiquitination of FANCD2 should be tightly regulated for cellular resistance to DNA crosslinking agents.

Interestingly, siRNA depletion of human homologues of Rad6, HHR6A and HHR6B in human cells (293T, MCF-7 and U2OS cells) inhibits monoubiquitination of FANCD2 in response to MMC or UV treatment, suggesting an interaction between RAD6 and the FA pathway [312]. The mechanism of this phenomenon is yet to be clarified. RAD6-RAD18 can monoubiquitinate the 9-1-1 checkpoint clamp complex [80], and the 9-1-1 complex is required for DNA damage-induced monoubiquitination of FANCD2 [93]. Therefore, this may be the link connecting HHR6 and FANCD2 monoubiquitination, although it has not been tested.

17.3.6 Localization of FA Proteins in Chromatin

Localization of FA proteins in chromatin seems to be important for their function. Monoubiquitinated FANCD2 is mainly localized in chromatin fraction and purified FANCD2 protein has direct DNA binding activity with specificity for dsDNA ends and Holliday junctions [210], suggesting a direct role of FANCD2 in DNA repair in the chromatin.

Until recently, the FA core complex was believed to be a soluble nuclear complex whose primary function was to monoubiquitinate FANCD2, but recent findings support the idea that the FA complex proteins move in and out of the chromatin, depending on the state of DNA damage and interact with UBE2T and FANCD2 in chromatin fraction [4].

There are many observations which support this idea. First, FANCA, FANCC and FANCG proteins are localized to chromatin in G1/S/G2 phases and their chromatin localization increases after treatment with MMC [179, 218, 260]. Second, FANCE and FANCC form nuclear foci and colocalize with FANCD2 after DNA damage [169, 206]. FANCG also forms ICL-induced nuclear foci and colocalizes with BRCA2 and RAD51 [113]. Third, FANCE interacts with BRCA2 and FANCD2 in chromatin [283]. Fourth, as described above, FANCM has a helicase motif, a degenerate nuclease motif, DNA-stimulated ATPase and translocase activity, and it obviously interacts with DNA [176].

According to a recent paper using chicken DT 40 cells [4], FA core complex is constitutively assembled and stable throughout the cell cycle, but accumulates in chromatin fraction in S phase compared to in G1 phase. After MMC treatment, more FA core complex accumulates in chromatin. In contrast, UBE2T is constitutively present in the chromatin fraction. FANCL itself (which is required for FA core complex formation) is required for chromatin localization of FANCD2 after MMC treatment, but ubiquitin ligase activity of FANCL or monoubiquitination of FANCD2 itself is not required for chromatin localization of FANCD2 in response to MMC treatment [4]. These observations lead to a model where in response to DNA damage, the FA core complex and FANCD2 accumulate in the chromatin fraction, interact with UBE2T, and then, FANCD2 is monoubiquitinated by UBE2T and the FA core complex in the chromatin [4].

FANCM and its partner FAAP24 play a critical role in the recruitment of the FA core complex to chromatin [128, 182]. In HeLa cells, FANCM is exclusively localized to chromatin fraction and undergo cell cycle-dependent phosphorylation and dephosphorylation. When FANCM (or its partner, FAAP24) is depleted by siRNA, accumulation of FA core complex to chromatin becomes impaired [128].

17.3.7 Interaction of FA Proteins and Non-FA Proteins Involved in DNA Repair and DNA Damage Response

Interaction of the BLM complex and the FA core complex may have important functional significance in DNA repair. BLM helicase is a part of BRAFT complex [174] and ICL-induced nuclear foci formation of BLM is dependent on FANCC [103, 214], FANCG [214], and FANCD2 [103]. ICL-induced phosphorylation of BLM is dependent on FANCC and FANCG, but not on FANCD2 [214]. BLM colocalizes with FANCD2 in ICL-induced nuclear foci and can be co-immunoprecipitated with monoubiquitinated FANCD2 [103, 214]. BLM-deficient cells are sensitive to ICL to some extent [214]. BLM has 3' to 5' DNA helicase activity and is associated with RPA and topoisomerase III α, which can break and rejoin DNA to alter its topology. The BLM complex stabilizes the stalled replication forks and allows subsequent homologous recombination-dependent processes to restart replication [19, 44]. The FA core complex and the BLM complex may cooperate at the stalled replication forks to help restart arrested replication forks [261, 282].

The NBS1-MRE11-RAD50 complex also interacts with FA proteins [188, 213] and this interaction seems to be functionally important. MRE11 has 3'-5' dsDNA exonuclease and ssDNA and dsDNA endonuclease activity [246]. NBS1 deficient cells and MRE11 deficient cells are sensitive to ICL [188]. FANCD2 and NBS1 partially colocalize in MMC-induced nuclear foci and are co-immunoprecipitated [188]. The FA core complex is required for phosphorylation of NBS1 and nuclear foci formation of MRE11 in response to ICL specifically [213].

Interaction between the XPF/ERCC complex and the FA proteins may have some functional significance. XPF is an endonuclease required for incision of ICL and forms a heterodimer with ERCC1. XPF deficient cells and ERCC1 deficient cells are hypersensitive to ICL. XPF colocalizes with FANCA in ICL-induced nuclear foci [243]. A related endonuclease, MUS81-Eme1, which is required for cellular resistance to ICL, may also have interactions with the FA pathway. In ERCC1-depleted cells, monoubiquitination of FANCD2 protein in response to MMC treatment is decreased and the localization of FANCD2 to nuclear foci is eliminated, suggesting a role of ERCC1 in the activation of the FA pathway in addition to its role in incision of ICL [170].

In a yeast two-hybrid screen using FANCD2 as bait, Tip60 was identified as a FANCD2-interacting protein [99]. Tip60 is a histone acetyltransferase and a component of a chromatin remodeling complex [116, 117]. It is implicated in double strand break repair. Chromatin remodeling complexes have an important role in DNA damage response and DNA repair [63]. Tip60 acetylates not only histones [139], but also p53 [252] and ATM [248]. The acetylation of ATM at lysine 3016 by Tip60 plays a critical role in the activation of ATM kinase in response to DNA damage [248, 249]. Tip60 and FANCD2 can be co-immunoprecipitated, but depletion of Tip60 does not affect monoubiquitination or nuclear foci formation of FANCD2 [99]. Tip60-depletion sensitizes wild type fibroblasts to MMC, but does not sensitize FANCC (or FANCA)-deficient fibroblasts, suggesting that Tip60, FANCA, and FANCC may work in a common pathway for cellular resistance to MMC [99]. Whether Tip60 acetylates FA proteins and whether FA proteins affect Tip60 functions have not been tested.

As described above, FANCJ interacts with the mismatch repair complex composed of PMS2 and MLH1 and this interaction is critical for cellular resistance to ICL [212].

How the FA proteins and these factors related to the FA pathway (helicases, nucleases, a topoisomerase, DNA polymerases, a recombinase, mismatch repair proteins and other DNA binding proteins) assemble in chromatin and repair damaged DNA lesions (especially ICL) is one of the most important questions in the FA research field [126, 194].

17.4 Cellular Defects in FA

17.4.1 Homologous Recombination

Increasing evidence suggests that FA cells are defective in homologous recombination (HR), although there is some inconsistency among reports (Table 17.2).

First, FANCD1/BRCA2-deficient cells and FANCN/PALB2-deficient cells are consistently reported to show a severe decrease in overall HR efficiency [184, 189, 244, 296].

Mild HR defects in other FA gene deficient cells (FA-A [189, 237, 306], C [103, 195], D2 [189, 236, 237, 302], G [189, 301], I [236, 237], J [155] and L [232]) have been reported. However, normal HR efficiency in FA-C [26], D2 [204], J [26] and M [182] cells have also been reported. Overall, there seems to be a mild defect of HR in FA cells.

Table 17.2 Defect of homologous recombination/single strand annealing in Fanconi anemia cells

Subtype	Cell type	Efficiency of HR	SCE spontaneous	SCE ICL-induced	SSA	References
FA-A	Human FA-A fibroblast (SV40 transformed)	Decreased			Decreased	[189]
	Fanca(-/-) mouse fibroblast	Decreased			Decreased	[306]
	Human DR-U2OS cells (FANCA siRNA)	Decreased				[237]
FA-B		Not reported				
FA-C	chicken DT40	Decreased	Increased	Poor response		[195]
	chicken DT40	Decreased	Increased	Poor response		[103]
	chicken DT40	Normal	Increased			[26]
FA-D1	Human Capan1 pancreatic cancer (*BRCA2*-mutated)	Decreased				[184]
	Mouse ES Cells with a BRCA2 Exon 27 Deletion	Decreased			Increased	[184, 244]
	Chinese hamster cell (VC-8) (*BRCA2*-mutated)	Decreased			Increased	[189, 306]
FA-D2	Human FA-D2 fibroblast (SV40 transformed)	Decreased			Decreased	[189]
	Human FA-D2 fibroblast (SV40 transformed)	Normal				[204]
	Chicken DT40	Decreased	Increased	Poor response		[302]
	Human DR-U2OS cells (FANCD2 siRNA)	Decreased				[236, 237]
FA-E		Not reported				
FA-F		Not reported				
FA-G	Human FA-G fibroblast (SV40 transformed)	Decreased			Decreased	[189]
	Chicken DT40	Decreased				[301]
FA-I	Human DR-U2OS cells (FANCI siRNA)	Decreased				[236, 237]
FA-J	Human MCF7 breast cancer (FANCJ siRNA)	Decreased				[155]
	Chicken DT40	Normal	Increased			[26]
FA-L	Chicken DT40	Decreased				[232]
FA-M	Chicken DT40	Normal				[182]
FA-N	Human DR-U2OS cells (FANCN/PALB2 siRNA)	Decreased				[296]

HR, monologous recombination; SCE, sister chromatid exchange; SSA, single strand annealing.
Adapted from Taniguchi [257] with modifications.

Sister chromatid exchange (SCE) is a recombination repair event and formation of SCE is considered to reflect efficiency of recombination repair. Spontaneous SCE is elevated in FANCC [103, 195], FANCD2 [302], or FANCJ [26] deficient chicken DT40 cells. These are intriguing findings, because SCE is thought to be mediated by HR and in human FA cells, spontaneous SCE is not elevated. What causes this discrepancy between human cells and chicken DT40 cells is not known. One speculative explanation is that in FA cells, a sub-pathway of HR which is not required for spontaneous SCE may be affected [302] and that something which masks (or compensates) increased SCE in human FA cells may be missing in chicken DT40 cells. Interestingly, in DT40 cells, increased SCE in FANCC deficient cells is epistatic with BLM deficient cells, indicating a functional linkage between FANCC and BLM in suppressing SCE [302].

Single-strand annealing (SSA) is a DNA double strand break repair pathway involving sequence homology, but is different from HR [244]. The annealing of complementary single strands formed after resection at a double strand break leads to SSA. While BRCA2 deficient cells show increased SSA [189, 244], BRCA1, FANCA, FANCG or FANCD2 deficient cells show reduced SSA [189, 244, 306]. These findings suggest that FA proteins (other than BRCA2) have a role in common steps for both SSA and HR.

FA cells (except for FA-D1 and probably FA-N) are generally not hypersensitive to ionizing radiation (IR) [65], but clinically some FA patients show unexpected toxicity due to radiation therapy [5]. In the absence of DNA-PK, which is an essential component of non-homologous end joining (NHEJ) repair, *Fancd2* knockout sensitizes fibroblasts and mice to IR [109], suggesting that NHEJ repair may mask the IR-hypersensitivity of FA cells. This may explain reported inconsistent IR sensitivities in FA.

Common fragile sites are chromosomal loci that preferentially exhibit gaps and breaks when cells have been cultured under replicative stress. FA cells (FA-A, FA-B and FA-D2) [111], ATR-deficient cells [34], CHK1-deficient cells [67], and BRCA1-deficient cells [10] show increased common fragile site instability, suggesting that these proteins have an important role in dealing with replicative stress.

17.4.2 Translesion Synthesis

Several lines of evidence implicate FA proteins in the regulation of translesion synthesis (TLS). TLS is a DNA damage tolerance mechanism that allows the replicative bypass of damaged nucleotides [143, 144].

TLS DNA polymerases (REV1 and REV3) are implicated in the FA pathway. Chicken DT40 cells deficient in REV1 or REV3 show crosslinker hypersensitivity that is epistatic to FANCC-deficient cells [195]. REV1 and FANCD2 colocalize in nuclear foci upon replication arrest [195]. These findings suggest that FA proteins and error prone TLS polymerases work in a common pathway. Consistently, FA cells are reported to be hypomutable [101, 102, 195, 209].

Recently, Mirchandani et al. reported the connection between FA proteins and REV1 [180]. A shuttle vector-based mutagenesis assay was developed to measure the mutation frequencies of FA cells and it was found that FANCA and FANCG are required for efficient point mutagenesis, but FANCD2 is not [180]. They also found that the recruitment of the error-prone TLS polymerase Rev1 into nuclear foci depends on an intact FA core complex, but not on FANCD2 or FANCI. For this FA core complex-dependent recruitment of REV1, BRCT domain of REV1 protein is required, but UBM domain of REV1 is not required [180]. In contrast, the UBM domain is required for the interaction with monoubiquitinated PCNA [292]. REV1-depleted cells show crosslinker-induced DNA breaks and radial chromosomes, similar to those seen in FA cells [180]. Taken together, these findings suggest that the FA core complex regulates REV1 independent of FANCD2 and FANCI. This is an example of another function of the FA core complex outside of monoubiquitination of FANCD2 and FANCI.

17.4.3 Function of FA Proteins in Intra S Phase Cell Cycle Checkpoints

In addition to its role in cellular resistance to ICL, the FANCD2 protein has a role in IR-induced intra S phase checkpoint [106, 188, 255, 283]. After IR exposure, an IR-activated signaling kinase ATM phosphorylates FANCD2 on several residues including serine 222. ATM phosphorylates and activates many proteins involved in cell cycle checkpoint responses and other cellular processes including p53, CHK2, NBS1, SMC1, BRCA1 and FANCD2 [233].

Normal cells stop synthesizing DNA in response to exposure to IR. This response is called intra S phase checkpoint. Cells deficient in ATM or some substrates of ATM, such as CHK2, NBS1, SMC1, BRCA1, and FANCD2, do not stop synthesizing DNA after exposure to IR. This phenomenon is called radioresistant DNA synthesis (RDS). FANCD1/BRCA2-deficient cells [135, 283] and FANCN/PALB2-deficient cells [296] also show the RDS phenotype.

IR-inducible ATM-dependent phosphorylation of FANCD2 on serine 222 is required for establishment of IR-induced intra-S phase checkpoint, but is not required for ICL resistance [255]. This phosphorylation is also dependent on NBS1 [188]. In contrast, monoubiquitination of FANCD2 on lysine 561 is required for ICL resistance, but not for intra-S phase checkpoint. Therefore, FANCD2 has two independent functions resulting from two independent post-translational modifications. Intriguingly, primary fibroblasts from Fancd2 knockout mice do not show the RDS phenotype [107]. The reason for this discrepancy is unknown.

It has been reported that FA cells fail to arrest DNA synthesis in response to ICL whereas normal cells arrest DNA replication, although in this study isogenic pairs of FA-deficient and FA-proficient cells were not used [228]. This suggests that FA cells have a defect in ICL-induced S phase checkpoint function, which is similar to a defect in IR-induced intra S phase checkpoint function seen in FANCD2 deficient cells. For the establishment of ICL-induced intra S phase checkpoint, the ICL-induced ATR-dependent phosphorylation of NBS1 and FANCD2 is required [215].

17.4.4 Notch-HES1 Pathway and the FA Core Complex

Perhaps, one of the most surprising recent findings related to the FA pathway is interaction of the FA pathway and the Notch-HES1 pathway [268]. HES1 (*H*airy *E*nhancer of *S*plit *1*) is a downstream effecter of the Notch signaling pathway and a transcriptional repressor. Tremblay et al. reported that HES1 directly interacts with the FA core complex [268]. HES1 is required for proper nuclear localization and stability of some members of the FA core complex, FANCD2 monoubiquitination and nuclear foci formation, and cellular resistance to MMC. HES1 itself forms MMC-induced nuclear foci, which do not colocalize with FANCD2 foci. HES1 foci formation is FA core complex dependent [268]. These findings indicate that the transcriptional repressor HES1 regulates the FA core complex directly and suggest that the FA core complex may regulate HES1 function. This interaction may explain some of the clinical phenotype of FA, such as defect of hematopoiesis, which cannot be explained by defect of DNA repair.

17.4.5 Other Functions of FA Proteins and Other Proteins Interacting with FA Proteins

Although nuclear functions (especially functions in chromatin) of FA proteins seem important, there are many reports describing functions of FA proteins outside of nucleus (reviewed in [21]).

FA cells are hypersensitive to oxygen and FA proteins have been implicated in the handling of oxidative stress [120]. Interaction of FANCC protein and some cytoplasmic proteins involved in handling of reactive oxygen species has been reported [51]. FANCC has also been implicated in JAK/STAT signaling and apoptotic signaling [27, 73, 207, 208, 226, 314].

Some FA proteins seem to have functions even in mitochondria. FANCG protein in mitochondria interacts with the mitochondrial peroxidase peroxiredoxin-3 (PRDX3) [186]. PRDX3 is deregulated in FA cells, including cleavage by a calpainlike cysteine protease and mislocalization. FA-G cells demonstrate distorted mitochondrial structures and mitochondrial extracts from FA-G cells have a significant decrease in thioredoxin-dependent peroxidase activity. FA-A and FA-C cells also have a similar defect.

Other proteins that interact with FA proteins include α-Spectrin II, FAZF, SNX5, NADPH cytochrome P450 reductase, a molecular chaperone GRP94, cdc2, GSTP1,

CYP2E1, BRG1, STAT1, Hsp70, PKR, menin, and many other proteins identified in yeast two hybrid screens (summarized in reference [222]).

17.5 FA Animal Models

17.5.1 Mouse Models

So far, generation of *Fanca*[39, 291], *Fancc*[36, 287], *Fancg*[132, 305], *Fancd2*[107], *Fanca-Fancc* double [198], *Fancd1/Brca2* [191], and *Fancl/Pog* knockout mice [2] have been reported.

Generally, these mice show a milder phenotype compared to human FA patients. For example, FA mice do not spontaneously develop bone marrow failure or anemia (except for mouse models of the FA-D1 group [191]), and develop bone marrow failure only after treatment with MMC [31]. Cells from FA knockout mice show increased cellular sensitivity to ICL agents and the FA mice generally show defective germ cell development.

Cancer susceptibility has been reported in some of the murine models of FA. *Fancd2* knockout mice develop tumors including adenocarcinoma and lymphoma [107]. *Fanca* knockout mice develop lymphoma, sarcoma, and ovarian granulosa cell tumor [291]. *Fancc* knockout mice develop mammary adenocarcinoma and histiocytic sarcoma [33]. *Fancc* knockout reduces latency to tumor development of *p53–/–* and *p53+/–* mice [79]. Heterozygosity for *p53* (*p53+/–*) accelerates epithelial tumor formation in *Fancd2* knockout mice [108]. Many conditional *Brca1* or *Brca2* deficient mice models with cancer susceptibility have been published (reviewed in reference [23, 72, 185]). Those results demonstrate the importance of the FA genes in tumor suppression.

FA mice are useful tools to test the applicability of new treatment for FA patients. FA mice have been used for testing gene therapy [81, 95, 97, 197, 223, 299, 300], ex vivo manipulation of hematopoietic stem/progenitor cells [152], cytokine treatment [32] and small molecules [313]. For example, Zhang et al. reported that tempol (a nitroxide antioxidant and a superoxide dismutase mimetics) [240] delays tumor development in *Fancd2–/– Trp53+/–* mice [313]. This finding suggests that tempol may be a useful tumor-preventive agent for FA patients.

17.5.2 Other Models

Drosophila, *C elegans*, *Xenopus* and zebrafish homologs of FA genes have been identified. The FA gene network is conserved from zebrafish [267] and *Xenopus* [238] to human. In *Drosophila,* only *FANCD2*, *FANCL* and *FANCM* homologs have been reported [166]. In *C. elegans*, *FANCD1/BRCA2 (BRC-2)*, *FANCD2 (FCD-2)*, *FANCJ (DOG-1)*and putative *FANCM* homologs have been reported [47, 308]. These proteins may constitute the minimal FA pathway machinery.

*Drosophila fancd2*and *fancl*mutants and *C elegans fancd2*and *fancj (dog-1)*mutants, show hypersensitivity to ICL agents, consistent with phenotype in vertebrates [47, 59, 142, 166, 167, 308].

A *fancd2*-deficient zebrafish model has been reported [157]. *fancd2*-deficient zebrafish embryos develop similar defects to those found in patients with FA, such as shortened body length, microcephaly (small head) and microphthalmia (small eyes) [157].

Xenopus egg extract provides an extremely powerful approach in the study of regulation of activation of the FA proteins by various DNA structures or during DNA replication [238, 239, 245]. Such assays could also be used for screening drugs that modulate the FA pathway.

17.6 The FA-BRCA Pathway in Human Cancer in the General (Non-FA) Population

Since FA patients have cancer susceptibility and FA knockout mice develop tumors, the FA-BRCA pathway seems important for tumor suppression. Therefore, many researchers have investigated whether inactivation of this pathway in human cancer in the general (non-FA) population occurs. Because integrity of the FA-BRCA pathway is critical for cellular resistance to ICL agents and many ICL agents are used as anti-cancer drugs, inactivation of the pathway in cancer may be related to chemosensitivity of cancer cells.

Recent reports describing abnormalities of the FA genes (except for *BRCA2*) in cancer in the general population are summarized in Table 17.3[49, 60, 100, 133, 145, 153, 168, 177, 190, 193, 205, 217, 256, 258, 263, 266, 269, 273–275, 285, 286, 294, 311]. Abnormalities of *BRCA1/2* in cancer have been reviewed elsewhere [270].

17.6.1 FANCF Methylation in Ovarian Cancer

Patients with ovarian carcinoma usually respond well to initial platinum (cisplatin or its derivative, carboplatin) containing chemotherapy, but eventually develop resistance. This acquired platinum resistance is a major clinical challenge in the treatment of women with advanced stage ovarian carcinoma [1]. Understanding the molecular basis of platinum sensitivity and acquired resistance is critical to the management of this disease. The FA-BRCA pathway is known to be critical for cellular resistance to ICL agents, including cisplatin and carboplatin. Therefore, it is reasonable to ask whether alterations of the pathway are involved in platinum sensitivity and resistance of ovarian cancer.

Indeed, methylation of the promoter region of the *FANCF* gene was found in a wide variety of human cancer, including ovarian cancer (Table 17.3). At least in two ovarian cancer cell lines (TOV-21G and 2008), *FANCF* methylation causes suppression of FANCF expression and increased sensitivity to MMC and cisplatin

Table 17.3 Abnormalities of Fanconi anemia genes in human cancer in the general population

Gene	Type of abnormalities	Tissue	Frequency in clinical samples	Cell lines with FA abnormalities	References
FANCF	Methylation	Ovarian cancer	4/19 cases (21%)	TOV-21G, 2008, C13*, OAW42	[256]
	Methylation	Stage III/IV epithelial ovarian tumors	0/? cases (0%)		[258]
	Methylation	Ovarian cancer	5/18 cases (28%)	OVCAR3	[285]
	Methylation	Ovarian cancer	7/53 cases (13.2%)		[153]
	Methylation	Granulosa cell tumors of the ovary	6/25 cases (24%)		[60]
	Methylation	Breast cancer	13/75 cases (17%)		[205]
	Methylation	Breast cancer	1/120 cases (0.8%)		[286]
	Methylation	Non-small cell lung cancer	22/158 cases (14%)		[168]
	Methylation	Head & neck squamous cell carcinoma	13/89 cases (15%)		[168]
	Methylation	Cervical cancer	27/91 cases (30%)	SiHa, ME-180, SW756	[190]
	Methylation	Testicular germ cell tumor (non-seminoma)	4/60 cases (6.7%)		[133]
	Methylation	Sporadic adult acute myeloid leukemia	0/36 cases (0%)	CHRF-288	[263]
	Methylation	Acute myeloid leukemia	0/139 cases (0%)		[100]
	Methylation	Sporadic childhood acute leukemia	0/81 cases (0%)		[177]
	Methylation	Bladder cancer	1/41 cases (2.4%)	BFTC909, 639 V, VMCub1	[193]
	Decreased expression (unknown mechanism)	Breast cancer		UACC812	[275]

Table 17.3 (continued)

Gene	Type of abnormalities	Tissue	Frequency in clinical samples	Cell lines with FA abnormalities	References
FANCA	Reduced expression (heterozygous deletion)	Acute myeloid leukemia (adult)	4/101cases (4.0%)		[266]
	Point mutation	Acute myeloid leukemia (adult)	6/79 cases (7.6%)		[49]
	Functional abnormality (unknown mechanism)	Acute myeloid leukemia (adult)		UoC-M1	[145]
	Methylation	Acute myeloid leukemia	0/139 cases (0%)		[100]
FANCB	Methylation	Sporadic childhood leukemia	0/81 cases (0%)		[177]
	Methylation	Acute myeloid leukemia	0/139 cases (0%)		[100]
FANCC	Point mutation/frameshift + LOH	Pancreatic cancer	2/22 cases (LOH at 9q22.3)		[273]
	Large homozygous deletion	Pancreatic cancer		PL11	[274]
	Methylation	Acute myeloid leukemia	1/143 cases (0.7%)		[100]
	Methylation	Acute lymphocytic leukemia	3/97 cases (3%)		[100]
FANCD2	Methylation	Acute myeloid leukemia	0/139 cases (0%)		[100]
	Absent protein expression (IHC)	Sporadic breast cancer	18/96 cases (19%)		[272]
	Absent protein expression (IHC)	*BRCA1*-mutated breast cancer	2/21 cases (10%)		[272]
FANCE	Methylation	Acute myeloid leukemia	0/139 cases (0%)		[100]
FANCG	Point mutation (nonsense) + LOH	Pancreatic cancer	0/22 cases (LOH at 9p13)	Hs766T	[273]
	Absent or low protein expression (IHC)	Oral squamous cell carcinoma (younger*)	30/37 (81%)		[269]
	Absent or low protein expression (IHC)	Oral squamous cell carcinoma (older**)	15/40 (36%)		[269]
	Methylation	Acute myeloid leukemia	0/139 cases (0%)		[100]
FANCJ/BRIP1	Methylation	Acute myeloid leukemia	0/139 cases (0%)		[100]
FANCL	Methylation	Acute myeloid leukemia	0/143 cases (0%)		[100]
	Methylation	Acute lymphocytic leukemia	1/97 cases (1%)		[100]
	Lack of protein expression (unknown mechanism)	Anaplastic lung carcinoma		Calu-6	[311]

Table 17.3 (continued)

Gene	Type of abnormalities	Tissue	Frequency in clinical samples	Cell lines with FA abnormalities	References
FANCM	Methylation	Acute myeloid leukemia	0/139 cases (0%)		[100]
FANCN/PALB2	Methylation	Inherited breast cancer	2/8 cases (25%)		[217]
	Methylation	Sporadic breast cancer	4/60 cases (6.7%)		[217]
	Methylation	Inherited ovarian cancer	0/9 cases (0%)		[217]
	Methylation	Sporadic ovarian cancer	4/53 cases (7.5%)		[217]
Unknown	Lack of monoubiquitination of FANCD2	Head & neck squamous cell carcinoma		FaDu	[275]
FA genes	mRNA downregulation of FA genes	Oral tongue squamous cell carcinoma	29/44 cases (66%)		[294]

IHC, immunohistochemistry; LOH, loss of heterozygosity.
* 40 years or younger.
** 60 years or older.
Adapted from Taniguchi [257] with modifications.

[256]. In a cisplatin-resistant derivative (C13*) of 2008 generated by selection in the presence of cisplatin in vitro, *FANCF* was partially demethylated and FANCF expression was restored, suggesting that reactivation of the FA-BRCA pathway can be a mechanism for acquired resistance to cisplatin. Therefore, at least in the cell lines, the integrity of the pathway is a critical determinant of cisplatin resistance.

The significance of *FANCF*methylation in clinical samples of ovarian cancer is inconsistent among reports. In the initial report, *FANCF*methylation was detected in 21% of a relatively small number of samples [256], but one subsequent report with a larger number of samples failed to detect *FANCF* methylation in stage III/IV epithelial ovarian tumors [258]. Two subsequent reports support the initial report. Wang et al. and Lim et al. reported that *FANCF*methylation was found in 28% (5 out of 18 cases) and 13.2% (7 out of 53 cases) of ovarian cancer clinical samples, respectively [153, 285]. In addition, *FANCF* methylation was found in 24% of clinical samples of a relatively rare subtype of non-epithelial ovarian tumor, granulosa cell tumors [60].

Lim et al. reported that *FANCF* methylation in ovarian cancer is not associated with overall survival, and that unmethylated *FANCF* is associated with better progression free survival, unexpectedly [153]. This may reflect the fact that cisplatin sensitivity and resistance mechanisms are multiple and not simple in clinical situations. In fact, the FA-BRCA pathway can be disrupted by some other mechanisms (for example, *BRCA1/2* mutations, *BRCA1* promoter methylation) in ovarian cancer. In addition, the defect of the FA-BRCA pathway in cancer is reversible during chemotherapy. Methylated *FANCF* can be demethylated and activated at least in a cell line model [256] and mutated *BRCA1/2* can be genetically reverted by secondary mutations of *BRCA1/2* themselves and then restore normal BRCA1/2 function [68, 227, 250]. This also complicates the situation.

17.6.2 FANCF Methylation in Other Tumors

FANCF methylation seems to have some clinical significance in lung cancer. First, for lung adenocarcinoma, *FANCF* methylation is a significant predictor of poor survival [168]. Second, for non small cell lung cancer, *FANCF* methylation is associated with a shorter duration of tobacco use [168]. Third, in lung squamous cell carcinoma, increased *p16INK4a* homozygous deletion occurred at higher frequency in *FANCF* methylated cases [136].

In head and neck squamous cell carcinoma, *FANCF* methylation is associated with a greater number of years of alcohol drinking [168]. In cervical cancer, *FANCF* methylation was frequently observed in 30% of cases and patients younger than 45 years of age showed a higher frequency of *FANCF* methylation [190]. In a small subset (6.7%) of germ cell testicular cancer (non-seminoma type), *FANCF* methylation was found [133]. In breast cancer, *FANCF* methylation was reported to be found in 17% (13 out of 75 cases) [205], but it turned out to be rare (0.8%, 1 out of 120 cases) in a later report [286]. In sporadic adult acute myeloid leukemia (AML)

[263] as well as sporadic childhood leukemia [177], *FANCF* methylation has not been found, while one *FANCF*-methylated AML cell line has been documented [263]. In bladder cancer clinical samples, *FANCF* methylation is rare, while one *FANCF*-methylated bladder cancer cell line has been documented [193]. In summary, *FANCF* methylation can be found in a wide variety of human cancer, but in certain types of cancer it is very rare.

Melphalan is an ICL agent and FA cells are sensitive to it. It is used as an anticancer drug and is a key drug in the treatment of multiple myeloma. A comparison of a melphalan-sensitive myeloma cell line and its melphalan-resistant derivative identified *FANCF* as a gene expressed significantly higher in the resistant derivative [98]. Consistently, modulation of *FANCF* expression in a myeloma cell line by siRNA or by overexpression of *FANCF* causes melphalan sensitivity or resistance, respectively [37]. However, whether the FA pathway is actually inactivated in clinical samples of myeloma remains to be tested and *FANCF* methylation in myeloma has not been reported.

17.6.3 Other FA Genes

van Der Heijden et al. reported that inherited and somatic mutations of *FANCC* and *FANCG* are present in a subset of young-onset pancreatic cancer [273]. Consistently, Couch et al. reported that two germline truncating *FANCC* mutations associated with loss of heterozygosity in tumor samples were identified in young onset pancreatic cancer cases [50], suggesting that inherited mutations in *FANCC* may cause predisposition to pancreatic cancer.

Alterations of FA genes have been analyzed in acute leukemia samples in non-FA patients and were found to be rare. *FANCA* abnormalities are involved in a small subset of AML. *FANCA* missense mutations were found in 7.6% of adult AML, although functional relevance of most of these mutations is unknown [49]. Heterozygous deletion of *FANCA* with reduced expression of *FANCA* was identified in 4% of adult AML patients [266].

Methylation of most of FA genes (*FANCA*, *FANCB*, *FANCC*, *FANCD1/BRCA2*, *FANCD2*, *FANCE*, *FANCF*, *FANCG*, *FANCJ/BRIP1*, *FANCL* and *FANCM*) has been analyzed in a large number of leukemia samples [100, 177], but methylation was not found in *FANCA*, *FANCB*, *FANCD1/BRCA2*, *FANCD2*, *FANCE*, *FANCF*, *FANCG*, *FANCJ/BRIP1* or *FANCM*. Promoter methylation of *FANCC* was found in one AML (out of 143) and three acute lymphoblastic leukemia (ALL) (out of 97) samples, while that of FANCL was found in only one ALL samples but not in any AML samples [100].

In 66% of oral tongue squamous cell carcinoma clinical samples, downregulation of at least one FA gene at mRNA level was observed (*FANCB*, *FANCF*, *FACNJ* and *FANCM* were most commonly affected by downregulation) [294], although the clinical implication of the finding is yet to be determined and the functional significance is not clear. Another group reported lower *FANCA* mRNA expression in oral squamous cell carcinoma (and nondysplastic mucosa) of young patients compared with older patients [269]. The same group reported that FANCG protein expression examined by immunohistochemistry is absent or low in 81% of oral squamous cell carcinoma samples of young patients compared with 36% of older patient tumors [269]. Although interpretation of this data is difficult, it suggests that the FA pathway is deregulated in a subset of sporadic oral squamous cell carcinomas.

In breast cancer, FANCD2 protein expression was analyzed. By immunohistochemistry, 18 of 96 (19%) sporadic breast cancers and two of 21 (10%) BRCA1-related breast cancers were completely FANCD2-negative and high FANCD2 expression appeared to be prognostically unfavorable for overall survival [272]. The mechanism of the lack of FANCD2 protein expression is unknown.

As described above, inherited mono-allelic mutations in *FANCJ/BRIP1/BACH1* and *FANCN/PALB2* have been implicated in breast cancer predisposition [30, 69, 70, 78, 220, 231, 265]. Disease-associated *FANCN/PALB2* point mutation in a multigenerational prostate cancer family has been reported [69].

FANCN/PALB2 can be epigenetically silenced. Promoter hypermethylation of *FANCN/PALB2* was found in 2 of 8 inherited breast tumors, 4 of 60 sporadic breast tumors, 0 of 9 inherited ovarian tumors, and 4 of 53 sporadic ovarian tumors [217]. Intriguingly, the two inherited breast cancer cases with *PALB2* methylation are *BRCA2* mutation carriers [217], although whether wild type *BRCA2* was lost in the tumors was not described.

FANCD1/BRCA2 and *BRCA1* are famous breast/ovarian cancer susceptibility genes and their abnormality in inherited and sporadic cancer has been extensively reviewed elsewhere [270, 293].

In summary, although defects of individual FA genes are rather infrequent in human cancer in the general population, substantial proportion of human cancer may have defects in the FA-BRCA pathway in total.

17.7 Implication of the FA-BRCA Pathway in Cancer Therapy

17.7.1 Exploiting the Defects of the FA-BRCA Pathway in Cancer Cells

At the cellular level, cancer cells with defective FA-BRCA pathway are sensitive to ICL agents, such as cisplatin, carboplatin, MMC, melphalan and cyclophosphamide [37, 82, 83, 256]. This is also true in in vivo mouse xenograft model [276]. Whether this can be applied to human cancer in clinical situation is an important issue, but has not been investigated extensively.

For *BRCA1/2*-mutated ovarian cancer, this seems to be the case. As described above, platinum compounds (cisplatin or its derivative, carboplatin) are key drugs for the treatment of ovarian cancer. At least 10% of all epithelial ovarian cancers are hereditary, with mutations in *BRCA1/2*explaining 90% of these cases [25]. Approximately 6% of all ovarian carcinomas are associated with germline mutations in *BRCA1* and 4% with germline mutations in *BRCA2*. Carcinomas from *BRCA1/2*mutation carriers almost invariably show loss of heterozygosity at the *BRCA1/2* locus and are considered to be BRCA1/2 deficient [46, 92, 192]. BRCA1 and BRCA2 regulate homologous recombination [183, 184], which is required for repair of DNA lesions caused by ICL agents and BRCA1/2-deficient cancer cells are hypersensitive to ICL agents including cisplatin [18, 77, 310]. Therefore, cisplatin or its derivative carboplatin are logical choices for the treatment of *BRCA1/2*-mutated neoplasms [271]. Consistently, patients developing *BRCA1/2*-mutated ovarian carcinoma have a better prognosis compared with non-carriers, particularly if they receive platinum-based therapy [16, 24, 35, 77].

At the cellular level, cancer cells with defective FA-BRCA pathway are sensitive to poly(ADP-ribose) polymerase (PARP) inhibitors [11, 28, 74, 171]. PARP1 is a critical component of the major short-patch base excision repair (BER) pathway, and inhibition of BRCA1/2 and inhibition of BER are considered to be synthetic lethal [11]. PARP inhibitors selectively kill BRCA1/2-deficient cancer cells [28, 74] and are expected to become a therapeutic option for patients with *BRCA1/2*-mutated cancers [271]. Not only BRCA1/2-deficiency, but also deficiency of RAD51, RAD54, DSS1, RPA1, NBS1, ATR, ATM, CHK1, CHK2, FANCD2, FANCA, or FANCC induces PARP inhibitor sensitivity [171]. Therefore, PARP inhibitors may be effective for the treatment of cancer with defective FA-BRCA pathway due to hypermethylation of *FANCF* or mutations of any FA genes, but this has not been tested clinically.

FA pathway deficient cells (FANCC or FANCG-deficient cells) are reported to be sensitive to ATM inhibition [127], suggesting that pharmaceutical inhibition of ATM kinase by ATM inhibitors such as KU-55933 can have a role in the treatment of FA pathway-deficient human cancers.

17.7.2 Functional Restoration of the FA-BRCA Pathway as a Mechanism of Acquired Drug Resistance

In several genetic disorders such as FA, spontaneous genetic alterations that compensate for inherited disease-causing mutations have been described [105]. These alterations in FA include secondary genetic changes of one of the mutated FA alleles, such as back-mutations to wild-type, compensatory mutations in *cis,* intragenic crossovers, and gene conversion. Cells derived from FA patients are usually hypersensitive to ICL agents, but FA cells with secondary genetic changes compensating for the inherited FA mutations show resistance to ICL agents [89, 90, 96, 115, 159, 165, 241, 281, 297]. It is conceivable that similar things happen in cancer cells.

If the mechanism of initial sensitivity to chemotherapy is the defect of the function of the FA-BRCA pathway, the mechanism of acquired resistance may be due to restored function of the pathway. Indeed, secondary mutations of *BRCA1/2* that restore the wild-type *BRCA1/2* reading frame were observed in platinum-resistant recurrent *BRCA1/2*-mutated ovarian carcinoma [68, 227, 250]. Secondary mutations of *BRCA2* were also observed in a *BRCA2*-mutated pancreatic cancer cell line Capan-1 after in vitro selection in the presence of cisplatin and these cisplatin-resistant Capan-1 clones with secondary *BRCA2* mutations were cross-resistant to a PARP inhibitor [227]. Similarly, secondary mutations of *BRCA2* were observed in Capan-1 after in vitro selection in the presence of a PARP inhibitor [68]. This data indicates that platinum (and PARP inhibitor) resistance of *BRCA1/2*-mutated carcinomas can be mediated by secondary intragenic mutations in *BRCA1/2* that restore the wild-type *BRCA1/2* reading frame. Furthermore, secondary mutations in *BRCA2* occurred in a MMC-resistant human AML cell line with biallelic *BRCA2* mutations derived from an FA (D1 subtype) patient [115] and *BRCA2*-mutated Chinese hamster fibroblast lines [288], suggesting that secondary mutations in *BRCA2* may have a general role in resistance to ICL agents in cells derived from various organs.

Similarly, in vitro cisplatin selection of a cisplatin-sensitive *FANCF*-methylated ovarian cancer cell line led to a cisplatin-resistant *FANCF*-demethylated clone with restored FA-BRCA pathway [256]. Whether this happens in clinical situation has not been tested.

17.7.3 The FA-BRCA Pathway as a Drug Target

One possible therapeutic approach for DNA crosslinker-resistant cancers is sensitization of tumor to DNA crosslinkers using a small-molecule inhibitor of the FA-BRCA pathway. As a proof-of-principle, it has been shown that suppression of the FA-BRCA pathway by an adenovirus overexpressing dominant-negative FANCA or siRNA against some FA-BRCA genes sensitizes tumor cells to DNA crosslinkers [15, 75, 161].

As described above, the FA-BRCA pathway involves formation of a multi-subunit protein complex harboring E3 ligase activity, several enzymes (ubiquitin ligase and conjugating enzyme, deubiquitinating enzyme, kinase, ATPase/DNA translocase and ATPase/helicase) and many protein–protein or protein–DNA interactions, which can be potential targets for small molecule inhibitors.

A high-throughput cell-based screening assay for small molecule inhibitors of the FA-BRCA pathway using inhibition of DNA damage-induced FANCD2 nuclear foci formation as a readout has been developed and a partial result of the screening focusing on one inhibitor, curcumin (diferuloylmethane) [38], has been published [42]. Curcumin inhibits FANCD2 monoubiquitination and nuclear foci formation and sensitizes an ovarian cancer cell line to cisplatin in an FA pathway-dependent manner [42]. However, in order to establish curcumin as a useful cisplatin chemosensitizer, a detailed isobologram analysis of combinations of curcumin with cisplatin, in vivo studies using mouse models, and identification of the target of curcumin in the FA-BRCA pathway are required.

17.8 Concluding Remarks

Identification and characterization of the 13 identified FA proteins has greatly enhanced our understanding of the FA-BRCA pathway, has allowed new diagnostic approaches to FA and suggests novel treatment options for FA patients and patients with cancer in the general population.

Critical questions remain for the FA-BRCA pathway. First, how this pathway regulates DNA repair has not been elucidated. Although FANCD2 monoubiquitination is a required event in this pathway, little is known about the downstream function of this post-translational modification. How FA proteins and other proteins cooperate to repair ICL lesions is still a central question in the field. Second, not all of the components of the pathway are identified yet and the signaling of the pathway has not been elucidated completely. Third, some of the clinical manifestations of FA cannot be explained by the current understanding of the FA-BRCA pathway. Why FA patients develop aplastic anemia and some congenital defects has not been understood. Forth, effective prevention of tumors and aplastic anemia in FA patients has not been established. Finally, further examination of the role of the FA-BRCA pathway in the cause and treatment of cancers in the general population is warranted. The importance of the FA-BRCA pathway in ICL sensitivity in cancer cells in vitro has been well characterized. What is still missing is an analysis of the correlation between disruption of the FA-BRCA pathway and response to DNA crosslinking agents in human primary tumors. The effort to solve these questions will lead to the development of novel diagnostic and therapeutic techniques beneficial for FA patients and to the better understanding of the pathogenesis of cancer in general population.

Acknowledgments We thank E. Villegas, C. Jacquemont, J. Huang and K, Dhillon for critical reading of the manuscript.

References

1. Agarwal R, Kaye SB (2003) Ovarian cancer: strategies for overcoming resistance to chemotherapy. Nat Rev Cancer 3:502–516
2. Agoulnik AI, Lu B, Zhu Q et al. (2002) A novel gene, Pog, is necessary for primordial germ cell proliferation in the mouse and underlies the germ cell deficient mutation, gcd. Hum Mol Genet 11:3047–3053
3. Akkari YM, Bateman RL, Reifsteck CA et al. (2001) The 4 N Cell Cycle Delay in Fanconi Anemia Reflects Growth Arrest in Late S Phase. Mol Genet Metab 74:403–412
4. Alpi A, Langevin F, Mosedale G et al. (2007) UBE2T, the Fanconi anemia core complex, and FANCD2 are recruited independently to chromatin: a basis for the regulation of FANCD2 monoubiquitination. Mol Cell Biol 27:8421–8430
5. Alter BP (2002) Radiosensitivity in Fanconi's anemia patients. Radiother Oncol 62:345–347
6. Alter BP (2003) Cancer in Fanconi anemia, 1927–2001. Cancer 97:425–440
7. Alter BP (2003) Inherited bone marrow failure syndromes. In: *Nathan and Oski's Hematology of Infancy and Childhood.* Edited by Nathan DG, Orkin SH, Ginsburg D, Look AT, Oski FA, vol. 1, 6th edn. Philadelphia, PA: Saunders; 280–365

8. Ameziane N, Errami A, Leveille F et al. (2008) Genetic subtyping of Fanconi anemia by comprehensive mutation screening. Hum Mutat 29:159–166
9. Andreassen PR, D'Andrea AD, Taniguchi T (2004) ATR couples FANCD2 monoubiquitination to the DNA-damage response. Genes Dev 18:1958–1963
10. Arlt MF, Xu B, Durkin SG et al. (2004) BRCA1 is required for common-fragile-site stability via its G2/M checkpoint function. Mol Cell Biol 24:6701–6709
11. Ashworth A (2008) A synthetic lethal therapeutic approach: poly(ADP) ribose polymerase inhibitors for the treatment of cancers deficient in DNA double-strand break repair. J Clin Oncol 26:3785–3790
12. Auerbach AD (1993) Fanconi anemia diagnosis and the diepoxybutane (DEB) test. Experimental Hematology 21:731–733
13. Auerbach AD, Buchwald M, Joenje H (2001) Fanconi Anemia. In: *The Metabolic and Molecular Bases of Inherited Disease.* Edited by Scriver CR, Sly WS, Childs B, Beaudet AL, Valle D, Kinzler KW, Vogelstein B, vol. 1, 8th edn. New York: McGraw-Hill; 753–768
14. Bagby Jr GC (2003) Genetic basis of Fanconi anemia. Curr Opin Hematol 10:68–76
15. Bartz SR, Zhang Z, Burchard J et al. (2006) Small interfering RNA screens reveal enhanced cisplatin cytotoxicity in tumor cells having both BRCA network and TP53 disruptions. Mol Cell Biol 26:9377–9386
16. Ben David Y, Chetrit A, Hirsh-Yechezkel G et al. (2002) Effect of BRCA Mutations on the Length of Survival in Epithelial Ovarian Tumors. J Clin Oncol 20:463–466
17. Berwick M, Satagopan JM, Ben-Porat L et al. (2007) Genetic heterogeneity among Fanconi anemia heterozygotes and risk of cancer. Cancer Res 67:9591–9596
18. Bhattacharyya A, Ear US, Koller BH et al. (2000) The breast cancer susceptibility gene BRCA1 is required for subnuclear assembly of Rad51 and survival following treatment with the DNA cross-linking agent cisplatin. J Biol Chem 275:23899–23903
19. Bjergbaek L, Cobb JA, Tsai-Pflugfelder M et al. (2005) Mechanistically distinct roles for Sgs1p in checkpoint activation and replication fork maintenance. EMBO J 24: 405–417
20. Blom E, van de Vrugt HJ, de Vries Y et al. (2004) Multiple TPR motifs characterize the Fanconi anemia FANCG protein. DNA Repair (Amst) 3:77–84
21. Bogliolo M, Cabre O, Callen E et al. (2002) The Fanconi anaemia genome stability and tumour suppressor network. Mutagenesis 17:529–538
22. Bogliolo M, Lyakhovich A, Callen E et al. (2007) Histone H2AX and Fanconi anemia FANCD2 function in the same pathway to maintain chromosome stability. Embo J 26:1340–1351
23. Bouwman P, Jonkers J (2008) Mouse models for BRCA1 associated tumorigenesis: from fundamental insights to preclinical utility. Cell Cycle, advanced on line publication
24. Boyd J, Sonoda Y, Federici MG et al. (2000) Clinicopathologic features of BRCA-linked and sporadic ovarian cancer. JAMA 283:2260–2265
25. Boyd J (2003) Specific keynote: hereditary ovarian cancer: what we know. Gynecol Oncol 88:S8–10; discussion S11–13
26. Bridge WL, Vandenberg CJ, Franklin RJ et al. (2005) The BRIP1 helicase functions independently of BRCA1 in the Fanconi anemia pathway for DNA crosslink repair. Nat Genet 37:953–957
27. Brodeur I, Goulet I, Tremblay CS et al. (2004) Regulation of the Fanconi anemia group C protein through proteolytic modification. J Biol Chem 279:4713–4720
28. Bryant HE, Schultz N, Thomas HD et al. (2005) Specific killing of BRCA2-deficient tumours with inhibitors of poly(ADP-ribose) polymerase. Nature 434:913–917
29. Cantor SB, Bell DW, Ganesan S et al. (2001) BACH1, a novel helicase-like protein, interacts directly with BRCA1 and contributes to its DNA repair function. Cell 105:149–160
30. Cao AY, Huang J, Hu Z et al. (2008) The prevalence of PALB2 germline mutations in BRCA1/BRCA2 negative Chinese women with early onset breast cancer or affected relatives. Breast Cancer Res Treat, advanced on line publication

31. Carreau M, Gan OI, Liu L et al. (1998) Bone marrow failure in the Fanconi anemia group C mouse model after DNA damage. Blood 91:2737–2744
32. Carreau M, Liu L, Gan OI et al. (2002) Short-term granulocyte colony-stimulating factor and erythropoietin treatment enhances hematopoiesis and survival in the mitomycin C-conditioned Fancc(–/–) mouse model, while long-term treatment is ineffective. Blood 100:1499–1501
33. Carreau M (2004) Not-so-novel phenotypes in the Fanconi anemia group D2 mouse model. Blood 103:2430
34. Casper AM, Nghiem P, Arlt MF et al. (2002) ATR regulates fragile site stability. Cell 111:779–789
35. Cass I, Baldwin RL, Varkey T et al. (2003) Improved survival in women with BRCA-associated ovarian carcinoma. Cancer 97:2187–2195
36. Chen M, Tomkins DJ, Auerbach W et al. (1996) Inactivation of Fac in mice produces inducible chromosomal instability and reduced fertility reminiscent of Fanconi anaemia. Nature Genetics 12:448–451
37. Chen Q, Van der Sluis PC, Boulware D et al. (2005) The FA/BRCA pathway is involved in melphalan-induced DNA interstrand cross-link repair and accounts for melphalan resistance in multiple myeloma cells. Blood 106:698–705
38. Cheng AL, Hsu CH, Lin JK et al. (2001) Phase I clinical trial of curcumin, a chemopreventive agent, in patients with high-risk or pre-malignant lesions. Anticancer Res 21:2895–2900
39. Cheng NC, van De Vrugt HJ, van Der Valk MA et al. (2000) Mice with a targeted disruption of the Fanconi anemia homolog fanca. Hum Mol Genet 9:1805–1811
40. Cheung I, Schertzer M, Rose A et al. (2002) Disruption of dog-1 in *Caenorhabditis elegans* triggers deletions upstream of guanine-rich DNA. Nat Genet 31:405–409
41. Chini CC, Chen J (2004) Claspin, a regulator of Chk1 in DNA replication stress pathway. DNA Repair (Amst) 3:1033–1037
42. Chirnomas D, Taniguchi T, de la Vega M et al. (2006) Chemosensitization to cisplatin by inhibitors of the Fanconi anemia/BRCA pathway. Mol Cancer Ther 5:952–961
43. Ciccia A, Ling C, Coulthard R et al. (2007) Identification of FAAP24, a Fanconi anemia core complex protein that interacts with FANCM. Mol Cell 25:331–343
44. Cobb JA, Bjergbaek L, Shimada K et al. (2003) DNA polymerase stabilization at stalled replication forks requires Mec1 and the RecQ helicase Sgs1. Embo J 22:4325–4336
45. Cohn MA, Kowal P, Yang K et al. (2007) A UAF1-containing multisubunit protein complex regulates the Fanconi anemia pathway. Mol Cell 28:786–797
46. Collins N, McManus R, Wooster R et al. (1995) Consistent loss of the wild type allele in breast cancers from a family linked to the BRCA2 gene on chromosome 13q12-13. Oncogene 10:1673–1675
47. Collis SJ, Barber LJ, Ward JD et al. (2006) *C. elegans* FANCD2 responds to replication stress and functions in interstrand cross-link repair. DNA Repair (Amst) 5:1398–1406
48. Collis SJ, Barber LJ, Clark AJ et al. (2007) HCLK2 is essential for the mammalian S-phase checkpoint and impacts on Chk1 stability. Nat Cell Biol 9:391–401
49. Condie A, Powles RL, Hudson CD et al. (2002) Analysis of the Fanconi anaemia complementation group A gene in acute myeloid leukaemia. Leuk Lymphoma 43:1849–1853
50. Couch FJ, Johnson MR, Rabe K et al. (2005) Germ line Fanconi anemia complementation group C mutations and pancreatic cancer. Cancer Res 65:383–386
51. Cumming RC, Lightfoot J, Beard K et al. (2001) Fanconi anemia group C protein prevents apoptosis in hematopoietic cells through redox regulation of GSTP1. Nat Med 7: 814–820
52. D'Andrea AD, Grompe M (2003) The Fanconi anaemia/BRCA pathway. Nat Rev Cancer 3:23–34
53. Daniels MJ, Wang Y, Lee M et al. (2004) Abnormal cytokinesis in cells deficient in the breast cancer susceptibility protein BRCA2. Science 306:876–879

54. Davies AA, Masson JY, McIlwraith MJ et al. (2001) Role of BRCA2 in control of the RAD51 recombination and DNA repair protein. Mol Cell 7:273–282
55. Davies OR, Pellegrini L (2007) Interaction with the BRCA2 C terminus protects RAD51-DNA filaments from disassembly by BRC repeats. Nat Struct Mol Biol 14:475–483
56. de Winter JP, Waisfisz Q, Rooimans MA et al. (1998) The Fanconi anaemia group G gene FANCG is identical with XRCC9. Nat Genet 20:281–283
57. de Winter JP, Leveille F, van Berkel CG et al. (2000) Isolation of a cDNA representing the fanconi anemia complementation group E gene. Am J Hum Genet 67:1306–1308
58. de Winter JP, Rooimans MA, van Der Weel L et al. (2000) The Fanconi anaemia gene FANCF encodes a novel protein with homology to ROM. Nat Genet 24:15–16
59. Dequen F, St-Laurent JF, Gagnon SN et al. (2005) The *Caenorhabditis elegans* FancD2 ortholog is required for survival following DNA damage. Comp Biochem Physiol B Biochem Mol Biol 141:453–460
60. Dhillon VS, Shahid M, Husain SA (2004) CpG methylation of the FHIT, FANCF, cyclin-D2, BRCA2 and RUNX3 genes in Granulosa cell tumors (GCTs) of ovarian origin. Mol Cancer 3:33
61. Digweed M, Rothe S, Demuth I et al. (2002) Attenuation of the formation of DNA-repair foci containing RAD51 in Fanconi anaemia. Carcinogenesis 23:1121–1126
62. Dorsman JC, Levitus M, Rockx D et al. (2007) Identification of the Fanconi anemia complementation group I gene, FANCI. Cell Oncol 29:211–218
63. Downs JA, Nussenzweig MC, Nussenzweig A (2007) Chromatin dynamics and the preservation of genetic information. Nature 447:951–958
64. Dronkert ML, Kanaar R (2001) Repair of DNA interstrand cross-links. Mutat Res 486: 217–247
65. Duckworth-Rysiecki G, Taylor AM (1985) Effects of ionizing radiation on cells from Fanconi's anemia patients. Cancer Res 45:416–420
66. Dufour C, Svahn J (2008) Fanconi anaemia: new strategies. Bone Marrow Transplant 41 Suppl 2:S90–95
67. Durkin SG, Arlt MF, Howlett NG et al. (2006) Depletion of CHK1, but not CHK2, induces chromosomal instability and breaks at common fragile sites. Oncogene 25:4381–4388
68. Edwards SL, Brough R, Lord CJ et al. (2008) Resistance to therapy caused by intragenic deletion in BRCA2. Nature 451:1111–1115
69. Erkko H, Xia B, Nikkila J et al. (2007) A recurrent mutation in PALB2 in Finnish cancer families. Nature 446:316–319
70. Erkko H, Dowty JG, Nikkila J et al. (2008) Penetrance analysis of the PALB2 c.1592delT founder mutation. Clin Cancer Res 14:4667–4671
71. Esashi F, Galkin VE, Yu X et al. (2007) Stabilization of RAD51 nucleoprotein filaments by the C-terminal region of BRCA2. Nat Struct Mol Biol 14:468–474
72. Evers B, Jonkers J (2006) Mouse models of BRCA1 and BRCA2 deficiency: past lessons, current understanding and future prospects. Oncogene 25:5885–5897
73. Fagerlie SR, Koretsky T, Torok-Storb B et al. (2004) Impaired type I IFN-induced Jak/STAT signaling in FA-C cells and abnormal CD4+ Th cell subsets in Fancc–/– mice. J Immunol 173:3863–3870
74. Farmer H, McCabe N, Lord CJ et al. (2005) Targeting the DNA repair defect in BRCA mutant cells as a therapeutic strategy. Nature 434:917–921
75. Ferrer M, de Winter JP, Mastenbroek DC et al. (2004) Chemosensitizing tumor cells by targeting the Fanconi anemia pathway with an adenovirus overexpressing dominant-negative FANCA. Cancer Gene Ther 11:539–546
76. Ferrer M, Rodriguez JA, Spierings EA et al. (2005) Identification of multiple nuclear export sequences in Fanconi anemia group A protein that contribute to CRM1-dependent nuclear export. Hum Mol Genet 14:1271–1281
77. Foulkes WD (2006) BRCA1 and BRCA2: chemosensitivity, treatment outcomes and prognosis. Fam Cancer 5:135–142

78. Foulkes WD, Ghadirian P, Akbari MR et al. (2007) Identification of a novel truncating PALB2 mutation and analysis of its contribution to early-onset breast cancer in French-Canadian women. Breast Cancer Res 9:R83
79. Freie B, Li X, Ciccone SL et al. (2003) Fanconi anemia type C and p53 cooperate in apoptosis and tumorigenesis. Blood 102:4146–4152
80. Fu Y, Zhu Y, Zhang K et al. (2008) Rad6-Rad18 mediates a eukaryotic SOS response by ubiquitinating the 9-1-1 checkpoint clamp. Cell 133:601–611
81. Galimi F, Noll M, Kanazawa Y et al. (2002) Gene therapy of Fanconi anemia: preclinical efficacy using lentiviral vectors. Blood 100:2732–2736
82. Gallmeier E, Calhoun ES, Rago C et al. (2006) Targeted disruption of FANCC and FANCG in human cancer provides a preclinical model for specific therapeutic options. Gastroenterology 130:2145–2154
83. Gallmeier E, Kern SE (2007) Targeting Fanconi anemia/BRCA2 pathway defects in cancer: the significance of preclinical pharmacogenomic models. Clin Cancer Res 13:4–10
84. Garcia-Higuera I, Kuang Y, Denham J et al. (2000) The fanconi anemia proteins FANCA and FANCG stabilize each other and promote the nuclear accumulation of the fanconi anemia complex. Blood 96:3224–3230
85. Garcia-Higuera I, Taniguchi T, Ganesan S et al. (2001) Interaction of the Fanconi anemia proteins and BRCA1 in a common pathway. Mol Cell 7:249–262
86. Gari K, Decaillet C, Stasiak AZ et al. (2008) The fanconi anemia protein FANCM can promote branch migration of Holliday junctions and replication forks. Mol Cell 29: 141–148
87. Godthelp BC, Wiegant WW, Waisfisz Q et al. (2006) Inducibility of nuclear Rad51 foci after DNA damage distinguishes all Fanconi anemia complementation groups from D1/BRCA2. Mutat Res 594:39–48
88. Gordon SM, Alon N, Buchwald M (2005) FANCC, FANCE, and FANCD2 form a ternary complex essential to the integrity of the Fanconi anemia DNA damage response pathway. J Biol Chem 280:36118–36125
89. Gregory JJ, Wagner JE, Verlander PC et al. (2001) Somatic mosaicism in Fanconi anemia: evidence of genotypic reversion in lymphohematopoietic stem cells. Proc Natl Acad Sci USA 98:2532–2537
90. Gross M, Hanenberg H, Lobitz S et al. (2002) Reverse mosaicism in Fanconi anemia: natural gene therapy via molecular self-correction. Cytogenet Genome Res 98:126–135
91. Gudmundsdottir K, Ashworth A (2006) The roles of BRCA1 and BRCA2 and associated proteins in the maintenance of genomic stability. Oncogene 25:5864–5874
92. Gudmundsson J, Johannesdottir G, Bergthorsson JT et al. (1995) Different tumor types from BRCA2 carriers show wild-type chromosome deletions on 13q12-q13. Cancer Res 55: 4830–4832
93. Guervilly JH, Mace-Aime G, Rosselli F (2008) Loss of CHK1 function impedes DNA damage-induced FANCD2 monoubiquitination but normalizes the abnormal G2 arrest in Fanconi anemia. Hum Mol Genet 17:679–689
94. Gupta R, Sharma S, Sommers JA et al. (2007) FANCJ (BACH1) helicase forms DNA damage inducible foci with replication protein A and interacts physically and functionally with the single-stranded DNA-binding protein. Blood 110:2390–2398
95. Gush KA, Fu KL, Grompe M et al. (2000) Phenotypic correction of Fanconi anemia group C knockout mice. Blood 95:700–704
96. Hamanoue S, Yagasaki H, Tsuruta T et al. (2006) Myeloid lineage-selective growth of revertant cells in Fanconi anaemia. Br J Haematol 132:630–635
97. Haneline LS, Li X, Ciccone SL et al. (2003) Retroviral-mediated expression of recombinant Fancc enhances the repopulating ability of Fancc–/– hematopoietic stem cells and decreases the risk of clonal evolution. Blood 101:1299–1307
98. Hazlehurst LA, Enkemann SA, Beam CA et al. (2003) Genotypic and phenotypic comparisons of de novo and acquired melphalan resistance in an isogenic multiple myeloma cell line model. Cancer Res 63:7900–7906

99. Hejna J, Holtorf M, Hines J et al. (2008) Tip60 is required for DNA interstrand cross-link repair in the Fanconi anemia pathway. J Biol Chem 283:9844–9851
100. Hess CJ, Ameziane N, Schuurhuis GJ et al. (2008) Hypermethylation of the FANCC and FANCL promoter regions in sporadic acute leukaemia. Cell Oncol 30:299–306
101. Hinz JM, Nham PB, Salazar EP et al. (2006) The Fanconi anemia pathway limits the severity of mutagenesis. DNA Repair (Amst) 5:875–884
102. Hinz JM, Nham PB, Urbin SS et al. (2007) Disparate contributions of the Fanconi anemia pathway and homologous recombination in preventing spontaneous mutagenesis. Nucleic Acids Res 35:3733–3740
103. Hirano S, Yamamoto K, Ishiai M et al. (2005) Functional relationships of FANCC to homologous recombination, translesion synthesis, and BLM. Embo J 24: 418–427
104. Hirsch B, Shimamura A, Moreau L et al. (2004) Association of biallelic BRCA2/FANCD1 mutations with spontaneous chromosomal instability and solid tumors of childhood. Blood 103:2554–2559
105. Hirschhorn R (2003) In vivo reversion to normal of inherited mutations in humans. J Med Genet 40:721–728
106. Ho GP, Margossian S, Taniguchi T et al. (2006) Phosphorylation of FANCD2 on two novel sites is required for mitomycin C resistance. Mol Cell Biol 26:7005–7015
107. Houghtaling S, Timmers C, Noll M et al. (2003) Epithelial cancer in Fanconi anemia complementation group D2 (Fancd2) knockout mice. Genes Dev 17:2021–2035
108. Houghtaling S, Granville L, Akkari Y et al. (2005) Heterozygosity for p53 (Trp53+/–) accelerates epithelial tumor formation in fanconi anemia complementation group D2 (Fancd2) knockout mice. Cancer Res 65:85–91
109. Houghtaling S, Newell A, Akkari Y et al. (2005) Fancd2 functions in a double strand break repair pathway that is distinct from non-homologous end joining. Hum Mol Genet 14: 3027–3033
110. Howlett NG, Taniguchi T, Olson S et al. (2002) Biallelic inactivation of BRCA2 in Fanconi anemia. Science 297:606–609
111. Howlett NG, Taniguchi T, Durkin SG et al. (2005) The Fanconi anemia pathway is required for the DNA replication stress response and for the regulation of common fragile site stability. Hum Mol Genet 14:693–701
112. Huang TT, Nijman SM, Mirchandani KD et al. (2006) Regulation of monoubiquitinated PCNA by DUB autocleavage. Nat Cell Biol 8:339–347
113. Hussain S, Witt E, Huber PA et al. (2003) Direct interaction of the Fanconi anaemia protein FANCG with BRCA2/FANCD1. Hum Mol Genet 12:2503–2510
114. Hussain S, Wilson JB, Medhurst AL et al. (2004) Direct interaction of FANCD2 with BRCA2 in DNA damage response pathways. Hum Mol Genet 13:1241–1248
115. Ikeda H, Matsushita M, Waisfisz Q et al. (2003) Genetic Reversion in an Acute Myelogenous Leukemia Cell Line from a Fanconi Anemia Patient with Biallelic Mutations in BRCA2. Cancer Res 63:2688–2694
116. Ikura T, Ogryzko VV, Grigoriev M et al. (2000) Involvement of the TIP60 histone acetylase complex in DNA repair and apoptosis. Cell 102:463–473
117. Ikura T, Tashiro S, Kakino A et al. (2007) DNA damage-dependent acetylation and ubiquitination of H2AX enhances chromatin dynamics. Mol Cell Biol 27:7028–7040
118. Jacquemont C, Taniguchi T (2006) Disruption of the Fanconi anemia pathway in human cancer in the general population. Cancer Biol Ther 5:1637–1639
119. Jacquemont C, Taniguchi T (2007) Proteasome function is required for DNA damage response and fanconi anemia pathway activation. Cancer Res 67:7395–7405
120. Joenje H, Arwert F, Eriksson AW et al. (1981) Oxygen-dependence of chromosomal aberrations in Fanconi's anaemia. Nature 290:142–143
121. Joenje H, Patel KJ (2001) The emerging genetic and molecular basis of Fanconi anaemia. Nat Rev Genet 2:446–457
122. Kalb R, Neveling K, Hoehn H et al. (2007) Hypomorphic mutations in the gene encoding a key Fanconi anemia protein, FANCD2, sustain a significant group of FA-D2 patients with severe phenotype. Am J Hum Genet 80:895–910

123. Kannouche PL, Wing J, Lehmann AR (2004) Interaction of human DNA polymerase eta with monoubiquitinated PCNA: a possible mechanism for the polymerase switch in response to DNA damage. Mol Cell 14:491–500
124. Karin M, Ben-Neriah Y (2000) Phosphorylation meets ubiquitination: the control of NF-[kappa]B activity. Annu Rev Immunol 18:621–663
125. Karppinen SM, Vuosku J, Heikkinen K et al. (2003) No evidence of involvement of germline BACH1 mutations in Finnish breast and ovarian cancer families. Eur J Cancer 39: 366–371
126. Kennedy RD, D'Andrea AD (2005) The Fanconi Anemia/BRCA pathway: new faces in the crowd. Genes Dev 19:2925–2940
127. Kennedy RD, Chen CC, Stuckert P et al. (2007) Fanconi anemia pathway-deficient tumor cells are hypersensitive to inhibition of ataxia telangiectasia mutated. J Clin Invest 117:1440–1449
128. Kim JM, Kee Y, Gurtan A et al. (2008) Cell cycle-dependent chromatin loading of the Fanconi anemia core complex by FANCM/FAAP24. Blood 111:5215–5222
129. Kobayashi J, Antoccia A, Tauchi H et al. (2004) NBS1 and its functional role in the DNA damage response. DNA Repair (Amst) 3:855–861
130. Koepp DM, Schaefer LK, Ye X et al. (2001) Phosphorylation-dependent ubiquitination of cyclin E by the SCFFbw7 ubiquitin ligase. Science 294:173–177
131. Komori K, Fujikane R, Shinagawa H et al. (2002) Novel endonuclease in Archaea cleaving DNA with various branched structure. Genes Genet Syst 77:227–241
132. Koomen M, Cheng NC, van de Vrugt HJ et al. (2002) Reduced fertility and hypersensitivity to mitomycin C characterize Fancg/Xrcc9 null mice. Hum Mol Genet 11:273–281
133. Koul S, McKiernan JM, Narayan G et al. (2004) Role of promoter hypermethylation in cisplatin treatment response of male germ cell tumors. Mol Cancer 3:16
134. Kowal P, Gurtan AM, Stuckert P et al. (2007) Structural determinants of human FANCF protein that function in the assembly of a DNA damage signaling complex. J Biol Chem 282:2047–2055
135. Kraakman-van der Zwet M, Overkamp WJ, van Lange RE et al. (2002) Brca2 (XRCC11) deficiency results in radioresistant DNA synthesis and a higher frequency of spontaneous deletions. Mol Cell Biol 22:669–679
136. Kraunz KS, Nelson HH, Lemos M et al. (2006) Homozygous deletion of p16(INK4a) and tobacco carcinogen exposure in nonsmall cell lung cancer. Int J Cancer 118:1364–1369
137. Krogan NJ, Lam MH, Fillingham J et al. (2004) Proteasome involvement in the repair of DNA double-strand breaks. Mol Cell 16:1027–1034
138. Kruisselbrink E, Guryev V, Brouwer K et al. (2008) Mutagenic capacity of endogenous G4 DNA underlies genome instability in FANCJ-defective *C. elegans*. Curr Biol 18: 900–905
139. Kusch T, Florens L, Macdonald WH et al. (2004) Acetylation by Tip60 is required for selective histone variant exchange at DNA lesions. Science 306:2084–2087
140. Kutler DI, Singh B, Satagopan J et al. (2003) A 20-year perspective on the International Fanconi Anemia Registry (IFAR). Blood 101:1249–1256
141. Kutler DI, Wreesmann VB, Goberdhan A et al. (2003) Human papillomavirus DNA and p53 polymorphisms in squamous cell carcinomas from Fanconi anemia patients. J Natl Cancer Inst 95:1718–1721
142. Lee KY, Yang I, Park JE et al. (2007) Developmental stage- and DNA damage-specific functions of *C. elegans* FANCD2. Biochem Biophys Res Commun 352:479–485
143. Lehmann AR (2006) Translesion synthesis in mammalian cells. Exp Cell Res 312: 2673–2676
144. Lehmann AR, Niimi A, Ogi T et al. (2007) Translesion synthesis: Y-family polymerases and the polymerase switch. DNA Repair (Amst) 6:891–899
145. Lensch MW, Tischkowitz M, Christianson TA et al. (2003) Acquired FANCA dysfunction and cytogenetic instability in adult acute myelogenous leukemia. Blood 102:7–16

146. Leveille F, Blom E, Medhurst AL et al. (2004) The Fanconi anemia gene product FANCF is a flexible adaptor protein. J Biol Chem 279:39421–39430
147. Leveille F, Ferrer M, Medhurst AL et al. (2006) The nuclear accumulation of the Fanconi anemia protein FANCE depends on FANCC. DNA Repair (Amst) 5:556–565
148. Levitus M, Rooimans MA, Steltenpool J et al. (2004) Heterogeneity in Fanconi anemia: evidence for 2 new genetic subtypes. Blood 103:2498–2503
149. Levitus M, Waisfisz Q, Godthelp BC et al. (2005) The DNA helicase BRIP1 is defective in Fanconi anemia complementation group J. Nat Genet 37:934–935
150. Levran O, Attwooll C, Henry RT et al. (2005) The BRCA1-interacting helicase BRIP1 is deficient in Fanconi anemia. Nat Genet 37:931–933
151. Li J, Zou C, Bai Y et al. (2006) DSS1 is required for the stability of BRCA2. Oncogene 25:1186–1194
152. Li X, Le Beau MM, Ciccone S et al. (2005) Ex vivo culture of Fancc–/– stem/progenitor cells predisposes cells to undergo apoptosis, and surviving stem/progenitor cells display cytogenetic abnormalities and an increased risk of malignancy. Blood 105:3465–3471
153. Lim SL, Smith P, Syed N et al. (2008) Promoter hypermethylation of FANCF and outcome in advanced ovarian cancer. Br J Cancer 98:1452–1456
154. Ling C, Ishiai M, Ali AM et al. (2007) FAAP100 is essential for activation of the Fanconi anemia-associated DNA damage response pathway. Embo J 26:2104–2114
155. Litman R, Peng M, Jin Z et al. (2005) BACH1 is critical for homologous recombination and appears to be the Fanconi anemia gene product FANCJ. Cancer Cell 8:255–265
156. Liu N, Lamerdin JE, Tucker JD et al. (1997) The human XRCC9 gene corrects chromosomal instability and mutagen sensitivities in CHO UV40 cells. Proceedings of the National Academy of Sciences of the United States of America 94:9232–9237
157. Liu TX, Howlett NG, Deng M et al. (2003) Knockdown of zebrafish Fancd2 causes developmental abnormalities via p53-dependent apoptosis. Dev Cell 5:903–914
158. Lo Ten Foe JR, Rooimans MA, Bosnoyan-Collins L et al. (1996) Expression cloning of a cDNA for the major Fanconi anaemia gene, FAA. Nat Genet 14:320–323
159. Lo Ten Foe JR, Kwee ML, Rooimans MA et al. (1997) Somatic mosaicism in Fanconi anemia: molecular basis and clinical significance. Eur J Hum Genet 5:137–148
160. Lomonosov M, Anand S, Sangrithi M et al. (2003) Stabilization of stalled DNA replication forks by the BRCA2 breast cancer susceptibility protein. Genes Dev 17:3017–3022
161. Lyakhovich A, Surralles J (2007) FANCD2 depletion sensitizes cancer cells repopulation ability in vitro. Cancer Lett 256:186–195
162. Machida YJ, Machida Y, Chen Y et al. (2006) UBE2T is the E2 in the Fanconi anemia pathway and undergoes negative autoregulation. Mol Cell 23:589–596
163. Maizels N (2006) Dynamic roles for G4 DNA in the biology of eukaryotic cells. Nat Struct Mol Biol 13:1055–1059
164. Maizels N (2008) Genomic stability: FANCJ-dependent G4 DNA repair. Curr Biol 18: R613–614
165. Mankad A, Taniguchi T, Cox B et al. (2006) Natural gene therapy in monozygotic twins with Fanconi anemia. Blood 107:3084–3090
166. Marek LR, Bale AE (2006) Drosophila homologs of FANCD2 and FANCL function in DNA repair. DNA Repair (Amst) 5:1317–1326
167. Marek LR, Kottemann MC, Glazer PM et al. (2008) MEN1 and FANCD2 mediate distinct mechanisms of DNA crosslink repair. DNA Repair (Amst) 7:476–486
168. Marsit CJ, Liu M, Nelson HH et al. (2004) Inactivation of the Fanconi anemia/BRCA pathway in lung and oral cancers: implications for treatment and survival. Oncogene 23:1000–1004
169. Matsushita N, Kitao H, Ishiai M et al. (2005) A FancD2-monoubiquitin fusion reveals hidden functions of Fanconi anemia core complex in DNA repair. Mol Cell 19:841–847
170. McCabe KM, Hemphill A, Akkari Y et al. (2008) ERCC1 is required for FANCD2 focus formation. Mol Genet Metab, advanced on line publication

171. McCabe N, Turner NC, Lord CJ et al. (2006) Deficiency in the Repair of DNA Damage by Homologous Recombination and Sensitivity to Poly(ADP-Ribose) Polymerase Inhibition. Cancer Res 66:8109–8115
172. Medhurst AL, Laghmani el H, Steltenpool J et al. (2006) Evidence for subcomplexes in the Fanconi anemia pathway. Blood 108:2072–2080
173. Meetei AR, de Winter JP, Medhurst AL et al. (2003) A novel ubiquitin ligase is deficient in Fanconi anemia. Nat Genet 35:165–170
174. Meetei AR, Sechi S, Wallisch M et al. (2003) A multiprotein nuclear complex connects Fanconi anemia and Bloom syndrome. Mol Cell Biol 23:3417–3426
175. Meetei AR, Levitus M, Xue Y et al. (2004) X-linked inheritance of Fanconi anemia complementation group B. Nat Genet 36:1219–1224
176. Meetei AR, Medhurst AL, Ling C et al. (2005) A human ortholog of archaeal DNA repair protein Hef is defective in Fanconi anemia complementation group M. Nat Genet 37: 958–963
177. Meyer S, White DJ, Will AM et al. (2006) No evidence of significant silencing of Fanconi genes FANCF and FANCB or Nijmegen breakage syndrome gene NBS1 by DNA hypermethylation in sporadic childhood leukaemia. Br J Haematol 134:61–63
178. Mi J, Qiao F, Wilson JB et al. (2004) FANCG is phosphorylated at serines 383 and 387 during mitosis. Mol Cell Biol 24:8576–8585
179. Mi J, Kupfer GM (2005) The Fanconi anemia core complex associates with chromatin during S phase. Blood 105:759–766
180. Mirchandani KD, McCaffrey RM, D'Andrea AD (2008) The Fanconi anemia core complex is required for efficient point mutagenesis and Rev1 foci assembly. DNA Repair (Amst) 7:902–911
181. Montes de Oca R, Andreassen PR, Margossian SP et al. (2005) Regulated interaction of the Fanconi anemia protein, FANCD2, with chromatin. Blood 105:1003–1009
182. Mosedale G, Niedzwiedz W, Alpi A et al. (2005) The vertebrate Hef ortholog is a component of the Fanconi anemia tumor-suppressor pathway. Nat Struct Mol Biol 12:763–771
183. Moynahan ME, Chiu JW, Koller BH et al. (1999) Brca1 controls homology-directed DNA repair. Mol Cell 4:511–518
184. Moynahan ME, Pierce AJ, Jasin M (2001) BRCA2 is required for homology-directed repair of chromosomal breaks. Mol Cell 7:263–272
185. Moynahan ME (2002) The cancer connection: BRCA1 and BRCA2 tumor suppression in mice and humans. Oncogene 21:8994–9007
186. Mukhopadhyay SS, Leung KS, Hicks MJ et al. (2006) Defective mitochondrial peroxiredoxin-3 results in sensitivity to oxidative stress in Fanconi anemia. J Cell Biol 175: 225–235
187. Murzin AG, Brenner SE, Hubbard T et al. (1995) SCOP: a structural classification of proteins database for the investigation of sequences and structures. J Mol Biol 247: 536–540
188. Nakanishi K, Taniguchi T, Ranganathan V et al. (2002) Interaction of FANCD2 and NBS1 in the DNA damage response. Nat Cell Biol 4:913–920
189. Nakanishi K, Yang YG, Pierce AJ et al. (2005) Human Fanconi anemia monoubiquitination pathway promotes homologous DNA repair. Proc Natl Acad Sci USA 102:1110–1115
190. Narayan G, Arias-Pulido H, Nandula SV et al. (2004) Promoter hypermethylation of FANCF: disruption of Fanconi Anemia-BRCA pathway in cervical cancer. Cancer Res 64:2994–2997
191. Navarro S, Meza NW, Quintana-Bustamante O et al. (2006) Hematopoietic dysfunction in a mouse model for Fanconi anemia group D1. Mol Ther 14:525–535
192. Neuhausen SL, Marshall CJ (1994) Loss of heterozygosity in familial tumors from three BRCA1-linked kindreds. Cancer Res 54:6069–6072
193. Neveling K, Kalb R, Florl AR et al. (2007) Disruption of the FA/BRCA pathway in bladder cancer. Cytogenet Genome Res 118:166–176

194. Niedernhofer LJ, Lalai AS, Hoeijmakers JH (2005) Fanconi Anemia (Cross) linked to DNA Repair. Cell 123:1191–1198
195. Niedzwiedz W, Mosedale G, Johnson M et al. (2004) The Fanconi anaemia gene FANCC promotes homologous recombination and error-prone DNA repair. Mol Cell 15: 607–620
196. Nijman SM, Huang TT, Dirac AM et al. (2005) The deubiquitinating enzyme USP1 regulates the Fanconi anemia pathway. Mol Cell 17:331–339
197. Noll M, Bateman RL, D'Andrea AD et al. (2001) Preclinical protocol for in vivo selection of hematopoietic stem cells corrected by gene therapy in Fanconi anemia group C. Mol Ther 3:14–23
198. Noll M, Battaile KP, Bateman R et al. (2002) Fanconi anemia group A and C double-mutant mice. Functional evidence for a multi-protein Fanconi anemia complex. Exp Hematol 30:679–688
199. Nookala RK, Hussain S, Pellegrini L (2007) Insights into Fanconi Anaemia from the structure of human FANCE. Nucleic Acids Res 35:1638–1648
200. O'Driscoll M, Ruiz-Perez VL, Woods CG et al. (2003) A splicing mutation affecting expression of ataxia-telangiectasia and Rad3-related protein (ATR) results in Seckel syndrome. Nat Genet 33:497–501
201. Oda T, Hayano T, Miyaso H et al. (2007) Hsp90 regulates the Fanconi anemia DNA damage response pathway. Blood 109:5016–5026
202. Oestergaard VH, Langevin F, Kuiken HJ et al. (2007) Deubiquitination of FANCD2 is required for DNA crosslink repair. Mol Cell 28:798–809
203. Offit K, Levran O, Mullaney B et al. (2003) Shared genetic susceptibility to breast cancer, brain tumors, and Fanconi anemia. J Natl Cancer Inst 95:1548–1551
204. Ohashi A, Zdzienicka MZ, Chen J et al. (2005) Fanconi anemia complementation group D2 (FANCD2) functions independently of BRCA2- and RAD51-associated homologous recombination in response to DNA damage. J Biol Chem 280:14877–14883
205. Olopade OI, Wei M (2003) FANCF methylation contributes to chemoselectivity in ovarian cancer. Cancer Cell 3:417–420
206. Pace P, Johnson M, Tan WM et al. (2002) FANCE: the link between Fanconi anaemia complex assembly and activity. EMBO J 21:3414–3423
207. Pang Q, Fagerlie S, Christianson TA et al. (2000) The Fanconi anemia protein FANCC binds to and facilitates the activation of STAT1 by gamma interferon and hematopoietic growth factors. Mol Cell Biol 20:4724–4735
208. Pang Q, Christianson TA, Keeble W et al. (2002) The anti-apoptotic function of Hsp70 in the PKR-mediated death signaling pathway requires the Fanconi anemia protein, FANCC. J Biol Chem 277:49638–49643
209. Papadopoulo D, Guillouf C, Mohrenweiser H et al. (1990) Hypomutability in Fanconi anemia cells is associated with increased deletion frequency at the HPRT locus. Proc Natl Acad Sci USA 87:8383–8387
210. Park WH, Margossian S, Horwitz AA et al. (2005) Direct DNA binding activity of the Fanconi anemia D2 protein. J Biol Chem 280:23593–23598
211. Pellegrini L, Yu DS, Lo T et al. (2002) Insights into DNA recombination from the structure of a RAD51-BRCA2 complex. Nature 420:287–293
212. Peng M, Litman R, Xie J et al. (2007) The FANCJ/MutLalpha interaction is required for correction of the cross-link response in FA-J cells. EMBO J 26:3238–3249
213. Pichierri P, Averbeck D, Rosselli F (2002) DNA cross-link-dependent RAD50/MRE11/NBS1 subnuclear assembly requires the Fanconi anemia C protein. Hum Mol Genet 11:2531–2546
214. Pichierri P, Franchitto A, Rosselli F (2004) BLM and the FANC proteins collaborate in a common pathway in response to stalled replication forks. Embo J 23:3154–3163
215. Pichierri P, Rosselli F (2004) The DNA crosslink-induced S-phase checkpoint depends on ATR-CHK1 and ATR-NBS1-FANCD2 pathways. Embo J 23:1178–1187

216. Pickart CM, Cohen RE (2004) Proteasomes and their kin: proteases in the machine age. Nat Rev Mol Cell Biol 5:177–187
217. Potapova A, Hoffman AM, Godwin AK et al. (2008) Promoter hypermethylation of the PALB2 susceptibility gene in inherited and sporadic breast and ovarian cancer. Cancer Res 68:998–1002
218. Qiao F, Moss A, Kupfer GM (2001) Fanconi anemia proteins localize to chromatin and the nuclear matrix in a dna damage- and cell cycle-regulated manner. J Biol Chem 276: 23391–23396
219. Qiao F, Mi J, Wilson JB et al. (2004) Phosphorylation of fanconi anemia (FA) complementation group G protein, FANCG, at serine 7 is important for function of the FA pathway. J Biol Chem 279:46035–46045
220. Rahman N, Seal S, Thompson D et al. (2007) PALB2, which encodes a BRCA2-interacting protein, is a breast cancer susceptibility gene. Nat Genet 39:165–167
221. Reid S, Schindler D, Hanenberg H et al. (2007) Biallelic mutations in PALB2 cause Fanconi anemia subtype FA-N and predispose to childhood cancer. Nat Genet 39:162–164
222. Reuter TY, Medhurst AL, Waisfisz Q et al. (2003) Yeast two-hybrid screens imply involvement of Fanconi anemia proteins in transcription regulation, cell signaling, oxidative metabolism, and cellular transport. Exp Cell Res 289:211–221
223. Rio P, Segovia JC, Hanenberg H et al. (2002) In vitro phenotypic correction of hematopoietic progenitors from Fanconi anemia group A knockout mice. Blood 100:2032–2039
224. Rosenberg PS, Greene MH, Alter BP (2003) Cancer incidence in persons with Fanconi anemia. Blood 101:822–826
225. Rosenberg PS, Alter BP, Socie G et al. (2005) Secular trends in outcomes for fanconi anemia patients who receive transplants: implications for future studies. Biol Blood Marrow Transplant 11:672–679
226. Saadatzadeh MR, Bijangi-Vishehsaraei K, Hong P et al. (2004) Oxidant hypersensitivity of Fanconi anemia type C-deficient cells is dependent on a redox-regulated apoptotic pathway. J Biol Chem 279:16805–16812
227. Sakai W, Swisher EM, Karlan BY et al. (2008) Secondary mutations as a mechanism of cisplatin resistance in BRCA2-mutated cancers. Nature 451:1116–1120
228. Sala-Trepat M, Rouillard D, Escarceller M et al. (2000) Arrest of S-phase progression is impaired in fanconi anemia cells. Exp Cell Res 260:208–215
229. Scully R, Xie A, Nagaraju G (2004) Molecular functions of BRCA1 in the DNA damage response. Cancer Biol Ther 3:521–527
230. Seal S, Barfoot R, Jayatilake H et al. (2003) Evaluation of Fanconi Anemia genes in familial breast cancer predisposition. Cancer Res 63:8596–8599
231. Seal S, Thompson D, Renwick A et al. (2006) Truncating mutations in the Fanconi anemia J gene BRIP1 are low-penetrance breast cancer susceptibility alleles. Nat Genet 38: 1239–1241
232. Seki S, Ohzeki M, Uchida A et al. (2007) A requirement of FancL and FancD2 monoubiquitination in DNA repair. Genes Cells 12:299–310
233. Shiloh Y (2003) ATM and related protein kinases: safeguarding genome integrity. Nat Rev Cancer 3:155–168
234. Shimamura A, D'Andrea AD (2003) Subtyping of Fanconi anemia patients: implications for clinical management. Blood 102:3459
235. Sigismund S, Polo S, Di Fiore PP (2004) Signaling through monoubiquitination. Curr Top Microbiol Immunol 286:149–185
236. Sims AE, Spiteri E, Sims RJ, 3rd et al. (2007) FANCI is a second monoubiquitinated member of the Fanconi anemia pathway. Nat Struct Mol Biol 14:564–567
237. Smogorzewska A, Matsuoka S, Vinciguerra P et al. (2007) Identification of the FANCI protein, a monoubiquitinated FANCD2 paralog required for DNA repair. Cell 129: 289–301

238. Sobeck A, Stone S, Costanzo V et al. (2006) Fanconi anemia proteins are required to prevent accumulation of replication-associated DNA double-strand breaks. Mol Cell Biol 26: 425–437
239. Sobeck A, Stone S, Hoatlin ME (2007) DNA structure-induced recruitment and activation of the Fanconi anemia pathway protein FANCD2. Mol Cell Biol 27: 4283–4292
240. Soule BP, Hyodo F, Matsumoto K et al. (2007) The chemistry and biology of nitroxide compounds. Free Radic Biol Med 42:1632–1650
241. Soulier J, Leblanc T, Larghero J et al. (2005) Detection of somatic mosaicism and classification of Fanconi anemia patients by analysis of the FA/BRCA pathway. Blood 105:1329–1336
242. Spardy N, Duensing A, Charles D et al. (2007) The human papillomavirus type 16 E7 oncoprotein activates the Fanconi anemia (FA) pathway and causes accelerated chromosomal instability in FA cells. J Virol 81:13265–13270
243. Sridharan D, Brown M, Lambert WC et al. (2003) Nonerythroid alphaII spectrin is required for recruitment of FANCA and XPF to nuclear foci induced by DNA interstrand cross-links. J Cell Sci 116:823–835
244. Stark JM, Pierce AJ, Oh J et al. (2004) Genetic steps of mammalian homologous repair with distinct mutagenic consequences. Mol Cell Biol 24:9305–9316
245. Stone S, Sobeck A, van Kogelenberg M et al. (2007) Identification, developmental expression and regulation of the Xenopus ortholog of human FANCG/XRCC9. Genes Cells 12:841–851
246. Stracker TH, Theunissen JW, Morales M et al. (2004) The Mre11 complex and the metabolism of chromosome breaks: the importance of communicating and holding things together. DNA Repair (Amst) 3:845–854
247. Strathdee CA, Gavish H, Shannon WR et al. (1992) Cloning of cDNAs for Fanconi's anaemia by functional complementation. Nature 356:763–767
248. Sun Y, Jiang X, Chen S et al. (2005) A role for the Tip60 histone acetyltransferase in the acetylation and activation of ATM. Proc Natl Acad Sci USA 102:13182–13187
249. Sun Y, Xu Y, Roy K et al. (2007) DNA damage-induced acetylation of lysine 3016 of ATM activates ATM kinase activity. Mol Cell Biol 27:8502–8509
250. Swisher EM, Sakai W, Karlan BY et al. (2008) Secondary BRCA1 mutations in BRCA1-mutated ovarian carcinomas with platinum resistance. Cancer Res 68:2581–2586
251. Syljuasen RG, Sorensen CS, Hansen LT et al. (2005) Inhibition of human Chk1 causes increased initiation of DNA replication, phosphorylation of ATR targets, and DNA breakage. Mol Cell Biol 25:3553–3562
252. Tang Y, Luo J, Zhang W et al. (2006) Tip60-dependent acetylation of p53 modulates the decision between cell-cycle arrest and apoptosis. Mol Cell 24:827–839
253. Taniguchi T, D'Andrea AD (2002) The Fanconi anemia protein, FANCE, promotes the nuclear accumulation of FANCC. Blood 100:2457–2462
254. Taniguchi T, Garcia-Higuera I, Andreassen PR et al. (2002) S-phase-specific interaction of the Fanconi anemia protein, FANCD2, with BRCA1 and RAD51. Blood 100: 2414–2420
255. Taniguchi T, Garcia-Higuera I, Xu B et al. (2002) Convergence of the Fanconi anemia and ataxia telangiectasia signaling pathways. Cell 109:459–472
256. Taniguchi T, Tischkowitz M, Ameziane N et al. (2003) Disruption of the Fanconi anemia-BRCA pathway in cisplatin-sensitive ovarian tumors. Nat Med 9:568–574
257. Taniguchi T, D'Andrea AD (2006) Molecular pathogenesis of Fanconi anemia: recent progress. Blood 107:4223–4233
258. Teodoridis JM, Hall J, Marsh S et al. (2005) CpG island methylation of DNA damage response genes in advanced ovarian cancer. Cancer Res 65:8961–8967
259. The-Fanconi-anaemia/breast-cancer-consortium (1996) Positional cloning of the Fanconi anaemia group A gene. Nat Genet 14:324–328

260. Thomashevski A, High AA, Drozd M et al. (2004) The Fanconi anemia core complex forms four complexes of different sizes in different subcellular compartments. J Biol Chem 279:26201–26209
261. Thompson LH, Hinz JM, Yamada NA et al. (2005) How Fanconi anemia proteins promote the four Rs: replication, recombination, repair, and recovery. Environ Mol Mutagen 45: 128–142
262. Timmers C, Taniguchi T, Hejna J et al. (2001) Positional cloning of a novel Fanconi anemia gene, FANCD2. Mol Cell 7:241–248
263. Tischkowitz M, Ameziane N, Waisfisz Q et al. (2003) Bi-allelic silencing of the Fanconi anaemia gene FANCF in acute myeloid leukaemia. Br J Haematol 123:469–471
264. Tischkowitz M, Dokal I (2004) Fanconi anaemia and leukaemia – clinical and molecular aspects. Br J Haematol 126:176–191
265. Tischkowitz M, Xia B, Sabbaghian N et al. (2007) Analysis of PALB2/FANCN-associated breast cancer families. Proc Natl Acad Sci USA 104:6788–6793
266. Tischkowitz MD, Morgan NV, Grimwade D et al. (2004) Deletion and reduced expression of the Fanconi anemia FANCA gene in sporadic acute myeloid leukemia. Leukemia 18: 420–425
267. Titus TA, Selvig DR, Qin B et al. (2006) The Fanconi anemia gene network is conserved from zebrafish to human. Gene 371:211–223
268. Tremblay CS, Huang FF, Habi O et al. (2008) HES1 is a novel interactor of the Fanconi anemia core complex. Blood, advanced online publication
269. Tremblay S, Pintor Dos Reis P, Bradley G et al. (2006) Young patients with oral squamous cell carcinoma: study of the involvement of GSTP1 and deregulation of the Fanconi anemia genes. Arch Otolaryngol Head Neck Surg 132:958–966
270. Turner N, Tutt A, Ashworth A (2004) Hallmarks of 'BRCAness' in sporadic cancers. Nat Rev Cancer 4:814–819
271. Tutt AN, Lord CJ, McCabe N et al. (2005) Exploiting the DNA repair defect in BRCA mutant cells in the design of new therapeutic strategies for cancer. Cold Spring Harb Symp Quant Biol 70:139–148
272. van der Groep P, Hoelzel M, Buerger H et al. (2008) Loss of expression of FANCD2 protein in sporadic and hereditary breast cancer. Breast Cancer Res Treat 107:41–47
273. van Der Heijden MS, Yeo CJ, Hruban RH et al. (2003) Fanconi anemia gene mutations in young-onset pancreatic cancer. Cancer Res 63:2585–2588
274. van der Heijden MS, Brody JR, Gallmeier E et al. (2004) Functional defects in the Fanconi anemia pathway in pancreatic cancer cells. Am J Pathol 165:651–657
275. van Der Heijden MS, Brody JR, Kern SE (2004) Functional screen of the Fanconi anemia pathway in cancer cells by fancd2 immunoblot. Cancer Biol Ther 3:534–537
276. van der Heijden MS, Brody JR, Dezentje DA et al. (2005) In vivo therapeutic responses contingent on Fanconi anemia/BRCA2 status of the tumor. Clin Cancer Res 11: 7508–7515
277. van Zeeburg HJ, Snijders PJ, Joenje H et al. (2004) Re: human papillomavirus DNA and p53 polymorphisms in squamous cell carcinomas from Fanconi anemia patients. J Natl Cancer Inst 96:968
278. Vandenberg CJ, Gergely F, Ong CY et al. (2003) BRCA1-independent ubiquitination of FANCD2. Mol Cell 12:247–254
279. Venkitaraman AR (2004) Tracing the network connecting BRCA and Fanconi anaemia proteins. Nat Rev Cancer 4:266–276
280. Wagner JE, Tolar J, Levran O et al. (2004) Germline mutations in BRCA2: shared genetic susceptibility to breast cancer, early onset leukemia, and Fanconi anemia. Blood 103: 3226–3229
281. Waisfisz Q, Morgan NV, Savino M et al. (1999) Spontaneous functional correction of homozygous fanconi anaemia alleles reveals novel mechanistic basis for reverse mosaicism. Nat Genet 22:379–383

282. Wang W (2007) Emergence of a DNA-damage response network consisting of Fanconi anaemia and BRCA proteins. Nat Rev Genet 8:735–748
283. Wang X, Andreassen PR, D'Andrea AD (2004) Functional interaction of monoubiquitinated FANCD2 and BRCA2/FANCD1 in chromatin. Mol Cell Biol 24:5850–5862
284. Wang X, Kennedy RD, Ray K et al. (2007) Chk1-mediated phosphorylation of FANCE is required for the Fanconi anemia/BRCA pathway. Mol Cell Biol 27:3098–3108
285. Wang Z, Li M, Lu S et al. (2006) Promoter hypermethylation of FANCF plays an important role in the occurrence of ovarian cancer through disrupting Fanconi anemia-BRCA pathway. Cancer Biol Ther 5:256–260
286. Wei M, Xu J, Dignam J et al. (2008) Estrogen receptor alpha, BRCA1, and FANCF promoter methylation occur in distinct subsets of sporadic breast cancers. Breast Cancer Res Treat 111:113–120
287. Whitney MA, Royle G, Low MJ et al. (1996) Germ cell defects and hematopoietic hypersensitivity to gamma-interferon in mice with a targeted disruption of the Fanconi anemia C gene. Blood 88:49–58
288. Wiegant WW, Overmeer RM, Godthelp BC et al. (2006) Chinese hamster cell mutant, V-C8, a model for analysis of Brca2 function. Mutat Res 600:79–88
289. Wilson JB, Yamamoto K, Marriott AS et al. (2008) FANCG promotes formation of a newly identified protein complex containing BRCA2, FANCD2 and XRCC3. Oncogene 27: 3641–3652
290. Wilson JH, Elledge SJ (2002) Cancer. BRCA2 enters the fray. Science 297:1822–1823
291. Wong JC, Alon N, McKerlie C et al. (2003) Targeted disruption of exons 1 to 6 of the Fanconi Anemia group A gene leads to growth retardation, strain-specific microphthalmia, meiotic defects and primordial germ cell hypoplasia. Hum Mol Genet 12:2063–2076
292. Wood A, Garg P, Burgers PM (2007) A ubiquitin-binding motif in the translesion DNA polymerase Rev1 mediates its essential functional interaction with ubiquitinated proliferating cell nuclear antigen in response to DNA damage. J Biol Chem 282:20256–20263
293. Wooster R, Weber BL (2003) Breast and ovarian cancer. N Engl J Med 348:2339–2347
294. Wreesmann VB, Estilo C, Eisele DW et al. (2007) Downregulation of fanconi anemia genes in sporadic head and neck squamous cell carcinoma. ORL J Otorhinolaryngol Relat Spec 69:218–225
295. Wu Y, Shin-ya K, Brosh RM, Jr. (2008) FANCJ helicase defective in Fanconia anemia and breast cancer unwinds G-quadruplex DNA to defend genomic stability. Mol Cell Biol 28:4116–4128
296. Xia B, Sheng Q, Nakanishi K et al. (2006) Control of BRCA2 cellular and clinical functions by a nuclear partner, PALB2. Mol Cell 22:719–729
297. Xia B, Dorsman JC, Ameziane N et al. (2007) Fanconi anemia is associated with a defect in the BRCA2 partner PALB2. Nat Genet 39:159–161
298. Xue Y, Li Y, Guo R et al. (2008) FANCM of the Fanconi anemia core complex is required for both monoubiquitination and DNA repair. Hum Mol Genet 17:1641–1652
299. Yamada K, Olsen JC, Patel M et al. (2001) Functional correction of Fanconi anemia group C hematopoietic cells by the use of a novel lentiviral vector. Mol Ther 3:485–490
300. Yamada K, Ramezani A, Hawley RG et al. (2003) Phenotype correction of Fanconi anemia group a hematopoietic stem cells using lentiviral vector. Mol Ther 8:600–610
301. Yamamoto K, Ishiai M, Matsushita N et al. (2003) Fanconi anemia FANCG protein in mitigating radiation- and enzyme-induced DNA double-strand breaks by homologous recombination in vertebrate cells. Mol Cell Biol 23:5421–5430
302. Yamamoto K, Hirano S, Ishiai M et al. (2005) Fanconi anemia protein FANCD2 promotes immunoglobulin gene conversion and DNA repair through a mechanism related to homologous recombination. Mol Cell Biol 25:34–43
303. Yamashita T, Barber DL, Zhu Y et al. (1994) The Fanconi anemia polypeptide FACC is localized to the cytoplasm. Proc Natl Acad Sci USA 91:6712–6716

304. Yang H, Jeffrey PD, Miller J et al. (2002) BRCA2 function in DNA binding and recombination from a BRCA2-DSS1-ssDNA structure. Science 297:1837–1848
305. Yang Y, Kuang Y, De Oca RM et al. (2001) Targeted disruption of the murine Fanconi anemia gene, Fancg/Xrcc9. Blood 98:3435–3440
306. Yang YG, Herceg Z, Nakanishi K et al. (2005) The Fanconi anemia group A protein modulates homologous repair of DNA double-strand breaks in mammalian cells. Carcinogenesis 26:1731–1740
307. Yin J, Sobeck A, Xu C et al. (2005) BLAP75, an essential component of Bloom's syndrome protein complexes that maintain genome integrity. EMBO J 24:1465–1476
308. Youds JL, Barber LJ, Ward JD et al. (2008) DOG-1 is the *Caenorhabditis elegans* BRIP1/FANCJ homologue and functions in interstrand cross-link repair. Mol Cell Biol 28:1470–1479
309. Yu X, Chini CC, He M et al. (2003) The BRCT domain is a phospho-protein binding domain. Science 302:639–642
310. Yuan SS, Lee SY, Chen G et al. (1999) BRCA2 is required for ionizing radiation-induced assembly of Rad51 complex in vivo. Cancer Res 59:3547–3551
311. Zhang J, Wang X, Lin CJ et al. (2006) Altered expression of FANCL confers mitomycin C sensitivity in Calu-6 lung cancer cells. Cancer Biol Ther 5:1632–1636
312. Zhang J, Zhao D, Wang H et al. (2008) FANCD2 monoubiquitination provides a link between the HHR6 and FA-BRCA pathways. Cell Cycle 7:407–413
313. Zhang QS, Eaton L, Snyder ER et al. (2008) Tempol protects against oxidative damage and delays epithelial tumor onset in Fanconi anemia mice. Cancer Res 68:1601–1608
314. Zhang X, Li J, Sejas DP et al. (2004) The Fanconi anemia proteins functionally interact with the protein kinase regulated by RNA (PKR). J Biol Chem 279:43910–43919
315. Zou L, Elledge SJ (2003) Sensing DNA damage through ATRIP recognition of RPA-ssDNA complexes. Science 300:1542–1548
316. Zou L, Liu D, Elledge SJ (2003) Replication protein A-mediated recruitment and activation of Rad17 complexes. Proc Natl Acad Sci USA 100:13827–13832

Chapter 18
BRCA1 and BRCA2: Role in the DNA Damage Response, Cancer Formation and Treatment

Kienan Savage and D. Paul Harkin

Abstract BRCA1 and BRCA2 are highly penetrant breast and ovarian cancer susceptibility genes that are mutated in a significant proportion of familial breast and ovarian cancer syndromes. Both of these genes are tumour suppressors, the products of which play vital roles in the cellular response to DNA damage. These proteins function in a number of cellular pathways in order to maintain genomic stability including DNA damage signaling, DNA repair, cell cycle regulation, protein ubiquitination, chromatin remodeling, transcriptional regulation and apoptosis. This chapter will discuss the functions of these proteins and how they relate to tumour development, and therapy.

18.1 Introduction

Approximately 10% of women diagnosed with breast cancer report a strong family history, prompting the search for breast cancer susceptibility genes. Using linkage analysis, the first breast cancer susceptibility gene BRCA1, was mapped to chromosome 17q21 [1]. Miki et al. subsequently cloned the gene in 1994 and identified a number of mutations in familial breast and ovarian cancer cases [2, 3]. To date over eight hundred distinct and clinically relevant mutations have been identified within the BRCA1 gene (BIC database). However, breast cancer families with a high incidence of male breast cancer did not have mutations in the BRCA1 gene, suggesting the existence of another susceptibility gene. Indeed in 1994 Wooster et al. linked the BRCA2 gene to chromosome 13 and cloned it shortly after [4]. Together mutations within BRCA1 and BRCA2 may account for up to 85% of inherited breast cancer cases [5]. Moreover, carriers of mutations within the BRCA1 or BRCA2 genes have up to an 80% life time risk of developing breast cancer – lifetime risk is generally defined as up to 70 years of age [6]. In addition, BRCA1 carriers have up to a 40% lifetime risk of ovarian cancer, whilst BRCA2 mutations confer approximately

K. Savage (✉)
Centre for Cancer Research and Cell Biology, Queen's University Belfast, Belfast, BT9 7BL, UK
e-mail: k.savage@qub.ac.uk

K.K. Khanna, Y. Shiloh (eds.), *The DNA Damage Response: Implications on Cancer Formation and Treatment*, DOI 10.1007/978-90-481-2561-6_18,

20% lifetime risk of ovarian cancer [6]. Although BRCA1/2 mutations account for a relatively low proportion of breast cancer cases as a whole, it has been found that between 50 and 70% of sporadic breast tumours have lost at least one BRCA1 allele, suggesting that loss of BRCA1 may also be important for sporadic breast cancer development [7]. In addition, hypermethylation of BRCA1 promoters and loss of BRCA1 transcript has also been found in a significant proportion of sporadic breast and ovarian tumours [8]. Moreover, BRCA1 and BRCA2 mutation carriers tend to develop early onset breast and ovarian cancers (<50 years of age), thus the years of life lost due to BRCA1/2 related cancers is disproportionately high in comparison to sporadic cancers [9].

Since their identification much attention has been focused on understanding the functions of BRCA1 and BRCA2. Both BRCA1 and BRCA2 play a role in maintaining genomic stability, namely cells deficient in either of these proteins display gross genetic instability.

BRCA2 plays a pivotal role in homologous recombination mediated repair of DNA double strand breaks, however little else is know about its function. Many functions have been identified for BRCA1 including DNA damage signaling, DNA repair, cell cycle regulation, protein ubiquitination, chromatin remodeling, transcriptional regulation and apoptosis. The functions of BRCA1 and BRCA2 will now be discussed in further detail.

18.2 BRCA1 Structure and Function

BRCA1 is located on chromosome 17q21.3, spans approximately 80 kb of genomic DNA, and is composed of 24 coding exons [3]. The BRCA1 gene encodes an 1863 amino acid protein, shares limited homology with other known proteins but does contain two functional domains. At its N-terminus BRCA1 contains a series of eight conserved Cys3-His-Cys4 motif repeats known as the RING domain, a catalytic domain involved in protein ubiquitination and protein:protein interactions [10]. BRCA1's RING domain interacts with a number of proteins (Fig. 18.1), in particular it binds with the *B*RCA1 *A*ssociated *R*ING *D*omain protein 1 (BARD1) forming the BRCA1/BARD1 heterodimer a known E3 ubiquitin ligase complex (this will be discussed in further detail later in this chapter) [11, 12]. Interestingly, cancer predisposing mutations have been found at five of the conserved cystine residues of the RING domain, suggesting that functional inactivation of the RING domain may in part contribute to cancer predisposition [13]. It was later discovered that interaction with BARD1 is also required for BRCA1 protein stability, leading to the finding that BARD1 mutations also predispose carriers to breast and ovarian cancer [14].

The C-terminus of BRCA1 contains two conserved *BR*CA1 *C-t*erminal (BRCT) domains. These domains contain distinct hydrophobic/acidic patches and have been shown to have transactivation activity [15–17]. Since their discovery within the

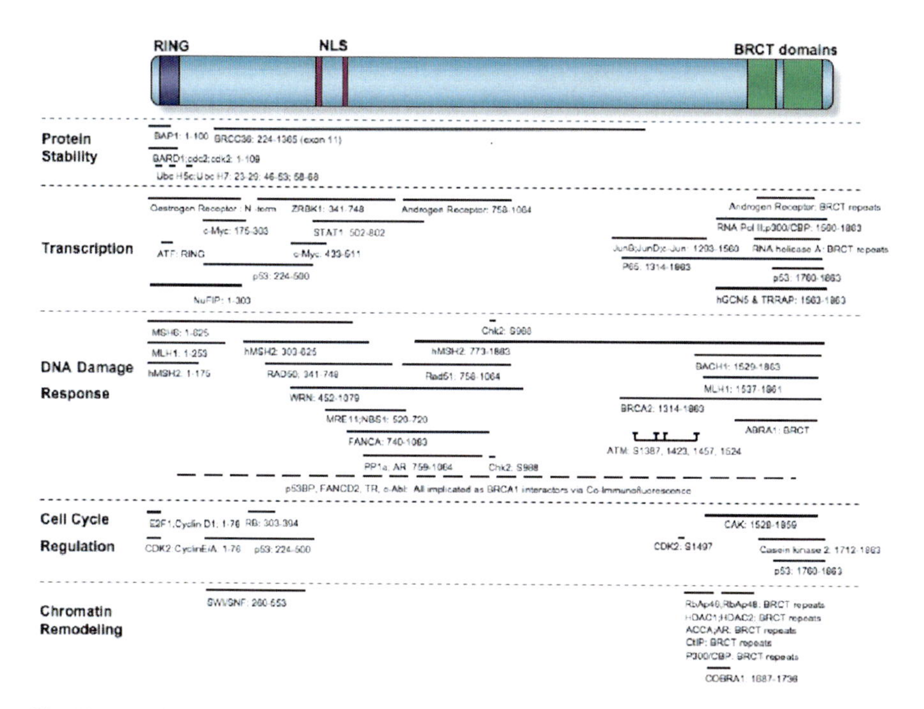

Fig. 18.1 BRCA1 Interacting proteins and their functions. Schematic diagram of BRCA1 including the RING, nuclear localisation signals (NLS), and the BRCT domains. Interacting proteins are shown underneath their known BRCA1 interacting region and are grouped according to their known/suspected function

BRCA1 protein, BRCT domains have been found in many proteins involved in the DNA damage response and have more recently been shown as phospho-specific protein interaction motifs. BRCT domains are found in a number of proteins involved in the DNA damage response pathway such as MDC1, 53BP1 and XRCC1 [18]. Not surprisingly a number of cancer predisposing mutations have also been found within the BRCT domain which affect both the transactivation activity of BRCA1 and its phospho-specific interaction with other proteins [15, 19].

BRCA1 also contains two nuclear localisation signals at amino acids 503–509 and 606–615 respectively, which target BRCA1 to the nucleus [20]. In addition, a nuclear export signal at amino acids 81–99, is involved in nuclear cytoplasmic shuttling [21].

Given that BRCA1 contains only one catalytic domain it has long been thought that it may carry out its diverse range of functions via its capacity to interact with a wide and varied array of proteins. Indeed, more than fifty BRCA1 interacting proteins have been identified with functions in a variety of pathways including cell cycle regulation, DNA damage signaling, DNA repair, ubiquitination and transcriptional regulation (Fig. 18.1).

18.2.1 BRCA1 and DNA Repair

BRCA1 was first implicated in the DNA damage response pathway due to its co-localisation in nuclear foci with the Rad51 protein, a protein essential for DNA double strand break repair [22]. In addition, Scully et al. showed that BRCA1 and RAD51 do not co-localise with the *p*roliferating *c*ell *n*uclear *a*ntigen protein (PCNA) localized at active replication forks, demonstrating that in unperturbed cells BRCA1 was not localised to active replication forks. This study went on to show that in response to treatment with DNA damaging agents [hydroxyurea (HU), mitomycin-C (MMC), ultraviolet (UV), or ionizing radiation (IR)] BRCA1 and Rad51 foci re-localise to PCNA containing foci, indicating BRCA1's involvement in post-replicative DNA repair. In addition, numerous studies have shown that BRCA1 deficient cells are hypersensitive to DNA double strand breaking agents such as ionizing radiation and DNA alkylating agents and exhibit impaired DNA double strand break repair kinetics, further suggesting an important role for BRCA1 in DNA repair [23].

There are three major cellular mechanisms for repair of DNA double strand breaks (DSBs), single strand annealing (SSA), non-homologous end joining (NHEJ) and homologous recombination (HR).

NHEJ, the most common form of repair in the human cell, occurs during G1 and M phases of the cell cycle. It is a relatively error prone repair mechanism involving resection and processing of broken DNA ends, after which the ends are directly ligated together. This often results in the removal or addition of bases at broken ends in order to produce ligatable DNA ends. In addition, in the case of multiple or complex breaks, non-contiguous DNA sequences are often ligated together forming non-homologous chromosomes and/or acentric and dicentric chromosomes which cause further problems during mitosis. Alternatively, SSA involves annealing of short sequences of homology which flank break sites, that are revealed after DSB resection. Due to the recognition of homologous sequences this process is less likely to lead to ligation of non-contiguous DNA sequences. However, bases lying between the homologous sequences are often deleted, and thus SSA is also relatively error prone. In contrast, HR occurs only during S- and G2 phases of the cell cycle when homologous sister chromosomes are present and in close proximity. HR is activated by DNA damage and involves the *A*taxia *t*elangiectasia *m*utated (ATM) kinase and *R*ad50:mre11:nbs1 (MRN) mediated processing and resection of broken DNA ends to form free 3` single stranded DNA (ssDNA) ends. These free ssDNA regions are then bound by the single stranded binding protein RPA, forming a substrate for loading and binding of the Rad51 recombinase. Rad51 then catalyses invasion of the homologous sister chromatid forming a holiday junction between the two chromosomes (Fig. 18.4). The intact sister chromatid then acts as a primer for DNA synthesis allowing homology directed repair of breaks. The resulting holiday structure is then resolved and the DNA ends ligated together to form two intact sister chromosomes. The finding that BRCA1 and RAD51 co-localise at sites of DNA repair suggested a direct role for BRCA1 in homologous recombination mediated DNA repair. Indeed BRCA1 deficient cells display a defect in homologous recombination [24]. In addition, Rad51 repair foci do not form in BRCA1 deficient cells suggesting a role for BRCA1 in regulation of Rad51 function [25]. Interestingly, as well as mediating broken DNA end processing in response to DNA damage, the ATM kinase activates the checkpoint kinase Chk2, which in turn phosphorylates BRCA1 on Ser-988. This phosphorylation event is required for BRCA1 mediated Rad51 foci formation and efficient homologous recombination, suggesting that BRCA1 may function as a DNA damage activated scaffolding protein facilitating Rad51 recruitment to broken DNA ends [26].

BRCA1 has also been implicated in the NHEJ mediated DSB repair pathway, however this remains contentious with numerous studies, presenting conflicting findings using both in-vitro and in-vivo assays. Another study has implicated BRCA1 in the SSA pathway, demonstrating a 50–100 fold reduction in micro-homology mediated end-joining in mouse embryonic fibroblasts depleted of BRCA1, however the role played by BRCA1 in this repair pathway remains to be defined [27]. In addition BRCA1 has been implicated in the nucleotide excision repair (NER) pathway due to its interaction with the NER proteins Msh2 and Msh6 [28]. This pathway is utilized to repair DNA base modifications/adducts such as those caused by UV irradiation and DNA alkylating agents. The process involves the excision of a 24–32 base fragment of DNA surrounding the adducted base followed by re-synthesis and ligation using the complementary strand as a template [29]. Nucleotide excision repair is also utilized to repair actively transcribed DNA templates, known as transcription coupled repair (TCR), a process in which both Msh2 and Msh6 are involved [29]. In addition, BRCA1 is required for transcription coupled repair of oxidative-8-oxoguanine lesions [30]. A mechanism for BRCA1 in this pathway has been suggested involving the BRCA1/BARD1 ubiquitin ligase mediated degradation of RNA Pol II, thereby allowing access for the repair machinery (this mechanism is discussed in more detail later in this chapter) [31]. Another mechanism through which BRCA1 may contribute to NER is via the transactivation of the XPC (xeroderma pigmentosum C) and DDB2 (damaged DNA binding) proteins both of which are involved in NER (BRCA1s involvement in transcriptional activation is discussed later in this chapter) [32].

18.2.2 BRCA1, DNA Damage Signaling and Cell Cycle Arrest

The DNA damage response is a complex network of signal transduction pathways initiated by the upstream kinases Ataxia telangiectasia mutated (ATM), ATM and rad3 related kinase (ATR) and the DNA dependent protein kinase (DNA-PK). These kinases are activated by different kinds of genotoxic stress, after which they phosphorylate a number of downstream target proteins which propagate the damage response signal resulting in the activation of a number of cellular processes including cell cycle arrest, DNA repair, transcriptional regulation, translation regulation, and apoptosis.

Initial studies on BRCA1 function revealed that BRCA1 expression was regulated throughout the cell cycle with levels of the protein highest during S and G2/M phases [33]. These studies also demonstrated a phosphorylation dependent shift in BRCA1 mobility during the S and G2/M phases of the cell cycle. Subsequent studies then revealed that BRCA1 phosphorylation was not only regulated by the cell cycle but was induced by DNA damaging agents such as IR, UV, HU, MMS and hydrogen peroxide [34]. Various research groups then demonstrated that BRCA1 phosphorylation was mediated by the DNA damage activated kinases ATM and ATR [35]. Traditionally, ATM and ATR are activated in response to different genetic lesions with ATM activated by DNA double strand breaks whereas ATR is activated by single stranded lesions and stalled replication forks such as those caused by chemotherapeutic DNA cross-linking agents and platinum compounds [35, 36]. Activated ATM and ATR then phosphorylate a number of downstream mediator and effector proteins, such as Chk1, Chk2 and MDC1, which propagate the DNA damage signal and initiate multiple DNA damage response functions from cell cycle arrest and DNA repair to apoptosis. Interestingly, Gatei et al. found that BRCA1 was phosphorylated at distinct and overlapping sites by ATM and ATR in response to different DNA damaging agents [36]. In particular, serines-1423 and -1524 were phosphorylated by either ATM or ATR in response to IR or UV, whereas serine-1387 was phosphorylated by ATM in response to IR but not by ATR after UV. Conversely, serine-1457 is phosphorylated predominantly by ATR in response to UV but not by ATM after IR. This finding suggested that BRCA1 may play a role in regulating or fine tuning the cellular response to different types of genetic lesions (Fig. 18.2).

Zhong et al. also discovered that BRCA1 associates with the Mre11/Rad50/Nbs1 (MRN) complex, a complex known to participate in both HR, NHEJ and cell cycle arrest in response to DNA damage [37]. In addition, Wang et al. published a seminal paper demonstrating that BRCA1 forms a multiprotein super complex with the MSH2, MSH6, MLH1, and BLM DNA repair proteins, the damage associated replication factor C (RFC), ATM and the MRN complex [28]. This complex was found to be involved in the recognition and repair of abnormal DNA structures and so was named the BRCA1 *a*ssociated genome *s*urveillance *c*omplex or BASC. These studies not only advanced the understanding of BRCA1's role in DNA repair but also implicated BRCA1 in cell cycle regulation in response to DNA damage. Indeed, the BRCA1 deficient cell line HCC1937 lacks functional G1/S, intra-S and G2/M cell cycle checkpoints [23, 38, 39]. In addition, ectopic expression of functional BRCA1 in this cell line restores both Intra-S and G2/M checkpoint function [38]. Although restoration of BRCA1 function in this cell line did not rescue the G1/S checkpoint, it is most likely due to a homozygous mutation within p53 a known regulator of the G1/S checkpoint [40]. Fabbro and co-workers clarified the function of BRCA1 in the G1/S checkpoint using siRNA depletion of BRCA1 in a p53 wild-type background [41]. This study demonstrated that BRCA1 acts as a scaffolding protein facilitating the ATM dependent phosphorylation of p53 in response to DNA damage leading to p53 mediated induction of the cyclin dependent kinase inhibitor p21, thereby causing G1/S arrest. Interestingly, this study also showed that the prior phosphorylation of BRCA1 on serines 1423 and -1524 by ATM are required

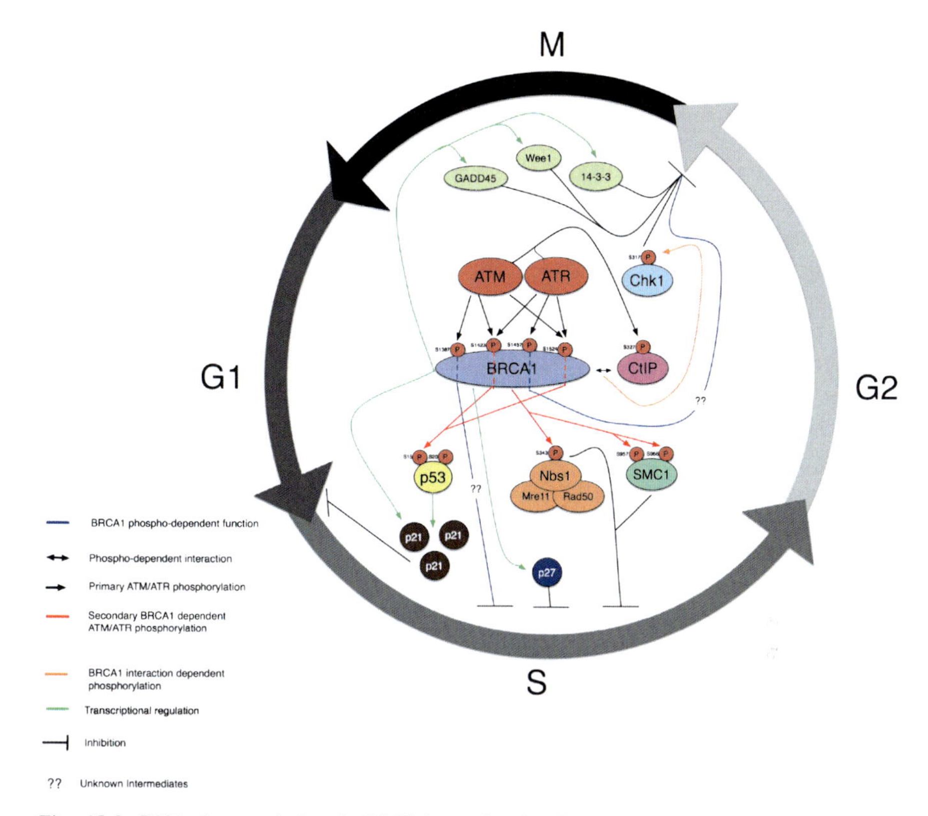

Fig. 18.2 DNA damage induced, BRCA1 regulated cell cycle arrest. Diagram representing BRCA1 regulated cell cycle regulation pathways in response to DNA damage

for this scaffolding function (Fig. 18.2). In contrast, BRCA1 is not required for a UV induced G1/S checkpoint arrest suggesting an alterative BRCA1 independent pathway for UV induced G1/S arrest, most likely through the ATR/Chk1 pathway. Similarly, BRCA1 is required for the DNA damage induced phosphorylation of another BRCA1 interacting protein, CtIP, an interaction required for phosphorylation and activation of the checkpoint kinase Chk1, thereby activating the G2/M checkpoint (Fig. 18.2) [42–44]. Xu et al. also demonstrated that ATM mediated phosphorylation of BRCA1 on serines-1387 and -1423 are required for IR induced intra-S and G2/M checkpoint arrest respectively, however the mechanistic action of these phosphorylation events in checkpoint activation remains unclear [38, 45]. It is probable that like the p53 induced G1/S checkpoint, the ATM/ATR dependent phosphorylation of BRCA1 allows BRCA1 to act as a scaffold/adapter protein allowing the downstream phosphorylation of further ATM/ATR target proteins which facilitate checkpoint arrest. In support of this multiple studies have demonstrated that BRCA1 facilitates the ATM/ATR dependent phosphorylation of a number of targets such as Nbs1 and Smc1, the phosphorylation of which are required for intra-S checkpoint arrest (Fig. 18.2) [42, 46].

In concert with DNA damage signal transduction BRCA1 regulates cell cycle arrest via transcriptional regulation of a number of cell cycle regulation genes. For example, BRCA1 directly transactivates the cyclin dependent kinase inhibitor p21 [47]. BRCA1 has also been shown to induce the S-phase inhibitor p27 and the G2/M inhibitors GADD45, Wee1, and 14-3-3 [44, 48]. In addition other studies have found that BRCA1 transcriptionally represses cyclin B, and PLK1 both of which are required for G2/M transition [49].

It is clear that BRCA1 plays an adapter role in the DNA damage signal transduction pathway, fine tuning the response to different genotoxic stimuli. This has become more evident from immunoaffinity chromatography studies which have identified at least five distinct DNA damage induced BRCA1 super complexes. Consistent with a role in the transcriptional regulation of cell cycle associated genes, BRCA1 has been identified as part of the RNA Pol II holoenzyme both in unperturbed cells and cells treated with DNA damaging agents. Another complex involved in G2/M checkpoint activation, contained BRCA1 the MRN complex and the transcriptional co-repressor CtIP [50]. A further complex, most likely the previously identified BASC complex, contained BRCA1/BRCA2 and Rad51 [50]. An additional complex containing BRCA1, BACH1 (BRCA1 associated helicase) and TopBP1 was identified to be involved in S-Phase checkpoint activation via inhibition of replication origin firing and DNA repair [50]. A similar study has identified yet another BRCA1 containing complex induced by treatment with the replication blocking drug, Hydroxyurea (HU). This complex was named the HUIC for *HU i*nduced *c*omplex and has been suggested to repress mRNA processing in HU treated cells. However, the exact constituents of this complex remain unclear [51].

Recently, a number of studies have identified another distinct BRCA1 containing complex. These studies identified BRCA1 and it's binding partner BARD1 in complex with a novel phosphoprotein named Abraxas (Abra1) and the Ubiquitin Interaction Motif (UIM) containing protein RAP80, along with the known BRCA1/BARD1 interacting proteins BRCC36 and BRCC45 [52]. These studies went on to show that RAP80, through it's UIM binds to Lys-63 linked polyubiquitin chains at sites of DNA damage/repair thereby mediating recruitment of BRCA1 to DNA break sites [52]. In addition, these studies showed that formation of the BRCA1/Abra1/Rap80 complex is required for efficient G2/M cell cycle arrest, resistance to DNA damage and efficient DNA repair in response to ionizing radiation [52]. Additional studies then demonstrated that phosphorylated MDC1, a DNA damage signal transducer protein, localized at break sites in response to DNA damage, interacts with and recruits the E3 ubiquitin ligase RNF8. RNF8 in turn recruits the E2 ubiquitin enzyme UBC13 which in conjunction with RNF8 polyubiquitinates histones H3 and H2AX at break sites in response to DNA damage. RNF8/UBC13 mediated ubiquitination conjugation at break sites is required for BRCA1/BARD1/Abra1/RAP80/BRCC36/BRCC45 foci formation, most likely due to binding and recruitment of RAP80 via its UIM to the ubiquitin conjugates [53].

BRCA1 also interacts with a number of the Fanconi Anemia proteins. These proteins were first identified as products of genes mutated in various complementation groups of the rare disease Fanconi Anemia (FA). FA is characterized by progressive bone marrow failure, congenital abnormalities and severe susceptibility to cancers [54]. To date twelve FA genes (termed the FANC genes) have been characterized. Eight of the FANC proteins (A, B, C, E, F, G, L, M) form a complex known as the core ubiquitin ligase complex [55]. FANCD2 and FANCD1 then form an additional complex which functions downstream of the core ubiquitin ligase complex [56]. Interestingly, recent work on identifying the gene responsible for FA complementation group D1 has identified it as the BRCA2 gene [57]. The exact role of the remaining FANC proteins (I, J) is unclear but it is postulated that they play accessory roles in one of the two main FANC complexes [54]. Like BRCA1 deficient cells, deficiency of any one of the FANC proteins is characterized by genomic instability and sensitivity to DNA cross-linking agents such as mitomycin C (MMC) and Cisplatin [54]. During S-phase of the cell cycle and in response to DNA cross-linker mediated damage the core FANC complex monoubiquitinates FANCD2. Activation of the core complex requires functional ATR and Chk1 kinases which are activated by replication fork stalling, caused by the advancing replication machinery encountering DNA damage such as cross-links, suggesting that the FANC pathway responds to stalled replication forks [58, 59]. Monoubiquitinated FANCD2 (Ub-FANCD2) then forms a complex with BRCA1, BRCA2 and Rad51 and localises to repair foci at sites of DNA damage [60, 61]. Repair of DNA crosslinks at stalled replication forks, involves excision of damaged bases leading to the generation of DNA double strand break intermediates, which can then be repaired by homologous recombination [59]. Localization of Ub-FANCD2 at repair foci with BRCA1, BRCA2 and Rad51 suggests that it may play a role in HR mediated repair of DSB intermediates. Indeed, FANCD2 is required for BRCA2 localisation to repair foci, an essential requirement for HR mediated DSB repair (The role of BRCA2 in HR will be discussed later in this chapter). Concurrently, BRCA1 is required for Ub-FANCD2 localisation to repair foci, suggesting a pivotal role for BRCA1 in the FANCD2/BRCA2 pathway. However, the exact role of Ub-FANCD2 and the FANC core complex in the BRCA1 repair pathway remains unclear. FANCD2 is also phosphorylated at ser-222 in response to DNA DSBs, an event required for an efficient intra-S phase cell cycle arrest however the role of BRCA1 in this pathway, if any, has not been identified [62].

In addition to regulating cell cycle checkpoints in response to DNA damage, BRCA1 has also been implicated in the mitotic spindle checkpoint. This checkpoint functions at the metaphse/anaphase boundary within mitosis and ensures that mitotic spindles are correctly attached to chromosomes, thus allowing equal segregation of sister chromosomes. Failure of one or more spindles to attach to chromosomes leads to activation of this checkpoint, which is maintained until attachment is achieved. The checkpoint functions through the action of the regulatory proteins Mad1, Mad2, Bub1, BubR1, Bub3, and Mps1 which inhibit the anaphase-promoting complex causing metaphase/anaphase arrest [63]. BRCA1s involvement in this checkpoint was first suggested by experiments demonstrating a requirement for BRCA1 for mitotic arrest in cells treated with spindle poisons such as paclitaxel and vinorelbine [64, 65]. One mechanism through which BRCA1 may participate in this checkpoint is via the transcriptional regulation of both BubR1 and Mad2 [66]. In addition, BRCA1 has been shown to degrade γ-tubulin causing inhibition of microtubule nucleation, providing an addition mechanism through which BRCA1 may participate in this checkpoint. Interestingly, a recent study has shown that in response to treatment with spindle poisons Chk2 is activated, and phosphorylates BRCA1 on ser-988. This study also demonstrated that this Chk2 dependent phosphorylation of BRCA1, is required for BRCA1's ability to inhibit microtubule nucleation. This is an intriguing finding and the role played by Chk2 in other DNA damage mediated cell cycle checkpoints suggests some degree of degeneracy between the spindle checkpoint and DNA damage response pathways. However, despite the known cell cycle arrest function of Chk2, and the requirement for activating phosphorylations from upstream kinases such as ATM and ATR it is surprising that activation of other cell cycle checkpoints in response to spindle poisons is not observed.

18.2.3 BRCA1 Ubiquitination and the DNA Damage Response

The RING domain of BRCA1 mediates interaction with BARD1 forming an active E3 ubiquitin ligase complex [11, 67]. Purified endogenous BRCA1 has also been found in complex with the E2 ubiquitin conjugating enzymes UbcH5 and UbcH7 [68]. In addition early studies investigating BRCA1/BARD1 ligase activity identified a pathogenic mutation (C61G) within the BRCA1 RING domain, which abolished the BRCA1/BARD1 E3 ligase activity. These finding suggested a physiological role for the BRCA1/BARD1 ligase complex and spurred the hunt for BRCA1/BARD1 ubiquitination targets [11].

Although traditionally associated with protein degradation, protein ubiquitination also plays a role in proteolysis independent post-translational modifications. Many studies have identified a number of BRCA1/BARD1 ubiquitination targets including histones H2A and H2AX, nucleophosmin/B23 (NM1), γ-tubulin, the RNA polymerase subunits RNA Pol II and RPB8, ERα, CtIP, the progesterone receptor (PR-A) and the TFIIE transcription factor. However, it is well accepted that the BRCA1/BARD1 complex is a very promiscuous ubiquitin ligase in vitro and has been known to ubiquitinate a number of targets in-vitro which are not *bona-fide, in-vivo* targets. For example, FANCD2, was long regarded as a BRCA1 ubiquitination substrate which, when ubiquitinated in response to DNA damage localises to BRCA1 containing nuclear repair foci [60]. However recent reports have demonstrated that monoubiquitination of FANCD2 occurs independently of BRCA1, consequently FANCD2 is no longer considered a target of the BRCA1/BARD1 ubiquitin ligase [55, 69, 70]. Nonetheless, although not a target of BRAC1:BARD1 ubiquitination, multiple studies have shown that BRCA1 is still required for FANCD2 localisation to sites of DNA damage/repair in response to DNA damage [60, 70, 71].

Thus due to the promiscuity of the BRCA1/BARD1 ligase, identification of target substrates remains difficult. One reason for this difficulty is the lack of understanding of the type of ubiquitination events carried out by BRCA1. Polyubiquitin chains are formed by linkage of single ubiquitin molecules through any one of seven different Lys residues within each molecule. The most common form of ubiquitin linkage is via Lys-48, however initial studies which identified the BRCA1/BARD1 complex as an autoubiquitination substrate, demonstrated that Ubiquitination chain linkage occurs via Lys-6 [72–74]. Interestingly this autoubiquitination event stimulates BRCA1/BARD1 E3 ligase activity approximately 20-fold forming a positive feedback loop [72]. However, a number of studies have identified different ubiquitin chain linkage conformations for BRCA1/BARD1 substrates creating contention in the field. Shedding light on this, a recent study using a structure based, yeast two hybrid assay identified six previously unidentified BRCA1/BARD1 interacting E2 ubiquitination proteins; UbcH6, Ube2e2, Ubc13, Ube2k and Ube2w [75]. In addition this study demonstrated that BRCA1/BARD1 can in-fact catalyse monoubiquitination events as well as direct the synthesis of Lys-6, Lys-48 and Lys-63 polyubiquitin chains dependent upon the E2 conjugation enzyme it interacts with.

Despite the confusion regarding the validity of BRCA1/BARD1 ubiquitination targets many of the target ubiquitination events have been linked to BRCA1 functions. For instance Lee et al. found that in response to DNA damage RNA PolII is ubiquitinated and degraded, thereby inhibiting further transcription of damaged mRNA templates [76]. Further studies found that the DNA damage dependent ubiquitination and degradation of RNA Pol II was BRCA1/BARD1 dependent [31]. This provides a mechanism in which RNA Pol II and 3′-RNA processing machinery stalled at sites of DNA damage can be targeted by BRCA1/BARD1 for degradation, thus allowing access for repair machinery, thereby linking BRCA1s ubiquitin ligase activity with its transcription coupled repair function.

Notwithstanding the lack of clarity with respect to BRCA1/BARD1 ubiquitination targets, the physiological importance of the BRCA1/BARD1 ligase activity has been clearly demonstrated [73]. Using an antibody directed against Lys-6 polyubiquitin chains, Morris and Solomon visualised ubiquitination events at stalled replication forks and DNA double strand break sites in response to HU or IR [73]. Notably these ubiquitination events co-localise with DNA damage induced BRCA1 foci, and are abolished by siRNA mediated depletion of BRCA1 or BARD1. Following this another study by the same group used this technology to demonstrate that cancer associated pathogenic mutations found in the BRCA1 N-terminus confer a defect in DNA damage induced ligase activity due to defective formation of BRCA1/BARD1/E2 ubiquitin ligase complexes [77].

Nonetheless, how BRCA1/BARD1 mediated ubiquitination at sites of DNA damage is initiated, and the downstream targets of this event remain unknown. It is possible that these targets may include histones, H3 and H2AX, which as mentioned earlier are postulated to recruit BRCA1 to sites of DNA damage via the RNF8/Ubc13 pathway. This would function to stabilize BRCA1 binding/foci formation at DNA break sites thereby facilitating efficient checkpoint activation and DNA repair.

18.2.4 BRCA1 and Transcriptional Regulation

Upon cloning BRCA1, Miki et al. first suggested transcriptional regulation as a potential BRCA1 function due to the high number of acidic amino acid residues in its C-terminus [78]. BRCA1 transactivation activity was then demonstrated with experiments using the C-terminus of BRCA1 fused to the GAL4 DNA binding domain [17]. In addition, tumour derived mutations in the C-terminus of BRCA1 abolished GAL4-BRCA1 mediated transactivation and growth suppression, suggesting a physiological role for BRCA1 mediated transactivation [17, 79]. In support of BRCA1 as a transcriptional regulator, BRCA1 was co-purified in a complex with the RNA Pol II holoenzyme via interaction with RNA helicase A and transcriptional activation by this complex required BRCA1 [80]. BRCA1 is also able to bind DNA directly, prompting the idea that BRCA1 itself may directly bind to DNA promoters and through interaction with the core transcriptional machinery directly affect transcription [17]. However, most recent evidence suggests that BRCA1 does not bind to DNA in a sequence specific manner.

It is now widely accepted that BRCA1 affects transcription in a number of ways. Firstly, BRCA1 is known to regulate signaling pathways that affect transcription factor activation. For example, as mentioned earlier BRCA1 phosphorylation in response to DNA damage is required for ATM mediated p53 signaling and activation, thus regulating p53 dependent p21 transcription [41]. In addition BRCA1 appears to modulate which p53 target genes are transactivated in response to DNA damage. Maclachlan et al. showed that although a large subset of p53 responsive genes are prop-apoptotic, the presence of functional BRCA1 mediated the p53 specific upregulation of DNA repair and cell cycle arrest genes, but not pro-apoptotic genes [81].

The estrogen response signaling pathway is also known to be affected by BRCA1 at multiple different levels. Estrogens translocate into the nucleus where they bind to the estrogen receptor (such as ERα) stimulating receptor dimerisation and activation. Activated estrogen receptors then bind estrogen response elements proximal to or within promoters and stimulate transcription of target genes. Estrogen in general, promotes proliferation and drives expression of a number of proliferative genes associated with cancer, such as Cyclins E & D, the *p*rogesterone *r*eceptor (PR), *e*pithelial *g*rowth *f*actor (EGF) and the *v*ascular *e*ndothelial *g*rowth *f*actor (VEGF) [82]. The link between estrogen signaling and the development of breast cancer is evidenced by the potent efficacy of treatment of this disease with anti-estrogen drugs such as tamoxifen. BRCA1 is known to bind to ERα and is reported to repress both ligand dependent and independent ERα signaling [83]. Concurrent with this, BRCA1 severely represses expression of a number of endogenous ERα regulated genes such as VEGF, a protein implicated in tumour growth and angiogenesis [84]. In addition, wildtype BRCA1 inhibits proliferative signaling to the extra-cellular signal-related kinase (ERK) by a membrane bound isoform of ERα. Another BRCA1 binding protein known as COBRA1, for *co*-factor of *BR*CA1, a subunit of the negative elongation factor complex (NELF), also binds to ERα and inhibits ERα mediated transcription by stalling RNA PolII mediated transcriptional elongation [85].

BRCA1 has also been associated directly with DNA promoters. Although BRCA1 has been shown to directly bind to DNA, it appears to have a preference for branched/abnormal DNA structures, consistent with its role in DNA repair and genomic surveillance [28, 86]. Thus, it is more likely that BRCA1 is recruited to promoters via interaction with sequence specific transcription factors. Consistent with this, we have recently shown that BRCA1 is recruited to the ERα promoter by the *oct*amer binding protein Oct-1, where it is required for ERα expression [87]. Interestingly, we showed that RNA Pol II is present on the ERα promoter irrespective of BRCA1 promoter occupancy, suggesting at least in the case of ERα, that BRCA1 does not recruit the core transcriptional machinery. Thus, exactly how BRCA1 stimulates transcription from the ERα promoter is unknown though it is probably via recruitment of secondary factors and/or chromatin remodeling factors. Indeed BRCA1 is known to bind to a number of chromatin remodeling factors such as the *h*istone *deac*etylases HDAC1 and HDAC2, and the BRG1 subunit of the SWI/SNF chromatin remodeling complex [88]. Although SWI/SNF mediated chromatin remodeling requires only the BRG1-BAF155 minimal complex transcriptional activation requires an activation domain [89]. Thus, it is thought that BRCA1 may provide this activation domain. BRCA1 also interacts with hGCN5 and TRRAP, forming part of a *h*istone *a*ctyl*t*ransferase (HAT) complex which requires BRCA1 for transcriptional activation [90]. In addition to regulating ERα expression, BRCA1 and Oct-1 are known to co-operate in the transactivation of the DNA damage induced cell cycle inhibitor GADD45 [91]. Interestingly, Oct1 is known to be stabilised via a post-transcriptional mechanism and has a higher affinity for DNA in response to DNA damage [92]. Whether BRCA1 plays a role in this response remains unknown.

In addition to Oct-1 and p53 mediated BRCA1 transactivation, BRCA1 functions as a transcriptional co-activator with a number of other transcription factors such as the *s*ignal *t*ransducer and *a*ctivator of *t*ranscription protein, Stat-1. BRCA1 co-activates a number of Stat-1 regulated genes such as the pro-apoptotic genes IRF7, MxA 2,5 OAS and ISG54 in response to interferon-γ stimulation [93].

Although BRCA1 transactivates GADD45 in response to DNA damage it has also been reported to repress GADD45 transcription in unperturbed cells, a mechanism which also requires the DNA damage response protein, CtIP [94, 95]. The BRCA1/CtIP/Zbrk1 complex has since been shown to repress over a dozen genes including the DNA repair genes RFC1, and HMGA2 [96]. Zbrk1 mediated transcriptional repression requires the Kap1 co-repressor, which recruits the histone methyltransferase SETDB1, and the methylhistone binding protein HP1, causing chromatin compaction and transcriptional repression. Exactly how BRCA1 is involved in this repression mechanism remains unknown, however Zbrk1 mediated Gadd45 repression does not occur in the absence of BRCA1 [95]. Recently, work from Jeffrey Parvin's group has provided a mechanism through which BRCA1/BARD1 bound to DNA promoters ubiquitinates the transcription pre-initiation complex, thereby inhibiting stable association of the TFIIE and TFIIH subunits and inhibiting transcription initiation, providing a mechanism through which BRCA1 mediates repression when recruited to promoters by sequence specific transcription factors such as Zbrk1 [97]. BRCA1 also acts as a transcriptional co-repressor in a complex with c-Myc and the c-Myc interacting protein Nmi (*n-myc-i*nteracting protein), where it represses a number of cancer associated genes such as hTERT, and psoriasin (S100A7) [98].

The finding that BRCA1 forms part of, and is required for, both the Gadd45 pre-damage transcriptional repression complex (BRCA1/Zbrk1) and the post-damage transcriptional activation complex (BRCA1/Oct1) suggests that BRCA1 may receive and/or transduce some sort of "switch" signal to promoter binding proteins and/or the repression/activation machinery in response to DNA damage. One report has shown that in response to treatment with UV radiation or the DNA crosslinking agent methyl methanosulfonate (MMS) Zbrk1 is degraded through the ubiquitin/proteasome degradation pathway, thereby leading to Gadd45 de-repression. However, Zbrk1 degradation, was BRCA1 independent ruling out BRCA1/BARD1 dependent ubiquitination mediated degradation of Zbrk1. Another mechanism via which BRCA1 repression/activation switching could be regulated is via sumolation of BRCA1 itself. Park et al. reported that BRCA1 sumolation induces release of BRCA1 from promoters allowing the deacetylation of promoter bound histones thereby causing transcriptional repression [99]. It may be that a combination of DNA damage induced phosphorylation and sumolation events regulates the switch between transcriptional repression and activation. A prime example of this is the DNA damage induced de-repression of Kap1 targets. In response to DNA DSBs activated ATM phoshorylates Kap-1 on serine-824, inducing desumolation of Kap1 by the SENP1 desumolase, mediating de-repression of Kap-1 target genes [100, 101]. Conceivably DNA damage induced phosphorylation of BRCA1 may regulate BRCA1 sumolation thereby causing dissociation of BRCA1 repressor complexes (such as BRCA1/Zbrk1) and allowing association of BRCA1 activation complexes (such as BRCA1/Oct-1).

18.3 BRCA2 Structure and Function

The BRCA2 gene is comprised of 27 exons, is located on chromosome 13q12-q13 and encodes a 3418 amino acid protein which, like BRCA1, shares no homology with known proteins [4, 102]. Close inspection of the BRCA2 protein revealed a series of eight repeated motifs since named the BRC motif. Each motif is approximately 30 amino acids in length and is situated between amino acids 990 and 2100 [103]. Although little homology exists between BRCA2 orthologues, the BRC repeats remain highly conserved among different species indicating that they play an essential role in BRCA2 function [104]. Of critical importance to the function of BRCA2 in HR mediated DSB repair was the finding that the BRC motifs directly interact with Rad51 (Fig. 18.3) [105, 106]. Indeed initial research on BRCA2, followed that of BRCA1, quickly demonstrating that BRCA2 deficient cells also exhibit gross genomic instability and are sensitive to a wide range of DNA damaging agents [105]. Also in agreement with BRCA1 research, BRCA2 deficient cells are deficient in HR mediated DSB repair and are unable to generate Rad51 foci at DNA break sites. Pellegrini et al. shed light on this observation by solving the crystal structure of the BRCA2 BRC4 domain bound to Rad51 [107, 108]. This revealed that a region of 28 amino acids within BRC4 mimics the interaction interface between two Rad51 molecules preventing them from oligomerising. Pellegrini et al. then suggested that BRCA2 may act like a molecular velcro strip binding multiple Rad51 molecules and holding them in a monomeric state primed for loading onto ssDNA at processed DSBs (Fig. 18.4). Consistent with the critical role played by the BRCA2 BRC domains, a number of cancer causing mutations have been found throughout the BRC domains [107]. An additional Rad51 binding site has been identified in the extreme C-terminus of BRCA2 [109]. It has since been shown that the C-terminal egion of BRCA2 stabalises Rad51 filament formation on ssDNA, suggesting that perhaps the c-terminal Rad51 binding region of BRCA2 may regulate the BRC domain mediated loading of Rad51 onto ssDNA, faciliating HR mediated repair (Fig. 18.4) [110]. In addition, the C-terminal Rad51 interaction is disrupted by phosphorylation of BRCA2 on Ser-3291 by the cell cycle regulating, cyclin dependent kinases (CDKs). The CDK phosphorylation of BRCA2 was found to be low throughout S-phase of the cell cycle and increased through G2-M, providing a mechanism for cell cycle specific Rad51 loading onto ssDNA and regulation of homologous recombination [109].

The C-terminal region of BRCA2 is highly structured containing at least five separate domains. The first domain is a series of α-helices spanning 190 amino acids and is known as the helical domain, predicted to be involved in protein–protein interactions [111]. Following the helical domain are three distinct oligonucleotide/oligosaccharide binding folds (OB fold) of approximately 100 amino acids, named OB1, OB2 and OB3. OB folds are found in a number of single strand DNA (ssDNA) binding proteins such as RPA and hSSB1 and are involved in directly binding to single stranded DNA (Fig. 18.3) [112]. Interestingly, the BRCA2 OB2 contains a 130 amino acid insertion, which forms a protruding tower like structure, consisting of a pair of anti-parallel helices supporting a 3 headed helix at its end [111]. Consistent with OB fold containing proteins Yang et al. demonstrated that the C-terminal region binds single stranded DNA with high affinity. However, this study also found that the tower like structure within OB2 binds to double stranded DNA (dsDNA), suggesting that BRCA2 may localize Rad51 to double stranded-single stranded DNA junctions found at processed DSBs. The BRCA2 OB folds also bind the DSS1 protein, the product of a gene formally mapped to a region lost in the *d*eleted in *s*plit foot/*s*plit hand syndrome [111, 113]. Although the functional relevance of this interaction remains unclear the crystal structure of the DSS1 bound BRCA2 C-terminus suggests that DSS1 may mimic small regions of single stranded DNA, thereby may play a role in regulating the DNA binding of BRCA2 and thus efficient BRCA2 mediated DNA repair [114]. In addition, cancer associated mutations within BRCA2 have been found at DSS1 interacting residues suggesting a physiological role for DSS1 in HR [111].

As mentioned earlier, BRCA2 was recently identified as the gene mutated in Fanoni Anemia complementation group D1 and plays a role in the FANC pathway mediated repair of DNA cross-link induced damage [57]. Howlett et al. recognised that the cellular phenotype of FANCD1 cells such as genomic instability and sensitivity to DNA crosslinking agents overlapped those seen in BRCA2 deficient cells, prompting them to sequence BRCA2 in FANCD1 patients. This identified a number of bialellic mutations within the BRCA2 gene in FANCD1 patient derived cell lines. One particular patient was identified with two different truncating mutations within BRCA2 (one on each allele). Interestingly, both of these mutations have been identified in breast/ovarian cancer kindreds. This finding was surprising as early studies using mouse knockouts had suggested that homozygous mutations within BRCA2 resulted in embryonic lethality. However, further mouse studies demonstrated that although mice harbouring homozygous mutations resulting in loss of the majority of BRCA2 died early in embryogenesis, mice harbouring "less severe" or hypormorphic mutations were able to survive to adulthood. In retrospect these mice displayed many of the Fanconi anemia phenotypes such as small gonads, skeletal defects and sensitivity to DNA cross-linking agents [115].

Since this study another BRCA2 interacting protein PALB2, for *p*artner *a*nd *l*ocaliser of *B*RCA2, has also been identified as a gene mutated in FA patients causing the FA complementation group N related disease (Fig. 18.3) [116]. This was prompted by the finding that PALB2, originally identified in an immunoaffinity purified BRCA2 complex, is required for BRCA2 protein stability and thus function. Due to this important function it is not surprising that mutations within PALB2 have also been identified in families suffering from inherited breast cancer [117]. However, other than the stabilisation of BRCA2 protein no specific functions have been assigned to PALB2.

18.3.1 BRCA2 and Cell Cycle Regulation

In contrast to BRCA1, BRCA2 does not seem to play a role in cell cycle checkpoint responses to DNA damage and although a number of phosphorylation sites within BRCA2 have been identified the functional consequences of these phosphorylation events remains to be elucidated [118].

Although it does not appear to play a role in DNA damage induced cell cycle regulation BRCA2, like BRCA1, has been implicated in the mitotic spindle checkpoint, through its interaction with and phosphorylation by the spindle checkpoint kinase BubR1. However, the significance of this event with respect to the spindle checkpoint remains unknown [119]. In addition, inhibition of BRCA2 or another interacting protein BRAF35 (*BR*CA2 *a*ssociated *f*actor *35*) by injection of antibodies against either protein blocked entry into mitosis, implicating both of these proteins in the mitotic checkpoint [120]. BRAF35 was identified as part of a large 2MDa complex containing BRCA2 and has been found to bind to branched DNA structures, implicating it in recognition of/or binding to damaged DNA, implicating it in a BRCA2 related repair complex [120].

18.3.2 BRCA2 Chromatin Remodeling and Transcriptional Regulation

BRCA2 has also been implicated in chromatin remodeling through the discovery of the BRCA2 interacting protein EMSY. EMSY interacts with a 12 amino acid region of the BRCA2 N-terminus, and along with BRCA2 is localised to repair foci in response to DNA damage (Fig. 18.3) [121]. Although the exact function of EMSY is not well understood it has been shown to bind the heterochromatin protein 1-β (HP1-β) and two chromatin associated proteins BS69 and a novel unnamed protein. All of these proteins belong to a family of proteins containing domains known as "royal family" domains, which are a collection of different structural domains known to recognize histones and their modified variants (e.g. acetylated and methylated histones) [121]. Not surprisingly many royal family domain containing proteins form part of chromatin remodeling complexes, thus implicating BRCA2 in this process. The biological function of BRCA2 in chromatin remodeling is not defined though there are two major trains of thought. One is that BRCA2, like BRCA1 may play a role in transcriptional regulation. BRCA2 has been linked to transcriptional regulation via interaction with the chromatin remodeling histone acetyltransferase P/CAF. It has also been shown to interact with the androgen receptor (AR) and the AR interacting protein GRIP-1 and together with P/CAF, BRCA2 appears to positively regulate transcription of AR target genes [122]. The second function that BRCA2 mediated chromatin remodeling may carry out is regulation of chromatin condensation, perhaps contributing to chromatin decondensation in response to DNA damage thus allowing access to DNA repair machinery, similar to the role played by the nuclear co-repressor Kap1 in response to DNA damage [101, 121, 123]. Nonetheless extensive work will be required to define the exact role if any, played by BRCA2/EMSY in chromatin remodeling and/or transcription (Fig. 18.3).

Although the role of BRCA2 in homologous recombination mediated repair has been well defined it is clear that much of the role of BRCA2 remains unresolved. It is likely that BRCA2 has many as yet undiscovered functions, which like

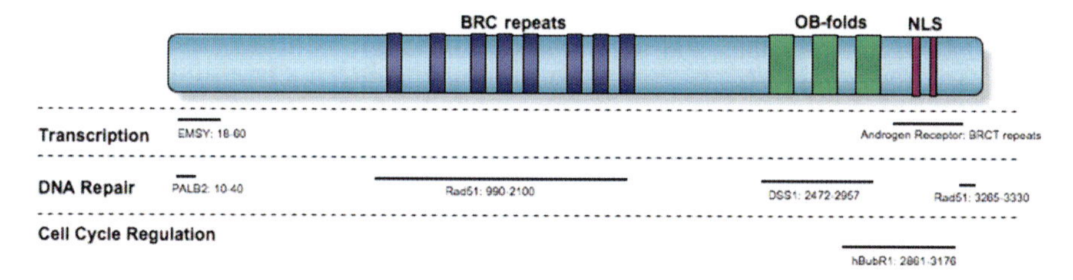

Fig. 18.3 BRCA2 Interacting proteins and their functions. Schematic diagram of BRCA2 including the BRC repeats, nuclear localisation signals (NLS), and the OB-folds. Interacting proteins are shown underneath their known BRCA2 interacting region and are grouped according to their known/suspected function

BRCA1 may be carried out by unidentified binding partners. Unfortunately unlike BRCA1 the size of the BRCA2 protein has hampered many studies to identify new binding partners, thus the understanding of BRCA2 function is progressing slowly.

18.4 Tissue Specificity of BRCA1 and BRCA2 Related Cancers

Perhaps the most perplexing feature of BRCA1 and BRCA2 related tumourigenesis is the observed tissue specificity. Given that these proteins are ubiquitously expressed in most tissues and appear to play an essential role fundamental to all cells, why do defects in these genes specifically cause breast and ovarian cancer? Initial theories hypothesised that the highly proliferative nature of breast and ovarian tissues during puberty and menstruation may increase the mutation rate within these tissues thereby increasing the rate of loss of BRCA1 or BRCA2 in these tissues. However, this theory is not supported by the lack of lymphocytic cancers in BRCA1/2 carriers. It has also been suggested that breast and ovarian tissues are more dependent on homologous recombination (HR) mediated DNA repair due to a higher rate of DNA damage in these tissues due to DNA adducts generated by estrogen metabolites (Fig. 18.4). Interestingly evidence that breast tissues may be more dependent on HR mediated DNA repair comes from the observation that atomic bomb survivors exposed to low dose ionizing radiation (which requires HR for error free repair), have an increased risk of breast cancer in comparison to other cancers [124]. In addition, BRCA1 has also been demonstrated to transcriptionally suppress the expression of aromatase, the protein responsible for estrogen production, in ovarian granulosa and breast adipose cells, the cells responsible for the bulk of estrogen production [125]. Thus, the combination of decreased DNA adducts due to reduced estrogen synthesis, along with increased proliferative signaling due to lack of BRCA1 mediated-suppression of ER levels and ER-mediated signaling, in addition to defective HR mediated repair may contribute to breast and ovarian specific tumourigenesis in BRCA1 mutated cells. However this model does not explain the tissue specificity of tumours in BRCA2 mutation carriers. Nonetheless, the finding that BRCA2 potentiates androgen receptor mediated anti-proliferative signaling may provide a link between BRCA2 and hormone responsive tissue specific tumours of the breast and ovary. BRCA1 has also been implicated in X-linked gene dosage compensation via stabilization of XIST, a non-coding RNA, which coats and inactivates one of the inactive X-chromosomes in somatic cells [126]. However, this finding is contentious as a number of groups have found that XIST stabilization and X-inactivation mediated dosage compensation occur independently of BRCA1 [127]. In addition, although these studies have assigned a "female specific" function to BRCA1 it is not clearly understood how this may affect breast and ovarian tissues or tumourigenesis.

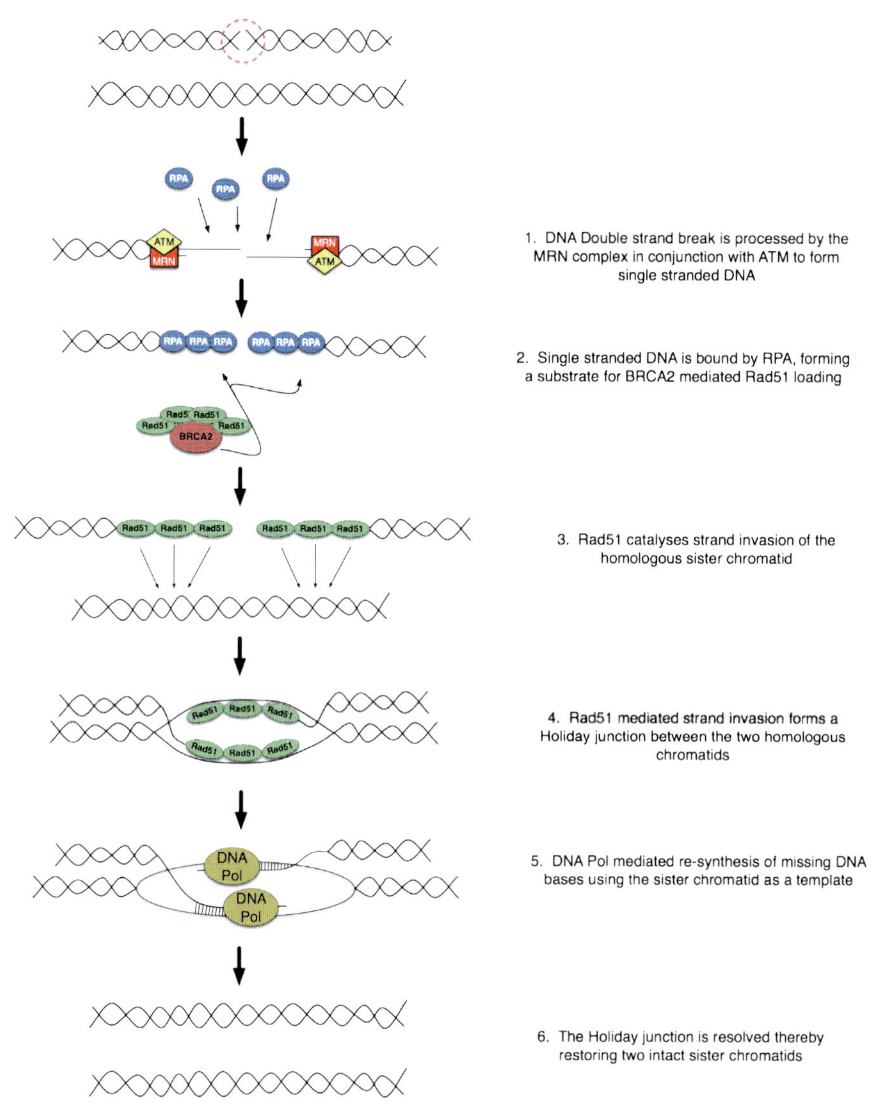

Fig. 18.4 Homologous recombination mediated double strand break repair. During S and G2 phases of the cell cycle DNA double strand breaks are processed in conjunction with ATM by the Mre11/Rad50/Nbs1 (MRN) complex to form single stranded DNA (ssDNA) ends. ssDNA is then bound by RPA, facilitating the BRCA2 mediated loading of Rad51 onto the ssDNA. Rad51 then catalyses invasion of the sister chromatid which is then used as a primer for DNA Pol mediated repair of damaged DNA. The resulting holiday junction is then resolved forming two intact sister chromatids

18.5 BRCA1, BRCA2 and Cancer Treatment

Although the current standard therapies for breast cancer are relatively effective estimates suggest that treatment failure occurs in approximately 30% of cases. Currently treatment plans are based on levels of response predictive biomarkers such as hormone receptors for response to tamoxifen, and HER2 receptor status for response to anti-herceptins such as trastuzamab. The important role played by BRCA1 and BRCA2 in the development of hereditary breast cancer and observations that both genes are epigenetically inactivated in a large proportion of sporadic breast and ovarian cancers has focused recent research on the molecular pathology of BRCA1/2 related tumours and the role of these proteins in regulating response to therapeutic treatment.

Normal breast tissue is made up of two distinct cell types, an inner luminal layer and an outer basal layer. BRCA1 related tumours exhibit a basal-like phenotype (characterized by expression of basal markers such as cyokeratin 5/6, 14 and 17) tend to be ER, PR and Her2 negative (triple negative) and have a poorer prognosis compared to ER, PR Her2 positive tumours. In contrast BRCA2 tumours do not appear to correlate with a particular tumour subtype. Pre-clinical studies have demonstrated that BRCA1 promotes resistance to DNA damaging chemotherapeutics such as cisplatin, and etoposide but in contrast confers sensitivity to spindle poisons such as paclitaxel and vinorelbine, thereby acting as a differential modulator of response to chemotherapy [65]. Many clinical studies investigating the function of BRCA1 in predicting response to chemotherapy have been difficult to interpret as they have been retrospective and have included both BRCA1 and BRCA2 mutant patients. In addition, most patients have been treated with a variety of different chemotherapeutic agents, making interpretation of the specific role of BRCA1 in differential modulation of therapeutic response difficult. Nonetheless, a number of studies have demonstrated a predictive role for BRCA1 and BRCA2 in response to chemotherapy. Chappuis et al. evaluated response to anthracycline based neoadjuvant chemotherapy in 38 patients with locally advanced breast cancer (seven of these patients carried germline mutations in BRCA1 and four had germline mutations in BRCA2). Ten of the eleven patients carrying BRCA1/2 mutations demonstrated a complete clinical response in comparison to only eight of twenty-seven non-carrier patients [128]. Multiple other clinical studies have reported similar responses to DNA damaging agents in breast cancer patients carrying BRCA1/2 mutations [129]. Additionally, a retrospective clinical study from our own laboratory has found that levels of BRCA1 mRNA in patient derived tumours predicts for overall survival in sporadic ovarian carcinoma [130]. In particular, patients with low and intermediate intra-tumoral levels of BRCA1 mRNA have an increased overall survival following treatment with platinum-based chemotherapy in comparison with patients with high levels of intra-tumoural BRCA1 mRNA. In addition, high intra-tumoral levels of BRCA1 mRNA appear to correlate with an increase in overall survival following taxane treatment [130]. These studies have important clinical implications and may influence the future cancer treatment of BRCA1 mutation carriers. In addition, BRCA1 mRNA and/or protein expression levels may provide important therapeutic response predictions in sporadic ovarian cancer patients, thereby influencing individual patient treatment plans.

Results of studies like these as well as the knowledge of the important role played by BRCA1 and BRCA2 in DNA repair has inspired recent research focusing on exploiting the DNA repair defect in BRCA1/2 deficient tumours. Clinical trials are now underway investigating the efficacy of carboplatin, a DNA alkylating agent, verses the spindle poison, docetaxel, in BRCA1 and BRCA2 mutation carriers with metastatic breast cancer and patients with metastatic or recurrent triple negative breast cancer. (Further information is available at http://clinicaltrials.gov)

In addition, recent work from Alan Ashowrth's laboratory using a synthetic lethal approach has demonstrated that BRCA1 and BRCA2 deficient tumours are highly sensitive to PARP inhibitors. PARP1 is a member of the base excision repair (BER) pathway and is involved in the BER mediated repair of single strand DNA breaks. Inhibition of PARP1 leads to lack of repair of ssDNA breaks which when met by the advancing replication fork during DNA replication, cause fork stalling leading to replication fork collapse and the formation of DNA DSBs [131]. Farmer et al. hypothesized that depletion of PARP1 in a BRCA1 or BRCA2 deficient background may lead to an increase in un-repaired DNA DSBs ultimately leading to gross chromosomal aberrations and cellular death. Depletion of PARP1 using targeted siRNA caused death in BRCA1 and BRCA2 deficient cells, and BRCA1/2 deficient cells were exquisitely sensitive to chemical inhibition of PARP1 using small molecular inhibitors [131]. This study provided a mechanism through which the tumour specific loss of BRCA1 or BRCA2 could be exploited to cause tumour specific death via PARP1 inhibition with minimal toxicity to normal tissues, which contain a functional copy of the BRCA1/2 gene. This seminal study has lead to the development of PARP inhibitors as a new class of chemotherapeutic drug, which are currently being trialed in phase two studies investigating their efficacy in the treatment of both hereditary and sporadic breast and ovarian cancers.

Although BRCA1 and BRCA2 play a role in predicting chemotherapeutic response, recent evidence has also shown that they may also play a direct role in resistance to chemotherapy. Edwards et al. generated PARP inhibitor resistant clones of the BRCA2 mutant tumour derived cell line CAPAN1, which contains the protein truncating, frame shift mutation C6174delT [132]. Edwards et al. noticed that in contrast to parental CAPAN1 cells PARP inhibitor resistant clones were able to generate Rad51 foci, and were no-longer sensitive to DNA damaging agents. They then went on to find that the PARP inhibitor resistant clones had intragenic deletions within BRCA2 resulting in restoration of the BRCA2 open reading frame (ORF). Reconstitution of parental CAPAN1 cells via ectopic expression of BRCA2 revertant alleles resulted in resistance to PARP inhibitors demonstrating that resistance to PARP inhibitors may be mediated by reversion mutations of BRCA2. Finally this study then went on to identify similar BRCA2 reversion mutations in carboplatin resistant ovarian tumours. Similarly, a study by Sakai et al. found BRCA2 reversion mutations which restored the ORF of truncated BRCA2 in a cisplatin resistant, BRCA2 mutant breast cancer cell line [133]. The same group have also found ORF restoring, reversion mutations within BRCA1 in platinum resistant,

BRCA1 mutated, ovarian tumours, thereby also implicating a role for BRCA1 in tumour chemoresistance [134].

Collectively these studies demonstrate the importance of BRCA1 and BRCA2 function in the development, treatment and prognosis of breast and ovarian cancers.

18.6 Conclusion

Since the initial discovery of BRCA1 and BRCA2 considerable progress has been made in understanding the cellular functions of these proteins. The role of BRCA1 in DNA damage signaling and cell cycle arrest is well established however its role in other pathways such as DNA repair and chemotherapeutic response remain poorly defined. The vast majority of BRCA1 functions are effected through its interaction with a large variety of proteins, as well as transcriptionally regulating the expression of a number of target proteins. Thus, in order to understand BRCA1 function clearly the identification of novel BRCA1 interacting proteins and target genes is vital. In addition, although BRCA1/BARD clearly play a role in DNA damage induced ubiquitination, targets proteins in this pathway remain elusive. Thus identifying these targets will likely add considerably to the understanding of BRCA1 function. Concurrently, although the role of BRCA2 in DNA repair is well understood, its role in other cellular processes such as, transcriptional regulation and the spindle checkpoint remain unclear. Therefore, focusing efforts on understanding these functions of BRCA2 via identification of additional interacting proteins or transcriptional targets and their possible role in the DNA damage induced modulation of BRCA2 function is critical in elucidating the role of BRCA2 in these processes.

Another major challenge is defining the role of BRCA1 and BRCA2 in breast and ovarian specific tumourigenesis. In this regard understanding the tissue specific differences in response to DNA damage stimuli or indeed the type of DNA damage induced in different tissues may aid to define this process more clearly.

Finally, by understanding the role of BRCA1 and BRCA2 in the DNA damage response, it is likely that response to various cancer therapeutics will be better predicted thus allowing for more defined cancer regimens.

References

1. Hall, J. M., Lee, M. K., and Newmann, B., Linkage of early-onset breast cancer to chromosome 17q21. *Science* **250**, 1684 (1990).
2. Futreal, P. A. et al., BRCA1 mutations in primary breast and ovarian carcinomas. *Science* **266** (5182), 120 (1994).
3. Miki, Y. et al., A strong candidate for the breast cancer susceptibility gene BRCA1. *Science* **266**, 66 (1994).
4. Wooster, R. et al., Identification of the breast cancer susceptibility gene, BRCA2. *Nature* **378**, 789 (1995); Wooster, R. et al., Localisation of a breast cancer susceptibility gene, BRCA2, to chromosome 13q12–13. *Science* **265**, 2088 (1994).
5. Ford, D. et al., Genetic heterogeneity and penetrance analysis of the BRCA1 and BRCA2 genes in breast cancer families. The Breast Cancer Linkage Consortium. *Am J Hum Genet* **62** (3), 676 (1998).

6. Risch, H. A. et al., Prevalence and penetrance of germline BRCA1 and BRCA2 mutations in a population series of 649 women with ovarian cancer. *Am J Hum Genet* **68** (3), 700 (2001).
7. Welcsh, P. L. and King, M. C., BRCA1 and BRCA2 and the genetics of breast and ovarian cancer. *Hum Mol Genet* **10** (7), 705 (2001).
8. Esteller, Manel et al., Promoter hypermethylation and BRCA1 inactivation in sporadic breast and ovarian tumors. *J Natl Cancer Inst* **92** (7), 564 (2000).
9. Blackwood, M. A. and Weber, B. L., BRCA1 and BRCA2: from molecular genetics to clinical medicine. *J Clin Oncol* **16** (5), 1969 (1998); Tan, D. S. P., Marchio, C., and Reis-Filho, J. S., Hereditary breast cancer: from molecular pathology to tailored therapies. *J Clin Pathol* **61** (10), 1073 (2008).
10. Paterson, J. W., BRCA1: a review of structure and putative functions. *Dis Markers* **13** (4), 261 (1998).
11. Hashizume, R. et al., The RING heterodimer BRCA1-BARD1 is a ubiquitin ligase inactivated by a breast cancer-derived mutation. *J Biol Chem* **276** (18), 14537 (2001).
12. Ruffner, H. et al., Cancer-predisposing mutations within the RING domain of BRCA1: loss of ubiquitin protein ligase activity and protection from radiation hypersensitivity. *Proc Natl Acad Sci USA* **98**, 5134 (2001).
13. Brzovic, P. S., Meza, J. E., King, M.-C., and Klevit, R. E., BRCA1 RING domain cancer-predisposing mutations: structural consequences and effects on protein–protein interactions. *J Biol Chem* **8**, 833 (2001).
14. Irminger-Finger, I. and Jefford, C. E., Is there more to BARD1 than BRCA1? *Nat Rev Cancer* **6** (5), 382 (2006); Sauer, M. K. and Andrulis, I. L., Identification and characterization of missense alterations in the BRCA1 associated RING domain (BARD1) gene in breast and ovarian cancer. *J Med Genet* **42** (8), 633 (2005); Thai, T. H. et al., Mutations in the BRCA1-associated RING domain (BARD1) gene in primary breast, ovarian and uterine cancers. *Hum Mol Genet* **7** (2), 195 (1998).
15. Chapman, M. S. and Verma, I. M., Transcriptional activation by BRCA1. *Nature* **382** (6593), 678 (1996).
16. Hu, Y.-F., Miyake, T., Ye, Q., and Li, R., Characterization of a novel *trans*-activation domain of BRCA1 that functions in concert with the BRCA1 C-terminal (BRCT) domain. *J Biol Chem* **275**, 40910 (2000).
17. Monteiro, A. N., August, A., and Hanafusa, H., Evidence for a transcriptional activation function of BRCA1 C-terminal region. *Proc Natl Acad Sci USA* **93** (24), 13595 (1996).
18. Yu, Xiaochun et al., The BRCT domain is a phospho-protein binding domain. *Science* **302** (5645), 639 (2003); Manke, I. A., Lowery, D. M., Nguyen, A., and Yaffe, M. B., BRCT repeats as phosphopeptide-binding modules involved in protein targeting. *Science* **302** (5645), 636 (2003).
19. Monteiro, A. N. A., August, A., and Hanafusa, H.,Evidence for a transcriptional activation function of BRCA1 C-terminal region. *Proc Natl Acad Sci USA* **93**, 13595 (1996); Williams, R. S. and Glover, J. N. M., Structural consequences of a cancer-causing BRCA1-BRCT missense mutation. *J Biol Chem* **278**, 2630 (2002).
20. Chen, C.-F. et al., The nuclear localisation sequences of the BRCA1 protein interact with the importin-α subunit of the nuclear transport signal. *J Biol Chem* **271** (51), 32863 (1996); Thakur, S. et al., Localisation of BRCA1 and a splice variant identifies the nuclear localisation signal. *Mol Cell Biol* **17**, 444 (1997).
21. Rodriguez, J. A. and Henderson, B. R., Identification of a functional nuclear export sequence in BRCA1. *J Biol Chem* **275** (49), 38589 (2000).
22. Scully, R. et al., Association of BRCA1 with Rad51 in mitotic and meiotic cells. *Cell* **88** (2), 265 (1997); Scully, R. et al., Location of BRCA1 in human breast and ovarian cancer cells. *Science* **272**, 123 (1996).
23. Scully, R. et al., Genetic analysis of BRCA1 function in a defined tumor cell line. *Mol Cell* **4** (6), 1093 (1999).
24. Moynahan, M. E., Chiu, J. W., Koller, B. H., and Jasin, M., Brca1 controls homology-directed DNA repair. *Mol Cell* **4** (4), 511 (1999); Snouwaert, J. N. et al., BRCA1 deficient embryonic stem cells display a decreased homologous recombination frequency and an increased

frequency of non-homologous recombination that is corrected by expression of a brca1 transgene. *Oncogene* **18** (55), 7900 (1999).
25. Bhattacharyya, A. et al., The breast cancer susceptibility gene BRCA1 is required for subnuclear assembly of Rad51 and survival following treatment with the DNA cross-linking agent cisplatin. *J Biol Chem* **275** (31), 23899 (2000); Zhou, C., Huang, P., and Liu, J., The carboxyl-terminal of BRCA1 is required for subnuclear assembly of RAD51 after treatment with cisplatin but not ionizing radiation in human breast and ovarian cancer cells. *Biochem Biophys Res Commun* **336** (3), 952 (2005).
26. Lee, J. S. et al., hCds1-mediated phosphorylation of BRCA1 regulates the DNA damage response. *Nature* **404** (6774), 201 (2000); Zhang, J. et al., Chk2 phosphorylation of BRCA1 regulates DNA double-strand break repair. *Mol Cell Biol* **24** (2), 708 (2004).
27. Zhong, Q., Chen, C.-F., Chen, P.-L., and Lee, W.-H., BRCA1 Facilitates microhomology-mediated end joining of DNA double strand breaks. *J Biol Chem* **277** (32), 28641 (2002).
28. Wang, Y. et al., BASC, a super complex of BRCA1-associated proteins involved in the recognition and repair of aberrant DNA structures. *Genes Dev* **14** (8), 927 (2000).
29. Wood, R. D., Nucleotide excision repair in mammalian cells. *J Biol Chem* **272** (38), 23465 (1997).
30. Le Page, F. et al., BRCA1 and BRCA2 are necessary for the transcription-coupled repair of the oxidative 8-oxoguanine lesion in human cells. *Cancer Res* **60** (19), 5548 (2000).
31. Kleiman, F. E. et al., BRCA1/BARD1 inhibition of mRNA 3' processing involves targeted degradation of RNA polymerase II. *Genes Dev* **19** (10), 1227 (2005); Starita, L. M. et al., BRCA1/BARD1 ubiquitinate phosphorylated RNA polymerase II. *J Biol Chem* **280** (26), 24498 (2005).
32. Hartman, A.-R. and Ford, J. M., BRCA1 induces DNA damage recognition factors and enhances nucleotide excision repair. *Nat Genet* **32**, 180 (2002).
33. Chen, Y. et al., BRCA1 is a 220-kDa nuclear phosphoprotein that is expressed and phosphorylated in a cell cycle-dependent manner. *Cancer Res* **56** (14), 3168 (1996); Thomas, J. E. et al., Subcellular localisation and analysis of apparent 180-kDa and 220-kDa proteins of the breast cancer susceptibility gene, BRCA1. *J Biol Chem* **271** (45), 28630 (1996).
34. Scully, R. et al., Dynamic changes of BRCA1 subnuclear location and phosphorylation state are initiated by DNA damage. *Cell* **90** (3), 425 (1997); Thomas, J. E. et al., Induction of phosphorylation on BRCA1 during the cell cycle and after DNA damage. *Cell Growth Differ* **8** (7), 801 (1997).
35. Gatei, M. et al., Role for ATM in DNA damage-induced phosphorylation of BRCA1. *Cancer Res* **60** (12), 3299 (2000); Tibbetts, R. S. et al., Functional interactions between BRCA1 and the checkpoint kinase ATR during genotoxic stress. *Genes Dev* **14** (23), 2989 (2000).
36. Gatei, M. et al., Ataxia telangiectasia mutated (ATM) kinase and ATM and Rad3 related kinase mediate phosphorylation of Brca1 at distinct and overlapping sites. In vivo assessment using phospho-specific antibodies. *J Biol Chem* **276** (20), 17276 (2001).
37. Zhong, Q. et al., Association of BRCA1 with the hRAd50-hMre11-p95 complex and the DNA damage response. *Science* **285**, 747 (1999).
38. Xu, B., Kim, St., and Kastan, M. B., Involvement of Brca1 in S-phase and G(2)-phase checkpoints after ionizing irradiation. *Mol Cell Biol* **21** (10), 3445 (2001).
39. Xu, X. et al., Centrosome amplification and a defective G2-M cell cycle checkpoint induce genetic instability in BRCA1 exon 11 isoform-deficient cells. *Mol Cell* **3**, 389 (1999).
40. Tomlinson, G. E. et al., Characterization of a breast cancer cell line derived from a germ-line BRCA1 mutation carrier. *Cancer Res* **58** (15), 3237 (1998).
41. Fabbro, M. et al., BRCA1-BARD1 complexes are required for p53Ser-15 phosphorylation and a G1/S arrest following ionizing radiation-induced DNA damage. *J Biol Chem* **279** (30), 31251 (2004).
42. Foray, N. et al., A subset of ATM- and ATR-dependent phosphorylation events requires the BRCA1 protein. *Embo J* **22** (11), 2860 (2003).
43. Li, Shang et al., Binding of CtIP to the BRCT repeats of BRCA1 involved in the transcription regulation of p21 is disrupted upon DNA damage. *J Biol Chem* **274** (16), 11334 (1999).

44. Yarden, R. I. et al., BRCA1 regulates the G2/M checkpoint by activating Chk1 kinase upon DNA damage. *Nat Genet* **30** (3), 285 (2002).
45. Xu, B., O'Donnell, A. H., Kim, S.-T., and Kastan, M. B., Phosphorylation of serine 1387 in Brca1 is specifically required for the ATM-mediated S-phase checkpoint after ionizing irradiation. *Cancer Res* **62**, 4588 (2002).
46. Kim, S.-T., Xu, B., and Kastan, M. B., Involvement of the cohesin protein, Smc1, in ATM-dependent and independent responses to DNA damage. *Genes Dev* **16** (5), 560 (2002); Kitagawa, R., Bakkenist, C. J., McKinnon, P. J., and Kastan, M. B., Phosphorylation of SMC1 is a critical downstream event in the ATM-NBS1-BRCA1 pathway. *Genes Dev* **18** (12), 1423 (2004); Yazdi, P. T. et al., SMC1 is a downstream effector in the ATM/NBS1 branch of the human S-phase checkpoint. *Genes Dev* **16** (5), 571 (2002).
47. Somasundaram, K. et al., Arrest of the cell cycle by the tumour-suppressor BRCA1 requires the CDK-inhibitor p21$^{WAF1/CIP1}$. *Nature* **389**, 187 (1997).
48. MacLachlan, T. K. et al., BRCA1 effects on the cell cycle and the DNA damage response are linked to altered gene expression. *J Biol Chem* **275** (4), 2777 (1999); Mullan, P. B. et al., BRCA1 and GADD45 mediated G2/M cell cycle arrest in response to antimicrotubule agents. *Oncogene* **20**, 6123 (2001); Williamson, E. A., Dadmanesh, F., and Koeffler, H. P., BRCA1 transactivates the cyclin-dependent kinase inhibitor p27^{Kip1}. *Oncogene* **21**, 3199 (2002).
49. MacLachlan, T. K. et al., BRCA1 effects on the cell cycle and the DNA damage response are linked to altered gene expression. *J Biol Chem* **275** (4), 2777 (2000); Ree, A. H. et al., Repression of mRNA for the PLK cell cycle gene after DNA damage requires BRCA1. *Oncogene* **22** (55), 8952 (2003).
50. Greenberg, R. A. et al., Multifactorial contributions to an acute DNA damage response by BRCA1/BARD1-containing complexes. *Genes Dev* **20** (1), 34 (2006).
51. Chiba, N. and Parvin, J. D., Redistribution of BRCA1 among four different protein complexes following replication blockage. *J Biol Chem* **276** (42), 38549 (2001).
52. Kim, H., Chen, J., and Yu, X., Ubiquitin-binding protein RAP80 mediates BRCA1-dependent DNA damage response. *Science* **316** (5828), 1202 (2007); Sobhian, B. et al., RAP80 targets BRCA1 to specific ubiquitin structures at DNA damage sites. *Science* **316** (5828), 1198 (2007); Wang, B. et al., Abraxas and RAP80 form a BRCA1 protein complex required for the DNA damage response. *Science* **316** (5828), 1194 (2007).
53. Huen, M. S. et al., RNF8 transduces the DNA-damage signal via histone ubiquitylation and checkpoint protein assembly. *Cell* **131** (5), 901 (2007); Kolas, N. K. et al., Orchestration of the DNA-damage response by the RNF8 ubiquitin ligase. *Science* **318** (5856), 1637 (2007); Mailand, N. et al., RNF8 ubiquitylates histones at DNA double-strand breaks and promotes assembly of repair proteins. *Cell* **131** (5), 887 (2007); Wang, B. and Elledge, S. J., Ubc13/Rnf8 ubiquitin ligases control foci formation of the Rap80/Abraxas/Brca1/Brcc36 complex in response to DNA damage. *PNAS* **104** (52), 20759 (2007).
54. Kennedy, R. D. and D'Andrea, A. D., The Fanconi anemia/BRCA pathway: new faces in the crowd. *Genes Dev* **19** (24), 2925 (2005).
55. Meetei, A. R. et al., A novel ubiquitin ligase is deficient in Fanconi anemia. *Nat Genet* **35** (2), 165 (2003).
56. Wang, X., Andreassen, P. R., and D'Andrea, A. D., Functional interaction of monoubiquitinated FANCD2 and BRCA2/FANCD1 in chromatin. *Mol Cell Biol* **24** (13), 5850 (2004).
57. Howlett, N. G. et al., Biallelic inactivation of BRCA2 in Fanconi anemia. *Science* **297** (5581), 606 (2002).
58. Andreassen, P. R., D'Andrea, A. D., and Taniguchi, T., ATR couples FANCD2 monoubiquitination to the DNA-damage response. *Genes Dev* **18** (16), 1958 (2004); Wang, X. et al., Chk1-mediated phosphorylation of FANCE is required for the Fanconi anemia/BRCA pathway. *Mol Cell Biol* **27** (8), 3098 (2007).
59. Andreassen, P. R., Ho, G. P. H., and D'Andrea, A. D., DNA damage responses and their many interactions with the replication fork. *Carcinogenesis* **27** (5), 883 (2006).

60. Garcia-Higuera, I. et al., Interaction of the Fanconi anemia proteins and BRCA1 in a common pathway. *Mol Cell* **7** (2), 249 (2001).
61. Taniguchi, T. et al., S-phase-specific interaction of the Fanconi anemia protein, FANCD2, with BRCA1 and RAD51. *Blood* **100** (7), 2414 (2002).
62. Nakanishi, K. et al., Interaction of FANCD2 and NBS1 in the DNA damage response. *Nat Cell Biol* **4** (12), 913 (2002).
63. May, K. M. and Hardwick, K. G., The spindle checkpoint. *J Cell Sci* **119** (20), 4139 (2006).
64. Lafarge, S., Sylvain, V., Ferrara, M., and Bignon, Y. J., Inhibition of BRCA1 leads to increased chemoresistance to microtubule-interfering agents, an effect that involves the JNK pathway. *Oncogene* **20** (45), 6597 (2001); Tassone, P. et al., BRCA1 expression modulates chemosensitivity of BRCA1-defective HCC1937 human breast cancer cells. *Br J Cancer* **88** (8), 1285 (2003).
65. Quinn, J. E. et al., BRCA1 functions as a differential modulator of chemotherapy-induced apoptosis. *Cancer Res* **63** (19), 6221 (2003).
66. Bae, I. et al., BRCA1 regulates gene expression for orderly mitotic progression. *Cell Cycle* **4** (11), 1641 (2005); Wang, R.-H., Yu, H., and Deng, C.-X., A requirement for breast-cancer-associated gene 1 (BRCA1) in the spindle checkpoint. *PNAS* **101** (49), 17108 (2004).
67. Lorick, K. L. et al., RING fingers mediate ubiquitin-conjugating enzyme (E2)-dependent ubiquitination. *Proc Natl Acad Sci USA* **96**, 11354 (1999).
68. Brzovic, P. S. et al., Binding and recognition in the assembly of an active BRCA1/BARD1 ubiquitin-ligase complex. *PNAS* **100** (10), 5646 (2003).
69. Meetei, A. R., Yan, Z., and Wang, W., FANCL replaces BRCA1 as the likely ubiquitin ligase responsible for FANCD2 monoubiquitination. *Cell Cycle* **3** (2), 179 (2004).
70. Vandenberg, C. J. et al., BRCA1-independent ubiquitination of FANCD2. *Mol Cell* **12** (1), 247 (2003).
71. Taniguchi, T. et al., S-phase-specific interaction of the Fanconi anemia protein, FANCD2, with BRCA1 and RAD51. *Blood* **100** (7), 2414 (2002); Taniguchi, T. et al., Convergence of the fanconi anemia and ataxia telangiectasia signaling pathways. *Cell* **109** (4), 459 (2002); Taniguchi, T. et al., Disruption of the Fanconi anemia-BRCA pathway in cisplatin-sensitive ovarian tumors. *Nat Med* **9** (5), 568 (2003).
72. Mallery, D. L., Vandenberg, C. J., and Hiom, K., Activation of the E3 ligase function of the BRCA1/BARD1 complex by polyubiquitin chains. *Embo J* **21** (24), 6755 (2002).
73. Morris, J. R. and Solomon, E., BRCA1: BARD1 induces the formation of conjugated ubiquitin structures, dependent on K6 of ubiquitin, in cells during DNA replication and repair. *Hum Mol Genet* **13** (8), 807 (2004).
74. Wu-Baer, F., Lagrazon, K., Yuan, W., and Baer, R., The BRCA1/BARD1 heterodimer assembles polyubiquitin chains through an unconventional linkage involving lysine residue K6 of ubiquitin. *J Biol Chem* **278** (37), 34743 (2003).
75. Christensen, D. E., Brzovic, P. S., and Klevit, R. E., E2-BRCA1 RING interactions dictate synthesis of mono- or specific polyubiquitin chain linkages. *Nat Struct Mol Biol* **14** (10), 941 (2007).
76. Lee, K.-B., Wang, D., Lippard, S. J., and Sharp, P. A., Transcription-coupled and DNA damage-dependent ubiquitination of RNA polymerase II in vitro. *PNAS* **99** (7), 4239 (2002).
77. Morris, J. R. et al., Genetic analysis of BRCA1 ubiquitin ligase activity and its relationship to breast cancer susceptibility. *Hum Mol Genet* **15** (4), 599 (2006).
78. Miki, Y. et al., A strong candidate for the breast and ovarian cancer susceptibility gene BRCA1. *Science* **266** (5182), 66 (1994).
79. Humphrey, J. S. et al., Human BRCA1 inhibits growth in yeast: potential use in diagnostic testing. *PNAS* **94** (11), 5820 (1997).
80. Anderson, S. F. et al., BRCA1 protein is linked to the RNA polymerase II holoenzyme complex via RNA helicase A. *Nat Genet* **19** (3), 254 (1998); Scully, R. et al., BRCA1 is a component of the RNA polymerase II holoenzyme. *Proc Natl Acad Sci USA* **94**, 5605 (1997).

81. MacLachlan, T. K., Takimoto, R., and El-Deiry, W. S., BRCA1 directs a selective p53-dependent transcriptional response towards growth arrest and DNA repair targets. *Mol Cell Biol* **22** (12), 4280 (2002).
82. Heldring, N. et al., Estrogen receptors: how do they signal and what are their targets. *Physiol Rev* **87** (3), 905 (2007).
83. Fan, S. et al., Role of direct interaction in BRCA1 inhibition of estrogen receptor activity. *Oncogene* **20**, 77 (2001); Zheng, L. et al., BRCA1 mediates ligand-independent transcriptional repression of the estrogen receptor. *Proc Natl Acad Sci USA* **98** (17), 9587 (2001).
84. Kawai, H. et al., Direct interaction between BRCA1 and the estrogen receptor regulates vascular endothelial growth factor (VEGF) transcription and secretion in breast cancer cells. *Oncogene* **21**, 7730 (2002).
85. Aiyar, S. E. et al., Attenuation of estrogen receptor {alpha}-mediated transcription through estrogen-stimulated recruitment of a negative elongation factor. *Genes Dev* **18** (17), 2134 (2004).
86. Paull, T. T. et al., From the cover: direct DNA binding by Brca1. *PNAS* **98** (11), 6086 (2001).
87. Hosey, A. M. et al., Molecular basis for estrogen receptor {alpha} deficiency in BRCA1-linked breast cancer. *J Natl Cancer Inst* **99** (22), 1683 (2007).
88. Chen, G.-C. et al., Rb-associated protein 46 (RbAp46) inhibits transcriptional transactivation mediated by BRCA1. *Biochem Biophys Res Commun* **284**, 507 (2001); Yarden, R. I. and Brody, L. C., BRCA1 interacts with components of the histone deacetylase complex. *Proc Natl Acad Sci USA* **96** (9), 4983 (1999).
89. Kadam, S. et al., Functional selectivity of recombinant mammalian SWI/SNF subunits. *Genes Dev* **14** (19), 2441 (2000).
90. Oishi, H. et al., An hGCN5/TRRAP histone acetyltransferase complex co-activates BRCA1 transactivation function through histone modification. *J Biol Chem* **281** (1), 20 (2006).
91. Fan, W. et al., BRCA1 regulates GADD45 through its interactions with the OCT-1 and CAAT motifs. *J Biol Chem* **277**, 8061 (2002).
92. Zhao, H. et al., Activation of the transcription factor Oct-1 in response to DNA damage. *Cancer Res* **60** (22), 6276 (2000).
93. Andrews, H. N. et al., BRCA1 regulates the interferon gamma –mediated apoptotic response. *J Biol Chem* **277** (29), 26225 (2002).
94. Li, S. et al., Binding of CtIP to the BRCT repeats of BRCA1 involved in the transcription regulation of p21 is disrupted upon DNA damage. *J Biol Chem* **274**, 11334 (1999); Yu, X. et al., The C-terminal (BRCT) domains of BRCA1 interact in vivo with CtIP, a protein implicated in the CtBP pathways of transcriptional repression. *J Biol Chem* **273** (39), 25388 (1998).
95. Zheng, L. et al., Sequence-specific transcriptional corepressor function for BRCA1 through a novel zinc finger protein, ZBRK1. *Mol Cell* **6**, 757 (2000).
96. Furuta, S. et al., Removal of BRCA1/CtIP/ZBRK1 repressor complex on ANG1 promoter leads to accelerated mammary tumor growth contributed by prominent vasculature. *Cancer Cell* **10** (1), 13 (2006).
97. Horwitz, A. A. et al., A mechanism for transcriptional repression dependent on the BRCA1 E3 ubiquitin ligase. *PNAS* **104** (16), 6614 (2007).
98. Kennedy, R. D. et al., BRCA1 and c-Myc associate to transcriptionally repress psoriasin, a DNA damage-inducible gene. *Cancer Res* **65** (22), 10265 (2005); Li, H., Lee, T.-H., and Avraham, H., A novel tricomplex of BRCA1, Nmi, and c-Myc inhibits c-Myc-induced human telomerase reverse transcriptase gene (hTERT) promoter activity in breast cancer. *J Biol Chem* **277** (23), 20965 (2002).
99. Park, M. A., Seok, Y.-J., Jeong, G., and Lee, J.-S., SUMO1 negatively regulates BRCA1-mediated transcription, via modulation of promoter occupancy. *Nucleic Acids Res* **36** (1), 263 (2008).
100. Li, C. J. et al., Dynamic redistribution of calmodulin in HeLa cells during cell division as revealed by a GFP-calmodulin fusion protein technique. *J Cell Sci* **112** (Pt 10), 1567 (1999);

Li, X. et al., Role for KAP1 serine 824 phosphorylation and sumoylation/desumoylation switch in regulating KAP1-mediated transcriptional repression. *J Biol Chem* **282** (50), 36177 (2007).
101. Ziv, Y. et al., Chromatin relaxation in response to DNA double-strand breaks is modulated by a novel ATM- and KAP-1 dependent pathway. *Nat Cell Biol* **8** (8), 870 (2006).
102. Tavtigian, S. V. et al., The complete BRCA2 gene and mutations in chromosome 13q-linked kindreds. *Nat Genet* **12** (3), 333 (1996).
103. Bork, P., Blomberg, N., and Nilges, M., Internal repeats in the BRCA2 protein sequence. *Nat Genet* **13** (1), 22 (1996).
104. Bignell, G. et al., The BRC repeats are conserved in mammalian BRCA2 proteins. *Hum Mol Genet* **6** (1), 53 (1997).
105. Sharan, S. K. et al., Embryonic lethality and radiation hypersensitivity mediated by Rad51 in mice lacking Brca2. *Nature* **386**, 804 (1997).
106. Wong, A. K. C. et al., RAD51 interacts with the evolutionarily conserved BRC motifs in the human breast cancer susceptibility gene BRCA2. *J Biol Chem* **272** (51), 31941 (1997).
107. Pellegrini, L. et al., Insights into DNA recombination from the structure of a RAD51-BRCA2 complex. *Nature* **420** (6913), 287 (2002).
108. Venkitaraman, A. R., Functions of BRCA1 and BRCA2 in the biological response to DNA damage. *J Cell Sci* **114** (Pt 20), 3591 (2001).
109. Esashi, F. et al., CDK-dependent phosphorylation of BRCA2 as a regulatory mechanism for recombinational repair. *Nature* **434** (7033), 598 (2005).
110. Esashi, F. et al., Stabilization of RAD51 nucleoprotein filaments by the C-terminal region of BRCA2. *Nat Struct Mol Biol* **14** (6), 468 (2007).
111. Yang, H. et al., BRCA2 function in DNA binding and recombination from a BRCA2-DSS1-ssDNA structure. *Science* **297** (5588), 1837 (2002).
112. Bochkarev, A. and Bochkareva, E., From RPA to BRCA2: lessons from single-stranded DNA binding by the OB-fold. *Curr Opin Struct Biol* **14** (1), 36 (2004); Richard, D. J. et al., Single-stranded DNA-binding protein hSSB1 is critical for genomic stability. *Nature* **453** (7195), 677 (2008).
113. Marston, N. J. et al., Interaction between the product of the breast cancer susceptibility gene BRCA2 and DSS1, a protein functionally conserved from yeast to mammals. *Mol Cell Biol* **19** (7), 4633 (1999).
114. Gudmundsdottir, K. et al., DSS1 is required for RAD51 focus formation and genomic stability in mammalian cells. *EMBO Rep* **5** (10), 989 (2004).
115. Connor, F. et al., Tumorigenesis and a DNA repair defect in mice with a truncating Brca2 mutation. *Nat Genet* **17** (4), 423 (1997); Yu, V. P. C. C. et al., Gross chromosomal rearrangements and genetic exchange between nonhomologous chromosomes following BRCA2 inactivation. *Genes Dev* **14** (11), 1400 (2000).
116. Reid, S. et al., Biallelic mutations in PALB2 cause Fanconi anemia subtype FA-N and predispose to childhood cancer. *Nat Genet* **39** (2), 162 (2007); Xia, B. et al., Fanconi anemia is associated with a defect in the BRCA2 partner PALB2. *Nat Genet* **39** (2), 159 (2007).
117. Erkko, H. et al., A recurrent mutation in PALB2 in Finnish cancer families. *Nature* **446** (7133), 316 (2007); Potapova, A. et al., Promoter hypermethylation of the PALB2 susceptibility gene in inherited and sporadic breast and ovarian cancer. *Cancer Res* **68** (4), 998 (2008); Rahman, N. et al., PALB2, which encodes a BRCA2-interacting protein, is a breast cancer susceptibility gene. *Nat Genet* **39** (2), 165 (2007); Tischkowitz, M. et al., Analysis of PALB2/FANCN-associated breast cancer families. *PNAS* **104** (16), 6788 (2007).
118. Matsuoka, S. et al., ATM and ATR substrate analysis reveals extensive protein networks responsive to DNA damage. *Science* **316** (5828), 1160 (2007); Patel, K. J. et al., Involvement of Brca2 in DNA repair. *Mol Cell* **1** (3), 347 (1998).
119. Futamura, M. et al., Potential role of BRCA2 in a mitotic checkpoint after phosphorylation by hBUBR1. *Cancer Res* **60** (6), 1531 (2000).

120. Marmorstein, L. Y. et al., A human BRCA2 complex containing a structural DNA binding component influences cell cycle progression. *Cell* **104** (2), 247 (2001).
121. Hughes-Davies, L. et al., EMSY links the BRCA2 pathway to sporadic breast and ovarian cancer. *Cell* **115** (5), 523 (2003).
122. Shin, S. and Verma, I. M., BRCA2 cooperates with histone acetyltransferases in androgen receptor-mediated transcription. *PNAS* **100** (12), 7201 (2003).
123. Haber, D. A., The BRCA2-EMSY connection: implications for breast and ovarian tumorigenesis. *Cell* **115** (5), 507 (2003).
124. Land, C. E. et al., Incidence of female breast cancer among atomic bomb survivors, Hiroshima and Nagasaki, 1950–1990. *Radiat Res* **160** (6), 707 (2003); Land, C. E., Tokunaga, M., Tokuoka, S., and Nakamura, N., Early-onset breast cancer in A-bomb survivors. *Lancet* **342** (8865), 237 (1993).
125. Hu, Y. et al., Modulation of aromatase expression by BRCA1: a possible link to tissue-specific tumor suppression. *Oncogene* **24** (56), 8343 (2005); Lu, M. et al., BRCA1 negatively regulates the cancer-associated aromatase promoters I.3 and II in breast adipose fibroblasts and malignant epithelial cells. *J Clin Endocrinol Metab* **91** (11), 4514 (2006).
126. Ganesan, S. et al., Association of BRCA1 with the inactive X chromosome and XIST RNA. *Philos Trans R Soc Lond B Biol Sci* **359** (1441), 123 (2004); Ganesan, S. et al., BRCA1 supports XIST RNA concentration on the inactive X chromosome. *Cell* **111** (3), 393 (2002).
127. Pageau, G. J., Hall, L. L., and Lawrence, J. B., BRCA1 does not paint the inactive X to localize XIST RNA but may contribute to broad changes in cancer that impact XIST and Xi heterochromatin. *J Cell Biochem* **100** (4), 835 (2007); Xiao, C. et al., The XIST noncoding RNA functions independently of BRCA1 in X inactivation. *Cell* **128** (5), 977 (2007).
128. Chappuis, P. O. et al., A significant response to neoadjuvant chemotherapy in BRCA1/2 related breast cancer. *J Med Genet* **39** (8), 608 (2002).
129. Delaloge, S. et al., BRCA1 germ-line mutation: predictive of sensitivity to anthracyclin alkylating agents regimens but not to taxanes? *ASCO Meet Abstr* **26** (15 Suppl), 574 (2008); Goffin, J. R. et al., Impact of germline BRCA1 mutations and overexpression of p53 on prognosis and response to treatment following breast carcinoma: 10-year follow up data. *Cancer* **97** (3), 527 (2003); Kirova, Y. M. et al., Risk of breast cancer recurrence and contralateral breast cancer in relation to BRCA1 and BRCA2 mutation status following breast-conserving surgery and radiotherapy. *Eur J Cancer* **41** (15), 2304 (2005).
130. Quinn, J. E. et al., BRCA1 mRNA expression levels predict for overall survival in ovarian cancer after chemotherapy. *Clin Cancer Res* **13** (24), 7413 (2007).
131. Farmer, H. et al., Targeting the DNA repair defect in BRCA mutant cells as a therapeutic strategy. *Nature* **434** (7035), 917 (2005).
132. Edwards, S. L. et al., Resistance to therapy caused by intragenic deletion in BRCA2. *Nature* **451** (7182), 1111 (2008).
133. Sakai, W. et al., Secondary mutations as a mechanism of cisplatin resistance in BRCA2-mutated cancers. *Nature* **451** (7182), 1116 (2008).
134. Swisher, E. M. et al., Secondary BRCA1 mutations in BRCA1-mutated ovarian carcinomas with platinum resistance. *Cancer Res* **68** (8), 2581 (2008).

Index

K.K. Khanna, Y. Shiloh (eds.), *The DNA Damage Response: Implications on Cancer Formation and Treatment*, DOI 10.1007/978-90-481-2561-6_BM2,